Special Functions 2000:
Current Perspective and Future Directions

NATO Science Series

A Series presenting the results of scientific meetings supported under the NATO Science Programme.

The Series is published by IOS Press, Amsterdam, and Kluwer Academic Publishers in conjunction with the NATO Scientific Affairs Division

Sub-Series

I. **Life and Behavioural Sciences**	IOS Press
II. **Mathematics, Physics and Chemistry**	Kluwer Academic Publishers
III. **Computer and Systems Science**	IOS Press
IV. **Earth and Environmental Sciences**	Kluwer Academic Publishers

The NATO Science Series continues the series of books published formerly as the NATO ASI Series.

The NATO Science Programme offers support for collaboration in civil science between scientists of countries of the Euro-Atlantic Partnership Council. The types of scientific meeting generally supported are "Advanced Study Institutes" and "Advanced Research Workshops", and the NATO Science Series collects together the results of these meetings. The meetings are co-organized bij scientists from NATO countries and scientists from NATO's Partner countries – countries of the CIS and Central and Eastern Europe.

Advanced Study Institutes are high-level tutorial courses offering in-depth study of latest advances in a field.
Advanced Research Workshops are expert meetings aimed at critical assessment of a field, and identification of directions for future action.

As a consequence of the restructuring of the NATO Science Programme in 1999, the NATO Science Series was re-organized to the four sub-series noted above. Please consult the following web sites for information on previous volumes published in the Series.

http://www.nato.int/science
http://www.wkap.nl
http://www.iospress.nl
http://www.wtv-books.de/nato-pco.htm

Series II: Mathematics, Physics and Chemistry – Vol. 30

Special Functions 2000: Current Perspective and Future Directions

edited by

Joaquin Bustoz

Department of Mathematics,
Arizona State University,
Tempe, Arizona, U.S.A.

Mourad E.H. Ismail

Department of Mathematics,
University of South Florida,
Tampa, Florida, U.S.A.

and

Sergei K. Suslov

Department of Mathematics,
Arizona State University,
Tempe, Arizona, U.S.A.

Kluwer Academic Publishers

Dordrecht / Boston / London

Published in cooperation with NATO Scientific Affairs Division

Proceedings of the NATO Advanced Study Institute on
Special Functions 2000: Current Perspective and Future Directions
Tempe, Arizona, U.S.A.
29 May–9 June 2000

A C.I.P. Catalogue record for this book is available from the Library of Congress.

ISBN 0-7923-7119-4 (HB)
ISBN 0-7923-7120-8 (PB)

Published by Kluwer Academic Publishers,
P.O. Box 17, 3300 AA Dordrecht, The Netherlands.

Sold and distributed in North, Central and South America
by Kluwer Academic Publishers,
101 Philip Drive, Norwell, MA 02061, U.S.A.

In all other countries, sold and distributed
by Kluwer Academic Publishers,
P.O. Box 322, 3300 AH Dordrecht, The Netherlands.

Printed on acid-free paper

Printed in the Netherlands.

TABLE OF CONTENTS

PREFACE

This volume contains the Proceedings of the NATO Advanced Study Institute *Special Functions 2000: Current Perspective and Future Directions* held at Arizona State University, Tempe, Arizona, USA from May 29 to June 9, 2000.

Special Functions 2000 comprised a NATO Advanced Study Institute (ASI) and an NSF Research Conference with a Computer Algebra component. The NATO ASI had a strong instructional component and summarized results in special functions and their diverse applications. The detailed program can be found on the web site: http://math.la.asu.edu/~sf2000/index.html. The Advanced Study Institute and the NSF Conference brought together researchers in main areas of special functions and applications to present recent developments in the theory, review the accomplishments in the past decades, and chart directions for research in the future.

Some of the topics covered are orthogonal polynomials and special functions in one and several variables, asymptotics, continued fractions, applications to number theory, combinatorics and mathematical physics, integrable systems, harmonic analysis and quantum groups, and Painlevé classification.

The Organizing Committee consisted of Joaquin Bustoz (Chair), Mourad Ismail, Tom Koornwinder, Vyacheslav Spiridonov (Director from the Partner country), Sergei Suslov (Director from the NATO country), and Luc Vinet.

We gratefully acknowledge the generous support provided by NATO Scientific and Environmental Affairs Division, Arizona State University, National Science Foundation, Centre de Recherches Mathématiques, The Fields Institute for Research in Mathematical Sciences, Pacific Institute for the Mathematical Sciences, and Wolfram Research.

Finally we wish to thank the distinguished lecturers at the ASI and hope that this volume will stimulate further significant advances in Special Functions and their applications.

Joaquin Bustoz, Mourad Ismail, Sergei Suslov

FOREWORD

The subject of Special Functions was an active research area during the last two decades of the 20th century. Much of this activity was initiated by George Andrews in his work in number theory, and Dick Askey in his research in orthogonal polynomials, and in a fruitful collaboration between these two mathematicians. The contribution of Andrews and Askey has created the current scientific school in the area of special functions and related topics; their students and collaborators have produced a substantial amount of interesting work. One of the landmarks is the discovery of the fundamental Askey-Wilson polynomials, and of many interesting results around them.

The main idea of the NATO Advanced Study Institute *Special Functions 2000: Current Perspective and Future Directions* was to summarize the important results that have been discovered during the current renaissance period of the theory of special functions and their diverse applications.

In order to achieve this goal the Organizing Committee has invited leading experts in the field to deliver lectures on such topics as orthogonal polynomials and special functions in one and several variables, asymptotics, continued fractions, applications to number theory, combinatorics and mathematical physics, integrable systems, harmonic analysis and quantum groups, and Painlevé classification.

The plenary talks were given in the form of broadly accessible survey lectures or series of such lectures. They can be roughly combined in the following overlapping categories:

Single Variable Theory

George Andrews, Lecture I: *q-Series in the twentieth century*; Lecture II: *Some recent developments in q-series*.

Richard Askey, *The last 100 years of special functions*.

Bruce C. Berndt, *Flowers which we cannot yet see growing in Ramanujan's garden of hypergeometric series, elliptic functions, and q's*.

Mourad E. H. Ismail, Lectures I–III: *Lectures on q-orthogonal polynomials*.

Olav Njåstad, *Orthogonal rational functions and continued fractions*.

Riemann Hilbert-Problems

Percy Deift, Lectures I–II: *Riemann-Hilbert Problems.*
Walter Van Assche, *Riemann-Hilbert problems for multiple orthogonal polynomial.*

Painlevé Classification

Alexander Kitaev, Lectures I–II: *Special functions of the isomonodromy type. The Painlevé equations.*

Bispecrtal Problem, Factorization Method and Related Topics

Alberto Grünbaum, *The bispectral problem.*
Luc Haine, *Krall's problem: some new perspectives.*
Vyacheslav P. Spiridonov, *The factorization method, self-similar potentials and quantum algebras.*

The reader will find many of these lectures, originally delivered at the Advanced Study Institute and then prepared for publication, in this volume.

In addition, research presentations in the form of invited one hour talks were given by A. Garsia, D. Leites, M. Noumi, G. Olshanski, H. Rosengren, P. Terwilliger, and J. Wimp. Contributed talks were presented by the participants in the afternoon parallel sessions. Many of them will appear in the forthcoming special issue of the Rocky Mountain Journal of Mathematics.

A Mini-Program: "Computer Algebra Systems and Special Functions on the Web" was organized during the evening hours. It included presentations by I. Gessel, C. Krattenthaler, L. Littlejohn, D. Lozier, O. Marichev and M. Trott, P. Paule, and A. Riese.

In the last day of the two week meeting, there was an interesting discussion of the future directions and open problems in the various areas of special functions, organized by Dick Askey. All this together gives a rather complete account of the current status of the research in special functions and their applications. We hope that the contents of this volume will inspire future contributions.

In spite of the intense schedule there was an opportunity for participants to meet, socialize and exchange ideas on an informal basis. Social programs included a welcome party, banquet and trip to the Grand Canyon.

The meeting was considered a success by the participants. The hot weather did not prevent, but rather, stimulated discussion of 'hot mathematics' at the beautiful surroundings of the campus of Arizona State University. The Advanced Study Institute has demonstrated once again that Special Functions are one of the main streams of research in mathematical sciences.

Sergei Suslov

BAILEY'S TRANSFORM, LEMMA, CHAINS AND TREE

GEORGE E. ANDREWS
Department of Mathematics
The Pennsylvania State University
University Park, PA 16802 USA

Abstract. In this paper we shall provide a brief survey of the work begun by L. J. Rogers and W. N. Bailey which has led to an iterative method for producing infinite chains of q-series identities. Apart from providing the reader with leads to the study of previous accomplishments, we shall emphasize the importance of examination of the seminal works in order to discern topics open to further development. This will lead us directly to a new construct: the Bailey tree.

1. Introduction

Bailey chains have their genesis in the work of L. J. Rogers. He first proved the following identities in 1894 [30], and they have become known as the Rogers-Ramanujan identities

$$\sum_{n=0}^{\infty} \frac{q^{n^2}}{(q;q)_n} = \frac{1}{(q,q^4;q^5)_\infty} \tag{1.1}$$

and

$$\sum_{n=0}^{\infty} \frac{q^{n^2+n}}{(q;q)_n} = \frac{1}{(q^2,q^3;q^5)_\infty}, \tag{1.2}$$

where

$$(A)_n = (A;q)_n = \prod_{j=0}^{n-1}(1-Aq^j), \tag{1.3}$$

1

J. Bustoz et al. (eds.), Special Functions 2000, 1–22.
© 2001 *Kluwer Academic Publishers. Printed in the Netherlands.*

and

$$(A_1, A_2, \ldots, A_r; q)_n = \prod_{j=1}^{r} (A_j; q)_n. \tag{1.4}$$

We shall provide a proof of (1.1) and (1.2) in Section 2. It is essentially Bressoud's simplification of Rogers' second proof [19]. This will provide us with the simplest possible example of a Bailey chain (cf. [5], [7; Ch.3]).

1.1. DEFINITION

A General Bailey Chain is an infinite sequence $(i \geq 0)$ of pairs of sequences of rational functions $\{\alpha_n^{(i)}\}_{n \geq 0}$ and $\{\beta_n^{(i)}\}_{n \geq 0}$ of two variables a and q and possibly others

$$(\alpha_n, \beta_n) \to (\alpha_n', \beta_n') \to (\alpha_n'', \beta_n'') \to \cdots. \tag{1.5}$$

There is an identity independent of i connecting $\alpha_n^{(i)}$ and $\beta_n^{(i)}$, say

$$\beta_n^{(i)} = F_n(\alpha_0^{(i)}, \alpha_1^{(i)}, \alpha_2^{(i)}, \ldots, \alpha_n^{(i)}); \qquad (i \geq 0) \tag{1.6}$$

furthermore there are given rules of construction

$$\beta_n^{(i)} = G_n(\beta_0^{(i-1)}, \ldots, \beta_n^{(i-1)}) \qquad (i \geq 1) \tag{1.7}$$

and

$$\alpha_n^{(i)} = H_n(\alpha_0^{(i-1)}, \ldots, \alpha_r^{(i-1)}) \qquad (i \geq 1), \tag{1.8}$$

again G_n and H_n are independent of i.

Put in this abstract setting, it is hard to discern the relationship of this iterative process with the Rogers-Ramanujan identities (1.1) and (1.2). This will become clear in Section 2. Section 3 will look at the overarching method dubbed by L. J. Slater, the Bailey transform [38; p. 58]. Section 4 will provide a summary of how this idea has been broadly applied. In Section 5, we consider the Milne-Lilly extensions [25]–[27], and other extensions [12]. We also return to a look at Rogers' second paper [31] with an eye to extensions as described in [8]. In Section 6, we reconsider Bailey's first papers on this topic; this leads us to new Bailey chains in Section 7 and eventually to the Bailey tree in Section 8.

2. The Bressoud-Rogers Proof

In this section we will provide the simplest known example of a non-trivial Bailey chain.

Theorem 1 *Let*

$$\alpha_n^{(0)} = \frac{(1 - aq^{2n})(a;q)_n(-1)^n q^{\binom{n}{2}}}{(1-a)(q;q)_n} \tag{2.1}$$

and

$$\beta_n^{(0)} = \delta_{n,0}. \tag{2.2}$$

Furthermore for $i > 0$

$$\alpha_n^{(i)} = a^n q^{n^2} \alpha_n^{(i-1)} \tag{2.3}$$

and

$$\beta_n^{(i)} = \sum_{j=0}^{n} \frac{q^{j^2} a^j \beta_j^{(i-1)}}{(q;q)_{n-i}}. \tag{2.4}$$

Then $\{\alpha_n^{(i)}\}_{n\geq 0}$, $\{\beta_n^{(i)}\}_{n\geq 0}$ $(i \geq 0)$ form a Bailey Chain.

2.1. PROOF

The G_n and H_n of (1.7) and (1.8) are clearly defined by (2.3) and (2.4). We shall show that

$$F_n(x_0, x_1, \ldots, x_n) = \sum_{j=0}^{n} \frac{x_j}{(q;q)_{n-j}(aq;q)_{n+j}}. \tag{2.5}$$

The treatment of the assertion that

$$\beta_n^{(0)} = \sum_{j=0}^{n} \frac{\alpha_j^{(0)}}{(q;q)_{n_j}(aq;q)_{n+j}} \tag{2.6}$$

follows immediately from the fact (proved easily by induction on M) that

$$\sum_{j=0}^{M} \frac{(a;q)_j(1-aq^{2j})(q^{-n};q);q^{nj}}{(q;q)_j(aq^{n+1};q)_j(1-a)} = \frac{(aq;q)_M q^{nM}(q^{1-n};q)_M}{(q;q)_M(aq^{n+1};q)_M}. \tag{2.7}$$

Now we assume that (1.6) holds for $i-1$ with F_n given by (2.5). Hence

$$\beta_n^{(i)} = \sum_{j=0}^{n} \frac{q^{j^2} a^j \beta_j^{(i-1)}}{(q;q)_{n-j}}$$

$$= \sum_{j=0}^{n} \frac{q^{j^2} a^j}{(q;q)_{n-j}} \sum_{r=0}^{j} \frac{\alpha_r^{(i-1)}}{(q;q)_{j-r}(aq;q)_{j+r}}$$

$$= \sum_{r=0}^{n} \sum_{j=r}^{n} \frac{q^{j^2} a^j \alpha_r^{(i-1)}}{(q;q)_{n-j}(q;q)_{j-r}(aq;q)_{j+r}}$$

$$= \sum_{r=0}^{n} \sum_{j=0}^{n-r} \frac{q^{j^2+2jr} a^j \alpha_r^{(i)}}{(q;q)_{n-j-r}(q;q)_j (aq;q)_{j+2r}}$$

$$= \sum_{r=0}^{n} \frac{\alpha_r^{(i)}}{(q;q)_{n-r}(aq;q)_{2r}}$$

$$\times \sum_{j \geq 0} \frac{(q^{-n+r};q)_j (-1)^j q^{(n-r)j+\binom{j+1}{2}+2jr}}{(q;q)_j (aq^{2r+1};q)_j}$$

$$= \sum_{r=0}^{r} \frac{\alpha_r^{(i)}}{(q;q)_{n-r}(aq;q)_{2r}} \frac{1}{(aq^{2r+1};q)_{n-r}}$$

$$\text{(by [23]; p. 11, eq (1.5.2) with } b \to \infty)$$

$$= \sum_{r=0}^{n} \frac{\alpha_r^{(i)}}{(q;q)_{n-r}(aq;q)_{n+r}}, \tag{2.8}$$

which proves that (1.6) holds for all i. □

The first couple of instances of (1.6) reveal the power of Bailey Chains. When $i = 1$,

$$\beta_n^{(1)} = \frac{1}{(q;q)_n} \tag{2.9}$$

by (2.4), and

$$\alpha_n^{(1)} = \frac{(1 - aq^{2n})(a;q)_n (-a)^n q^{n(3n-1)/2}}{(1-a)(q;q)_n}. \tag{2.10}$$

Next,

$$\beta_n^{(2)} = \sum_{j=0}^{n} \frac{q^{j^2} q^j}{(q)_{n-j}(q)_j}, \tag{2.11}$$

and

$$\alpha_n^{(2)} = \frac{(1 - aq^{2n})(a;q)_n (-1)^n a^{2n} q^{n(5n-1)/2}}{(1-a)(q;q)_n}. \tag{2.12}$$

Now by Theorem 1, we know that

$$\beta_n^{(2)} = \sum_{r=0}^{n} \frac{\alpha_r^{(2)}}{(q;q)_{n-r}(aq;q)_{n+r}}, \tag{2.13}$$

and therefore

$$\beta_\infty^{(2)} = \frac{1}{(q;q)_\infty (aq;q)_\infty} \sum_{r=0}^{\infty} \alpha_r^{(2)}, \tag{2.14}$$

i.e. after multiplication by $(q;q)_\infty$, we obtain

$$\sum_{n=0}^{\infty} \frac{a^n q^{n^2}}{(q;q)_n} = \frac{1}{(aq;q)_\infty} \sum_{n=0}^{\infty} \frac{(a;q)_n (1 - aq^{2n})(-1)^n a^{2n} q^{n(5n-1)/2}}{(q;q)_n (1-a)}. \tag{2.15}$$

From this familiar formula and Jacobi's Triple Product Indentity [23; p. 12, eq. (1.6.1)] we obtain (1.1) by setting $a = 1$, and (1.2) by setting $a = q$.

This process was essentially carried out by D. Bressoud in [19] (cf. [17; Sec. 3.4]). It is a limiting case of the proof given in Chapter 3 of [7]. We go through it here to illustrate as cleanly as possible precisely how a Bailey Chain works.

3. The Bailey Transform

The actual transformation that we performed in (2.8) was treated in a fully abstract manner by Bailey in [15]. L. J. Slater [38; p. 58] has christened this the Bailey Transform:

If

$$\beta_n = \sum_{r=0}^{n} \alpha_r u_{n-r} v_{n+r}, \tag{3.1}$$

and

$$\gamma_n = \sum_{r=n}^{\infty} \delta_r u_{r-n} v_{r+n}, \tag{3.2}$$

then

$$\sum_{n=0}^{\infty} \alpha_n \gamma_n = \sum_{n=0}^{\infty} \beta_n \delta_n. \tag{3.3}$$

As Bailey notes [15; p. 1] the proof is a mere series rearrangement.

$$\sum_{n=0}^{\infty} \alpha_n \gamma_n = \sum_{n=0}^{\infty} \sum_{r=n}^{\infty} \alpha_n \delta_r u_{r-n} v_{r+n} \tag{3.4}$$

$$= \sum_{r=0}^{\infty} \sum_{n=0}^{r} \delta_r \alpha_n u_{r-n} v_{r+n} \tag{3.5}$$

$$= \sum_{r=0}^{\infty} \delta_r \beta_r . \tag{3.6}$$

In (2.8), the Bailey Transform underlies the process in the instance $u_n = 1/(q;q)_n$, $v_n = 1/(aq;q)_n$, $\delta_r = a^r q^{r^2}$, $\gamma_n = a^n q^{n^2}/(aq;q)_\infty$.

In Section 4, we shall examine other instances of the Bailey Transform that lead to other Bailey Chains. Indeed, one of the reasons for a separate section on the Bailey Transform is to distinguish it from Bailey's Lemma (an instance of the Bailey Transform [cf. (4.1)-(4.4) in Section 4]) and Bailey Chains (as defined in Section 1).

Section 4 will only consider the case $u_n = 1/(q;q)_n$, $v_n = 1/(aq;q)_n$, and in Section 5 we shall restrict ourselves to natural multidimensional extensions of this instance. However, there are several quite distinct instances of Bailey's Transform in the literature, and we shall touch briefly on a few.

D. Bressoud [18] was perhaps the first to obtain really striking variations. He chose

$$u_n = \frac{(\beta;q)_n}{(q;q)_n}, \quad v_n = \frac{(\alpha\beta;q)_n}{(q;q)_n}, \quad \text{and} \quad \delta_n = r^n . \tag{3.7}$$

Perhaps his most appealing application is the discovery of two new polynomial identities which imply the Rogers-Ramanujan identities:

$$\sum_{m=0}^{N} q^{m^2} \begin{bmatrix} N \\ m \end{bmatrix} = \sum_{m=-\infty}^{\infty} (-1)^m q^{m(5m+1)/2} \begin{bmatrix} 2N \\ N+2m \end{bmatrix} \tag{3.8}$$

and

$$(1-q^N) \sum_{n=0}^{\infty} q^{m^2+m} \begin{bmatrix} N-1 \\ m \end{bmatrix} = \sum_{m=-\infty}^{\infty} (-1)^m q^{m(5m+3)/2} \begin{bmatrix} 2N \\ N+2m+1 \end{bmatrix}, \tag{3.9}$$

where

$$\begin{bmatrix} A \\ B \end{bmatrix} = \begin{bmatrix} A \\ B \end{bmatrix}_q = \begin{cases} \frac{(q;q)_A}{(q;q)_B (q;q)_{A-B}} & \text{if } 0 \le A \le B \\ 0 & \text{otherwise.} \end{cases} \tag{3.10}$$

In [10], it was pointed out that one of D. B. Sears' most useful transformations [35; p. 167, Th. 3] was the following instance of the Bailey Transform:

$$u_n = \frac{(ab/e;q)_n}{(q;q)_n}, \quad v_n = 1, \quad \gamma_s = \sum_{t=0}^{\infty} \frac{(ab/e;q)_t \delta_{s+t}}{(q;q)_t}$$

$$\alpha_n = \frac{(e/a;q)_n (e/b;q)_n \left(\frac{ab}{e}\right)^n}{(q;q)_n (e;q)_n}, \quad \beta_n = \frac{(a;q)_n (b;q)_n}{(q;q)_n (e;q)_n}. \qquad (3.11)$$

Following up on this [10], one can obtain an $(r-1)$-fold expansion of

$$_{r+1}\phi_r \left(\begin{matrix} c_0, c_1, \ldots, c_r \\ f_1, \ldots, f_r \end{matrix} ; q, \frac{f_1 f_2 \cdots f_r}{c_0 c_1 \cdots c_r} \right)$$

$$:= \sum_{n \geq 0} \frac{(c_0, c_1, \ldots, c_r; q)_n}{(q, f_1, \ldots, f_r; q)_n} \left(\frac{f_1 f_2 \ldots f_r}{c_0 c_1 \ldots c_r} \right)^n. \qquad (3.12)$$

When $r = 2$, the result reduces to

$$_3\phi_2 \left(\begin{matrix} c_0, c_1, c_2 \\ f_1, f_2 \end{matrix} ; q, \frac{f_1 f_2}{c_0 c_1 c_2} \right)$$

$$= \frac{\left(\frac{f_2}{c_2}, \frac{f_1 f_2}{c_0 c_2}; q \right)_\infty}{\left(f_2, \frac{f_1 f_2}{c_0 c_1 c_2}; q \right)_\infty} \, _3\phi_2 \left(\begin{matrix} f_1/c_0, f_1/c_1, c_2 \\ f_1, \frac{f_1 f_2}{c_0 c_1} \end{matrix} ; q, f_2/c_2 \right) \qquad (3.13)$$

a familiar and useful formula [23, p. 62, eq. (3.2.7)].

Very much related to this is a series of papers by Schilling and Warnaar [32], [33], [34] in which they also concentrate on the γ_n and δ_n and consider multi-dimensional extensions thereof. Also a variation on γ_n and δ_n appears in Bailey's paper [15, Sec. 5].

In this same general area, but not explicitly following from the Bailey Transform is the work in [9] on a q-trinomial coefficient analog of Bailey's Lemma. In [39], O. Warnaar shows how this work is closely related to the classical Bailey Lemma.

4. One-Dimensional Bailey Chains

The proof of the Rogers-Ramanujan identities given in Section 2 is in fact a very special case of the following result known as Bailey's Lemma [5], [7; Ch. 3].

If for $n \geq 0$

$$\beta_n = \sum_{r=0}^{n} \frac{\alpha_r}{(q;q)_{n-r}(aq;q)_{n+r}}, \qquad (4.1)$$

then

$$\beta'_n = \sum_{r=0}^{n} \frac{\alpha'_r}{(q;q)_{n-r}(aq;q)_{n+r}} , \qquad (4.2)$$

where

$$\alpha'_r = \frac{(\rho_1;q)_r(\rho_2;q)_r(aq/\rho_1\rho_2)^r \alpha_r}{(aq/\rho_1;q)_r(aq/\rho_2;q)_r} \qquad (4.3)$$

and

$$\beta'_n = \sum_{j\geq 0} \frac{(\rho_1;q)_j(\rho_2;q)_j(aq/\rho_1\rho_2;q)_{n-j}(aq/\rho_1\rho_2)^j \beta_j}{(q;q)_{n-j}(aq/\rho_1;q)_n(aq/\rho_2;q)_n} . \qquad (4.4)$$

We shall suppress the proof of this; it is very similar to the proof of (2.8) (cf. [7]; Ch. 3). This, of course, can be iterated immediately to produce a Bailey Chain as described in Section 1.

This result as outlined (but not stated) by Bailey himself ([15]; Sec. 4) uses the Bailey Transform in the case where

$$u_n = \frac{1}{(q;q)_n}, \qquad v_n = \frac{1}{(aq;q)_n},$$

$$\delta_n = \frac{(\rho_1,\rho_2,q^{-N};q)_n q^n}{(\rho_1\rho_2 q^{-N}/a;q)_n},$$

$$\gamma_n = \frac{(aq/\rho_1, aq/\rho_2;q)_N}{(aq, aq/(\rho_1\rho_2);q)_N} \frac{(\rho_1,\rho_2,q^{-N};q)_n \left(\frac{-aq}{\rho_1\rho_2}\right)^n q^{nN-\binom{n}{2}}}{(aq/\rho_1, aq/\rho_2, aq^{N+1};q)_n}. \qquad (4.5)$$

Subsequently this Bailey Chain has been applied to prove many disparate results. Its original application [3] (cf. [7]; Ch. 3) was to provide an analog of Watson's q-analog of Whipple's Theorem [41] for

$$_{2k+r}\phi_{2k+3} \left(\begin{array}{c} a, q\sqrt{a}, -q\sqrt{a}, b_1, c_1, b_2, c_2, \ldots, b_k, c_k, q^{-N} \\ \sqrt{a}, -\sqrt{a}, \frac{aq}{b_1}, \frac{aq}{c_1}, \frac{aq}{b_2}, \frac{aq}{c_2}, \ldots, \frac{aq}{b_k}, \frac{aq}{c_k}, aq^{N+1} \end{array} ; q, \frac{a^k q^{k+N}}{b_1 c_1 \cdots b_k c_k} \right). \qquad (4.6)$$

This in turn can be specialized to the natural extension of the first Rogers-Ramanujan identity [2], [4; Ch. 7]:

$$\sum_{n_1 \geq n_2 \geq \cdots n_{k-1} \geq 0} \frac{q^{n_1^2 + n_2^2 + \cdots + n_{k-1}^2}}{(q;q)_{n_1-n_2}(q;q)_{n_2-n_3} \cdots (q;q)_{n_{k-2}-n_{k-1}}(q;q)_{n_{k-1}}}$$

$$= \prod_{\substack{n = 1 \\ n \not\equiv 0, \pm k \pmod{2k+1}}}^{\infty} \frac{1}{1 - q^n}. \tag{4.7}$$

But other results also follow such as [11; p. 392]

$$\sum_{n=0}^{\infty} \frac{q^{\binom{n+1}{2}}}{(-q; q)_n} = \sum_{\substack{n \geq 0 \\ |j| \leq n}} (-1)^{n+j} q^{n(3n+1)/2 - j^2} (1 - q^{2n+1}), \tag{4.8}$$

or [11; p. 404]

$$\sum_{n \geq 1} \frac{(-1)^n q^{n^2}}{(q; q^2)_n} = \sum_{n \geq 1} \sum_{j=0}^{2n-1} (-1)^n q^{n(3n-1) - j(j+1)/2} (1 + q^{2n}). \tag{4.9}$$

Further applications are given in [5], [6], [36] and [37].

5. Multidimensional Bailey Chains

Lilly and Milne were the first researchers [25], [26] and [27] to recognize that the concepts described in Sections 2–5 could be extended to higher dimensional series. In a substantial series of papers extending classical ordinary and q-hypergeometric functions to multiple series based on the unitary or symplectic groups, S. Milne and his collaborators laid the ground work for an extension of Bailey's Lemma and Bailey chains to these groups.

As an example of their achievement we state their generalization for the symplectic groups C_ℓ [26; pp. 494–495]:

Theorem 2 *Let $A = \{A_y\}$ and $B = \{B_y\}$ be sequences that satisfy*

$$B_N = \sum_{\substack{0 \leq y_k \leq N_k \\ k = 1, 2, \dots, \ell}} \left\{ \prod_{r,s=1}^{\ell} \left[\left(q \frac{x_r}{x_s} q^{y_r - y_s} \right)_{N_r - y_r}^{-1} (q x_r x_s q^{y_r + y_s})_{N_r - y_r}^{-1} \right] A_y \right\} \tag{5.1}$$

for every $N_i \geq 0$, $i = 1, 2, \dots, \ell$. (A and B form a C_ℓ Bailey pair). If we define

$$A'_N := \left\{ \prod_{k=1}^{\ell} \left[\frac{(\alpha x_k)_{N_k} (q x_k \beta^{-1})_{N_k}}{(\beta x_k)_{N_k} (q x_k \alpha^{-1})_{N_k}} \right] \left(\frac{\beta}{\alpha} \right)^{(N_1 + \dots + N_\ell)} A_N \right\} \tag{5.2}$$

and if we also define

$$
B'_N := \sum_{\substack{0 \le m_k \le N_k \\ k = 1, 2, \ldots, \ell}} \left\{ \left(\frac{\beta}{\alpha}\right)_{(N_1 + \cdots + N_\ell) - (m_1 + \cdots + m_\ell)} \right.
$$

$$
\times \prod_{k=1}^{\ell} \left[\frac{(\alpha x_k)_{m_k} (q x_k \beta^{-1})_{m_k}}{(\beta x_k)_{N_k} (q x_k \alpha^{-1})_{N_k}} \right] \left(\frac{\beta}{\alpha}\right)^{(m_1 + \cdots + m_\ell)}
$$

$$
\times \prod_{1 \le r < s \le \ell} [(q x_r x_s q^{m_r + m_s})^{-1} (q x_r x_3 q^{N_s - m_s})_{N_r - m_r}^{-1}]
$$

$$
\times \prod_{r,s=1}^{\ell} \left. \left(q \frac{x_r}{x_s} q^{m_r - m_s} \right)^{-1}_{N_r - m_r} B_m \right\}, \tag{5.3}
$$

then $A' = \{A'_y\}$ and $B' = \{B'_y\}$ also satisfy (5.1); i.e. they form a C_ℓ Bailey pair.

In [27], they consider the applications of their extensions and provide relevant generalizations of the q-analog of Whipple's transformation, the q-Dougall summation and other results.

In [12], Schilling, Warnaar and I provide an alternative approach to multiple series. It is not nearly as broad in scope as the Lilly-Milne work, but it does yield a variety of new series-product identities. The principal ideas of [12] culminate in the following formulation.

5.1. DEFINITION

(A_2 Bailey pair of type I). Denote $\alpha_k = \alpha_{k_1,k_2,k_3}$ and $\beta_L = \beta_{L_1,L_2}$, and let $\alpha = \{\alpha_k\}_{\substack{k_1 \ge k_2 \ge k_3 \\ k_1 + k_2 + k_3 = 0}}$ and $\beta = \{\beta_l\}_{L_1,L_2 \ge 0}$ be a pair of sequences that satisfies

$$
\beta_L = \sum_{\substack{k_1 \ge k_2 \ge k_3 \\ k_1 + k_2 + k_3 = 0}} \alpha_k \sum_r \frac{q^{r_1 r_{23}}}{(q)_{r_1}(q)_{r_2}(aq)_{r_3}(q)_{r_{12}}(q)_{r_{13}}(q)_{r_{23}}}, \tag{5.4}
$$

where $\sum_r$ denotes a sum over $r_1, \ldots, r_{23}$ such that

$$
r_1 + r_{12} + r_{13} = L_2 - k_1, \quad r_2 + r_{12} + r_{23} = L_2 - k_2, \quad r_3 + r_{13} + r_{23} = L_2 - k_3
$$

and

$$
r_1 + r_2 + r_3 = 2L_1 - L_2, \quad r_{12} + r_{13} + r_{23} = 2L_2 - L_1.
$$

Then (α, β) forms an A_2 Bailey pair of type I relative to α.

5.2. DEFINITION

(A_2 Bailey pair of type II). Denote $\alpha_k = \alpha_{k_1, k_2, k_3}$ and $\beta_L = \beta_{L_1, L_2}$, and let $\alpha = \{\alpha_k\}$ $\begin{matrix} k_1 \geq k_2 \geq k_3 \\ k_1 + k_2 + k_3 = 0 \end{matrix}$ and $\beta = \{\beta_L\}_{L_1, L_2 \geq 0}$ be a pair of sequences that satisfies

$$\beta_L = \sum_{\substack{k_1 \geq k_2 \geq k_3 \\ k_1 + k_2 + k_3 = 0}} \alpha_k$$

$$\times \frac{(aq)_{L_1+L_2}}{(aq)_{L_1+k_1}(aq)_{L_1+k_2}(q)_{L_1+k_3}(q)_{L_2-k_1}(q)_{L_2-k_2}(aq)_{L_2-k_3}}. \quad (5.5)$$

Then (α, β) forms an A_2 Bailey pair of type II relative to α.

With these definitions, the A_2 version of Bailey's lemma reads

Theorem 3 [12] *Let* (α, β) *form an A_2 bailey pair of type $T = I, II$ relative to a. Then the pair (α', β') given by*

$$\alpha'_k = a^{k_1 + k_2} q^{\frac{1}{2}(k_1^2 + k_2^2 + k_3^2)} \alpha_k,$$

$$\beta'_L = f_L^{(T)} \sum_{r_1=0}^{L_1} \sum_{r_2=0}^{L_2} \frac{a^{r_1} q^{r_1^2 - r_1 r_2 + r_2^2}}{(q)_{L_1-r_1}(q)_{L_2-r_2}} \beta_r \quad (5.6)$$

forms an A_2 Bailey pair of type II relative to a. Here $f_L^{(I)} = (aq)_{L_1+L_2}^{-1}$ and $f_L^{(II)} = 1$.

As examples of the series-product results that follow from this theorem, we include [12; Th. 5.2]

Theorem 4 *For* $|q| < 1$,

$$\sum_{r_1 r_2 \geq 0} \frac{q^{r_1^2 - r_1 r_2 + r_2^2}}{(q)_{r_1}} \begin{bmatrix} 2r_1 \\ r_2 \end{bmatrix} \quad (5.7)$$

$$= \prod_{n=1}^{\infty} \frac{1}{(1 - q^{7n-1})^2 (1 - q^{7n-3})(1 - q^{7n-4})(1 - q^{7n-6})^2},$$

$$\sum_{r_1 r_2 \geq 0} \frac{q^{r_1^2 - r_1 r_2 + r_2^2 + r_1 + r_2}}{(q)_{r_1}} \begin{bmatrix} 2r_1 \\ r_2 \end{bmatrix}$$

$$= \prod_{n=1}^{\infty} \frac{1}{(1-q^{7n-2})(1-q^{7n-3})^2(1-q^{7n-4})^2(1-q^{7n-5})}, \quad (5.8)$$

and

$$\sum_{r_1 r_2 \geq 0} \frac{q^{r_1^2 - r_1 r_2 + r_2^2 + r_1}}{(q)_{r_1}} \begin{bmatrix} 2r_1 + 1 \\ r_2 \end{bmatrix}$$

$$= \sum_{r_1, r_2 \geq 0} \frac{q^{r_1^2 - r_1 r_2 + r_2^2 + r_2}}{(q)_{r_1}} \begin{bmatrix} 2r_1 \\ r_2 \end{bmatrix} \quad (5.9)$$

$$= \prod_{n=1}^{\infty} \frac{1}{(1-q^{7n-1})(1-q^{7n-2})(1-q^{7n-3})}$$

$$\times \frac{1}{(1-q^{7n-4})(1-q^{7n-5})(1-q^{7n-6})}.$$

(the symbol $\begin{bmatrix} A \\ B \end{bmatrix}$ *is defined in (3.10)).*

In [8], I undertook an examination of the work of L. J. Rogers which inspired all of Bailey's insights. The main result there might be described as a "direct product" generalization of Bailey's Lemma and Bailey Chains [7; Ch. 3]. Its bilateral nature is dealt with in the $s = 1$ case in [16] and [29].

Theorem 5 *If for* $n_1, n_2, \ldots, n_s \geq 0$,

$$\beta_{n_1, n_2, \ldots, n_s} = \sum_{r_1 = -\infty}^{n_1} \sum_{r_2 = -\infty}^{n_2} \cdots \sum_{r_s = -\infty}^{n_s} \frac{\alpha_{r_1, r_2, \ldots, r_s}}{\prod_{j=1}^{s} (q; q)_{n_j - r_j} (a_j q; q)_{n_j + r_j}} \quad (5.10)$$

then

$$\beta'_{n_1, n_2, \ldots, n_s} = \sum_{r_1 = -\infty}^{n_1} \sum_{r_2 = -\infty}^{n_2} \cdots \sum_{r_s = -\infty}^{n_s} \frac{\alpha'_{r_1, r_2, \ldots, r_s}}{\prod_{j=1}^{s} (q; q)_{n_j - r_j} (a_j q; q)_{n_j + r_j}}, \quad (5.11)$$

where

$$\alpha'_{r_1, r_2, \ldots, r_s} = \left(\prod_{j=1}^{s} \frac{(\rho_j)_{r_j} (\sigma_j)_{r_j} (a_j q/(\rho_j \sigma_j))^{r_j}}{(a_j q/\rho_j)_{r_j} (a_j q/\sigma_j)_{r_j}} \right) \alpha_{r_1, r_2, \ldots, r_s} \quad (5.12)$$

and

$$\beta'_{n_1,n_2,\ldots,n_s} \tag{5.13}$$

$$= \prod_{j=1}^{s} \sum_{m_j=-\infty}^{n_j} \frac{(\rho_j)_{m_j} (\sigma_j)_{m_j} (a_j q/(\rho_j \sigma_j))_{n_j-m_j} \left(\frac{a_j q}{\rho_j \sigma_j}\right)^{m_j} \beta_{m_j,\ldots,m_s}}{(q)_{n_j-m_j}(a_j q/\rho_j)_{n_j}(a_j q/\sigma_j)_{n_j}}.$$

The instance of this theorem that is utilized in [8] is the

First Corollary *If for* $n_1, n_2, \ldots, n_s \geq 0$

$$\beta_{n_1,n_2,\ldots,n_s} = \sum_{r_1=-n_1}^{n_1} \sum_{r_2=-n_2}^{n_2} \cdots \sum_{r_s=-n_s}^{n_s} \frac{\alpha_{r_1,r_2,\ldots,r_s}}{\prod\limits_{j=1}^{s}(q;q)_{n_j-r_j}(q;q)_{n_j+r_j}}, \tag{5.14}$$

then

$$\sum_{n_1,\ldots,n_s \geq 0} q^{n_1^2+n_2^2+\cdots+n_s^2} \beta_{n_1,n_2,\ldots,n_s} \tag{5.15}$$

$$= \frac{1}{(q;q)_\infty^s} \sum_{m_1,m_2,\ldots,m_s=-\infty} q^{m_1^2+m_2^2+\cdots+m_s^2} \alpha_{m_1,m_2,\ldots,m_s}.$$

To illustrate the implications of this result for q-series identities, new pentagonal number theorems are proved:

$$\sum_{n=1}^{\infty} \frac{q^{2n^2}}{(q;q)_{2n}} = \frac{\displaystyle\sum_{n,m=-\infty}^{\infty} (-1)^{n+m} q^{n(3n-1)/2+m(3m-1)/2+nm}}{\displaystyle\prod_{n-1}^{\infty}(1-q^n)^2}. \tag{5.16}$$

and

$$\sum_{i,j,k\geq 0} \frac{q^{i^2+j^2+k^2}}{(q;q)_{i+j-k}(q;q)_{i+k-j}(q;q)_{j+k-i}} \tag{5.17}$$

$$= \frac{\displaystyle\sum_{n,m,p=-\infty}^{\infty} (-1)^{n+m+p} q^{n(3n-1)/2+m(3m-1)/2+p(3p-1)/2+nm+np+mp}}{\displaystyle\prod_{n=1}^{\infty}(1-q^n)^3}.$$

To further illustrate the application of these results we shall prove:

Theorem 6

$$\sum_{m,n\geq 0} \frac{(-1)^m q^{2m^2+2n^2}(-q;q^2)_{n-m}}{(q;q^2)_{n-m}(q^4;q^4)_m(q^4;q^4)_m} = \frac{\sum_{i,j=-\infty}^{\infty} (-1)^i q^{3i^2+3j^2+2ij}}{(q^2;q^2)_\infty^2}. \tag{5.18}$$

Proof. This result is a direct application of the First Corollary wherein we replace q by q^2, set $s = 2$, and

$$\alpha_{i,j} = (-1)^i q^{(i+j)^2}, \tag{5.19}$$

$$\beta_{m,n} = \frac{(-1)^m(-q;q^2)_{n-m}}{(q;q^2)_{n-m}(q^4;q^4)_m(q^4;q^4)_n}. \tag{5.20}$$

This means that we must prove that

$$\sum_{i,j=-\infty}^{\infty} (-1)^i q^{(i+j)^2} \begin{bmatrix} 2m \\ m+i \end{bmatrix}_{q^2} \begin{bmatrix} 2n \\ n+j \end{bmatrix}_{q^2}$$

$$= \frac{(-1)^m(-q;q^2)_{n-m}(q^2;q^4)_m(q^2;q^4)_n}{(q;q^2)_{n-m}}. \tag{5.21}$$

To see this, we note

$$\sum_{i,j} (-1)^i q^{(i+j)^2} \begin{bmatrix} 2m \\ m+i \end{bmatrix}_{q^2} \begin{bmatrix} 2n \\ n+j \end{bmatrix}_{q^2} \tag{5.22}$$

$$= \sum_j \begin{bmatrix} 2n \\ n+j \end{bmatrix}_{q^2} q^{j^2} \sum_i (-1)^k q^{i^2+2ij} \begin{bmatrix} 2m \\ m+i \end{bmatrix}_{q^2}$$

$$= \sum_j \begin{bmatrix} 2n \\ n+j \end{bmatrix}_{q^2} q^{j^2} (q^{2j+1},q^2)_m(q^{1-2j};q^2)_m$$

(by [4; p. 49, Ex. 1])

$$= \frac{(-1)^m(q;q^2)_{m-n}q^{(n+m)^2}}{(q;q^2)_{-m-n}} {}_2\phi_1 \left(\begin{matrix} q^{-4n}, q^{2m-2n+1} \\ q^{-2m-2n+1} \end{matrix} ; q^2, -q^{2n-2m+1} \right)$$

$$= \frac{(-1)^m(-q;q^2)_{n-m}(q^2;q^4)_m(q^2;q^4)_r}{(q;q^2)_{n-m}}$$

(by [1; eq. (1.7)], [23; Sec. 1.8]),

which is the desired result. □

6. A New Bailey Chain, the WP-Bailey Chain

The work described in Section 4 yields (among many other results) transformations for the series given in (4.6). The most famous of these are the case $k = 1$, a summation due to F. H. Jackson [24], and $k = 2$, Watson's q-analog of Whipple's theorem.

However, there is a classical transformation due to Bailey [23; p. 39, eq. (2.9.1)]) which is not a direct corollary of the Bailey Chains considered in Section 4.

$$
{}_{10}\phi_9 \left(\begin{array}{c} a, qa^{\frac{1}{2}}, -qa^{\frac{1}{2}}, b, c, d, e, f, \lambda aq^{n+1}/ef, q^{-n} \\ a^{\frac{1}{2}}, -a^{\frac{1}{2}}, aq/b, aq/c, aq/d, aq/e, aq/f, efq^{-n}/\lambda, aq^{n+1} \end{array} ; q, q \right)
$$

$$
= \frac{(aq, aq/ef, \lambda q/e, \lambda q/f; q)_n}{(aq/e, aq/f, \lambda q/ef, \lambda q; q)_n} {}_{10}\phi_9 \left(\begin{array}{c} \lambda, q\lambda^{\frac{1}{2}}, q\lambda^{\frac{1}{2}}, \lambda b/a, \lambda c/a, \lambda d/a, \\ \lambda^{\frac{1}{2}} - \lambda^{\frac{1}{2}}, aq/b, aq/c, aq/d, \end{array} \right.
$$

$$
\left. \begin{array}{c} e, f, \lambda aq^{n+1}/ef, q^{-n}, \\ \lambda q/e, \lambda q/f, efq^{-n}/a, \lambda q^{n+1} \end{array} ; q, q \right), \tag{6.1}
$$

where $\lambda = qa^2/bcd$.

If we examine the Bailey method for proving (6.1) (beautifully, presented in [23; Sec. 2.2 and Sec. 2.9]), we find that we can easily produce a new type of Bailey Chain. We shall call it a WP-Bailey Chain because both the series in (6.1) are examples of very well-poised basic hypergeometric series [23; p. 32].

Definition 6.1 *We say that two sequences $\alpha_n(a, k)$ and $\beta_n(a, k)$ form a WP-Bailey pair provided*

$$
\beta_n(a, k) = \sum_{i=0}^{n} \frac{(k/a; q)_{n-i}(k; q)_{n+i}}{(q; q)_{n-i}(aq; q)_{n+i}} \alpha_i(a, k). \tag{6.2}
$$

Note that if $k = 0$ then we have a Bailey pair as considered in Section 4.

Theorem 7 *If $\alpha_n(a, k)$ and $\beta_n(a, k)$ form a WP-Bailey pair, then so do $\alpha'_n(a, k)$ and $\beta'_n(a, k)$, where*

$$
\alpha'_n(a, k) = \frac{(\rho_1; q)_n(\rho_2; q)_n(aq/\rho_1\rho_2)^n}{(aq/\rho_1; q)_n(aq/\rho_2; q)_n} \alpha_n \left(a, \frac{k\rho_1\rho_2}{aq} \right) \tag{6.3}
$$

and

$$
\beta'_n(a, k) = \frac{\left(\frac{k\rho_1}{a}; q \right)_n \left(\frac{k\rho_2}{a} q \right)_n}{\left(\frac{aq}{\rho_1}, q \right)_n \left(\frac{aq}{\rho_2}, q \right)_n} \sum_{j=0}^{n} \frac{(\rho_1; q)_j(\rho_2; q)_j}{\left(\frac{k\rho_2}{a}; q \right)_j \left(\frac{k\rho_1}{a}; q \right)_j} \tag{6.4}
$$

$$
\times \; \frac{\left(1 - \frac{k\rho_1\rho_2 q^{2j-1}}{a}\right)\left(\frac{aq}{\rho_1\rho_2};q\right)_{n-j}(k;q)_{n+j}}{\left(1 - \frac{k\rho_1\rho_2}{aq}\right)(q;q)_{n-j}\left(\frac{k\rho_1\rho_2}{a};q\right)_{n+j}} \left(\frac{aq}{\rho_1\rho_2}\right)^j
$$

$$
\times \; \beta_j\left(a, \frac{k\rho_1\rho_2}{aq}\right).
$$

Remark Note that if we at $k = 0$, this result reduces to the assertions in (4.1)–(4.4).

Proof. We must show that $\alpha_n'(a, k)$ and $\beta_n'(a, k)$ as given in (6.3) and (6.4) actually satisfy (6.2). By (6.4) and (6.2)

$$
\beta_n'(a,k) = \frac{\left(\frac{k\rho_1}{a}, \frac{k\rho_2}{a};q\right)_n}{\left(\frac{aq}{\rho_1}, \frac{aq}{\rho_2};q\right)_n} \sum_{j=0}^{n} \frac{(\rho_1,\rho_2;q)_j}{\left(\frac{k\rho_2}{a}, \frac{k\rho_1}{a};q\right)_j}
$$

$$
\times \; \frac{\left(1 - \frac{k\rho_1\rho_2 q^{2j-1}}{a}\right)\left(\frac{aq}{\rho_1\rho_2};q\right)_{n-j}(k;q)_{n+j}}{\left(1 - \frac{k\rho_1\rho_2}{aq}\right)(q;q)_{n-j}\left(\frac{k\rho_1\rho_2}{a};q\right)_{n+j}} \left(\frac{aq}{\rho_1\rho_2}\right)^j
$$

$$
\times \; \sum_{i=0}^{j} \frac{\left(\frac{k\rho_1\rho_2}{a^2 q};q\right)_{j-i}\left(\frac{k\rho_1\rho_2}{aq};q\right)_{j+1} \alpha_i\left(a, \frac{k\rho_1\rho_2}{aq}\right)}{(q;q)_{j-i}(aq;q)_{j+i}}
$$

$$
= \; \frac{\left(\frac{k\rho_1}{a}, \frac{k\rho_2}{a};q\right)_n}{\left(\frac{aq}{\rho_1}, \frac{aq}{\rho_2};q\right)_n} \sum_{i=0}^{n}\sum_{j=0}^{n-i} \frac{(\rho_1,\rho_2;q)_{j+i}}{\left(\frac{k\rho_2}{a}, \frac{k\rho_1}{a};q\right)_{j+i}}
$$

$$
\times \; \frac{\left(1 - \frac{k\rho_1\rho_2 q^{2i+2j-1}}{a}\right)\left(\frac{aq}{\rho_1\rho_2};q\right)_{n-i-j}(k;q)_{n+i+j}\left(\frac{aq}{\rho_1\rho_2}\right)^{i+j}}{\left(1 - \frac{k\rho_1\rho_2}{aq}\right)(q;q)_{n-i-j}\left(\frac{k\rho_1\rho_2}{a};q\right)_{n+i+j}}
$$

$$
\times \; \frac{\left(\frac{k\rho_1\rho_2}{a^2 q};q\right)_j\left(\frac{k\rho_1\rho_2}{aq};q\right)_{j+2i} \alpha_i\left(a, \frac{k\rho_1\rho_2}{aq}\right)}{(q;q)_j(aq;q)_{j+2i}}
$$

$$
= \; \frac{\left(\frac{k\rho_1}{a}, \frac{k\rho_2}{a};q\right)_n}{\left(\frac{aq}{\rho_1}, \frac{aq}{\rho_2};q\right)_n} \sum_{i=0}^{n} \frac{(\rho_1,\rho_2;q)_i}{\left(\frac{k\rho_2}{a}, \frac{k\rho_1}{a};q\right)_i} \alpha_i\left(a, \frac{k\rho_1\rho_2}{aq}\right)
$$

$$
\times \; \frac{\left(\frac{aq}{\rho_1\rho_2};q\right)_{n-i}(k;q)_{n+i}\left(\frac{aq}{\rho_1\rho_2}\right)^i\left(1 - \frac{k\rho_1\rho_2 q^{2i-1}}{a}\right)\left(\frac{k\rho_1\rho_2}{aq};q\right)_{2i}}{(q;q)_{n-i}\left(\frac{k\rho_1\rho_2}{a};q\right)_{n+i}\left(1 - \frac{k\rho_1\rho_2}{aq}\right)(aq;q)_{2i}}
$$

$$\times \ {}_8\phi_7 \left(\begin{array}{c} \dfrac{k\rho_1\rho_2 q^{2i-1}}{a}, \ q^{i+1}\sqrt{\dfrac{k\rho_1\rho_2}{aq}}, \ -q^{i+1}\sqrt{\dfrac{k\rho_1\rho_2}{aq}}, \ \rho_1 q^i, \rho_2 q^i, \\[2ex] q^i\sqrt{\dfrac{k\rho_1\rho_2}{aq}}, \ -q^i\sqrt{\dfrac{k\rho_1\rho_2}{aq}}, \ \dfrac{k\rho_2 q^i}{a}, \ \dfrac{k\rho_1 q^i}{a}, \\[2ex] \dfrac{k\rho_1\rho_2}{a^2 q}, \ kq^{n+i}, \ q^{-n+i} \\[2ex] aq^{2i+1}, \ \dfrac{\rho_1\rho_2 q^{-n+i}}{a}, \ \dfrac{k\rho_1\rho_2 q^{n+i}}{a} \end{array} ; q, q \right)$$

$$= \frac{\left(\dfrac{k\rho_1}{a}, \dfrac{k\rho_2}{a}; q\right)_n}{\left(\dfrac{aq}{\rho_1}, \dfrac{aq}{\rho_2}; q\right)_n} \sum_{i=0}^{n} \frac{(\rho_1, \rho_2; q)_i}{\left(\dfrac{k\rho_2}{a}, \dfrac{k\rho_1}{a}; q\right)_i} \alpha_i\left(a, \frac{k\rho_1\rho_2}{aq}\right)$$

$$\times \ \frac{\left(\dfrac{aq}{\rho_1\rho_2}; q\right)_{n-i} (k; q)_{n+i} \left(\dfrac{aq}{\rho_1\rho_2}\right)^i \left(\dfrac{k\rho_1\rho_2}{a}; q\right)_{2i}}{(q; q)_{n-i} \left(\dfrac{k\rho_1\rho_2}{a}; q\right)_{n+i} (aq; q)_{2i}}$$

$$\times \ \frac{\left(\dfrac{k\rho_1\rho_2 q^{2i}}{a}, \dfrac{k}{a}, \dfrac{\rho_1 q^{-n}}{a}, \dfrac{\rho_2 q^{-n}}{a}; q\right)_{n-i}}{\left(\dfrac{k\rho_2 q^i}{a}, \dfrac{k\rho_1 q^i}{a}, \dfrac{\rho_1\rho_2 q^{-n+i}}{a}, \dfrac{q^{-n-i}}{a}; q\right)_{n-i}}$$

(by [23; p. 35, eq. (2.6.2)])

$$= \sum_{i=0}^{n} \frac{\left(\dfrac{k}{a}; q\right)_{n-i} (k; q)_{n+i}}{(q; q)_{n-i}(aq; q)_{n+i}} \frac{(\rho_1, \rho_2; q)_i}{\left(\dfrac{aq}{\rho_1}, \dfrac{aq}{\rho_2}; q\right)_i} \left(\frac{aq}{\rho_1\rho_2}\right)^i \alpha_i\left(a, \frac{k\rho_1\rho_2}{aq}\right)$$

$$= \sum_{i=0}^{n} \frac{\left(\dfrac{k}{a}; q\right)_{n-i} (k; q)_{n+i}}{(q; q)_{n-i}(aq; q)_{n+i}} \alpha_i'(a, k), \tag{6.5}$$

as desired. □

It is now a simple matter to begin a WP-Bailey chain by noting that

$$\beta_n(a, k) = \delta_{n,0}, \tag{6.6}$$

$$\alpha_n(a, k) = \frac{(a; q)_n (1 - aq^{2n})\left(\dfrac{a}{k}; q\right)_n}{(q; q)_n (1 - a)(kq; q)_n} \left(\frac{k}{a}\right)^n \tag{6.7}$$

forms a WP-Bailey pair. This follows immediately by setting $m = n$ in the following identity

$$\sum_{n=0}^{m} \frac{\left(\dfrac{k}{a}; q\right)_{n-i} (k; q)_{n+i} \alpha_i(a, k)}{(q; q)_{n-i}(aq; q)_{n+i}}$$

$$= \frac{\left(\frac{kq^{-m}}{a};q\right)_n (1-k)(kq^{m+1};q)_n (q^{1-n};q)_m q^{nm}}{(1-kq^n)(aq^{m+1};q)_n (q;q)_n (q;q)_m}. \tag{6.8}$$

Identity (6.8) follows immediately by mathematical induction on m.

Corollary (F. H. Jackson's q-analog of Dougall's theorem [24], [23; p. 35, eq. (2.6.2)])

$$_8\phi_7 \left(\begin{array}{c} a, q\sqrt{a}, -q\sqrt{a}, kq^n, \rho_1, \rho_2, \dfrac{a^2 q}{k\rho_1\rho_2}, q^{-n} \\ \sqrt{a}, -\sqrt{a}, \dfrac{aq^{1-n}}{k}, \dfrac{aq}{\rho_1}, \dfrac{aq}{\rho_2}, \dfrac{k\rho_1\rho_2}{a}, aq^{n+1} \end{array} ; q, q \right)$$

$$= \frac{\left(aq, \frac{aq^{1-n}}{\rho_1 k}, \frac{aq^{1-n}}{\rho_2 k}, \frac{aq}{\rho_1\rho_2}; q\right)_n}{\left(\frac{aq^{1-n}}{k}, \frac{aq}{\rho_1}, \frac{aq}{\rho_2}, \frac{aq^{1-n}}{k\rho_1\rho_2}; q\right)_n}. \tag{6.9}$$

Proof. This is just the assertion that the WP-Bailey pair $\alpha'_n(a,k)$ and $\beta'_n(a,k)$ defined from Theorem 7 satisfy (6.2) where $\alpha_n(a,k)$ and $\beta_n(a,k)$ are given by (6.6) and (6.7). $\square$

Corollary (W. N. Bailey's very well-poised $_{10}\phi_9$ identity [23; p. 38, eq. (2.9.1)]) *Equation (6.1) is valid.*

Proof. This is (after changes of variable), the assertion that the WP-Bailey pair $\alpha''_n(a,k)$, $\beta''_n(a,k)$ (with new parameters ρ_3 and ρ_4) defined from (6.3) and (6.4) satisfy (6.2) where $\alpha'_n(a,k)$ and $\beta'_n(a,k)$ are given in the previous corollary. $\square$

We close this section by remarking that the N-th entry in this WP-Bailey chain produces a representation of

$$_{2N+4}\phi_{2N+3} \left(\begin{array}{c} a, q\sqrt{a}, -a\sqrt{a}, \rho_1, \rho_2, \rho_3\rho_4, \ldots, \rho_{2N-3}, \rho_{2N-2}, \lambda q^n, q^{-n} \\ \sqrt{a}, -\sqrt{a}, \frac{aq}{\rho_1}, \frac{aq}{\rho_2} \cdots \frac{aq}{\rho_{2N-3}}, \frac{aq}{\rho_{2N-2}}, \frac{aq^{1-n}}{\lambda}, aq^{n+1} \end{array} ; q, q \right) \tag{6.10}$$

(where $a^N q^{N-1} = \rho_1\rho_2\cdots\rho_{2N-2}\lambda$) as an $(N-2)$ fold summation. Thus when $N = 4$ (the first new case) we find the very-well poised $_{12}\phi_{11}$ of the above type represented by a double sum.

7. An Alternative WP-Bailey Pair

In [13; Sec. 9] and [14], W. N. Bailey used ideas similar to those recounted up to this point in order to develop identities for nearly poised series. These are series in which one column of parameters does not satisfy the well-poised requirement.

When we translate Bailey's ideas into the world of Bailey chains, a most extraordinary thing happens. Namely, in contrast with Theorem 3, we find that from any given WP-Bailey pair we can construct a new WP-Bailey pair $\tilde{\alpha}_n(a,k)$, $\tilde{\beta}_n(a,k)$ which is not at all the pair given by (6.3) and (6.4).

Theorem 8 *If $\alpha_n(a,k)$ and $\beta_n(a,k)$ form a WP-Bailey pair, then so do $\tilde{\alpha}_n(a,k)$ and $\tilde{\beta}_n(a,k)$, where*

$$\tilde{\alpha}_n(a,k) = \frac{\left(\frac{qa^2}{k};q\right)_{2n}\left(\frac{k^2}{qa^2}\right)^n}{(k;q)_{2n}}\,\alpha_n\left(a,\frac{qa^2}{k}\right), \qquad (7.1)$$

$$\tilde{\beta}_n(a,k) = \sum_{j=0}^n \frac{\left(\frac{k^2}{a^2q};q\right)_{n-j}}{(q;q)_{n-j}}\left(\frac{k^2}{a^2q}\right)^j \beta_j\left(a,\frac{aq^2}{k}\right). \qquad (7.2)$$

Remark There is absolutely no intersection of the WP-Bailey pairs produced by Theorem 3 and those produced by Theorem 4.

Proof. We are given that the sequences $\alpha_n(a,k)$ and $\beta_n(a,k)$ satisfy (6.2). Hence

$$
\begin{aligned}
\tilde{\beta}_n(a,k) &= \sum_{j=0}^n \frac{\left(\frac{k^2}{a^2q};q\right)_{n-j}\left(\frac{k^2}{a^2q}\right)^j}{(q;q)_{n-j}} \sum_{i=0}^j \frac{\left(\frac{qa}{k};q\right)_{j-1}\left(\frac{qa^2}{k};q\right)_{j+i}}{(q;q)_{j-i}(aq;q)_{j+i}}\alpha_i\left(a,\frac{qa^2}{k}\right) \\
&= \sum_{i=0}^n \sum_{j=0}^{n-i} \frac{\left(\frac{k^2}{a^2q};q\right)_{n-i-j}\left(\frac{k^2}{a^2q}\right)^{j+i}\left(\frac{qa}{k};q\right)_j\left(\frac{aq^2}{k};q\right)_{j+2i}}{(q;q)_{n-i-j}(q;q)_j(aq;q)_{j+2i}}\alpha_i\left(a,\frac{qa^2}{k}\right) \\
&= \sum_{i=0}^n \frac{\left(\frac{k^2}{a^2q};q\right)_{n-i}\left(\frac{k^2}{a^2q}\right)^i\left(\frac{aq^2}{k};q\right)_{2i}\alpha_i\left(a,\frac{aq^2}{k}\right)}{(q;q)_{n-i}(aq;q)_{2i}} \\
&\quad \times \sum_{j=0}^{n-i} \frac{(q^{-n+i};q)_j}{\left(\frac{a^2q^{2-n+i}}{k^2};q\right)_j}\,q^j\,\frac{\left(\frac{aq}{k};q\right)_j\left(\frac{a^2q^{2i+1}}{k};q\right)_j}{(q;q)_j(aq^{2i+1};q)_j} \\
&= \sum_{i=0}^n \frac{\left(\frac{k^2}{a^2q};q\right)_{n-i}\left(\frac{k^2}{a^2q}\right)^i\left(\frac{qa^2}{k};q\right)_{2i}\alpha_i\left(a,\frac{qa^2}{k}\right)}{(q;q)_{n-i}(aq;q)_{2i}} \\
&\quad \times \frac{\left(\frac{k}{a};q\right)_{n-i}(kq^{2i};q)_{n-i}}{(aq^{2i+1};q)_{n-i}\left(\frac{k^2}{a^2q};q\right)_{n-i}}
\end{aligned}
$$

(by [23; p. 13, eq. (1.7.2)])

$$= \sum_{i=0}^{n} \frac{\left(\frac{k}{a};q\right)_{n-i}(k;q)_{n+i}}{(q;q)_{n-i}(aq;q)_{n+i}} ; \frac{\left(\frac{qa^2}{k};q\right)_{2i}\left(\frac{k^2}{a^2q}\right)^i \alpha_i\left(a;\frac{qa^2}{k}\right)}{(k;q)_{2i}}$$

$$= \sum_{i=0}^{n} \frac{\left(\frac{k}{a};q\right)_{n-i}(k;q)_{n+i}}{(q;q)_{n-i}(aq:q)_{n+i}} \widetilde{\alpha}_i(a,k), \tag{7.3}$$

as desired.   □

The most immediate application of Theorem 4 is the application in which the WP-Bailey pair $\alpha'_n(a,k)$ and $\beta'_n(a,k)$ of the first Corollary are taken as input. The resulting assertion that $\widetilde{\alpha}_n(a,k)$ and $\widetilde{\beta}_n(a,k)$ form a WP-Bailey pair is equivalent to the result of Bailey [13; p. 431]:

$$_5\phi_4\left(\begin{array}{c} a,\ b,\ c,\ d,\ q^{-N} \\ aq/b,\ aq/c,\ aq/d,\ a^2q^{-N}/k^2 \end{array}; q,\, q\right) \tag{7.4}$$

$$= \frac{(kq/a)_N(k^2q/a)_N}{(kq)_N(k^2q/a^2)_N}$$

$$\times {}_{12}\phi_{11}\left(\begin{array}{c} k,\ q\sqrt{k},-q\sqrt{k}, kb/a,\ kc/a,\ kd/a,\quad \sqrt{a},-\sqrt{a}, \\ \sqrt{k},\ -\sqrt{k},\ aq/b,\ aq/c,\ aq/d,\ kq/\sqrt{a},\ -kq/\sqrt{a}, \end{array}\right.$$
$$\left.\begin{array}{c} \sqrt{(aq)},\quad -\sqrt{(aq)},\ k^2q^{N+1}/a,\quad q^{-N}, \\ k\sqrt{(q/a)},\ -k\sqrt{(q/a)},\quad aq^{-N}/k,\ kq^{N+1} \end{array}; q,\, q\right).$$

8. Summary and the Bailey Tree

First it should be noted that this survey paper has shortchanged a number of papers that have made substantial contributions to Bailey chains. Among these are [16], [20], [21], [22], [29]. I would draw special attention to [40] wherein Warnaar gives an extended account of Bailey chains especially emphasizing their role in statistical mechanics.

Finally, I note again the surprise contained in Sections 7 and 8. The new, two parameter WP-Bailey pairs (which reduce to the original Bailey pairs when $k = 0$) generate new WP-Bailey pairs in two entirely different ways (Theorems 3 and 4). Thus at each step in the generation of a chain there are two offspring pairs. So instead of generating a chain, we are generating a binary tree. We shall wait to explore Bailey Trees subsequently.

The author was supported in part by NSF Grant DMS-9206993.

References

1. G. E. Andrews, *On the q-analog of Kummer's theorem and applications*, Duke Math. J. **40** (1973), 525–528.

2. G. E. Andrews, *An analytic generalization of the Rogers-Ramanujan identities for odd moduli*, Proc. Nat. Acad. Sci. USA **71** (1974), 4082–4085.

3. G. E. Andrews, *Problems and prospects for basic hypergeometric functions* in: "Theory and Applications of Special Functions", R. Askey, ed., Academic Press, New York, 1975, pp. 191–224.

4. G. E. Andrews, *The Theory of Partitions*, Encyl. of Math. and Its Appl., Vol. 2, Addison-Wesley, Reading, 1976; Reissued: Cambridge University Press, Cambridge, 1985 and 1998.

5. G. E. Andrews, *Multiple series Rogers-Ramanujan type identities*, Pacific J. Math. **114** (1984), 267–283.

6. G. E. Andrews, *The fifth and seventh order mock theta functions*, Trans. Amer. Math. Soc. **293** (1986), 113–134.

7. G. E. Andrews, *q-Series: Their Development*, CBMS Regional Conf. Lecture Series 66, Amer. Math. Soc., Providence, 1986.

8. G. E. Andrews, *Umbral calculus, Bailey chains and pentagonal number theorems*, J. Comb. Theory, Ser. A, **91** (2000), 464–475.

9. G. E. Andrews and A. Berkovich, *A trinomial analogue of Bailey's lemma and $N = 2$ superconformal invariance*, Comm. in Math. Phys. **192** (1998), 245–260.

10. G. E. Andrews and D. Bowman, *The Bailey transform and D. B. Sears*, Quaest. Math. **22** (1999), 19–26.

11. G. E. Andrews, F. J. Dyson and D. Hickerson, *Partitions and indefinite quadratic forms*, Invent. Math. **91** (1988), 391–407.

12. G. E. Andrews, A. Schilling and S. O. Warnaar, *An A_2 Bailey Lemma and Rogers-Ramanujan-type identities*, J. Amer. Math. Soc. **12** (1999), 677–702.

13. W. N. Bailey, *Some identities in combinatory analysis*, Proc. London Math. Soc. (2) **49** (1947), 421–435.

14. W. N. Bailey, *A transformation of nearly-poised basic hypergeometric series*, J. London Math. Soc. **22** (1947), 237–240.

15. W. N. Bailey, *Identities of the Rogers-Ramanujan type*, Proc. London Math. Soc. (2) **50** (1949), 1–10.

16. A. Berkovich, B. M. McCoy and A. Schilling, *$N = 2$ supersymmetry and Bailey pairs*, Phys. A **228** (1996), 33–62.

17. J. M. and P. B. Borwein, *Pi and the AGM*, Wiley, New York, 1987.

18. D. M. Bressoud, *Some identities for terminating q-series*, Math. Proc. Camb. Phil. Soc. **81** (1981), 211–223.

19. D. M. Bressoud, *An easy proof of the Rogers-Ramanujan identities*, J. Number Theory **16** (1983), 235–241.

20. D. M. Bressoud, *A matrix inverse*, Proc. Amer. Math. Soc. **88** (1983), 446–448.

21. D. M. Bressoud, *The Bailey lattice: an introduction*, in: "Ramanujan Revisited", G. E. Andrews et al, eds., Academic Press, New York 1988, pp. 57–67.

22. O. Foda and V.-H. Iuano, *Virasoro character identities from the Andrews-Bailey construction*, Int. J. Mod. Phys. A **12** (1997), 1651–1676.

23. G. Gasper and M. Rahman, *Basic Hypergeometric Series*, Encyl. of Math and Its Appl., Vol. 35, Cambridge University Press, Cambridge, 1990.

24. F. H. Jackson, *Summation of q-hypergeometric series*, Messenger of Math. **50** (1921), 101–112.

25. G. M. Lilly and S. C. Milne, *The A_ℓ and C_ℓ Bailey transform and lemma*, Bull. Amer. Math. Soc. (NS) **26** (1992), 258–263.

26. G. M. Lilly and S. C. Milne, *The C_ℓ Bailey transform and Bailey lemma*, Constr. Approx. **9** (1993), 473–500.

27. G. M. Lilly and S. C. Milne, *Consequences of the A_ℓ and C_ℓ Bailey transform and Bailey lemma*, Discr. Math. **139** (1995), 319–346.

28. P. Paule, *On identities of the Rogers-Ramanujan type*, J. Math. Anal. and Appl. **107** (1985), 225–284.

29. P. Paule, *The concept of Bailey chains*, Sem. Lothar. Combin. B, 18f (1987), 24.

30. L. J. Rogers, *Second memoir on the expansion of certain infinite products*, Proc. London Math. Soc. **25** (1894), 318–343.

31. L. J. Rogers, *On two theorems of combinatory analysis and allied identities*, Proc. London Math. Soc. (2) **16** (1917), 315–336.

32. A. Schilling and S. O. Warnaar, *A higher-level Bailey lemma*, Int. J. Mod. Phys., **B11** (1997), 189–195.

33. A. Schilling and S. O. Warnaar, *A higher level Bailey lemma, proof and application*, Ramanujan journal **2** (1998), 327–349.

34. A. Schilling and S. O. Warnaar, *Conjugate Bailey pairs*, to appear.

35. D. B. Sears, *On the transformation theory of basic hypergeometric functions*, Proc. London Math. Soc. (2) **53** (1951), 158–180.

36. L. J. Slater, *A new proof of Rogers' transformations of infinite series*, Proc. London Math. Soc. (2) **53** (1951), 460–475.

37. L. J. Slater, *Further identities of the Rogers-Ramanujan type*, Proc. London Math. Soc. (2) **54** (1952), 147–167.

38. L. J. Slater, *Generalized Hypergeometric Functions*, Cambridge University Press, Cambridge, 1966.

39. O. Warnaar, *A note on the trinomial analogue of Bailey's lemma*, J. Combin. Theory A **81** (1998), 114–118.

40. O. Warnaar, *50 years of Bailey's lemma*, Kerber Festschrift, to appear.

41. G. N. Watson, *A new proof of the Rogers-Ramanujan identities*, J. London Math. Soc., **4** (1929) pp. 4–9.

RIEMANN-HILBERT PROBLEMS FOR MULTIPLE ORTHOGONAL POLYNOMIALS

WALTER VAN ASSCHE
Department of Mathematics
Katholieke Universiteit Leuven
Celestijnenlaan 200B
B-3001 Leuven Belgium

JEFFREY S. GERONIMO
School of Mathematics
Georgia Institute of Technology
Atlanta, GA 30332 USA

AND

ARNO B. J. KUIJLAARS
Department of Mathematics
Katholieke Universiteit Leuven
Celestijnenlaan 200B
B-3001 Leuven Belgium

Abstract. In the early nineties, Fokas, Its and Kitaev observed that there is a natural Riemann-Hilbert problem (for 2×2 matrix functions) associated with a system of orthogonal polynomials. This Riemann-Hilbert problem was later used by Deift et al. and Bleher and Its to obtain interesting results on orthogonal polynomials, in particular strong asymptotics which hold uniformly in the complex plane. In this paper we will show that a similar Riemann-Hilbert problem (for $(r+1) \times (r+1)$ matrix functions) is associated with multiple orthogonal polynomials. We show how this helps in understanding the relation between two types of multiple orthogonal polynomials and the higher order recurrence relations for these polynomials. Finally we indicate how an extremal problem for vector potentials is important for the normalization of the Riemann-Hilbert problem. This ex-

J. Bustoz et al. (eds.), Special Functions 2000, 23–59.

tremal problem also describes the zero behavior of the multiple orthogonal polynomials.

1. Introduction

Recently it was observed that one can describe various aspects of the theory of orthogonal polynomials using a Riemann-Hilbert problem (Fokas, Its, Kitaev [12] [13] [16], Deift [6], Deift et al. [7] [8]). The Riemann-Hilbert problem is to find a complex 2×2 matrix valued function which is analytic in $\mathbb{C} \setminus \mathbb{R}$, having a prescribed growth as $z \to \infty$, which satisfies a jump condition when crossing the real line. The jump matrix contains the weight function w with respect to which the polynomials are orthogonal. In this paper we will show that an extension of this Riemann-Hilbert approach to $(r+1) \times (r+1)$ matrix valued functions describes certain polynomials which obey orthogonality conditions with respect to $r > 1$ weights on the real line. These polynomials are known as multiple orthogonal polynomials [2] [19] [21] [22] [26].

In this introduction we will explain the notion of multiple orthogonal polynomials. In Section 2 we describe the Riemann-Hilbert problem for type I multiple orthogonal polynomials and Section 3 gives the Riemann-Hilbert problem for type II multiple orthogonal polynomials. In Section 4 we will use these Riemann-Hilbert problems to give a relation between type I and type II multiple orthogonal polynomials. It is worth noting that one can also introduce various combinations of type I and type II multiple orthogonal polynomials, which can all be shown to be related. Section 5 shows that the type II multiple orthogonal polynomials satisfy some finite order recurrence relations, a (known) property which in this paper follows as a nice consequence of this Riemann-Hilbert approach. Finally, in Section 6 we show how to normalize the Riemann-Hilbert problem so that the growth condition for $z \to \infty$ is replaced by the condition that one obtains the identity matrix as $z \to \infty$. This normalization involves a vector of probability measures describing the asymptotic zero distribution of the multiple orthogonal polynomials, and this vector of measures solves an equilibrium problem in logarithmic potential theory. We will give a survey of how to obtain the asymptotic zero distribution for an Angelesco system (Section 6.1) and for a Nikishin system (Section 6.2) and indicate how this vector of equilibrium measures is used for the normalization of the Riemann-Hilbert problem. Strong asymptotics for multiple orthogonal polynomials has been obtained in a general setting for Angelesco systems [1] and Nikishin systems [3]. The Riemann-Hilbert approach gives an alternative way to study the asymptotics. This aspect, however, is outside the scope of this survey, but will be considered in forthcoming publications.

Multiple orthogonal polynomials are related to Hermite-Padé approximation to a system of Markov functions [21, Chapter 4]. Let $f_1, f_2, \ldots, f_r$ be r Markov functions, i.e.,

$$f_j(z) = \int_{\Delta_j} \frac{w_j(x)}{z - x}\, dx, \qquad z \notin \Delta_j,\; j = 1, 2, \ldots, r,$$

where each Δ_j is a real interval. We will think of the weights w_j as weights on the real line such that $w_j(x) = 0$ for $x \notin \Delta_j$. We also assume that the limits

$$f_j^+(x) = \lim_{\epsilon \to 0+} f_j(x + i\epsilon), \quad f_j^-(x) = \lim_{\epsilon \to 0+} f_j(x - i\epsilon),$$

exist, so that the Sokhotsky-Plemelj formula holds

$$f_j^+(x) - f_j^-(x) = -2\pi i w_j(x), \qquad x \in \mathbb{R}. \tag{1.1}$$

1.1. TYPE I MULTIPLE ORTHOGONAL POLYNOMIALS

Multiple orthogonal polynomials are also known as poly-orthogonal polynomials or Hermite-Padé polynomials. Good references are Aptekarev [2], Mahler [19], the book by Nikishin and Sorokin [21], Nuttall [22], and [26]. Here we briefly explain what we mean by type I and type II multiple orthogonal polynomials, how they arise from a problem of simultaneous rational approximation, and state the orthogonality conditions.

Let $\vec{n} = (n_1, n_2, \ldots, n_r)$ be a multi-index in $\mathbb{N}^r$. For type I Hermite-Padé approximation we look for a vector of polynomials $\vec{A}_{\vec{n}} = (A_{\vec{n},1}, A_{\vec{n},2}, \ldots, A_{\vec{n},r})$, where $A_{\vec{n},j}$ has degree $n_j - 1$, and a polynomial $B_{\vec{n}}$ such that

$$\sum_{j=1}^{r} A_{\vec{n},j}(z) f_j(z) - B_{\vec{n}}(z) = \mathcal{O}(1/z^{n_1 + n_2 + \cdots + n_r}), \qquad z \to \infty. \tag{1.2}$$

The type I vector polynomial $(A_{\vec{n},1}, A_{\vec{n},2}, \ldots, A_{\vec{n},r})$ satisfies a number of orthogonality conditions, namely

$$\int x^k \sum_{j=1}^{r} A_{\vec{n},j}(x) w_j(x)\, dx = 0, \qquad k = 0, 1, 2, \ldots, n_1 + n_2 + \cdots + n_r - 2. \tag{1.3}$$

This gives $n_1 + n_2 + \cdots + n_r - 1$ homogeneous equations for the $n_1 + n_2 + \cdots + n_r$ unknown coefficients, so that we can find the vector polynomial up to a common factor if the matrix of the linear system has full rank (the index $\vec{n}$ is called normal in this case). We will normalize the solution by imposing that

$$\int x^{n_1+n_2+\cdots+n_r-1} \sum_{j=1}^{r} A_{\vec{n},j}(x) w_j(x)\, dx = 1. \tag{1.4}$$

The solution will exist and be unique if the weights $w_1, w_2, \ldots, w_r$ or the functions $f_1, f_2, \ldots, f_r$ are sufficiently 'independent'. Some useful systems are

  − Angelesco systems: in this case $\overset{\circ}{\Delta}_i \cap \overset{\circ}{\Delta}_j = \emptyset$ whenever $i \neq j$;
  − AT systems: in this case $\Delta_j = \Delta$ for all j and

$$\begin{aligned} \{w_1, xw_1, \ldots, x^{n_1-1}w_1, w_2, xw_2, \ldots, x^{n_2-1}w_2, \\ \ldots, w_r, xw_r, \ldots, x^{n_r-1}w_r\} \end{aligned} \tag{1.5}$$

is a Chebyshev system for every multi-index $(n_1, n_2, \ldots, n_r)$, i.e., every linear combination of the basis functions in (1.5) has at most $n_1 + n_2 + \cdots + n_r - 1$ zeros on Δ.

Both systems guarantee the existence and uniqueness of the type I multiple orthogonal polynomials. Mixtures of Angelesco and AT systems have been discussed recently [15]. The remaining polynomial $B_{\vec{n}}$ in (1.2) is given by

$$B_{\vec{n}}(z) = \int \sum_{j=1}^{r} \frac{A_{\vec{n},j}(z) - A_{\vec{n},j}(x)}{z - x} w_j(x)\, dx.$$

1.2. TYPE II MULTIPLE ORTHOGONAL POLYNOMIALS

Type II Hermite-Padé approximation consists of finding a polynomial $P_{\vec{n}}$ of degree $n_1 + n_2 + \cdots + n_r$ and polynomials $Q_{\vec{n},j}$ ($j = 1, 2, \ldots, r$) such that

$$P_{\vec{n}}(z) f_j(z) - Q_{\vec{n},j}(z) = \mathcal{O}(1/z^{n_j+1}), \qquad z \to \infty, \; j = 1, 2, \ldots, r. \tag{1.6}$$

In this case we look for rational functions approximating $f_1, f_2, \ldots, f_r$ near infinity and with the same denominator $P_{\vec{n}}$. This denominator satisfies a number of orthogonality conditions and is known as a type II multiple orthogonal polynomial:

$$\int x^k P_{\vec{n}}(x) w_j(x)\, dx = 0, \qquad k = 0, 1, \ldots, n_j - 1, \; j = 1, 2, \ldots, r. \tag{1.7}$$

The orthogonality conditions are now distributed over the r weights. We have $n_1 + n_2 + \cdots + n_r$ homogeneous linear conditions for the $n_1 + n_2 + \cdots +$

$n_r + 1$ unknown coefficients of $P_{\vec{n}}$. If the matrix of this linear system has full rank, then the index $\vec{n}$ is called normal and the multiple orthogonal polynomial is unique up to a constant factor. We will normalize the polynomial by taking it to be monic, i.e., with leading coefficient one. Again existence and uniqueness is guaranteed for Angelesco systems and AT systems. The numerator polynomials will be given by

$$Q_{\vec{n},j}(z) = \int \frac{P_{\vec{n}}(z) - P_{\vec{n}}(x)}{z - x} w_j(x)\, dx.$$

2. Riemann-Hilbert Problems for Type I Multiple Orthogonal Polynomials

Fokas, Its, and Kitaev [16] [12] [13] observed that there is a natural Riemann-Hilbert problem associated with a system of orthogonal polynomials with weight function $w(x)$ on the real line. Deift has described this Riemann-Hilbert problem and some of its applications in his lecture notes [6]. He and his collaborators [6] [7] and Bleher and Its [4] have used this idea to obtain many interesting results on orthogonal polynomials and random matrices. Here we want to extend this idea to the multiple orthogonal polynomials described in the previous section.

Theorem 2.1 *Consider the following Riemann-Hilbert problem: determine an $(r+1) \times (r+1)$ matrix function $Y(z)$ such that*

1. *$Y(z)$ is analytic in $\mathbb{C} \setminus \mathbb{R}$.*
2. *On the real line there is the jump condition*

$$Y^+(x) = Y^-(x) \begin{pmatrix} 1 & 0 & \cdots & 0 & -2\pi i w_1(x) \\ 0 & 1 & \cdots & 0 & -2\pi i w_2(x) \\ \vdots & & \ddots & \vdots & \vdots \\ 0 & \cdots & 0 & 1 & -2\pi i w_r(x) \\ 0 & \cdots & 0 & 0 & 1 \end{pmatrix}, \qquad (2.1)$$

where $Y^{\pm}(x) = \lim_{\epsilon \to 0+} Y(x \pm i\epsilon)$.
3. *For $z \to \infty$ we have*

$$\lim_{z \to \infty} Y(z) \begin{pmatrix} z^{-n_1} & & & & 0 \\ & z^{-n_2} & & & \\ & & \ddots & & \\ & & & z^{-n_r} & \\ 0 & & & & z^{n_1 + n_2 + \cdots + n_r} \end{pmatrix} = I. \qquad (2.2)$$

If the indices $\vec{n}$ and $\vec{n} + \vec{e}_k$ ($k = 1, 2, \ldots, r$) are normal, where $\vec{e}_k = (0, \ldots, 0, 1, 0, \ldots, 0)$ with 1 on the kth position, then the solution is unique and given by

$$Y(z) = \begin{pmatrix} c_1^{-1}\vec{A}_{\vec{n}+\vec{e}_1}(z) & c_1^{-1}R_{\vec{n}+\vec{e}_1}(z) \\ c_2^{-1}\vec{A}_{\vec{n}+\vec{e}_2}(z) & c_2^{-1}R_{\vec{n}+\vec{e}_2}(z) \\ \vdots & \vdots \\ c_r^{-1}\vec{A}_{\vec{n}+\vec{e}_r}(z) & c_r^{-1}R_{\vec{n}+\vec{e}_r}(z) \\ \vec{A}_{\vec{n}}(z) & R_{\vec{n}}(z) \end{pmatrix} \tag{2.3}$$

where $\vec{A}_{\vec{n}}$ is the vector containing the type I multiple orthogonal polynomials,

$$R_{\vec{n}}(z) = \int \sum_{j=1}^{r} A_{\vec{n},j}(x) w_j(x) \frac{dx}{z - x},$$

and c_j is the leading coefficient of $A_{\vec{n}+\vec{e}_j, j}$.

Proof. Let us write the matrix Y as

$$Y = \begin{pmatrix} U & v \\ u^t & g \end{pmatrix}$$

where U is an $r \times r$ matrix function, v and u are column vector functions of size r (with u^t the transpose of u), and g is a complex function. The jump condition (2.1) then implies

$$U^+(x) = U^-(x),$$

so that U is analytic on the whole complex plane. The asymptotic condition (2.2) implies

$$\lim_{n \to \infty} U(z) \begin{pmatrix} z^{-n_1} & & & 0 \\ & z^{-n_2} & & \\ & & \ddots & \\ 0 & & & z^{-n_r} \end{pmatrix} = I,$$

so that each diagonal element $U_{k,k}(z)$ is a monic polynomial of degree n_k and each non-diagonal element $U_{k,j}(z)$ (with $k \neq j$) is a polynomial of degree at most $n_j - 1$. Here we used the fact that an entire function f for which $f(z)/z^n$ remains bounded as $z \to \infty$, is a polynomial of degree at most n (Liouville). For the vector u the jump condition (2.1) is

$$u^+(x) = u^-(x),$$

so that u is entire, and the asymptotic condition is

$$\lim_{n\to\infty} u_k(z)/z^{n_k} = 0, \qquad k = 1, \dots, r,$$

so that each $u_k(z)$ is a polynomial of degree at most $n_k - 1$. We will show that the polynomials in U and u satisfy the orthogonality conditions for type I multiple orthogonal polynomials. For this we use the jump condition (2.1) which for the vector v becomes

$$v_k^+(z) = v_k^-(z) - 2\pi i \sum_{j=1}^{r} U_{k,j}(x) w_j(x),$$

but then the Sokhotsky-Plemelj formula (1.1) implies that

$$v_k(z) = \int \sum_{j=1}^{r} \frac{U_{k,j}(x) w_j(x)}{z - x} \, dx, \qquad z \notin \mathbb{R}.$$

Use the expansion

$$\frac{1}{z - x} = \sum_{\ell=0}^{n-1} \frac{x^\ell}{z^{\ell+1}} + \frac{x^n}{z^n} \frac{1}{z - x}, \tag{2.4}$$

to find for any n

$$v_k(z) = \sum_{\ell=0}^{n-1} \frac{1}{z^{\ell+1}} \int x^\ell \sum_{j=1}^{r} U_{k,j}(x) w_j(x) \, dx + \frac{1}{z^n} \int x^n \sum_{j=1}^{r} \frac{U_{k,j}(x) w_j(x)}{z - x} \, dx. \tag{2.5}$$

The asymptotic condition (2.2) for v is

$$\lim_{z\to\infty} v_k(z) z^{n_1 + n_2 + \cdots + n_r} = 0, \qquad k = 1, 2, \dots, r,$$

which combined with (2.5) implies

$$\int x^\ell \sum_{j=1}^{r} U_{k,j}(x) w_j(x) \, dx = 0, \qquad \ell = 0, 1, \dots, n_1 + n_2 + \cdots + n_r - 1.$$

Finally for the function g the jump condition (2.1) becomes

$$g^+(x) = g^-(x) - 2\pi i \sum_{j=1}^{r} u_j(x) w_j(x),$$

so that the Sokhotsky-Plemelj formula (1.1) implies that

$$g(z) = \int \sum_{j=1}^{r} \frac{u_j(x)w_j(x)}{z-x}\,dx, \qquad z \notin \mathbb{R}.$$

Again, this can be written as

$$g(z) = \sum_{\ell=0}^{n-1} \frac{1}{z^{\ell+1}} \int x^\ell \sum_{j=1}^{r} u_j(x)w_j(x)\,dx + \frac{1}{z^n} \int x^n \sum_{j=1}^{r} \frac{u_j(x)w_j(x)}{z-x}\,dx. \tag{2.6}$$

The asymptotic condition (2.2) for g becomes

$$\lim_{z\to\infty} g(z)z^{n_1+n_2+\cdots n_r} = 1,$$

which combined with (2.6) gives

$$\int x^\ell \sum_{j=1}^{r} u_j(x)w_j(x)\,dx = \begin{cases} 0, & \text{if } \ell = 0,1,2,\ldots,n_1 + n_2 + \cdots + n_r - 2, \\ 1, & \text{if } \ell = n_1 + n_2 + \cdots + n_r - 1. \end{cases}$$

If we compare all these orthogonality conditions with the orthogonality conditions (1.3) and (1.4) of type I multiple orthogonal polynomials, then we see that $u(z) = A_{\vec{n}}(z)$. Similarly, the kth row of $U(z)$ satisfies the orthogonality conditions (1.3) for $A_{\vec{n}+\vec{e}_k}(z)$, and the requirement that $U_{k,k}(z)$ is a monic polynomial of degree n_k shows that the appropriate normalizing factor is c_k^{-1}, where c_k is the leading coefficient of $A_{\vec{n}+\vec{e}_k,k}(z)$. $\qquad\square$

3. Riemann-Hilbert Problems for Type II Multiple Orthogonal Polynomials

There is a similar Riemann-Hilbert problem for type II multiple orthogonal polynomials.

Theorem 3.1 *Consider the following Riemann-Hilbert problem: determine an $(r+1) \times (r+1)$ matrix function $Z(z)$ such that*

1. *$Z(z)$ is analytic in $\mathbb{C} \setminus \mathbb{R}$,*
2. *On the real line there is the jump condition*

$$Z^+(x) = Z^-(x) \begin{pmatrix} 1 & -2\pi i w_1(x) & -2\pi i w_2(x) & \cdots & -2\pi i w_r(x) \\ 0 & 1 & 0 & \cdots & 0 \\ 0 & 0 & 1 & \cdots & 0 \\ \vdots & \cdots & & \ddots & \vdots \\ 0 & \cdots & 0 & 0 & 1 \end{pmatrix}$$

$$\tag{3.1}$$

3. *For $z \to \infty$ we have*

$$\lim_{z\to\infty} Z(z) \begin{pmatrix} z^{-n_1-n_2-\cdots-n_r} & & & & 0 \\ & z^{n_1} & & & \\ & & z^{n_2} & & \\ & & & \ddots & \\ 0 & & & & z^{n_r} \end{pmatrix} = I. \tag{3.2}$$

If the indices $\vec{n}$ and $\vec{n} - \vec{e}_k$ $(k = 1, 2 \ldots, r)$ are normal, then the solution is unique and given by

$$Z(z) = \begin{pmatrix} P_{\vec{n}}(z) & \vec{R}_{\vec{n}}(z) \\ d_1 P_{\vec{n}-\vec{e}_1}(z) & d_1 \vec{R}_{\vec{n}-\vec{e}_1}(z) \\ d_2 P_{\vec{n}-\vec{e}_2}(z) & d_2 \vec{R}_{\vec{n}-\vec{e}_2}(z) \\ \vdots & \vdots \\ d_r P_{\vec{n}-\vec{e}_r}(z) & d_r \vec{R}_{\vec{n}-\vec{e}_r}(z) \end{pmatrix} \tag{3.3}$$

where $P_{\vec{n}}(z)$ is the type II multiple orthogonal polynomial and $\vec{R}_{\vec{n}} = (R_{\vec{n},1}, R_{\vec{n},2}, \ldots, R_{\vec{n},r})$ is the vector containing

$$R_{\vec{n},j}(z) = \int P_{\vec{n}}(x) w_j(x) \frac{dx}{z - x},$$

and

$$\frac{1}{d_j} = \int x^{n_j-1} P_{\vec{n}-\vec{e}_j}(x) w_j(x) \, dx.$$

Proof. We write the matrix Z as

$$\begin{pmatrix} h & u^t \\ v & U \end{pmatrix},$$

where h is a complex function, u and v are column vectors of size r, and U is an $r \times r$ matrix function. The jump condition (3.1) implies that $h^+(x) = h^-(x)$ for $x \in \mathbb{R}$, hence h is an entire function. The asymptotic condition (3.2) shows that

$$\lim_{z\to\infty} h(z)/z^{n_1+\cdots+n_r} = 1,$$

hence $h(z)$ is a monic polynomial of degree $n_1 + \cdots + n_r$. The jump condition (3.1) for u is

$$\begin{pmatrix} u_1^+(x) \\ u_2^+(x) \\ \vdots \\ u_r^+(x) \end{pmatrix} = \begin{pmatrix} -2\pi i h(x) w_1(x) + u_1^-(x) \\ -2\pi i h(x) w_2(x) + u_2^-(x) \\ \vdots \\ -2\pi i h(x) w_r(x) + u_r^-(x) \end{pmatrix},$$

hence for each $k \in \{1, 2, \ldots, r\}$, the Sokhotsky-Plemelj formula gives

$$u_k(z) = \int \frac{h(x)w_k(x)}{z - x} \, dx.$$

If we use (2.4), then this gives

$$u_k(z) = \sum_{\ell=0}^{n_k-1} \frac{1}{z^{\ell+1}} \int x^\ell h(x)w_k(x) \, dx + \frac{1}{z^{n_k}} \int x^{n_k} \frac{h(x)w_k(x)}{z - x} \, dx.$$

The asymptotic condition (3.2) gives

$$\lim_{z \to \infty} u_k(z) z^{n_k} = 0, \qquad k = 1, 2, \ldots, r,$$

hence one needs to have

$$\int x^\ell h(x)w_k(x) \, dx = 0, \qquad \ell = 0, 1, \ldots, n_k - 1,$$

so that the monic polynomial h of degree $n_1 + \ldots + n_r$ needs to satisfy the orthogonality conditions in (1.7). Hence h is the type II multiple orthogonal polynomial $P_{\vec{n}}$ and

$$u_k(x) = \int \frac{P_{\vec{n}}(x)w_k(x)}{z - x} \, dx.$$

This gives the first row of the matrix Z.

The jump condition (3.1) for the vector v is $v^+(x) = v^-(x)$, so that v is a vector of entire functions. The asymptotic condition (3.2) gives

$$\lim_{z \to \infty} v_k(z)/z^{n_1 + n_2 + \cdots + n_r} = 0,$$

so that each v_k $(k = 1, 2, \ldots, r)$ is a polynomial of degree $< n_1 + n_2 + \cdots + n_r$. The jump condition (3.1) for the kth row of U is

$$\begin{pmatrix} U_{k,1}^+(x) \\ U_{k,2}^+(x) \\ \vdots \\ U_{k,r}^+(x) \end{pmatrix}^t = \begin{pmatrix} -2\pi i v_k(x) w_1(x) + U_{k,1}^-(x) \\ -2\pi i v_k(x) w_2(x) + U_{k,2}^-(x) \\ \vdots \\ -2\pi i v_k(x) w_r(x) + U_{k,r}^-(x) \end{pmatrix}^t.$$

Hence, Sokhotsky-Plemelj gives

$$U_{k,j}(z) = \int \frac{v_k(x)w_j(x)}{z - x} \, dx.$$

We can expand this using (2.4) to find

$$U_{k,j}(z) = \sum_{\ell=0}^{n_k-2} \frac{1}{z^{\ell+1}} \int x^\ell v_k(x) w_j(x)\, dx + \frac{1}{z^{n_k-1}} \int x^{n_k-1} \frac{v_k(x) w_j(x)}{z-x}\, dx.$$

The asymptotic condition (3.2) implies

$$\lim_{z \to \infty} U_{k,j}(z) z^{n_j} = \begin{cases} 1 & \text{if } k = j, \\ 0 & \text{if } k \neq j. \end{cases}$$

Therefore we have

$$\int x^\ell v_k(x) w_j(x)\, dx = 0, \qquad \ell = 0, 1, \ldots, n_k - 1, \ j \neq k$$

$$\int x^\ell v_k(x) w_k(x)\, dx = 0, \qquad \ell = 0, 1, \ldots, n_k - 2,$$

$$\int x^{n_k-1} v_k(x) w_k(x)\, dx = 1.$$

This means that the polynomial v_k satisfies the orthogonality conditions of a type II multiple orthogonal polynomial with index $\vec{n} - \vec{e}_k$ and hence, since the index $\vec{n} - \vec{e}_k$ is normal, $v_k(x) = d_k P_{\vec{n}-\vec{e}_k}(x)$. Here the normalizing constant d_k is such that

$$\int x^{n_k-1} d_k P_{\vec{n}-\vec{e}_k}(x) w_k(x)\, dx = 1,$$

which gives the required result. $\qquad\qquad\square$

4. Relation between Type I and Type II Multiple Orthogonal Polynomials

Theorem 4.1 *Denote by Y the matrix function solving the Riemann-Hilbert problem for type I multiple orthogonal polynomials (Theorem 2.1), and by Z the matrix function solving the Riemann-Hilbert problem for type II multiple orthogonal polynomials (Theorem 3.1). Then, assuming $\vec{n}$ and $\vec{n} \pm \vec{e}_k$ ($k = 1, \ldots, r$) are normal indices,*

$$Z = \begin{pmatrix} \vec{0}^{\ t} & 1 \\ -I_r & \vec{0} \end{pmatrix} (Y^{-1})^t \begin{pmatrix} \vec{0} & -I_r \\ 1 & \vec{0}^{\ t} \end{pmatrix}, \tag{4.1}$$

where I_r is the identity matrix of order r and $\vec{0}$ is the column vector containing r zeros.

Proof. For simplification, we write $\vec{w} = -2\pi i(w_1, w_2, \ldots, w_r)^t$ for the column vector containing the weights. The jump condition for Y then becomes

$$Y^+ = Y^- \begin{pmatrix} I_r & \vec{w} \\ \vec{0}^{\,t} & 1 \end{pmatrix}.$$

The inverse of Y exists since $\det Y$ is analytic on $\mathbb{C} \setminus \mathbb{R}$, $\det Y^+ = \det Y^-$ on the real line, so there is no jump and $\det Y$ is an entire function, and the asymptotic condition tells us that $\det Y$ is bounded and equal to 1 as $z \to \infty$. Hence $\det Y = 1$ everywhere. Taking inverse and transpose gives

$$[(Y^+)^{-1}]^t = [(Y^-)^{-1}]^t \begin{pmatrix} I_r & \vec{0} \\ -\vec{w}^{\,t} & 1 \end{pmatrix}.$$

Multiplying to the left and right by the appropriate matrices and using

$$\begin{pmatrix} \vec{0} & -I_r \\ 1 & \vec{0}^{\,t} \end{pmatrix} \begin{pmatrix} \vec{0}^{\,t} & 1 \\ -I_r & \vec{0} \end{pmatrix} = I_{r+1}$$

gives

$$\begin{pmatrix} \vec{0}^{\,t} & 1 \\ -I_r & \vec{0} \end{pmatrix} [(Y^+)^{-1}]^t \begin{pmatrix} \vec{0} & -I_r \\ 1 & \vec{0}^{\,t} \end{pmatrix} = \begin{pmatrix} \vec{0}^{\,t} & 1 \\ -I_r & \vec{0} \end{pmatrix} [(Y^-)^{-1}]^t \begin{pmatrix} \vec{0} & -I_r \\ 1 & \vec{0}^{\,t} \end{pmatrix} \begin{pmatrix} 1 & \vec{w}^{\,t} \\ \vec{0} & I_r \end{pmatrix},$$

where the last jump matrix is the one for the Riemann-Hilbert problem for Z. The asymptotic condition for Y is

$$\lim_{z \to \infty} Y(z) \begin{pmatrix} z^{-n_1} & & & & 0 \\ & z^{-n_2} & & & \\ & & \ddots & & \\ & & & z^{-n_r} & \\ 0 & & & & z^{n_1+n_2+\cdots+n_r} \end{pmatrix} = I_{r+1},$$

so that

$$\lim_{z \to \infty} \begin{pmatrix} \vec{0}^{\,t} & 1 \\ -I_r & \vec{0} \end{pmatrix} [Y(z)^{-1}]^t \begin{pmatrix} \vec{0} & -I_r \\ 1 & \vec{0}^{\,t} \end{pmatrix}$$

$$\times \begin{pmatrix} \vec{0}^{\,t} & 1 \\ -I_r & \vec{0} \end{pmatrix} \begin{pmatrix} z^{n_1} & & & & 0 \\ & z^{n_2} & & & \\ & & \ddots & & \\ & & & z^{n_r} & \\ 0 & & & & z^{-(n_1+n_2+\cdots+n_r)} \end{pmatrix} \begin{pmatrix} \vec{0} & -I_r \\ 1 & \vec{0}^{\,t} \end{pmatrix} = I_{r+1}.$$

An easy calculation gives

$$
\begin{pmatrix} \vec{0}^{\,t} & 1 \\ -I_r & \vec{0} \end{pmatrix}
\begin{pmatrix} z^{n_1} & & & & 0 \\ & z^{n_2} & & & \\ & & \ddots & & \\ & & & z^{n_r} & \\ 0 & & & & z^{-(n_1+n_2+\cdots+n_r)} \end{pmatrix}
\begin{pmatrix} \vec{0} & -I_r \\ 1 & \vec{0}^{\,t} \end{pmatrix}
$$

$$
= \begin{pmatrix} z^{-(n_1+n_2+\cdots+n_r)} & & & & 0 \\ & z^{n_1} & & & \\ & & z^{n_2} & & \\ & & & \ddots & \\ 0 & & & & z^{n_r} \end{pmatrix},
$$

so that we have the asymptotic condition for the Riemann-Hilbert problem for Z. This means that

$$
\begin{pmatrix} \vec{0}^{\,t} & 1 \\ -I_r & \vec{0} \end{pmatrix} [Y(z)^{-1}]^t \begin{pmatrix} \vec{0} & -I_r \\ 1 & \vec{0}^{\,t} \end{pmatrix} = Z,
$$

hence proving the theorem. $\qquad\square$

A consequence of this theorem is a relationship between the multiple orthogonal polynomials of type I and type II, which is well-known (see, e.g., Mahler [19]). If we take the $(1,1)$-entry of Z then

$$
Z_{1,1} = P_{\vec{n}}(z).
$$

On the other hand, from (4.1) we also have

$$
Z_{1,1} = (Y^{-1})_{r+1,r+1},
$$

so that

$$
P_{\vec{n}}(z) = \frac{1}{c_1 c_2 \cdots c_r} \det \begin{pmatrix} \vec{A}_{\vec{n}+\vec{e}_1}(z) \\ \vec{A}_{\vec{n}+\vec{e}_2}(z) \\ \vdots \\ \vec{A}_{\vec{n}+\vec{e}_r}(z) \end{pmatrix}. \tag{4.2}
$$

5. Recurrence Relations

It is very well known that a system of orthogonal polynomials on the real line always satisfies a three-term recurrence relation (a second order linear recurrence relation). Multiple orthogonal polynomials also satisfy a number of recurrence relations. This was already known to Padé [23] and one can also find various recurrence relations in the work of Mahler [19]. In this section we show that such recurrences follow rather easily from the Riemann-Hilbert problem for these multiple orthogonal polynomials. In view of Theorem 4.1 it is sufficient to study the matrix Z for type II multiple orthogonal polynomials. We will use the $Z_{\vec{n}}$ to denote the matrix function for the multi-index $\vec{n}$ and assume that all indices involved are normal. Consider the matrix function

$$X_{\vec{n},k}(z) = Z_{\vec{n}}(z)Z_{\vec{n}-\vec{e}_k}^{-1}(z), \qquad k = 1, 2, \dots, r.$$

The inverse Z^{-1} always exists since $\det Z = 1$ in the complex plane. Clearly $X_{\vec{n},k}$ is analytic in $\mathbb{C} \setminus \mathbb{R}$ and since $Z_{\vec{n}}$ and $Z_{\vec{n}-\vec{e}_k}$ have the same jump condition on the real line, we see that $X_{\vec{n},k}$ has no jump condition and hence $X_{\vec{n},k}$ is an entire matrix function. For $z \to \infty$ it has the asymptotic behavior

$$X_{\vec{n},k}(z) = Z_{\vec{n}} \begin{pmatrix} z^{-n_1-n_2-\cdots-n_r} & & & & & \\ & z^{n_1} & & & & \\ & & \ddots & & & \\ & & & z^{n_k} & & \\ & & & & \ddots & \\ & & & & & z^{n_r} \end{pmatrix}$$

$$\times \begin{pmatrix} z & & & & & \\ & 1 & & & & \\ & & \ddots & & & \\ & & & 1/z & & \\ & & & & \ddots & \\ & & & & & 1 \end{pmatrix}$$

$$\times \begin{pmatrix} z^{n_1+\cdots+n_r-1} & & & & & \\ & z^{-n_1} & & & & \\ & & \ddots & & & \\ & & & z^{-n_k+1} & & \\ & & & & \ddots & \\ & & & & & z^{-n_r} \end{pmatrix} Z_{\vec{n}-\vec{e}_k}^{-1}$$

$$
= \left(I + \frac{A(\vec{n})}{z} + \mathcal{O}(\frac{1}{z^2})\right)
\begin{pmatrix}
z & & & & \\
& 1 & & & \\
& & \ddots & & \\
& & & 1/z & \\
& & & & \ddots & \\
& & & & & 1
\end{pmatrix}
$$

$$
\times \left(I + \frac{B(\vec{n} - \vec{e}_k)}{z} + \mathcal{O}(\frac{1}{z^2})\right), \tag{5.1}
$$

where $A_{i,j}(\vec{n})_{0 \le i,j \le r}$ and $B_{i,j}(\vec{n} - \vec{e}_k)_{0 \le i,j \le r}$ are matrices independent of z but depending on $\vec{n}$ and k. Hence $X_{\vec{n},k}$ is a matrix polynomial of degree 1. Comparing the coefficients of z and of z^0 in (5.1) shows that

$$
X_{\vec{n},k}(z) =
\begin{pmatrix}
z + A_{0,0} + B_{0,0} & B_{0,1} & B_{0,2} & \cdots & B_{0,k} & \cdots & B_{0,r} \\
A_{1,0} & 1 & & & & & \\
A_{2,0} & & \ddots & & & & \\
\vdots & & & 1 & & & \\
A_{k,0} & & & & 0 & & \\
\vdots & & & & & 1 & \\
A_{r,0} & & & & & & \ddots \\
& & & & & & & 1
\end{pmatrix}. \tag{5.2}
$$

Using this expression for X, the entry in the first row and first column in the equation

$$
X_{\vec{n},k}(z) Z_{\vec{n} - \vec{e}_k} = Z_{\vec{n}} \tag{5.3}
$$

gives

$$
\begin{aligned}
P_{\vec{n}}(z) &= [z + A_{0,0}(\vec{n}) + B_{0,0}(\vec{n} - \vec{e}_k)] P_{\vec{n} - \vec{e}_k}(z) \\
&\quad + \sum_{j=1}^{r} B_{0,j}(\vec{n}) d_j(\vec{n} - \vec{e}_k) P_{\vec{n} - \vec{e}_k - \vec{e}_j}(z).
\end{aligned} \tag{5.4}
$$

The entry in the first column and on row $j + 1$ of (5.3) gives

$$
\begin{aligned}
d_j(\vec{n}) P_{\vec{n} - \vec{e}_j}(z) &= A_{j,0}(\vec{n}) P_{\vec{n} - \vec{e}_k}(z) + d_j(\vec{n} - \vec{e}_k) P_{\vec{n} - \vec{e}_k - \vec{e}_j}(z), \quad j \neq k \\
d_k(\vec{n}) P_{\vec{n} - \vec{e}_k}(z) &= A_{k,0}(\vec{n}) P_{\vec{n} - \vec{e}_k}(z).
\end{aligned} \tag{5.5}
$$

The latter equation shows that $A_{k,0}(\vec{n}) = d_k(\vec{n})$. Summarizing we have

Theorem 5.1 *Type II multiple orthogonal polynomials satisfy the following recurrence relation*

$$(z + A_{0,0}(\vec{n} + \vec{e}_k) + B_{0,0}(\vec{n}))P_{\vec{n}}(z) = P_{\vec{n}+\vec{e}_k}(z) \tag{5.6}$$
$$- \sum_{j=1}^{r} B_{0,j}(\vec{n} + \vec{e}_k)d_j(\vec{n})P_{\vec{n}-\vec{e}_j}(z).$$

Proof. This follows by replacing $\vec{n}$ by $\vec{n} + \vec{e}_k$ in (5.4). $\square$

Corollary 1 *Type II multiple orthogonal polynomials also satisfy the recurrence relation*

$$(z - a_0(\vec{n}))P_{\vec{n}}(z) = P_{\vec{n}+\vec{e}_1}(z) + \sum_{j=1}^{r} a_j(\vec{n})P_{\vec{n}-\vec{e}_{r-j+1}-\cdots-\vec{e}_r}(z), \tag{5.7}$$

for certain coefficients $a_j(\vec{n})$ $(j = 0, 1, \ldots, r)$.

Proof. Take $k = 1$ in (5.6). Then

$$(z + A_{0,0}(\vec{n} + \vec{e}_1) + B_{0,0}(\vec{n}))P_{\vec{n}}(z) = P_{\vec{n}+\vec{e}_1}(z) \tag{5.8}$$
$$- \sum_{j=1}^{r} B_{0,j}(\vec{n} + \vec{e}_1)d_j(\vec{n})P_{\vec{n}-\vec{e}_j}(z).$$

If we use (5.5) with $k = j + 1$, then

$$P_{\vec{n}-\vec{e}_j}(z) = P_{\vec{n}-\vec{e}_{j+1}}(z) + c_j(\vec{n})P_{\vec{n}-\vec{e}_j-\vec{e}_{j+1}}(z), \tag{5.9}$$

with $c_j(\vec{n})$ some constant. More general, for $0 \leq k \leq r - j - 1$ we have

$$P_{\vec{n}-\vec{e}_j-\cdots-\vec{e}_{j+k}}(z) = P_{\vec{n}-\vec{e}_{j+1}-\cdots-\vec{e}_{j+k+1}}(z) + c_{j,k}(\vec{n})P_{\vec{n}-\vec{e}_j-\cdots-\vec{e}_{j+k+1}}(z),$$

which follows from (5.5) with $\vec{n}$ replaced by $\vec{n} - \vec{e}_{j+1} - \cdots - \vec{e}_{j+k}$ and k replaced by $j + k + 1$. Repeated application of these identities then shows that $P_{\vec{n}-\vec{e}_j}(z)$ is a linear combination of the form

$$P_{\vec{n}-\vec{e}_j}(z) = \sum_{i=j}^{r} \hat{c}_{i,j}(\vec{n})P_{\vec{n}-\vec{e}_i-\cdots-\vec{e}_r}(z).$$

If we insert this into (5.8), then we get the required recurrence relation. $\square$

6. Vector Potentials and g-Functions

The Riemann-Hilbert problems described in Theorems 2.1 and 3.1 are not normalized in the sense that conditions (2.2) and (3.2) impose some growth condition as $z \to \infty$. In order to normalize the Riemann-Hilbert problem, we would like to modify Y and Z in such a way that we get another Riemann-Hilbert problem with the same contours (but possibly different jump conditions) for which the solution tends to the identity matrix as $z \to \infty$. For the normalization we need to take into account the behavior of $Y(z)$ or $Z(z)$ for large z. This depends heavily on the distribution of the zeros of the multiple orthogonal polynomial. The zero distribution of orthogonal polynomials is usually given by an extremal problem in logarithmic potential theory [24] [25]. For multiple orthogonal polynomials one needs to study an extremal problem for vector potentials [14] [21]. Suppose μ and ν are two probability measures on the real line and define the *(logarithmic) energy* of μ by

$$I(\mu) = \iint \log \frac{1}{|x - y|} \, d\mu(x) \, d\mu(y), \tag{6.1}$$

and the *mutual energy* of μ and ν by

$$I(\mu, \nu) = \iint \log \frac{1}{|x - y|} \, d\mu(x) \, d\nu(y). \tag{6.2}$$

These quantities are bounded from below if the support of μ and ν is compact, but the energies can be $+\infty$. Observe that $I(\mu) = I(\mu, \mu)$. Let $C = (c_{i,j})_{i,j=1}^r$ be a positive definite matrix of order r. Then the extremal problem is to minimize

$$I(\vec{\mu}) = \sum_{i=1}^r \sum_{j=1}^r c_{i,j} I(\mu_i, \mu_j) \tag{6.3}$$

over all vectors $\vec{\mu} = (\mu_1, \dots, \mu_r)$ of probability measures μ_i supported on given compact sets Δ_i ($i = 1, 2, \dots, r$). Under fairly weak conditions on the sets Δ_i, this minimum is finite and attained at a (unique) vector of measures $\vec{\nu} = (\nu_1, \dots, \nu_r)$, which is called the equilibrium. The support of ν_i will be denoted by Δ_i^* and is a subset of Δ_i. The solution of this minimization problem can also be described using *(logarithmic) potentials*

$$U(x; \mu_i) = \int_{\Delta_i} \log \frac{1}{|x - y|} \, d\mu_i(y). \tag{6.4}$$

Indeed, the variational conditions are the following: there exist constants $F_1, \dots, F_r$ (Lagrange multipliers) such that

$$\sum_{i=1}^r c_{i,j} U(x; \nu_i) = F_j, \qquad x \in \Delta_j^*, \tag{6.5}$$

$$\sum_{i=1}^{r} c_{i,j} U(x; \nu_i) \geq F_j, \qquad x \in \Delta_j, \tag{6.6}$$

holds for $j = 1, 2, \ldots, r$. The precise form of the positive definite matrix C depends on the problem at hand.

6.1. ANGELESCO SYSTEMS

Suppose that each Δ_i is a finite interval and that the open intervals are disjoint: $\overset{\circ}{\Delta}_i \cap \overset{\circ}{\Delta}_j = \emptyset$ whenever $i \neq j$. In view of the connection between type I and type II multiple orthogonal polynomials, given by Theorem 4.1, we can limit the discussion to type II multiple orthogonal polynomials, which is the most convenient for Angelesco systems. Let $P_{\vec{n}}(x)$ be the type II multiple orthogonal polynomials for weights $(w_1, \ldots, w_r)$ defined on the intervals $(\Delta_1, \ldots, \Delta_r)$. The orthogonality relations on Δ_i imply that $P_{\vec{n}}$ has at least n_i zeros on $\overset{\circ}{\Delta}_i$. Indeed, suppose that $P_{\vec{n}}$ has $m < n_i$ sign changes on Δ_i at the points $x_1, \ldots, x_m$. Let $Q_m(x) = (x - x_1) \cdots (x - x_m)$. Then $P_{\vec{n}}(x) Q_m(x)$ does not change sign on $\overset{\circ}{\Delta}_i$, but since $m < n_i$, the orthogonality on Δ_i implies

$$\int_{\Delta_i} P_{\vec{n}}(x) Q_m(x) w_i(x) \, dx = 0,$$

which is not possible. This contradiction shows that $P_{\vec{n}}$ has at least n_i zeros on $\overset{\circ}{\Delta}_i$. But all these open intervals are disjoint and the degree of $P_{\vec{n}}$ is $n_1 + \cdots + n_r$. Hence $P_{\vec{n}}$ has precisely n_i simple zeros on each open interval $\overset{\circ}{\Delta}_i$. This means that

$$P_{\vec{n}}(x) = q_{n_1,1}(x) \cdots q_{n_r,r}(x),$$

where each $q_{n_i,i}$ has n_i simple zeros on $\overset{\circ}{\Delta}_i$. The orthogonality relation

$$\int_{\Delta_i} x^k q_{n_i,i}(x) \prod_{j \neq i} q_{n_j,j}(x) \, w_i(x) \, dx = 0, \qquad k = 0, \ldots, n_i - 1,$$

means that $q_{n_i,i}$ is the (monic) orthogonal polynomial of degree n_i for the weight

$$\prod_{j \neq i} |q_{n_j,j}(x)| \, w_i(x)$$

on Δ_i. This means that $q_{n_i,i}$ minimizes the integral

$$\int_{\Delta_i} |q(x)|^2 \prod_{j \neq i} |q_{n_j,j}(x)| \, w_i(x) \, dx \tag{6.7}$$

over all monic polynomials q of degree n_i. It is known that the distribution of the zeros of orthogonal polynomials with varying weights is given by the solution of an equilibrium problem with external field [24]. Suppose $\{x_{j,n_i}, \ j = 1, \ldots, n_i\}$ are the zeros of $q_{n_i,i}$, which are all on Δ_i, and let

$$\mu(q_{n_i,i}) = \frac{1}{n_i} \sum_{j=1}^{n_i} \delta(x_{j,n_i})$$

be the distribution of these n_i zeros, where $\delta(c)$ is the Dirac measure at c. Observe that we can write

$$|q_{n_i,i}(x)| = \prod_{j=1}^{n_i} |x - x_{j,n_i}| = \exp[-n_i U(x; \mu(q_{n_i,i}))].$$

If we set $\mu_i = \mu(q_{n_i,i})$, then the integrand in (6.7) becomes

$$\exp\left(-2n_i U(x; \mu_i) - \sum_{j \neq i} n_j U(x; \mu_j) + \log w_i(x)\right). \tag{6.8}$$

We want to minimize the integral of this on Δ_i over all probability measures μ_i supported on Δ_i. We will instead minimize the maximum on Δ_i over all probability measures μ_i supported on Δ_i. Assume that

$$\lim_{|\vec{n}| \to \infty} \frac{n_i}{|\vec{n}|} = p_i > 0, \qquad i = 1, \ldots, r,$$

where $p_1 + \cdots + p_r = 1$, and that $w_i(x) > 0$ on Δ_i, then, as each $n_i \to \infty$, minimizing the maximum of (6.8) on Δ_i is equivalent with maximizing

$$\inf_{x \in \Delta_i} \left(2p_i U(x; \mu_i) + \sum_{j \neq i} p_j U(x; \mu_j)\right).$$

The variational conditions for this extremal problem are

$$2p_i U(x; \nu_i) + \sum_{j \neq i} p_j U(x; \mu_j) = \ell_i, \qquad x \in \Delta_i^*,$$

$$2p_i U(x; \nu_i) + \sum_{j \neq i} p_j U(x; \mu_j) \geq \ell_i, \qquad x \in \Delta_i,$$

where ℓ_i is some constant and Δ_i^* is the support of the extremal measure ν_i. Now we have to do this for every $i \in \{1, 2, \ldots, r\}$, which finally gives us the variational conditions

$$2p_i U(x; \nu_i) + \sum_{j \neq i} p_j U(x; \nu_j) = \ell_i, \qquad x \in \Delta_i^*, \tag{6.9}$$

$$2p_i U(x; \nu_i) + \sum_{j \neq i} p_j U(x; \nu_j) \geq \ell_i, \qquad x \in \Delta_i, \qquad (6.10)$$

for $i = 1, 2, \ldots, r$. This equilibrium problem therefore corresponds to the vector equilibrium problem for the interaction matrix

$$C = \begin{pmatrix} 2p_1^2 & p_1 p_2 & p_1 p_3 & \cdots & p_1 p_r \\ p_2 p_1 & 2p_2^2 & p_2 p_3 & \cdots & p_2 p_r \\ \vdots & & \ddots & & \cdots \\ p_r p_1 & p_r p_2 & \cdots & p_r p_{r-1} & 2p_r^2 \end{pmatrix}, \qquad (6.11)$$

and Lagrange multipliers $F_i = p_i \ell_i$. If we denote the intervals by $\Delta_i = [a_i, b_i]$, then one knows [21] that the supports of ν_i are again intervals which we denote by $\Delta_i^* = [a_i^*, b_i^*] \subset [a_i, b_i]$.

The normalization of the Riemann-Hilbert problem for an Angelesco system now is as follows. Let $(\nu_1, \ldots, \nu_r)$ be the vector of equilibrium measures satisfying the variational conditions (6.9)–(6.10), and let $\vec{n}$ be a multi-index such that

$$n_k = np_k \in \mathbb{N}, \qquad k = 1, \ldots, r.$$

This can be done if all $p_k = a_k/b_k$ are rational and n is a multiple of the least common multiple of the denominators $b_1, \ldots, b_r$. Observe that $|\vec{n}| = n$. Define

$$g_k(z) = \int_{a_k^*}^{b_k^*} \log(z - y) \, d\nu_k(y), \qquad k = 1, \ldots, r. \qquad (6.12)$$

Then

$$\begin{aligned} g_k^+(x) &= -U(x; \nu_k), & x \geq b_k^*, \\ g_k^+(x) &= -U(x; \nu_k) + i\pi, & x \leq a_k^*, \\ g_k^+(x) &= -U(x; \nu_k) + \varphi_k(x), & x \in [a_k^*, b_k^*], \end{aligned}$$

where

$$\varphi_k(x) = i\pi \int_x^{b_k^*} d\nu_k(y).$$

Similarly

$$\begin{aligned} g_k^-(x) &= -U(x; \nu_k), & x \geq b_k^*, \\ g_k^-(x) &= -U(x; \nu_k) - i\pi, & x \leq a_k^*, \\ g_k^-(x) &= -U(x; \nu_k) - \varphi_k(x), & x \in [a_k^*, b_k^*]. \end{aligned}$$

It will be convenient to write the variational conditions (6.9)–(6.10) as

$$-p_k[g_k^+(x) + g_k^-(x)] - \sum_{j \neq k} p_j g_j^-(x) = \ell_k^*, \qquad x \in \Delta_k^*, \qquad (6.13)$$

$$-p_k[g_k^+(x) + g_k^-(x)] - \sum_{j \neq k} p_j g_j^-(x) - \ell_k^* \geq 0, \qquad x \in \Delta_k, \qquad (6.14)$$

where $\ell_k^* - \ell_k$ is an integer multiple of $i\pi/n$. If Z is the solution of the Riemann-Hilbert problem given by Theorem 3.1, then we define

$$M(z) = \begin{pmatrix} 1 \\ & e^{-n\ell_1^*} \\ & & \ddots \\ & & & e^{-n\ell_k^*} \\ & & & & \ddots \\ & & & & & e^{-n\ell_r^*} \end{pmatrix} Z(z)$$

$$\times \begin{pmatrix} e^{-n_1 g_1(z) - \cdots - n_r g_r(z)} \\ & e^{n_1 g_1(z)} \\ & & \ddots \\ & & & e^{n_k g_k(z)} \\ & & & & \ddots \\ & & & & & e^{n_r g_r(z)} \end{pmatrix}$$

$$\times \begin{pmatrix} 1 \\ & e^{n\ell_1^*} \\ & & \ddots \\ & & & e^{n\ell_k^*} \\ & & & & \ddots \\ & & & & & e^{n\ell_r^*} \end{pmatrix}. \qquad (6.15)$$

Then M is analytic on $\mathbb{C} \setminus \bigcup_{i=1}^r [a_i, b_i]$. Let us write the matrix product in (6.15) as

$$M(z) = LZ(z)G(z)L^{-1}.$$

Then

$$\lim_{z \to \infty} M(z) = L \left[\lim_{z \to \infty} Z(z)G(z) \right] L^{-1},$$

and since

$$\lim_{z \to \infty} (g_k(z) - \log z) = 0, \qquad k = 1, \ldots, r,$$

the growth condition (3.2) implies that

$$\lim_{z \to \infty} M(z) = I, \qquad (6.16)$$

which shows that the Riemann-Hilbert problem for M is normalized. In order to find the jump on the interval Δ_k, we compute

$$M^+(x) = LZ^+(x)G^+(x)L^{-1}.$$

Recall that the jump for the original Riemann-Hilbert problem on $\Delta_k = [a_k, b_k]$ is

$$Z^+(x) = Z^-(x) \begin{pmatrix} 1 & -2\pi i w_k(x)\vec{e}_k^{\,t} \\ 0 & I \end{pmatrix}, \qquad x \in [a_k, b_k],$$

where $\vec{e}_k^{\,t} = (0, 0, \dots, 0, 1, 0, \dots, 0)$ is the kth unit vector in $\mathbb{R}^r$. Hence we have

$$M^+(x) = L Z^-(x) \begin{pmatrix} 1 & -2\pi i w_k(x)\vec{e}_k^{\,t} \\ 0 & I \end{pmatrix} G^+(x) L^{-1}, \qquad x \in [a_k, b_k].$$

Use $M^-(x) = L Z^-(x) G^-(x) L^{-1}$, then

$$M^+(x) = M^-(x) L [G^-(x)]^{-1} \begin{pmatrix} 1 & -2\pi i w_k(x)\vec{e}_k^{\,t} \\ 0 & I \end{pmatrix} G^+(x) L^{-1},$$
$$x \in [a_k, b_k]. \quad (6.17)$$

First we compute the matrix product

$$[G^-(x)]^{-1} \begin{pmatrix} 1 & -2\pi i w_k(x)\vec{e}_k^{\,t} \\ 0 & I \end{pmatrix} G^+(x)$$

$$= \begin{pmatrix} e^{\sum_{j=1}^r n_j [g_j^-(x) - g_j^+(x)]} & 0 & \cdots & a(x) & \cdots & 0 \\ & e^{n_1 [g_1^+(x) - g_1^-(x)]} & & & & \\ & & \ddots & & & \\ & & & e^{n_k [g_k^+(x) - g_k^-(x)]} & & \\ & & & & \ddots & \\ & & & & & e^{n_r [g_r^+(x) - g_r^-(x)]} \end{pmatrix},$$

where

$$a(x) = -2\pi i w_k(x) e^{n_k [g_k^+(x) + g_k^-(x)] + \sum_{j \neq k} n_j g_j^-(x)}.$$

Observe that on Δ_k^* we have

$$e^{g_j^+(x) - g_j^-(x)} = \begin{cases} 1 & \text{if } j \neq k, \\ 2\varphi_k(x) & \text{if } j = k, \end{cases}$$

and that, taking into account (6.13) we also have

$$a(x) = -2\pi i w_k(x) e^{-n\ell_k^*}.$$

This means that the jump on Δ_k^* is given by

$$L [G^-(x)]^{-1} \begin{pmatrix} 1 & -2\pi i w_k(x)\vec{e}_k^{\,t} \\ 0 & I \end{pmatrix} G^+(x) L^{-1}$$

$$
= \begin{pmatrix}
e^{-2n_k\varphi_k(x)} & 0 & \cdots & 0 & -2\pi i w_k(x) & 0 & \cdots & 0 \\
 & 1 & & & & & & \\
 & & \ddots & & & & & \\
 & & & 1 & & & & \\
 & & & & e^{2n_k\varphi_k(x)} & & & \\
 & & & & & 1 & & \\
 & & & & & & \ddots & \\
 & & & & & & & 1
\end{pmatrix}, \qquad x \in \Delta_k^*.
$$
$$(6.18)$$

On $\Delta_k \setminus \Delta_k^*$ we have

$$
e^{g_j^+(x) - g_j^-(x)} = 1, \qquad j = 1, \ldots, r,
$$

and (6.14) gives

$$
a(x) = -2\pi i w_k(x) e^{-n\ell_k^* - n V_k(x)},
$$

where $V_k(x) \geq 0$. Hence on $\Delta_k \setminus \Delta_k^*$ the jump is

$$
L[G^-(x)]^{-1} \begin{pmatrix} 1 & -2\pi i w_k(x) \vec{e}_k^{\,t} \\ 0 & I \end{pmatrix} G^+(x) L^{-1}
$$

$$
= \begin{pmatrix}
1 & 0 & \cdots & 0 & -2\pi i w_k(x) e^{-n V_k(x)} & 0 & \cdots & 0 \\
 & \ddots & & & & & & \\
 & & & & & & & \\
 & & & 1 & & & & \\
 & & & & & & & \\
 & & & & & & \ddots & \\
 & & & & & & & 1
\end{pmatrix}, \qquad x \in \Delta_k \setminus \Delta_k^*.
$$
$$(6.19)$$

The normalized Riemann-Hilbert matrix $M(z)$ therefore has rather simple jumps on the intervals Δ_k. On the supports of the measures ν_k the jump (6.18) has oscillatory terms $e^{\pm 2n_k\varphi_k(x)}$ on the diagonal (recall that $\varphi_k(x)$ is purely imaginary), and on $\Delta_k \setminus \Delta_k^*$ the jump (6.19) is the identity matrix, except for one entry on the first row which decreases exponentially fast as $n \to \infty$ provided $V_k(x) > 0$, which will be typically the case. Such a Riemann-Hilbert problem with oscillatory and exponentially decreasing jumps can be analysed asymptotically by using the steepest descent method introduced by Deift and Zhou [9] [10]. We will apply this deepest descent method to some relevant cases in future contributions, since this would be outside of the scope of this survey.

6.2. NIKISHIN SYSTEMS

When all the weight functions $w_1, \ldots, w_r$ are defined on the same interval Δ_r, then one needs appropriate conditions on these weights in order to guarantee that a multi-index $\vec{n}$ is normal. An interesting construction was suggested by Nikishin [20], which in the book [21] is called an MT-system, but which is nowadays known as a Nikishin system. The construction is by induction. A Nikishin system of order 1 on Δ_1 consists of a weight function $w_{1,1}$ on an interval Δ_1 of the real line. A Nikishin system of order 2 on Δ_2 consists of weight functions $(w_{1,2}, w_{2,2})$ on an interval Δ_2, with $w_{1,2}$ a positive weight on Δ_2 and

$$w_{2,2}(x) = w_{1,2}(x) \int_{\Delta_1} \frac{w_{1,1}(t)}{x - t} \, dt, \qquad x \in \Delta_2, \qquad (6.20)$$

where $\overset{\circ}{\Delta}_1 \cap \overset{\circ}{\Delta}_2 = \emptyset$ and $w_{1,1}$ is a Nikishin system of order 1 on Δ_1. In general, a Nikishin system of order r on Δ_r consists of weights $(w_{1,r}, \ldots, w_{r,r})$ on Δ_r, such that $w_{1,r}$ is a positive weight on Δ_r and for $k = 2, 3, \ldots, r$

$$w_{k,r}(x) = w_{1,r}(x) \int_{\Delta_{r-1}} \frac{w_{k-1,r-1}(t)}{x - t} \, dt, \qquad x \in \Delta_r, \qquad (6.21)$$

where $\overset{\circ}{\Delta}_{r-1} \cap \overset{\circ}{\Delta}_r = \emptyset$ and $(w_{1,r-1}, \ldots, w_{r-1,r-1})$ is a Nikishin system of order $r - 1$ on Δ_{r-1}. Observe that each $w_{k,r}$ has constant sign on Δ_r since the intervals $\overset{\circ}{\Delta}_r$ and $\overset{\circ}{\Delta}_{r-1}$ are disjoint. For a Nikishin system of order r one knows that the weights $(w_{1,r}, \ldots, w_{r,r})$ form an AT-system on Δ_r for the multi-indices $\vec{n} = (n_1, \ldots, n_r)$ with

$$n_1 \geq n_2 \geq \cdots \geq n_r, \qquad (6.22)$$

so that these multi-indices are normal. It is still an open problem (except for $r = 2$ [11] [5]) whether or not every multi-index is normal, but one already knows that there are more normal indices than given by (6.22) [11]. For Nikishin systems it is more convenient to work with the type I multiple orthogonal polynomials. We will assume that (6.22) holds so that the multi-index $\vec{n}$ is normal.

We will assume that $w_{1,r}(x) > 0$ on Δ_r. Consider the function

$$L_{\vec{n}}(x) = A_{\vec{n},1}(x) + \sum_{j=2}^{r} A_{\vec{n},j}(x) \int_{\Delta_{r-1}} \frac{w_{j-1,r-1}(t)}{x - t} \, dt. \qquad (6.23)$$

Then

$$w_{1,r}(x) L_{\vec{n}}(x) = \sum_{j=1}^{r} A_{\vec{n},j}(x) w_{j,r}(x),$$

and $L_{\vec{n}}$ has exactly $|\vec{n}| - 1$ sign changes on $\overset{\circ}{\Delta}_r$. Indeed, since the weights form an AT-system on Δ_r, we already know that $L_{\vec{n}}$ has at most $|\vec{n}| - 1$ sign changes on $\overset{\circ}{\Delta}_r$. If there are $m < |\vec{n}| - 1$ sign changes on $\overset{\circ}{\Delta}_r$, then we can form the polynomial Q_m with zeros at these points and $L_{\vec{n}}Q_m$ will not change sign on Δ_r. But the orthogonality (1.3) gives

$$\int_{\Delta_r} w_{1,r}(x) L_{\vec{n}}(x) Q_m(x)\, dx = 0,$$

and this contradiction shows that there are exactly $|\vec{n}| - 1$ sign changes on $\overset{\circ}{\Delta}_r$. Denote by $H_{\vec{n}}$ the monic polynomial of degree $|\vec{n}| - 1$ with zeros at the points where $L_{\vec{n}}$ changes sign on $\overset{\circ}{\Delta}_r$. Then the ratio $L_{\vec{n}}(z)/H_{\vec{n}}(z)$ is analytic on $\mathbb{C} \setminus \Delta_{r-1}$ and of constant sign on Δ_r. It turns out that $H_{\vec{n}}$ is the monic orthogonal of degree $|\vec{n}| - 1$ on Δ_r for the weight function

$$\frac{|L_{\vec{n}}(x)|}{|H_{\vec{n}}(x)|}\, w_{1,r}(x), \tag{6.24}$$

because

$$\int_{\Delta_r} H_{\vec{n}}(x) x^k \frac{|L_{\vec{n}}(x)|}{|H_{\vec{n}}(x)|}\, w_{1,r}(x)\, dx = 0, \qquad k = 0, 1, \dots, |\vec{n}| - 2,$$

which follows from the orthogonality (1.3). This means that the zero distribution of $H_{\vec{n}}$ will be given by an equilibrium problem with an external field given by

$$Q_1(x) = -\lim_{|\vec{n}| \to \infty} \frac{1}{|\vec{n}|} \log \frac{|L_{\vec{n}}(x)|}{|H_{\vec{n}}(x)|}, \qquad x \in \Delta_r. \tag{6.25}$$

Observe that $L_{\vec{n}}(x)/H_{\vec{n}}(x)$ is analytic in $\overline{\mathbb{C}} \setminus \Delta_{r-1}$ and that

$$\frac{L_{\vec{n}}(z)}{H_{\vec{n}}(z)} = \mathcal{O}(z^{-(n_2 + \cdots + n_r)}), \qquad z \to \infty,$$

where we have taken into account (6.22). If we choose a contour Γ that goes around Δ_{r-1} counterclockwise but which does not contain any points of Δ_r, then the residue theorem applied to the domain outside Γ gives

$$\frac{1}{2\pi i} \int_\Gamma \frac{L_{\vec{n}}(z)}{H_{\vec{n}}(z)} z^k\, dz = 0, \qquad k = 0, 1, \dots, n_2 + \cdots + n_r - 2.$$

We can evaluate this contour integral as

$$\frac{1}{2\pi i} \int_\Gamma \frac{L_{\vec{n}}(z)}{H_{\vec{n}}(z)} z^k\, dz$$

$$= \frac{1}{2\pi i} \int_\Gamma \frac{A_{\vec{n},1}(z)}{H_{\vec{n}}(z)} z^k \, dz + \sum_{j=2}^{r} \frac{1}{2\pi i} \int_\Gamma \frac{A_{\vec{n},j}(z)}{H_{\vec{n}}(z)} z^k \int_{\Delta_{r-1}} \frac{w_{j-1,r-1}(t)}{z-t} \, dt \, dz.$$

The first integral on the right hand side vanishes by applying the Cauchy theorem for the domain inside Γ (since all the zeros of $H_{\vec{n}}$ are on Δ_r). Changing the order of integration for the other terms and using Cauchy's theorem for the domain inside Γ gives

$$\frac{1}{2\pi i} \int_\Gamma \frac{A_{\vec{n},j}(z)}{H_{\vec{n}}(z)} z^k \int_{\Delta_{r-1}} \frac{w_{j-1,r-1}(t)}{z-t} \, dt \, dz = \int_{\Delta_{r-1}} \frac{A_{\vec{n},j}(t)}{H_{\vec{n}}(t)} t^k w_{j-1,r-1}(t) \, dt$$

so that we get

$$\int_{\Delta_{r-1}} t^k \sum_{j=2}^{r} A_{\vec{n},j}(t) w_{j-1,r-1}(t) \frac{dt}{H_{\vec{n}}(t)} = 0,$$
$$k = 0, 1, \ldots, n_2 + \cdots + n_r - 2. \quad (6.26)$$

This means that $(A_{\vec{n},2}, \ldots, A_{\vec{n},r})$ is the type I multiple orthogonal polynomial of multi-index $(n_2, \ldots, n_r)$ for the Nikishin system $(w_{1,r-1}, \ldots, w_{r-1,r-1})/H_{\vec{n}}$ of order $r-1$ on Δ_{r-1}. This way we have reduced the problem by going down one order, and we can repeat the reasoning. For the external field Q_1 given by (6.25), we can use Cauchy's formula with a contour Γ going counterclockwise around Δ_{r-1}, but not around x, so that for $x \notin \Delta_{r-1}$

$$\frac{L_{\vec{n}}(x)}{H_{\vec{n}}(x)} = -\frac{1}{2\pi i} \int_\Gamma \frac{L_{\vec{n}}(z)}{H_{\vec{n}}(z)} \frac{dz}{z-x},$$

which gives

$$\frac{L_{\vec{n}}(x)}{H_{\vec{n}}(x)} = -\frac{1}{2\pi i} \int_\Gamma \frac{A_{\vec{n},1}(z)}{H_{\vec{n}}(z)} \frac{dz}{z-x}$$
$$- \sum_{j=2}^{r} \frac{1}{2\pi i} \int_\Gamma \frac{A_{\vec{n},j}(z)}{H_{\vec{n}}(z)} \int_{\Delta_{r-1}} \frac{w_{j-1,r-1}(t)}{z-t} \, dt \, \frac{dz}{z-x}.$$

The first integral on the right vanishes because of Cauchy's theorem, and by interchanging the order of integration, the remaining integrals give

$$\frac{L_{\vec{n}}(x)}{H_{\vec{n}}(x)} = \int_{\Delta_{r-1}} \frac{1}{x-t} \sum_{j=2}^{r} A_{\vec{n},j}(t) w_{j-1,r-1}(t) \frac{dt}{H_{\vec{n}}(t)}.$$

If we set

$$w_{1,r-1}(t) L_{n_2,\ldots,n_r}(t) = \sum_{j=2}^{r} A_{\vec{n},j}(t) w_{j-1,r-1}(t),$$

and define $H_{n_2,\ldots,n_r}$ to be the polynomial of degree $n_2 + \cdots + n_r - 1$ with the sign changes of $L_{n_2,\ldots,n_r}$ on Δ_{r-1}, then

$$\frac{L_{\vec{n}}(x)}{H_{\vec{n}}(x)} = \int_{\Delta_{r-1}} \frac{1}{x-t} L_{n_2,\ldots,n_r}(t) \frac{w_{1,r-1}(t)}{H_{\vec{n}}(t)} dt$$

$$= \frac{1}{H_{n_2,\ldots,n_r}(x)} \int_{\Delta_{r-1}} \frac{H_{n_2,\ldots,n_r}(x)}{x-t} L_{n_2,\ldots,n_r}(t) \frac{w_{1,r-1}(t)}{H_{\vec{n}}(t)} dt.$$

If we write

$$\frac{H_{n_2,\ldots,n_r}(x)}{x-t} = \frac{H_{n_2,\ldots,n_r}(x) - H_{n_2,\ldots,n_r}(t)}{x-t} + \frac{H_{n_2,\ldots,n_r}(t)}{x-t},$$

then the first term on the right hand side is a polynomial in t of degree at most $n_2 + \cdots + n_r - 2$, and hence the orthogonality gives

$$\frac{L_{\vec{n}}(x)}{H_{\vec{n}}(x)} = \frac{1}{H_{n_2,\ldots,n_r}(x)} \int_{\Delta_{r-1}} \frac{H_{n_2,\ldots,n_r}(t)}{x-t} L_{n_2,\ldots,n_r}(t) \frac{w_{1,r-1}(t)}{H_{\vec{n}}(t)} dt. \tag{6.27}$$

Observe that $w_{1,r-1}(t) L_{n_2,\ldots,n_r}(t) H_{n_2,\ldots,n_r}(t)/H_{\vec{n}}(t)$ does not change sign on Δ_{r-1} so that

$$\int_{\Delta_{r-1}} \frac{H_{n_2,\ldots,n_r}(t)}{x-t} L_{n_2,\ldots,n_r}(t) \frac{w_{1,r-1}(t)}{H_{\vec{n}}(t)} dt$$

is the Stieltjes transform (Markov function, Cauchy transform) of a positive weight, which we can turn into a probability weight after an appropriate normalization. The nth root of this Stieltjes function hence converges to 1 uniformly on compact subsets of $\mathbb{C} \setminus \Delta_{r-1}$. This means that the external field Q_1 in (6.25) depends only on the distribution of the zeros of $H_{n_2,\ldots,n_r}$. Let

$$\nu_k = \lim_{|\vec{n}| \to \infty} \frac{1}{n_k + \cdots + n_r - 1} \sum_{j=1}^{n_k+\cdots+n_r-1} \delta(x_{j,n_k,\ldots,n_r}), \qquad k = 1,\ldots,r, \tag{6.28}$$

where $x_{j,n_k,\ldots,n_r}$ are the zeros of $H_{n_k,\ldots,n_r}$ (which are all on Δ_{r+1-k}), and assume that

$$\lim_{|\vec{n}| \to \infty} \frac{n_k + \cdots + n_r}{n_1 + \cdots + n_r} = q_{k-1}, \qquad k = 1,\ldots,r, \tag{6.29}$$

with $q_0 = 1$. Then

$$Q_1(x) = -q_1 U(x;\nu_2), \qquad x \in \Delta_r. \tag{6.30}$$

Hence the distribution of the zeros of $H_{\vec{n}}$ is governed by the variational conditions

$$2U(x; \nu_1) - q_1 U(x; \nu_2) = \ell_1, \qquad x \in \Delta_r. \tag{6.31}$$

Now repeat the reasoning for the type I multiple orthogonal polynomials $(A_{\vec{n},2}, \ldots, A_{\vec{n},r})$ for the Nikishin system $(w_{1,r-1}, \ldots, w_{r-1,r-1})/H_{\vec{n}}$ on Δ_{r-1}. First of all, the polynomial $H_{n_2, \ldots, n_r}$ will be an orthogonal polynomial of degree $n_2 + \cdots + n_r - 1$ on Δ_{r-1} with weight function

$$\frac{|L_{n_2, \ldots, n_r}(x)|}{|H_{n_2, \ldots, n_r}(x)|} \frac{w_{1,r-1}(x)}{|H_{\vec{n}}(x)|}, \qquad x \in \Delta_{r-1}. \tag{6.32}$$

Hence the zero distribution ν_2 of $H_{n_2, \ldots, n_r}$ is governed by an equilibrium problem with external field

$$
\begin{aligned}
Q_2(x) &= -\lim_{|\vec{n}| \to \infty} \frac{1}{|\vec{n}|} \log \frac{|L_{n_2, \ldots, n_r}(x)|}{|H_{n_2, \ldots, n_r}(x)|} \frac{w_{1,r-1}(x)}{|H_{\vec{n}}(x)|}. \\
&= -q_2 U(x; \nu_3) - U(x; \nu_1), \qquad x \in \Delta_{r-1}, \tag{6.33}
\end{aligned}
$$

where the last equality follows because

$$
\begin{aligned}
&\frac{L_{n_2, \ldots, n_r}(x)}{H_{n_2, \ldots, n_r}(x)} \\
&= \frac{1}{H_{n_3, \ldots, n_r}(x)} \int_{\Delta_{r-2}} \frac{H_{n_3, \ldots, n_r}(t)}{x-t} L_{n_3, \ldots, n_r}(t) \frac{w_{1,r-2}(t)}{H_{n_2, \ldots, n_r}(t) H_{\vec{n}}(t)} \, dt.
\end{aligned}
$$

The variational condition hence becomes

$$2q_1 U(x; \nu_2) - q_2 U(x; \nu_3) - U(x; \nu_1) = \ell_2, \qquad x \in \Delta_{r-1}. \tag{6.34}$$

The remaining vector $(A_{\vec{n},3}, \ldots, A_{\vec{n},r})$ consists of the type I multiple orthogonal polynomial for the Nikishin system $(w_{1,r-2}, \ldots, w_{r-2,r-2})/H_{n_2, \ldots, n_r}$ of order $r - 2$ on Δ_{r-2}. In general the polynomial $H_{n_k, \ldots, n_r}$ with the sign changes of

$$L_{n_k, \ldots, n_r}(x) = A_{\vec{n},k}(x) + \sum_{j=k+1}^{r} A_{\vec{n},j}(x) \int_{\Delta_{r-k}} \frac{w_{j-k, r-k}(t)}{x-t} \, dt.$$

on Δ_{r-k+1} has a zero distribution ν_k which is given by the variational condition

$$2q_{k-1} U(x; \nu_k) - q_k U(x; \nu_{k+1}) - q_{k-2} U(x; \nu_{k-1}) = \ell_k, \qquad x \in \Delta_{r-k+1}. \tag{6.35}$$

and this for $k = 2, \dots, r-1$. For $k = 1$ one has (6.31), which corresponds to $q_{-1} = 0$, and for $k = r$ the polynomial H_{n_r} is precisely equal to the (monic version of) the polynomial $A_{\vec{n},r}$, which is orthogonal on Δ_1 with respect to the weight $w_{1,1}/H_{n_{r-1},n_r}$. Hence the zero distribution for ν_r is determined by

$$2q_{r-1}U(x; \nu_r) - q_{r-2}U(x; \nu_{r-1}) = \ell_r, \qquad x \in \Delta_1, \qquad (6.36)$$

which corresponds to (6.35) with $k = r$ and $q_r = 0$. The numbers ℓ_k ($k = 1, \dots, r$) in (6.35) are constants. All this means that the relevant vector potential problem for Nikishin systems has an interaction matrix of the form

$$C = \begin{pmatrix} 2q_0^2 & -q_0q_1 \\ -q_0q_1 & 2q_1^2 & -q_1q_2 \\ & -q_1q_2 & 2q_2^2 & -q_2q_3 \\ & & \ddots & \ddots & \ddots \\ & & & -q_{r-3}q_{r-2} & 2q_{r-2}^2 & -q_{r-2}q_{r-1} \\ & & & & -q_{r-2}q_{r-1} & 2q_{r-1}^2 \end{pmatrix}, \qquad (6.37)$$

and the Lagrange multipliers are $F_k = q_{k-1}\ell_k$. It turns out that for each $k = 1, \dots, r$ the support of ν_k is the full interval Δ_k (see, e.g., [15]).

We will consider the Riemann-Hilbert problem of Theorem 2.1 for the matrix function Y which is analytic in $\mathbb{C} \setminus \Delta_r$, with the jump condition

$$Y^+(x) = Y^-(x) \begin{pmatrix} 1 & 0 & \cdots & 0 & -2\pi i w_{r,r}(x) \\ 0 & 1 & \cdots & 0 & -2\pi i w_{r-1,r}(x) \\ \vdots & & \cdots & & \vdots \\ 0 & \cdots & 0 & 1 & -2\pi i w_{1,r}(x) \\ 0 & \cdots & 0 & 0 & 1 \end{pmatrix}, \qquad x \in \Delta_r, \qquad (6.38)$$

and growth condition

$$\lim_{z \to \infty} Y(z) \begin{pmatrix} z^{-n_r} \\ & \ddots \\ & & z^{-n_2} \\ & & & z^{-n_1} \\ & & & & z^{n_1 + \cdots + n_r} \end{pmatrix} = I. \qquad (6.39)$$

Observe that, compared with Theorem 2.1, we have reversed the order of the weights, which means that for the solution Y we must also reverse the order of the indices in the multi-index $\vec{n}$, i.e.,

$$Y(z) = \begin{pmatrix} c_r^{-1} A_{\vec{n}+\vec{e}_r,r}(z) & \cdots & c_r^{-1} A_{\vec{n}+\vec{e}_r,1}(z) & c_r^{-1} R_{\vec{n}+\vec{e}_r}(z) \\ \vdots & \cdots & \vdots & \vdots \\ c_1^{-1} A_{\vec{n}+\vec{e}_1,r}(z) & \cdots & c_1^{-1} A_{\vec{n}+\vec{e}_1,1}(z) & c_1^{-1} R_{\vec{n}+\vec{e}_1}(z) \\ A_{\vec{n},r}(z) & \cdots & A_{\vec{n},1}(z) & R_{\vec{n}}(z) \end{pmatrix} . \quad (6.40)$$

The normalization of this Riemann-Hilbert matrix Y for type I multiple orthogonal polynomials for a Nikishin system now goes as follows. Suppose

$$n_k = n(q_{k-1} - q_k) \in \mathbb{N},$$

where $q_0 = 1 \geq q_1 \geq \cdots \geq q_{r-1} \geq q_r = 0$ and $n = |\vec{n}|$. First we introduce the matrix

$$U_1(z) = Y(z) \begin{pmatrix} 1 & 0 & \cdots & 0 & \int_{\Delta_{r-1}} \frac{w_{r-1,r-1}(t)}{z-t} dt & 0 \\ & 1 & & 0 & \int_{\Delta_{r-1}} \frac{w_{r-2,r-1}(t)}{z-t} dt & 0 \\ & & \ddots & & \vdots & \vdots \\ & & & 1 & \int_{\Delta_{r-1}} \frac{w_{1,r-1}(t)}{z-t} dt & 0 \\ & & & & 1 & 0 \\ & & & & & 1 \end{pmatrix} . \quad (6.41)$$

Clearly U_1 is analytic in $\mathbb{C} \backslash (\Delta_r \cup \Delta_{r-1})$ and it has the same growth condition as Y as $z \to \infty$ because if we write (6.41) as $U_1(z) = Y(z) W_{r-1}(z)$, then

$$\begin{pmatrix} z^{n_r} & & & & \\ & \ddots & & & \\ & & z^{n_2} & & \\ & & & z^{n_1} & \\ & & & & z^{-n_1 - \cdots - n_r} \end{pmatrix} W_{r-1}(z)$$

$$\begin{pmatrix} z^{-n_r} & & & & \\ & \ddots & & & \\ & & z^{-n_2} & & \\ & & & z^{-n_1} & \\ & & & & z^{n_1 + \cdots + n_r} \end{pmatrix}$$

$$= \begin{pmatrix} 1 & & & \mathcal{O}(z^{-n_1+n_r-1}) & 0 \\ & \ddots & & \vdots & \\ & & 1 & \mathcal{O}(z^{-n_1+n_2-1}) & 0 \\ & & & 1 & 0 \\ & & & & 1 \end{pmatrix},$$

and since we assume (6.22) we see that this is $I + \mathcal{O}(1/z)$ as $z \to \infty$. The jump condition on Δ_r is

$$U_1^+(x) = Y^+(x)W_{r-1}(x) = Y^-(x)\begin{pmatrix} I_r & \vec{w}_r \\ \vec{0}\,^t & 1 \end{pmatrix}W_{r-1}(x),$$

where $\vec{w}_r{}^t = -2\pi i(w_{r,r}, \dots, w_{1,r})$. Since $U_1^-(x) = Y^-(x)W_{r-1}(x)$, we get

$$U_1^+(x) = U_1^-(x)[W_{r-1}(x)]^{-1}\begin{pmatrix} I_r & \vec{w}_r \\ \vec{0}\,^t & 1 \end{pmatrix}W_{r-1}(x).$$

Working out the matrix product gives

$$U_1^+(x) = U_1^-(x)\begin{pmatrix} 1 & 0 & \cdots & 0 & 0 \\ & \ddots & & & \vdots \\ & & 1 & 0 & 0 \\ & & & 1 & -2\pi i w_{1,r}(x) \\ & & & & 1 \end{pmatrix}, \qquad x \in \Delta_r, \tag{6.42}$$

because the entry on row $r-k+1$ ($k = 2, \dots, r$) and the last column is

$$-2\pi i w_{k,r}(x) + 2\pi i w_{1,r}(x)\int_{\Delta_{r-1}} \frac{w_{k-1,r-1}(t)}{x-t}\,dt,$$

which vanishes by (6.21). There will also be a jump on Δ_{r-1}, which is given by

$$U_1^+(x) = U_1^-(x)[W_{r-1}^-(x)]^{-1}W_{r-1}^+(x),$$

since Y itself has no jump on Δ_{r-1}. This matrix product is easily evaluated and if we use the Sokhotsky-Plemelj formula (1.1) then we find

$$U_1^+(x) = U_1^-(x)\begin{pmatrix} I_{r-1} & \vec{w}_{r-1} & \vec{0}_{r-1} \\ \vec{0}\,^t_{r-1} & 1 & 0 \\ \vec{0}\,^t_{r-1} & 0 & 1 \end{pmatrix}, \qquad x \in \Delta_{r-1}, \tag{6.43}$$

where $\vec{w}_{r-1}{}^t = -2\pi i(w_{r-1,r-1}(x), \dots, w_{1,r-1}(x))$. The next step is to consider

$$U_2(z) = U_1(z)\begin{pmatrix} 1 & & & \int_{\Delta_{r-2}} \frac{w_{r-2,r-2}(t)}{z-t}\,dt & 0 & 0 \\ & 1 & & \int_{\Delta_{r-2}} \frac{w_{r-3,r-2}(t)}{z-t}\,dt & 0 & 0 \\ & & \ddots & \vdots & & \\ & & 1 & \int_{\Delta_{r-2}} \frac{w_{1,r-2}(t)}{z-t}\,dt & 0 & 0 \\ & & & 1 & 0 & 0 \\ & & & & 1 & 0 \\ & & & & & 1 \end{pmatrix}, \tag{6.44}$$

and in general

$$U(z) = Y(z) \prod_{j=1}^{r-1} \begin{pmatrix} I_{r-j} & \int_{\Delta_{r-j}} \frac{\tilde{w}_{r-j}(t)}{z-t}\, dt & 0_{r-j,j} \\ 0_{1,r-j} & 1 & 0_{1,j} \\ 0_{j,r-j} & 0_{j,1} & I_j \end{pmatrix}, \qquad (6.45)$$

where the matrix product is taken from left ($j = 1$) to right ($j = r - 1$) and $0_{m,n}$ is a zero matrix with m rows and n columns. The matrix function U is then analytic on $\mathbb{C} \setminus \bigcup_{j=1}^{r} \Delta_j$. The growth condition is again

$$\lim_{z \to \infty} U(z) \begin{pmatrix} z^{-n_r} & & & 0 \\ & \ddots & & \\ & & z^{-n_1} & \\ 0 & & & z^{n_1+n_2+\cdots+n_r} \end{pmatrix} = I,$$

since

$$\begin{pmatrix} z^{n_r} & & & 0 \\ & \ddots & & \\ & & z^{n_1} & \\ 0 & & & z^{-n_1-\cdots-n_r} \end{pmatrix} \begin{pmatrix} I_{r-j} & \int_{\Delta_{r-j}} \frac{\tilde{w}_{r-j}(t)}{z-t}\, dt & 0_{r-j,j} \\ 0_{1,r-j} & 1 & 0_{1,j} \\ 0_{j,r-j} & 0_{j,1} & I_j \end{pmatrix}$$

$$\begin{pmatrix} z^{-n_r} & & & 0 \\ & \ddots & & \\ & & z^{-n_1} & \\ 0 & & & z^{n_1+\cdots+n_r} \end{pmatrix}$$

$$= \begin{pmatrix} 1 & & & \mathcal{O}(z^{-n_j+n_r-1}) \\ & \ddots & & \\ & & 1 & \mathcal{O}(z^{-n_j+n_{j+1}-1}) \\ & & & \ddots \\ & & & & 1 \end{pmatrix} = I + \mathcal{O}(1/z),$$

since we assume (6.22). The jump condition requires some computation, but the use of the Sokhotsky-Plemelj formula (1.1) and the relation (6.21) for the weight functions in a Nikishin system, gives for $x \in \mathbb{R}$

$$U^+(x) = U^-(x) \begin{pmatrix} 1 & -2\pi i w_{1,1}(x) & 0 & \cdots & 0 \\ & 1 & -2\pi i w_{1,2}(x) & \cdots & 0 \\ & & 1 & \ddots & \vdots \\ & & & 1 & -2\pi i w_{1,r}(x) \\ & & & & 1 \end{pmatrix}. \qquad (6.46)$$

Observe that there will only be jumps on the intervals Δ_k and that this jump matrix only contains the first weight function $w_{1,k}$ of each Nikishin system of order k. In case all the intervals Δ_k are disjoint, the jump will only contain one entry outside the diagonal on each interval Δ_k.

Let $(\nu_1, \ldots, \nu_r)$ be the vector of equilibrium measures for the extremal problem with interaction matrix C given by (6.37) and variational conditions (6.35). The normalization will be of the form

$$M(z)$$

$$= \begin{pmatrix} e^{n(\ell_1^* + \cdots + \ell_r^*)} & & & & \\ & \ddots & & & \\ & & e^{n(\ell_1^* + \ell_2^*)} & & \\ & & & e^{n\ell_1^*} & \\ & & & & 1 \end{pmatrix} U(z)$$

$$\times \begin{pmatrix} e^{-n q_{r-1} g_r(z)} & & & & \\ & \ddots & & & \\ & & e^{-n[q_{k-1} g_k(z) - q_k g_{k+1}(z)]} & & \\ & & & \ddots & \\ & & & & e^{-n[q_0 g_1(z) - q_1 g_2(z)]} \\ & & & & & e^{n q_0 g_1(z)} \end{pmatrix}$$

$$\times \begin{pmatrix} e^{-n(\ell_1^* + \cdots + \ell_r^*)} & & & & \\ & \ddots & & & \\ & & e^{-n(\ell_1^* + \ell_2^*)} & & \\ & & & e^{-n\ell_1^*} & \\ & & & & 1 \end{pmatrix} \tag{6.47}$$

where

$$g_k(z) = \int_{\Delta_{r-k+1}} \log(z - x)\, d\nu_k(x).$$

The numbers ℓ_k^* are obtained by rewriting the variational conditions (6.35) in the form

$$-q_{k-1}[g_k^+(x) + g_k^-(x)] + q_k g_{k+1}^-(x) + q_{k-2} g_{k-1}^+(x) = \ell_k^*, \qquad x \in \Delta_{r-k+1}, \tag{6.48}$$

where $\ell_k^* - \ell_k$ is an integer multiple of $i\pi/n$. Observe that

$$n[q_{k-1} g_k(z) - q_k g_{k+1}(z)] = n(q_{k-1} - q_k) \log z + \mathcal{O}(1/z), \qquad z \to \infty,$$

so that

$$e^{-n[q_{k-1} g_k(z) - q_k g_{k+1}(z)]} = z^{-n_k}[1 + \mathcal{O}(1/z)], \qquad k = 1, \ldots, r,$$

and

$$e^{nq_0 g_1(z)} = z^{n_1 + \cdots + n_r}[1 + \mathcal{O}(1/z)],$$

which means that

$$\lim_{z \to \infty} M(z) = I.$$

This shows that the matrix function M is normalized as $z \to \infty$. If we write (6.47) as $M(z) = L^{-1}U(z)G(z)L$, then the jump condition for this function on the real line is

$$M^+(x) = M^-(x)L^{-1}[G^-(x)]^{-1}W(x)G^+(x)L,$$

where W is the jump matrix for U given by (6.46). First we compute the matrix product

$$[G^-(x)]^{-1}W(x)G^+(x) = \begin{pmatrix} a_r(x) & b_r(x) & 0 & \cdots & 0 \\ & a_{r-1}(x) & b_{r-1}(x) & 0 & \vdots \\ & & \ddots & \ddots & 0 \\ & & & a_1(x) & b_1(x) \\ & & & & a_0(x) \end{pmatrix},$$

(6.49)

where for $k = 1, \ldots, r$

$$a_k(x) = e^{n[q_k(g_{k+1}^+(x) - g_{k+1}^-(x)) - q_{k-1}(g_k^+(x) - g_k^-(x))]},$$

$$b_k(x) = -2\pi i w_{1,r-k+1}(x) e^{n[-q_k g_{k+1}^-(x) + q_{k-1}(g_k^+(x) + g_k^-(x)) - q_{k-2} g_{k-1}^+(x)]},$$

and

$$a_0(x) = e^{n(g_1^+(x) - g_1^-(x))}.$$

We have

$$g_k^{\pm}(x) = \begin{cases} -U(x; \nu_k), & x > b_k, \\ -U(x; \nu_k) \pm i\pi, & x < a_k, \\ -U(x; \nu_k) \pm \varphi_k(x), & x \in [a_k, b_k], \end{cases}$$

where we define $\Delta_{r-k+1} = [a_k, b_k]$, and

$$\varphi_k(x) = i\pi \int_x^{b_k} d\nu_k(t).$$

With this information and (6.48) we find

$$a_k(x) = e^{2n[q_k \varphi_{k+1}(x) - q_{k-1} \varphi_k(x)]},$$

$$b_k(x) = -2\pi i w_{1,r-k+1}(x) e^{-n\ell_k^*},$$

for $k = 1, \ldots, r$, and

$$a_0(x) = e^{2n\varphi_1(x)}.$$

The jump for M thus becomes

$$L^{-1}[G^-(x)]^{-1}W(x)G^+(x)L$$

$$= \begin{pmatrix} a_r(x) & -2\pi i w_{1,1}(x) & 0 & \cdots & 0 \\ & a_2(x) & -2\pi i w_{1,2}(x) & 0 & 0 \\ & & \ddots & \ddots & \\ & & & a_1(x) & -2\pi i w_{1,r}(x) \\ & & & & a_0(x) \end{pmatrix}, \quad (6.50)$$

where each a_k is an oscillatory function. This means that this normalized Riemann-Hilbert problem now has oscillatory jumps on each of the intervals Δ_k, so that the steepest descent method of Deift and Zhou [9] [10] can be used for the asymptotic analysis of $M(z)$ as $|\vec{n}|$ tends to infinity. We will return to this in future work.

7. Conclusion

We have shown that one can find a Riemann-Hilbert problem for $(r + 1) \times (r + 1)$ matrix functions for studying multiple orthogonal polynomials of type I and type II (Hermite-Padé polynomials). This Riemann-Hilbert problem easily gives an important relationship between type I and type II multiple orthogonal polynomials (Section 4) and a recurrence relation for contiguous type II multiple orthogonal polynomials. For the asymptotic analysis of these Riemann-Hilbert problems as the multi-index $\vec{n}$ is large, it is more convenient to normalize the Riemann-Hilbert problem. We have worked out the proper normalization for two important systems of weights, namely for Angelesco systems (Section 6.1) and for Nikishin systems (Section 6.2). This involves the asymptotic zero distribution for multiple orthogonal polynomials, which is related to an extremal problem for vector potentials. This is explained in Section 6. Details of this equilibrium problem are given again for Angelesco systems and for Nikishin systems. We hope that the material in Section 6 is a useful survey of these interesting systems for multiple orthogonal polynomials. The normalized Riemann-Hilbert problems has oscillatory jumps on the real line, which allows the use of a steepest descent method introduced by Deift and Zhou, but a detailed account of this would be out of the scope of the present survey.

Acknowledgements. The authors like to thank Percy Deift and Alexander I. Aptekarev for their interest in this work and for many fruitful discussions. This work was supported by the project INTAS-2000-272.

References

1. A. I. Aptekarev, *Asymptotics of simultaneously orthogonal polynomials in the Angelesco case*, Mat. Sb. **136** (178) (1988), 56–84 (in Russian); Math. USSR Sb. **64** (1989), 57–84.
2. A. I. Aptekarev, *Multiple orthogonal polynomials*, J. Comput. Appl. Math. **99** (1998), 423–447.
3. A. I. Aptekarev, *Strong asymptotics of multiple orthogonal polynomials for Nikishin systems*, Mat. Sb. **190** No. 5 (1999), 3–44 (in Russian); Sbornik Math. **190** No. 5 (1999), 631–669.
4. P. Bleher, A. Its, *Semiclassical asymptotics of orthogonal polynomials, Riemann-Hilbert problem, and the universality in the matrix model*, Ann. of Math. **150** (1999), 185–266.
5. J. Bustamante, G. López Lagomasino, *Hermite-Padé approximation to a Nikishin type system of analytic functions*, Mat. Sb. **183** (1992), 117–138; translated in Russian Acad. Sci. Sb. Math. **77** (1994), 367–384.
6. P. Deift, *Orthogonal Polynomials and Random Matrices: A Riemann-Hilbert Approach*, Courant Lecture Notes **3**, Courant Institute, New York, 1999.
7. P. Deift, T. Kriecherbauer, K. T-R. McLaughlin, S. Venakides, X. Zhou, *Strong asymptotics of orthogonal polynomials with respect to exponential weights*, Commun. Pure Appl. Math. **52** (1999), 1491–1552.
8. P. Deift, T. Kriecherbauer, K. T-R. McLaughlin, S. Venakides, X. Zhou, *Uniform asymptotics for polynomials orthogonal with respect to varying exponential weights and applications to universality questions in random matrix theory*, Commun. Pure Appl. Math. **52** (1999), 1335–1425.
9. P. Deift, X. Zhou, *A steepest descent method for oscillatory Riemann-Hilbert problems*, Bull. Amer. Math. Soc. **26** (1992), 119–123.
10. P. Deift, X. Zhou, *A steepest descent method for oscillatory Riemann-Hilbert problems, Asymptotics for the MKdV equation*, Ann. of Math. (2) **137** (1993), 295–368.
11. K. Driver, H. Stahl, *Normality in Nikishin systems*, Indag. Math. (New Series) **5** (1994), 161–187.
12. A. S. Fokas, A. R. Its, A. V. Kitaev, *Discrete Painlevé equations and their appearance in quantum gravity*, Comm. Math. Phys. **142** (2) (1991), 313–344.
13. A. S. Fokas, A. R. Its, A. V. Kitaev, *The isomonodromy approach to matrix models in 2D quantum gravity*, Comm. Math. Phys. **147** (1992), 395–430.
14. A. A. Gonchar, E. A. Rakhmanov, *On the equilibrium problem for vector potentials*, Usp. Mat. Nauk **40**, No. 4 (244) (1985), 155–156; translated in Russ. Math. Surveys **40**, No. 4 (1985), 183–184.
15. A. A. Gonchar, E. A. Rakhmanov, V. N. Sorokin, *Hermite-Padé approximants for systems of Markov-type functions*, Mat. Sb. **188** (1997), 33–58 (in Russian); Russian Acad. Sci. Sb. Math. **188** (1997), 671–696.
16. A. R. Its, A. V. Kitaev, A. S. Fokas, *An isomonodromy approach to the theory of two-dimensional quantum gravity*, Uspekhi Mat. Nauk **45** No. 6 (276) (1990), 135–136; translated in Russ. Math. Surveys **45**, No. 6 (1990), 155–157.
17. V. A. Kalyagin, *On a class of polynomials defined by two orthogonality relations*, Mat. Sb. **110** (1979), 609–627 (in Russian); Math. USSR Sb. **38** (1981), 563–580.
18. V. A. Kaliaguine, A. Ronveaux, *On a system of classical polynomials of simultaneous orthogonality*, J. Comput. Appl. Math. **67** (1996), 207–217.
19. K. Mahler, *Perfect systems*, Comp. Math. **19** (1968), 95–166.
20. E. M. Nikishin, *On simultaneous Padé approximations*, Mat. Sb. **113** (155) (1980), 499–519; translated in Math. USSR Sb. **41** (1982), 409–425.
21. E. M. Nikishin, V. N. Sorokin, *Rational Approximations and Orthogonality*, Translations of Mathematical Monographs **92**, Amer. Math. Soc., Providence RI, 1991.
22. J. Nuttall, *Asymptotics of diagonal Hermite-Padé polynomials*, J. Approx. Theory **42** (1984), 299–386.

23. H. Padé, *Mémoire sur les développements en fractions continues de la fonction exponentielle pouvant servir d'introduction à la théorie des fractions continues algébriques*, Ann. Sci. École Norm. Sup. (3) **16** (1899), 395–426.
24. E. B. Saff, V. Totik, *Logarithmic Potentials with External Fields*, Grundlehren der Mathematischen Wissenschaften **316**, Springer-Verlag, Berlin, 1997.
25. H. Stahl, V. Totik, *General Orthogonal Polynomials*, Encyclopedia of Mathematics and its Applications, Cambridge University Press, 1992.
26. W. Van Assche, *Multiple orthogonal polynomials, irrationality and transcendence*, Contemporary Mathematics **236** (1999), 325–342.

FLOWERS WHICH WE CANNOT YET SEE GROWING IN RAMANUJAN'S GARDEN OF HYPERGEOMETRIC SERIES, ELLIPTIC FUNCTIONS, AND q'S

BRUCE C. BERNDT
Department of Mathematics
University of Illinois
1409 West Green Street
Urbana, IL 61801 USA

Abstract. Many of Ramanujan's ideas and theorems form the seeds of questions and problems, many of which remain unresolved or even to be thoroughly examined. This survey raises questions arising from Ramanujan's work on theta-functions and other q-series, with Gaussian hypergeometric functions making frequent appearances.

1. Introduction

In his lecture on June 1, 1987 at the Ramanujan centenary conference at the University of Illinois, Freeman Dyson [32] proclaimed, "That was the wonderful thing about Ramanujan. He discovered so much, and yet he left so much more in his garden for other people to discover." Our goal in this paper is to tell you about some of the flowers in Ramanujan's garden that have yet to be discovered. Of course, because we have not yet seen the flowers, we cannot describe their colors or fragrances for you. We cannot even identify their genus and species, and, in fact, we may need to concoct some new botanical names. The flowers may also be extremely difficult to find. What we can do is point out some signposts and provide a map with some pathways, some of which may lead you nowhere.

Many of our queries originate with Ramanujan's alternative theories of elliptic functions, and so we begin our stroll through Ramanujan's garden basking in the beauty of these flowers, the seeds of which were planted in Ramanujan's famous paper on modular equations and approximations to π [48], [49, pp. 23–39] in 1914, but which only germinated in his notebooks

J. Bustoz et al. (eds.), Special Functions 2000, 61–85.

[50]. These flowers were not seen by anyone but Ramanujan until Berndt, S. Bhargava, and F. G. Garvan [11] cultivated them in 1995. Since then, other gardeners have found further flowers, but Ramanujan has left a huge plot of land for still undiscovered related species to grow and be nurtured.

Arising out of both the classical and alternative theories are transformations between hypergeometric functions. In many cases, in particular with those transformations arising from modular equations, we ask if these are isolated, or are they really special cases of more general transformations yet to be discovered. Sections 2 and 3 give examples.

In Section 4, we turn our attention to the explicit evaluation of theta-functions, hypergeometric series, and elliptic integrals. An important and useful analogue of the classical class invariants of Weber and Ramanujan found in Ramanujan's lost notebook is discussed. This points the way to other analogues which have not yet sprung up in Ramanujan's garden.

In Section 5, the focus is on q-continued fractions, in particular, those which can be represented by a q-product, such as the Rogers–Ramanujan continued fraction, which has an extensive elegant theory. Less extensive theories have been developed for the Ramanujan–Gordon–Göllnitz and cubic continued fractions. There appear to be many garden paths needing development.

In the last garden plot, the flowers are really mysterious. In his lost notebook [51], Ramanujan left a very beautiful rare species of flower, so rare that, except for a few additional flowers found by Berndt, Chan, and S.–S. Huang [14], no one has found any others or really explained how they grow, or if they have relationships with other species. Ramanujan found several integrals of eta-functions that can be represented as incomplete elliptic integrals of the first kind. If you have not previously seen this species, you will definitely agree that it is an exotic one.

2. Ramanujan's Alternative Theories of Elliptic Functions

In his famous paper [48], [49, pp. 23–39], Ramanujan records several elegant series for $1/\pi$ and asserts, "There are corresponding theories in which q is replaced by one or other of the functions"

$$q_r := q_r(x) := \exp\left(-\pi \csc(\pi/r)\frac{{}_2F_1\left(\frac{1}{r}, \frac{r-1}{r}; 1; 1-x\right)}{{}_2F_1\left(\frac{1}{r}, \frac{r-1}{r}; 1; x\right)}\right), \qquad (2.1)$$

where $r = 3, 4$, or 6, and where ${}_2F_1$ denotes the classical Gaussian hypergeometric function. In the classical theory of elliptic functions, the variable q is given by q_2, and Ramanujan implies that most of his series for $1/\pi$ arise not out of the classical theory but out of new alternative theories wherein

q is replaced by either q_3, q_4, or q_6. Ramanujan gives no proofs of his series for $1/\pi$ or of any of his theorems in the "corresponding" or "alternative" theories. It was not until 1987 that J. M. and P. B. Borwein [21] proved Ramanujan's series formulas for $1/\pi$. In his second notebook [50, pp. 257–262], Ramanujan records without proofs his theorems in these new theories, which were first proved in 1995 by Berndt, Bhargava, and Garvan [11], who gave these theories the appellation, the theories of signature r ($r = 3, 4, 6$). An account of this work may also be found in Berndt's book [9, Chap. 33].

Define the complete elliptic integral $K(k)$ of the first kind by

$$K := K(k) := \int_0^{\pi/2} \frac{d\phi}{\sqrt{1 - k^2 \sin^2 \phi}} = \frac{\pi}{2} \; {}_2F_1\left(\frac{1}{2}, \frac{1}{2}; 1; k^2\right),$$
(2.2)

where the latter representation is achieved by expanding the integrand in a binomial series and integrating termwise. Here, k, $0 < k < 1$, is called the modulus. In the classical theory, the theta-functions

$$\varphi(q) := \sum_{n=-\infty}^{\infty} q^{n^2} \quad \text{and} \quad \psi(q) := \sum_{n=0}^{\infty} q^{n(n+1)/2}$$
(2.3)

play key roles. In particular, Jacobi's identity [7, p. 40, Entry 25(vii)]

$$\varphi^4(q) - \varphi^4(-q) = 16q\psi^4(q^2)$$
(2.4)

is crucially used in establishing the fundamental inversion formula [7, pp. 100–101]

$$z_2 := {}_2F_1\left(\frac{1}{2}, \frac{1}{2}; 1; x\right) = \frac{2}{\pi} K(k) = \varphi^2(q),$$
(2.5)

where $x = k^2$, $q := q_2$ is given by (2.1), and $K(k)$ is given by (2.2).

A common, sumptuous flower in the classical theory is the Dedekind eta-function, $\eta(\tau)$, defined by

$$\eta(\tau) := e^{2\pi i \tau/24}(q; q)_\infty =: q^{1/24} f(-q), \qquad q = e^{2\pi i \tau}, \quad \text{Im}\,\tau > 0,$$
(2.6)

where $f(-q)$ is the notation employed by Ramanujan. We close our brief description of the classical theory by noting the well-known formulas [7, p. 39, Entries 24 (ii), (iv)]

$$\varphi(q) = \frac{f^2(q)}{f(-q^2)} \quad \text{and} \quad \psi(q) = \frac{f^2(-q^2)}{f(-q)}.$$
(2.7)

In the cubic theory, or the theory of signature 3, for $\omega = \exp(2\pi i/3)$, define

$$a(q) := \sum_{m,n=-\infty}^{\infty} q^{m^2+mn+n^2}, \tag{2.8}$$

$$b(q) := \sum_{m,n=-\infty}^{\infty} \omega^{m-n} q^{m^2+mn+n^2}, \tag{2.9}$$

and

$$c(q) := \sum_{m,n=-\infty}^{\infty} q^{(m+1/3)^2+(m+1/3)(n+1/3)+(n+1/3)^2}. \tag{2.10}$$

The functions defined in (2.8)–(2.10) are the "cubic" theta-functions, first introduced by the Borweins [22], who proved that

$$a^3(q) = b^3(q) + c^3(q). \tag{2.11}$$

Ramanujan [50, p. 258] established the fundamental inversion formula

$$z_3 := {}_2F_1\left(\frac{1}{3}, \frac{2}{3}; 1; x\right) = a(q), \tag{2.12}$$

where $q := q_3$ is given by (2.1). This theorem was first proved in print by Berndt, Bhargava, and Garvan [11], [9, p. 99], with (2.11) being a necessary ingredient in their proof.

The analogues of (2.7) in the theory of signature 3 are given by [9, p. 109, Lemma 5.1]

$$b(q) = \frac{f^3(-q)}{f(-q^3)} \quad \text{and} \quad c(q) = 3q^{1/3}\frac{f^3(-q^3)}{f(-q)}. \tag{2.13}$$

In the theory of signature 4, or in the quartic theory, Berndt, Bhargava, and Garvan [11], [9, p. 146, eq. (9.7)] established a "transfer" principle of Ramanujan by which formulas in the theory of signature 4 can be derived from formulas in the classical theory. Taking the place of $a(q), b(q)$, and $c(q)$ in the cubic theory are the functions

$$A(q) := \varphi^4(q) + 16q\psi^4(q^2), \qquad B(q) := \varphi^4(q) - 16q\psi^4(q^2), \tag{2.14}$$

and

$$C(q) := 8\sqrt{q}\varphi^2(q)\psi^2(q^2), \tag{2.15}$$

(where φ and ψ are defined in (2.3)) which, by Jacobi's identity (2.4), satisfy the equality

$$A^2(q) = B^2(q) + C^2(q), \tag{2.16}$$

the quartic analogue of (2.4) and (2.11). Berndt, Chan, and Liaw [16] used (2.16) to establish the inversion formula

$$z_4 := {}_2F_1\left(\frac{1}{4}, \frac{3}{4}; 1; x\right) = \sqrt{A(q)}, \tag{2.17}$$

which is clearly an analogue of (2.5) and (2.12). They therefore were able to bypass the aforementioned transfer theorem and directly prove theorems in the quartic theory.

To again demonstrate the ubiquitous role of the Dedekind eta-function, we note that the quartic analogues of (2.7) and (2.13) are given by [16, Thm. 3.1]

$$B(q) = \left(\frac{f^2(-q)}{f(-q^2)}\right)^4 \quad \text{and} \quad C(q) = 8\sqrt{q}\left(\frac{f^2(-q^2)}{f(-q)}\right)^4. \tag{2.18}$$

Ramanujan's theory of elliptic functions of signature 6 is not as complete as those in the cubic and quartic theories [11], [9, pp. 161–164]. In particular, we have been unable to obtain a sextic analogue of (2.5), (2.12), and (2.17). The few results in the sextic theory recorded by Ramanujan in his notebooks were proved by using the transformation formula

$$\sqrt{1+2p}\,{}_2F_1(\tfrac{1}{6}, \tfrac{5}{6}; 1; \beta) = \sqrt{1+p+p^2}\,{}_2F_1(\tfrac{1}{2}, \tfrac{1}{2}; 1; \alpha),$$

where $0 < p < 1$ and α and β are given by

$$\alpha := \frac{p(2+p)}{1+2p} \quad \text{and} \quad \beta := \frac{27p^2(1+p)^2}{4(1+p+p^2)^3}.$$

Chan and Y. L. Ong [30] have established a few results pointing toward the beginnings of a theory of signature 7. The analogue of $\varphi(q)$ and $a(q)$ in the septic theory is given by

$$\sum_{m,n=-\infty}^{\infty} q^{m^2+mn+2n^2}.$$

However, we do not know any other theta functions in the septic theory.

No further alternative theories have been found. It may be that if other theories of signature n exist, then the class number of $\mathbf{Q}(\sqrt{-n})$ must be equal to 1. Observe that the discriminant of $m^2 + mn + 2n^2$ is -7 and that the class number of $\mathbf{Q}(\sqrt{-7})$ equals 1.

The alternative cubic theory appears to be the most interesting of the three known alternative theories for several reasons. In particular, the new theta functions $a(q)$, $b(q)$, and $c(q)$ arise, and the many results found by Ramanujan in the cubic theory do not seem to be reachable through the classical theory. If there is a complete cubic theory of elliptic functions analogous to the rich and beautiful classical theory, then it is necessary to seek the identities of the cubic analogues of the Jacobian elliptic functions and the cubic analogues of the elliptic integrals from which the Jacobian functions arise by inverting the elliptic integrals. Ramanujan bequeathed to us one extraordinarily beautiful flower found on page 257 in his second notebook [50], [11, Theorem 8.1], [9, p. 133] given below. Proofs have been given by Berndt, Bhargava, and Garvan [11] and by L.-C. Shen [57].

Theorem 1 *Let* $q := q_3$ *be defined by* (2.1), *and let* $z := z_3$ *be defined by* (2.12). *For* $0 \leq \varphi \leq \pi/2$, *define* $\theta = \theta(\varphi)$ *by*

$$\theta z = \int_0^{\varphi} {}_2F_1\left(\tfrac{1}{3}, \tfrac{2}{3}; \tfrac{1}{2}; x\sin^2 t\right) dt. \tag{2.19}$$

Then, for $0 \leq \theta \leq \pi/2$,

$$\varphi = \theta + 3\sum_{n=1}^{\infty} \frac{\sin(2n\theta)}{n(1 + 2\cosh(ny))} = \theta + 3\sum_{n=1}^{\infty} \frac{\sin(2n\theta)q^n}{n(1 + q^n + q^{2n})}, \tag{2.20}$$

where $q := e^{-y}$.

Recall that [6, p. 99, Entry 35(iii), Chap. 11]

$${}_2F_1\left(\tfrac{1}{2} + n, \tfrac{1}{2} - n; \tfrac{1}{2}; x^2\right) = (1 - x^2)^{-1/2}\cos\left(2n\,\sin^{-1}x\right), \tag{2.21}$$

where n is arbitrary. With $n = \frac{1}{6}$ in (2.21), we see that the integral in (2.19) is an analogue of the incomplete integral of the first kind, which arises from the case $n = 0$ in (2.21). Since ${}_2F_1\left(\tfrac{1}{3}, \tfrac{2}{3}; \tfrac{1}{2}; x\sin^2 t\right)$ is a nonnegative, monotonically increasing function on $[0, \pi/2]$, there exists a unique inverse function $\varphi = \varphi(\theta)$. Thus, (2.20) gives the "Fourier series" of this inverse function and is analogous to familiar Fourier series for the Jacobian elliptic functions (Whittaker and Watson [65, pp. 511–512]). The function φ may therefore be considered a cubic analogue of the Jacobian function sn.

When $\varphi = 0 = \theta$, (2.20) is trivial. When $\varphi = \pi/2$,

$$\int_0^\varphi {}_2F_1\left(\tfrac{1}{3}, \tfrac{2}{3}; \tfrac{1}{2}; x\sin^2 t\right) dt = \sum_{n=0}^\infty \frac{(\tfrac{1}{3})_n(\tfrac{2}{3})_n}{(\tfrac{1}{2})_n n!} x^n \int_0^{\pi/2} \sin^{2n} t\, dt$$

$$= \sum_{n=0}^\infty \frac{(\tfrac{1}{3})_n(\tfrac{2}{3})_n}{(\tfrac{1}{2})_n n!} x^n \frac{(\tfrac{1}{2})_n}{n!} \frac{\pi}{2}$$

$$= \tfrac{1}{2}\pi\, {}_2F_1\left(\tfrac{1}{3}, \tfrac{2}{3}; 1; x\right).$$

Thus, $\theta = \pi/2$, which is implicit in our statement of Theorem 1.

What are the cubic analogues of the other Jacobian elliptic functions, if they do indeed exist? Are there further integrals like (2.19) which when inverted yield these new cubic Jacobian functions? Some progress in these directions has been made in unpublished work by Y.-S. Choi.

In the classical theory, Pfaff's familiar quadratic transformation [6, p. 93]

$$ {}_2F_1\left(\frac{1}{2}, \frac{1}{2}; 1; 1 - \left(\frac{1-x}{1+x}\right)^2\right) = (1+x)\, {}_2F_1\left(\frac{1}{2}, \frac{1}{2}; 1; x^2\right) \tag{2.22}$$

is the key to proving the inversion formula (2.5), while in the cubic theory, Ramanujan's cubic transformation [22], [11], [9, p. 97, Cor. 2.4]

$$ {}_2F_1\left(\frac{1}{3}, \frac{2}{3}; 1; 1 - \left(\frac{1-x}{1+2x}\right)^3\right) = (1+2x)\, {}_2F_1\left(\frac{1}{3}, \frac{2}{3}; 1; x^3\right) \tag{2.23}$$

plays the leading role in proving (2.12). The first proof of (2.23) was by the Borweins [22], while another proof is given in [11], [9, p. 97, Cor. 2.4]. However, Chan's [25] two proofs are more natural. In the quartic theory, the transformation [50, p. 260], [9, p. 146, Thm. 9.4], [16, Thm. 2.1]

$$ {}_2F_1\left(\frac{1}{4}, \frac{3}{4}; 1; 1 - \left(\frac{1-x}{1+3x}\right)^2\right) = \sqrt{1+3x}\, {}_2F_1\left(\frac{1}{4}, \frac{3}{4}; 1; x^2\right) \tag{2.24}$$

is central. Are there any further transformations of the types (2.22)–(2.24)?

There are essentially but two quadratic transformations for ${}_2F_1$ [3, p. 125, Thm. 3.1.1; p. 127, Thm. 3.1.3]; (2.22) and (2.24) are special cases. The transformation (2.23) and is a special case of the cubic transformation [11], [9, p. 96, p. 96]

$$ {}_2F_1\left(c, c + \frac{1}{3}; \frac{3c+1}{2}; 1 - \left(\frac{1-x}{1+2x}\right)^3\right) \tag{2.25}$$

$$= (1+2x)^{3c}\, {}_2F_1\left(c, c + \frac{1}{3}; \frac{3c+5}{6}; x^3\right).$$

Further cubic transformations were found by Ramanujan [50], Berndt, Bhargava, and Garvan [11], [9, pp. 173–175], and Garvan [34]. E. Goursat [35] found many cubic transformations. How many distinct cubic transformations exist? Apparently no general transformations of order exceeding three have been found. We shall return to this question in the next section.

3. Modular Equations and Hypergeometric Transformations

We begin by reviewing the definition of a modular equation, as it was understood by Ramanujan. In the sequel, for brevity, set $q = q_2$ and $z = z_2$.

Let n denote a fixed positive integer, and suppose that

$$n\frac{{}_2F_1\left(\frac{1}{2},\frac{1}{2};1;1-k^2\right)}{{}_2F_1\left(\frac{1}{2},\frac{1}{2};1;k^2\right)} = \frac{{}_2F_1\left(\frac{1}{2},\frac{1}{2};1;1-\ell^2\right)}{{}_2F_1\left(\frac{1}{2},\frac{1}{2};1;\ell^2\right)}, \tag{3.1}$$

where $0 < k, \ell < 1$. Then a modular equation of degree n is a relation between the moduli k and ℓ which is implied by (3.1). Following Ramanujan, we put $\alpha = k^2$ and $\beta = \ell^2$, We often say that β has degree n, or degree n over α. The multiplier m is defined by

$$m := \frac{{}_2F_1\left(\frac{1}{2},\frac{1}{2};1;\alpha\right)}{{}_2F_1\left(\frac{1}{2},\frac{1}{2};1;\beta\right)}. \tag{3.2}$$

The fundamental relations (2.5) and (3.2) are the keys to deriving modular equations, for the most successful approach has been to establish theta-function identities and then convert them to modular equations by using a catalogue of formulas for theta-functions based on (2.5) and in terms of z, q, and x (or α or k.)

In view of (2.1), we could also define modular equations in the alternative theories. Indeed, Ramanujan did this and derived several modular equations in the cubic and quartic theories [11], [9, pp. 120–132, 153–161]. Further modular equations in the cubic theory were found by Chan and Liaw [28].

Ramanujan discovered several hundred modular equations, and although almost all of them have now been proved, for most of them, we have not discerned Ramanujan's methods. In particular, to prove some of them, we have had to resort to the theory of modular forms, a subject with which Ramanujan was apparently unfamiliar. See [7]–[9] for proofs. With the publication of his lost notebook [51], there appears a fragment in which Ramanujan summarizes by type many of his results on modular equations. Proofs for most of them can be found in [9], but some do not appear in his notebooks; the entire fragment is discussed in [10], [4, Chap. 9]. In this brief section, we focus our attention on Ramanujan's formulas for the multiplier m, defined in (3.2), which are conveniently gathered together in the aforementioned fragment. Almost all of Ramanujan's formulas of this sort are completely new, and moreover we do not know how Ramanujan derived them. Most of our proofs use the theory of modular forms. Note that, by (3.2), a formula for the multiplier yields a transformation formula for hypergeometric functions. For $n > 3$, these are thus higher order transformations for which we do not know more general transformations.

We now give three examples found in Ramanujan's second notebook. To repeat, by (3.2), each provides a transformation between hypergeometric functions.

Theorem 2 *If β and the multiplier m have degree 5, then*

$$m = \left(\frac{\beta}{\alpha}\right)^{1/4} + \left(\frac{1-\beta}{1-\alpha}\right)^{1/4} - \left(\frac{\beta(1-\beta)}{\alpha(1-\alpha)}\right)^{1/4}.$$

Theorem 3 *If β and the multiplier m have degree 7, then*

$$m^2 = \left(\frac{\beta}{\alpha}\right)^{1/2} + \left(\frac{1-\beta}{1-\alpha}\right)^{1/2} - \left(\frac{\beta(1-\beta)}{\alpha(1-\alpha)}\right)^{1/2} - 8\left(\frac{\beta(1-\beta)}{\alpha(1-\alpha)}\right)^{1/3}.$$

Theorem 4 *If β and the multiplier m have degree 13, then*

$$m = \left(\frac{\beta}{\alpha}\right)^{1/4} + \left(\frac{1-\beta}{1-\alpha}\right)^{1/4} - \left(\frac{\beta(1-\beta)}{\alpha(1-\alpha)}\right)^{1/4} - 4\left(\frac{\beta(1-\beta)}{\alpha(1-\alpha)}\right)^{1/6}.$$

Theorems 2 and 3 can be found in Entries 13(xii) and 19(v), respectively, in Chapter 19 [7, pp. 281–282, 314], and Theorem 4 coincides with Entry 8(iii) in Chapter 20 [7, p. 376] of Ramanujan's second notebook. See also [10, p. 72]. Are these transformations particular to only the function $_2F_1(\frac{1}{2}, \frac{1}{2}; 1; x)$, or are they special cases of more general transformations?

Similar questions can be asked in the alternative theories. Thus, in the cubic theory, the multiplier m is defined by

$$m := \frac{{}_2F_1\left(\frac{1}{3}, \frac{2}{3}; 1; \alpha\right)}{{}_2F_1\left(\frac{1}{3}, \frac{2}{3}; 1; \beta\right)}, \tag{3.3}$$

where β has degree n over α. We cite one example of Ramanujan from his second notebook [9, p. 124, Thm. 7.5]. If β has degree 9 and m is the multiplier, defined by (3.3), for modular equations of degree 9 in the theory of signature 3, then

$$m = 3\frac{1 + 2\beta^{1/3}}{1 + 2(1-\alpha)^{1/3}}.$$

We close this section with an open problem about a quasi-transformation formula given by Ramanujan on page 209 of his lost notebook in the pagination of [51]. Recall that $K(k)$ is defined by (2.2), and put $K' := K(k')$, where $k' := \sqrt{1-k^2}$; k' is called the complementary modulus. If $q = e^{-\pi K'/K}$ and $q' = e^{-\pi K/K'}$, then

$$\left\{\prod_{n=0}^{\infty}\left(\frac{1-(-1)^n q^{(2n+1)/2}}{1+(-1)^n q^{(2n+1)/2}}\right)^{2n+1}\right\}^{\log q}\left\{\prod_{n=1}^{\infty}\left(\frac{1-(-1)^n iq'^n}{1+(-1)^n iq'^n}\right)^{n}\right\}^{2\pi i}$$

$$= \exp\left(\frac{\pi^2}{4} - \frac{k\ _3F_2(1,1,1;\frac{3}{2},\frac{3}{2};k^2)}{_2F_1(\frac{1}{2},\frac{1}{2};1;k^2)}\right). \tag{3.4}$$

The formulation given in [51] is not clear. In particular, the definition of q is faded. Also, there is a wavy line to the right of (3.4), with the definition of q on the right of one indentation of the wave, and the definition of q' on the right of the other indentation of the wave. The claim (3.4) appears to be a modular-type transformation formula analogous to the transformation formula for theta functions. Thus, one naturally thinks of using the Poisson summation formula to effect a proof, but this and other attempts have so far been fruitless.

4. Explicit Evaluations

We now turn our attention to explicit evaluations. Observe, from (2.5), that an explicit evaluation of one of $\varphi(q)$, $K(k)$, or $_2F_1(\frac{1}{2},\frac{1}{2};1;k^2)$ yields an explicit evaluation of the other two functions. However, note that, for example, if one can determine $\varphi(q)$ for a certain specified value of q, it may be difficult to find the corresponding value of k. Historically, there have been several approaches to explicit evaluations, but the two most successful have been through the Chowla–Selberg formula [56] and its generalizations, and through singular moduli and class invariants; the methods are not unrelated. For evaluations via the former method, see, for example, papers of I. J. Zucker [71] and G. S. Joyce and Zucker [40]. In our opinion, the most successful approach has been through class invariants and singular moduli. This approach also leads to the evaluation of other theta-functions, the Rogers–Ramanujan continued fraction, and certain other q-continued fractions.

Before discussing class invariants, we briefly mention a recent wonderful result of A. van der Poorten and K. S. Williams [59]. Recall the definition of the Dedekind eta-function, $\eta(\tau)$, given by (2.6). Van der Poorten and

Williams evaluated $\eta((b + \sqrt{d})/(2a))$, where $d = b^2 - 4ac$ and d is a fundamental discriminant [59, Thm. 9.3]. They then showed that the Chowla–Selberg formula and the most far reaching generalization of the Chowla–Selberg formula, found in a paper by J. G. Huard, P. Kaplan, and Williams [39], follow as special cases. Their formulas are quite complicated and are expressed in terms of quantities appearing in the theory of positive definite binary quadratic forms. Specific determinations are therefore difficult. Now, class invariants, theta-functions, the Rogers–Ramanujan continued fraction, and many other functions in the theory of elliptic functions and modular forms can be expressed in terms of Dedekind eta-functions. The particular products and quotients arising in these contexts usually have more elegant representations, without the appearances of parameters from the theory of quadratic forms. Thus, it is usually best to directly address the specific evaluations of the relevant functions, instead of attempting to determine individual values of the relevant eta-functions.

To define Ramanujan's class invariants or the Weber–Ramanujan class invariants, first set

$$\chi(q) = (-q; q^2)_\infty. \tag{4.1}$$

If

$$q = \exp(-\pi\sqrt{n}), \tag{4.2}$$

where n is a positive rational number, the two *class invariants* G_n and g_n are defined by

$$G_n := 2^{-1/4} q^{-1/24} \chi(q) \quad \text{and} \quad g_n := 2^{-1/4} q^{-1/24} \chi(-q). \tag{4.3}$$

In the notation of Weber [64], $G_n =: 2^{-1/4}\mathfrak{f}(\sqrt{-n})$ and $g_n =: 2^{-1/4}\mathfrak{f}_1(\sqrt{-n})$. It is well known that G_n and g_n are algebraic; for example, see Cox's book [31, p. 214, Thm. 10.23; p. 257, Thm. 12.17]. In fact, they are frequently units [9, p. 184, Thm. 1.1]. As above, let $k := k(q)$, $0 < k < 1$, denote the modulus. The singular modulus k_n is defined by $k_n := k(e^{-\pi\sqrt{n}})$, where n is a natural number. Following Ramanujan, set $\alpha = k^2$ and $\alpha_n = k_n^2$. Class invariants and singular moduli are intimately related by the equalities [9, p. 185, eq. (1.6)]

$$G_n = \{4\alpha_n(1 - \alpha_n)\}^{-1/24} \quad \text{and} \quad g_n = \{4\alpha_n(1 - \alpha_n)^{-2}\}^{-1/24}. \tag{4.4}$$

Ramanujan calculated a total of 116 class invariants and over 30 singular moduli. See Berndt's book [9, Chap. 34] for a description of this work. By

(4.4), if one can find G_n or g_n for a particular value of n, then one can find α_n by solving a quadratic equation. However, generally, this simple device yields a rather ugly expression for α_n, and so other methods yielding more elegant representations of α_n are desirable. Some simple examples are given by

$$\alpha_5 = \frac{1}{2}\left(\frac{\sqrt{5}-1}{2}\right)^3\left(\frac{\sqrt{5}+1}{2} - \sqrt{\frac{\sqrt{5}+1}{2}}\right)^2, \tag{4.5}$$

$$\alpha_{10} = \left(\sqrt{10}-3\right)^2\left(3-2\sqrt{2}\right)^2, \quad G_5 = \left(\frac{1+\sqrt{5}}{2}\right)^{1/4}, \quad g_{10} = \sqrt{\frac{1+\sqrt{5}}{2}}.$$

Ramanujan left a huge, blooming, colorful garden of class invariants, singular moduli, and explicit evaluations. But it is now time to stroll down some paths where we have not yet been able to espy the flowers.

Define, for positive rational numbers k and n,

$$r_{k,n} := \frac{f\left(-e^{-2\pi\sqrt{n/k}}\right)}{k^{1/4}e^{-\pi\sqrt{n/k}(k-1)/12}f\left(-e^{-2\pi\sqrt{nk}}\right)} \tag{4.6}$$

and

$$s_{k,n} := \frac{f\left(e^{-\pi\sqrt{n/k}}\right)}{k^{1/4}e^{-\pi\sqrt{n/k}(k-1)/24}f\left(-(-1)^k e^{-\pi\sqrt{nk}}\right)}. \tag{4.7}$$

Our definition of $r_{k,n}$ is the same as that of J. Yi [68], but our definition of $s_{k,n}$ is slightly different from her definition of $r'_{k,n}$.

Suppose that $k = 2$. Then, by (4.6) and (4.4),

$$\begin{aligned}
r_{2,n} &= \frac{f\left(-e^{-2\pi\sqrt{n/2}}\right)}{2^{1/4}e^{-\pi\sqrt{n/2}/12}f\left(-e^{-2\pi\sqrt{2n}}\right)} \\
&= \frac{f\left(-e^{-\pi\sqrt{2n}}\right)}{2^{1/4}e^{-\pi\sqrt{2n}/24}f\left(-e^{-2\pi\sqrt{2n}}\right)} \\
&= 2^{-1/4}e^{\pi\sqrt{2n}/24}\left(e^{-\pi\sqrt{2n}}; e^{-2\pi\sqrt{2n}}\right)_\infty = g_{2n}.
\end{aligned}$$

Thus, $r_{2,n}$ is simply equal to the class invariant g_{2n}. Also, by (4.7) and (4.4),

$$\begin{aligned}
s_{2,2n} &= \frac{f\left(e^{-\pi\sqrt{n}}\right)}{2^{1/4}e^{-\pi\sqrt{n}/24}f\left(-e^{-2\pi\sqrt{n}}\right)} \\
&= 2^{-1/4}e^{\pi\sqrt{n}/24}\left(-e^{-\pi\sqrt{n}}; e^{-2\pi\sqrt{n}}\right)_\infty = G_n.
\end{aligned}$$

Hence, $s_{2,2n}$ is simply G_n.

On page 212 in his lost notebook in the pagination of [51], Ramanujan defined

$$\lambda_n := \frac{1}{3\sqrt{3}} \frac{f^6(q)}{\sqrt{q} f^6(q^3)},$$

(4.8)

where $q = e^{-\pi\sqrt{n/3}}$. Thus,

$$\lambda_n = s_{3,n}^6.$$

On the same page, Ramanujan provided a table of eleven recorded values of λ_n, and ten unrecorded values of λ_n. For example,

$$\lambda_{73} = \left(\sqrt{\frac{11 + \sqrt{73}}{8}} + \sqrt{\frac{3 + \sqrt{73}}{8}} \right)^6.$$

All twenty-one values and several more were established by Berndt, Chan, S.-Y. Kang, and L.-C. Zhang [15]. Several methods needed to be devised in order to calculate all the values indicated by Ramanujan. The definition of λ_n given by (4.8) suggests that the theory and calculation of λ_n are connected with Ramanujan's alternative cubic theory. This indeed is correct, and one of the methods used by these authors depends on modular equations in the theory of signature 3. They also showed that the alternative cubic theory and the theory of λ_n lead to new methods for calculating values of the modular j-invariant. Further applications for the values of λ_n have been developed by Chan, W.-C. Liaw, and V. Tan [29], Chan, A. Gee, and Tan [26], and by Berndt and Chan [13]. In particular, using the theory of λ_n, Berndt and Chan [13] derived new infinite series for $1/\pi$ which are unobtainable by previous methods. One of their series gives a (current) world record of approximately 73 or 74 digits of π per term.

In a very brief study of λ_n, K. G. Ramanathan [47] introduced the function

$$\mu_n := \frac{1}{3\sqrt{3}} \frac{f^6(-q^2)}{q f^6(-q^6)},$$

(4.9)

where $q = e^{-\pi\sqrt{n/3}}$. Thus, by (4.6),

$$\mu_n = r_{3,n}^6.$$

It is easy to show [15, Thm. 3.1] that λ_n and μ_n are related by the equality

$$\frac{\lambda_n}{\mu_n} = \left(\frac{G_{n/3}}{G_{3n}} \right)^6.$$

In 1828, N. H. Abel [1] established the beautiful formula

$$\frac{\eta^2(i\sqrt{5}/2)}{2\eta^2(2i\sqrt{5})} = \frac{\sqrt{5}+1}{2} + \sqrt{\frac{\sqrt{5}+1}{2}}. \tag{4.10}$$

Note that the left side of (4.10) equals $r_{4,5}^2$. This example was brought to our attention by F. Hajir and F. Rodriguez Villegas [37], who showed that a certain eta-quotient always generates a Kummer extension of the ring class field. We have established (4.10) by using basic formulas for $\eta(\tau)$ and $\eta(4\tau)$ (or for $f(-q)$ and $f(-q^4)$) [7, p. 124, Entries 12(ii), (iv)] along with the evaluation of α_5 given in (4.5). A systematic study of the functions $r_{4,n}$ and $s_{4,n}$ has not been undertaken.

A serious study of the case $k = 5$ should also prove fruitful. In particular, such values arise in the explicit evaluation of the Rogers–Ramanujan continued fraction, defined by

$$R(q) := \frac{q^{1/5}}{1} + \frac{q}{1} + \frac{q^2}{1} + \frac{q^3}{1} + \cdots. \tag{4.11}$$

For example [17], [4, Entry 2.2.7], the value,

$$5^{3/2} s_{5,7}^6 = e^{\pi\sqrt{7/5}} \frac{f^6(e^{-\pi\sqrt{7/5}})}{f^6(e^{-\pi\sqrt{35}})} = 125 \left(\frac{5+\sqrt{5}}{10}\right)^3,$$

is necessary in the evaluation,

$$R(-e^{-\pi\sqrt{7/5}}) = -\left(-5\sqrt{5} - 7 + \sqrt{35(5+2\sqrt{5})}\right)^{1/5},$$

one of the many explicit values for the Rogers–Ramanujan continued fraction found in Ramanujan's lost notebook [51, p. 210]. For further examples in the case $k = 5$ with applications to the evaluation of the Rogers–Ramanujan continued fraction, see papers by J. Yi [66] and K. R. Vasuki and M. S. Mahadeva Naika [60].

One of the primary tools for finding values of G_n, g_n, λ_n, and other quotients of eta-functions employs eta-function identities, a particular type of modular equation. This has been exploited, in particular, by Berndt, Chan, and Zhang [18], Berndt, Chan, Kang, and Zhang [15], and Yi [66]–[68]. Ramanujan discovered 23 of these beautiful identities; for proofs, see Berndt's book [8, pp. 204–244]. For example [8, p. 221, Entry 62], if

$$P := \frac{f(-q)}{q^{1/12}f(-q^3)} \quad \text{and} \quad Q := \frac{f(-q^5)}{q^{5/12}f(-q^{15})},$$

then

$$(PQ)^2 + 5 + \frac{9}{(PQ)^2} = \left(\frac{Q}{P}\right)^3 - \left(\frac{P}{Q}\right)^3. \tag{4.12}$$

Several others have been found by Yi [66], [68]. It is not difficult to discover such identities by using, for example, Garvan's *etamake* package on Maple V. One can always then rigorously prove them by using the theory of modular forms. But these are dull proofs; the proofs in [66] and most of them in [8] employ Ramanujan's more interesting ideas.

Since those working in special functions are most likely more interested in classical theta-functions than in the eta-function, especially because of the relations in (2.5), one might ask if it is feasible to calculate directly values of quotients of theta-functions, and if theta-analogues of modular equations like (4.12) exist. One can answer such a query in the affirmative, but not much is known at present. In his notebooks [50], Ramanujan recorded a couple examples like (4.12) with $f(-q)$ replaced by either $\varphi(q)$ or $\psi(q)$. For example [8, p. 235, Entry 67], if

$$P := \frac{\varphi(q)}{\varphi(q^5)} \quad \text{and} \quad Q := \frac{\varphi(q^3)}{\varphi(q^{15})},$$

then

$$PQ + \frac{5}{PQ} = \left(\frac{Q}{P}\right)^3 + 3\frac{Q}{P} + 3\frac{P}{Q} - \left(\frac{P}{Q}\right)^3.$$

J. Yi [68] is currently systematically searching for other examples. Also recall from (3.2) and (2.5) that the multiplier m can be represented as a quotient of the theta functions φ. Representations for m, such as those given in Theorems 2–4, can then be used to explicitly determine such quotients. In his notebooks, Ramanujan recorded several examples which were first proved by Berndt and Chan [12]. For example [12], [9, p. 327, Entry 4],

$$\frac{\varphi(e^{-3\pi})}{\varphi(e^{-\pi})} = \frac{1}{\sqrt[4]{6\sqrt{3}-9}}. \tag{4.13}$$

Since the value $\varphi(e^{-\pi}) = \pi^{1/4}/\Gamma(\frac{3}{4})$ is well-known [7, p. 103], (4.13) provides an explicit evaluation of $\varphi(e^{-3\pi})$.

Ramanujan's lost notebook also contains several theorems leading to the explicit evaluation of theta-functions. Most of these theorems have been proved by Soon–Yi Kang [41]. For example, if $k = R(q)R^2(q^2)$ [51, p. 56], [41], [4, Chap. 1], then

$$\frac{\varphi^2(-q)}{\varphi^2(-q^5)} = \frac{1-4k-k^2}{1-k^2} \quad \text{and} \quad \frac{\psi^2(q)}{\psi^2(q^5)} = \frac{1+k-k^2}{k}. \tag{4.14}$$

Thus, if we can evaluate $R(q)$ and $R(q^2)$ at the requisite values of q, then we can evaluate the quotients of theta-functions above. The introduction of the parameter k is an ingenious device of Ramanujan. It would seem that these ideas can be further exploited.

5. Continued Fractions

In Chapters 12 and 16 and the unorganized pages of his second notebook [50], [6, Chap. 12], [9, Chap. 32], Ramanujan recorded many beautiful continued fractions for quotients of gamma functions. For one of the simpler examples, we cite Entry 25 of Chapter 12 [6, p. 140]. If either n is an odd integer and x is any complex number, or if n is any complex number and Re $x > 0$, then

$$\frac{\Gamma\left(\frac{1}{4}(x+n+1)\right)\Gamma\left(\frac{1}{4}(x-n+1)\right)}{\Gamma\left(\frac{1}{4}(x+n+3)\right)\Gamma\left(\frac{1}{4}(x-n+3)\right)}$$
$$= \frac{4}{x} - \frac{n^2-1^2}{2x} - \frac{n^2-3^2}{2x} - \frac{n^2-5^2}{2x} - \cdots. \tag{5.1}$$

This particular result is, in fact, due to Euler [33, Sec. 67], but most, indeed, are original with Ramanujan. (For references to other proofs of (5.1), see [6, p. 141].) D. Masson [43], [44], [45] and Zhang [70] have shown that Ramanujan's continued fractions for quotients of gamma functions are related to contiguous relations of hypergeometric series. Ramanujan, in Chapter 12 and the unorganized pages in his second notebook [6, pp. 131–163], [9, pp. 50–66], worked out many beautiful limiting cases. For example, the continued fraction [6, p. 155],

$$\zeta(3) = 1 + \frac{1}{2\cdot 2} + \frac{1^3}{1} + \frac{1^3}{6\cdot 2} + \frac{2^3}{1} + \frac{1}{10\cdot 2} + \frac{3^3}{1} + \cdots,$$

was made famous by R. Apéry's [5] famous proof of the irrationality of $\zeta(3)$, where $\zeta(s)$ denotes the Riemann zeta-function. Ramanujan was not exhaustive in his search for elegant special cases. We recommend that the particular special and limiting cases be systematically unearthed and brought out into the sunlight.

It is natural to ask if q-analogues exist. G. N. Watson [63] and D. P. Gupta and Masson [36] have established q-analogues of Ramanujan's most general theorem on continued fractions for quotients of gamma functions [6, p. 163, Entry 40]. In some cases, we know q-continued fractions which share the same features, but which apparently are not q-analogues. For example, the extremely beautiful Entry 12 in Chapter 16 of the second notebook, which we now state, has some of the same salient features as (5.1). Suppose that $a, b,$ and q are complex numbers with $|ab| < 1$ and $|q| < 1$, or that $a = bq^{2m+1}$ for some integer m. Then

$$\frac{(a^2q^3; q^4)_\infty (b^2q^3; q^4)_\infty}{(a^2q; q^4)_\infty (b^2q; q^4)_\infty} = \cfrac{1}{1 - ab} + \cfrac{(a - bq)(b - aq)}{(1 - ab)(q^2 + 1)} + \cfrac{(a - bq^3)(b - aq^3)}{(1 - ab)(q^4 + 1)} + \cdots. \qquad (5.2)$$

This is one of my favorite flowers. If $|ab| > 1$, then the continued fraction in (5.2) is equivalent to

$$\frac{-1/(ab)}{1 - 1/(ab)} + \frac{(1/a - q/b)(1/b - q/a)}{(1 - 1/(ab))(q^2 + 1)} + \frac{(1/a - q^3/b)(1/b - q^3/a)}{(1 - 1/(ab))(q^4 + 1)} + \cdots,$$

which, by (5.2), converges to

$$-\frac{1}{ab} \frac{(q^3/a^2; q^4)_\infty (q^3/b^2; q^4)_\infty}{(q/a^2; q^4)_\infty (q/b^2; q^4)_\infty}.$$

Since the continued fraction in (5.2) diverges when $|ab| = 1$, the hypotheses on a and b in (5.2) cannot be further relaxed. However, if $|ab| < 1$ and $|q| > 1$, the continued fraction in (5.2) converges to

$$\frac{(a^2/q^3; 1/q^4)_\infty (b^2/q^3; 1/q^4)_\infty}{(a^2/q; 1/q^4)_\infty (b^2/q; 1/q^4)_\infty}.$$

Each beautiful symmetry yields a different view of the flower.

Thus, what are the q-analogues of Ramanujan's continued fractions for quotients of gamma functions? Also, try to find continued fractions, such as (5.2), which are "almost" q-analogues.

Several elegant q-continued fractions have representations as q-products. The most famous one, of course, is the Rogers–Ramanujan continued fraction defined by (4.11). We record several of them. For $|q| < 1$,

$$\frac{(-q^2; q^2)_\infty}{(-q; q^2)_\infty} = \cfrac{1}{1} + \cfrac{q}{1} + \cfrac{q^2 + q}{1} + \cfrac{q^3}{1} + \cfrac{q^4 + q^2}{1} + \cfrac{q^5}{1} + \cdots, \qquad (5.3)$$

$$\frac{(q^2; q^3)_\infty}{(q; q^3)_\infty} = \cfrac{1}{1} - \cfrac{q}{1 + q} - \cfrac{q^3}{1 + q^2} - \cfrac{q^5}{1 + q^3} - \cfrac{q^7}{1 + q^4} - \cdots, \qquad (5.4)$$

$$\frac{(q; q^2)_\infty}{(q^2; q^4)_\infty^2} = \cfrac{1}{1} + \cfrac{q}{1} + \cfrac{q + q^2}{1} + \cfrac{q^3}{1} + \cfrac{q^2 + q^4}{1} + \cdots, \qquad (5.5)$$

$$\frac{(-q^3;q^4)_\infty}{(-q;q^4)_\infty} = \frac{1}{1} + \frac{q}{1} + \frac{q^2+q^3}{1} + \frac{q^5}{1} + \frac{q^4+q^7}{1} + \cdots, \tag{5.6}$$

$$\frac{(q^3;q^4)_\infty}{(q;q^4)_\infty} = \frac{1}{1} - \frac{q}{1+q^2} - \frac{q^3}{1+q^4} - \frac{q^5}{1+q^6} - \cdots, \tag{5.7}$$

$$\frac{(q;q^5)_\infty(q^4;q^5)_\infty}{(q^2;q^5)_\infty(q^3;q^5)_\infty} = \frac{1}{1} + \frac{q}{1} + \frac{q^2}{1} + \frac{q^3}{1} + \cdots, \tag{5.8}$$

$$\frac{(q;q^6)_\infty(q^5;q^6)_\infty}{(q^3;q^6)_\infty^2} = \frac{1}{1} + \frac{q+q^2}{1} + \frac{q^2+q^4}{1} + \frac{q^3+q^6}{1} + \cdots, \tag{5.9}$$

$$\frac{(q;q^8)_\infty(q^7;q^8)_\infty}{(q^3;q^8)_\infty(q^5;q^8)_\infty} = \frac{1}{1} + \frac{q+q^2}{1} + \frac{q^4}{1} + \frac{q^3+q^6}{1} + \cdots. \tag{5.10}$$

In examining the bases on the left sides of (5.3)–(5.10), we find that q^7 is missing. Thus, we ask if there is a continued fraction representation for

$$\frac{(q;q^7)_\infty(q^2;q^7)_\infty(q^4;q^7)_\infty}{(q^3;q^7)_\infty(q^5;q^7)_\infty(q^6;q^7)_\infty}. \tag{5.11}$$

Now, D. Bowman and G. Choi [23] have recently found a G-continued fraction for

$$\frac{(q;q^7)_\infty(q^6;q^7)_\infty}{(q^3;q^7)_\infty(q^4;q^7)_\infty}.$$

Thus, maybe one must look for a G-continued fraction for (5.11).

The product representation for $q^{-1/5}R(q)$ given in (5.8) was originally discovered by L. J. Rogers [52] in 1894 and rediscovered by Ramanujan; see his notebooks [50, Chap. 16, Entry 38(iii)], [7, p. 79]. The function $R(q)$ possesses a very beautiful and extensive theory, almost all of which was found by Ramanujan. In particular, his lost notebook [51] contains an enormous amount of material on the Rogers–Ramanujan continued fraction. See papers by Berndt, S.-S. Huang, J. Sohn, and S. H. Son [19] and by Kang [41], [42] for proofs of many of these theorems. In fact, the first five chapters of the first volume by Andrews and Berndt [4] on Ramanujan's lost notebook are devoted to the Rogers–Ramanujan continued fraction.

The continued fraction in (5.10) is called the Ramanujan–Gordon–Göllnitz continued fraction. It also possesses a beautiful theory, much of it developed by Chan and Huang [27]. The continued fraction appears in both

Ramanujan's second notebook [50, p. 290], [9, p. 50] and lost notebook [51, p. 44], [4, Cor. 2.2.14]. However, the first published proof of (5.10) is by A. Selberg [54, eq. (53)], [55, pp. 18–19]. For references to further proofs, see [9] and [4].

The continued fraction in (5.9) has also been studied and is called Ramanujan's cubic continued fraction. The first appearance of (5.9) is in Chapter 19 of Ramanujan's third notebook [50], where several of its properties are given [7, pp. 345–346, Entry 1]. It also appears in his third notebook [9, p. 45, Entry 18]. The first proof of (5.9) is by Watson [62], and the second is by Selberg [54, p. 19], [55]. In a fragment published with his lost notebook [51, p. 366], Ramanujan writes "and many results analogous to the previous continued fraction," indicating that there is a theory of the cubic continued fraction similar to that for the Rogers–Ramanujan continued fraction. Piqued by this remark, Chan [24] developed an extensive theory of the cubic continued fraction.

All the remaining continued fractions cited above can be found in Ramanujan's notebooks or lost notebook. Theories for the continued fractions in (5.3) and (5.5)–(5.7) have not been developed, but the theories are probably less difficult than those for the three continued fractions discussed above. The proof of (5.4) is difficult [9, pp. 46–47], and we would expect that if this continued fraction has a theory comparable to those of (5.8)–(5.10), it will be challenging to discover.

Although theories for the Ramanujan–Gordon–Göllnitz and cubic continued fractions have been developed, there are many theorems for the Rogers–Ramanujan continued fraction for which analogues have not been found for the former two continued fractions and the other continued fractions cited above. We give three examples.

In his lost notebook [51, p. 50], Ramanujan examined the power series coefficients v_n defined by

$$C(q) := \frac{1}{q^{-1/5}R(q)} =: \sum_{n=0}^{\infty} v_n q^n, \qquad |q| < 1. \tag{5.12}$$

In particular, he derived identities for

$$\sum_{n=0}^{\infty} v_{5n+j} q^n, \qquad 0 \le j \le 4. \tag{5.13}$$

These were first proved by Andrews [2], who also showed that these coefficients have partition-theoretic interpretations. Let $B_{k,a}(n)$ denote the number of partitions of n of the form $n = b_1 + b_2 + \cdots + b_s$, where $b_i \ge b_{i+1}, b_i - b_{i+k-1} \ge 2$ and at most $a - 1$ of the b_i equal 1. Then, for example,

$$v_{5n} = B_{37,37}(n) + B_{37,13}(n - 4).$$

Do the power series coefficients of any of the other continued fractions cited above have partition-theoretic interpretations?

Two of the most important properties for $R(q)$ are given by

$$\frac{1}{R(q)} - 1 - R(q) = \frac{f(-q^{1/5})}{q^{1/5} f(-q^5)} \tag{5.14}$$

and

$$\frac{1}{R^5(q)} - 11 - R^5(q) = \frac{f^6(-q)}{q f^6(-q^5)}. \tag{5.15}$$

These equalities were found by Watson in Ramanujan's notebooks [50], [7, pp. 265–267] and proved by him [61] in order to establish claims about the Rogers–Ramanujan continued fraction communicated by Ramanujan in his first two letters to Hardy [61], [62]. Some of the continued fractions mentioned above have similar properties. However, in his lost notebook [51, p. 207], Ramanujan states two-variable generalizations which we offer in the next theorem.

Theorem 5 *If*

$$P = \frac{f(-\lambda^{10}q^7, -\lambda^{15}q^8) + \lambda q f(-\lambda^5 q^2, -\lambda^{20}q^{13})}{q^{1/5} f(-\lambda^{10}q^5, -\lambda^{15}q^{10})} \tag{5.16}$$

and

$$Q = \frac{\lambda f(-\lambda^5 q^4, -\lambda^{20}q^{11}) - \lambda^3 q f(-q, -\lambda^{25}q^{14})}{q^{-1/5} f(-\lambda^{10}q^5, -\lambda^{15}q^{10})}, \tag{5.17}$$

then

$$P - Q = 1 + \frac{f(-q^{1/5}, -\lambda q^{2/5})}{q^{1/5} f(-\lambda^{10}q^5, -\lambda^{15}q^{10})} \tag{5.18}$$

and

$$P^5 - Q^5 = 1 + 5PQ + 5P^2Q^2 + \frac{f(-q, -\lambda^5 q^2) f^5(-\lambda^2 q, -\lambda^3 q^2)}{q \, f^6(-\lambda^{10}q^5, -\lambda^{15}q^{10})}. \tag{5.19}$$

This theorem was first proved by Son [58]. His task was made more difficult because Ramanujan did not divulge the identities of P and Q given in (5.16) and (5.17), respectively. Although we forego all details, it is not difficult to show that if we set $\lambda = 1$ above, then (5.18) and (5.19) reduce to (5.14) and (5.15), respectively. Very few theorems in the theory of theta functions have two-variable generalizations such as this one. Further investigations along these lines will be difficult, but presumably worthwhile.

In his second notebook and especially in his lost notebook, Ramanujan recorded interesting claims about finite versions of the Rogers–Ramanujan continued fraction. For example, finite versions at roots of unity are examined, and relations with certain class invariants are found. Most of these assertions were proved in a paper by Huang [38]. See also a paper by Berndt, Huang, Sohn, and Son [19]. Finite versions of the other continued fractions cited above have not been examined.

On the other side of the coin, on page 45 in his lost notebook in the pagination of [51], Ramanujan states two very interesting asymptotic formulas for equivalent continued fractions of (5.4) and (5.7). We state one of the two examples, namely for (5.4), which arises when we set $e^x = 1/q$ below.

Let $\zeta(s)$ denote the Riemann zeta-function, and let

$$L(s, \chi) = \sum_{n=1}^{\infty} \chi(n) n^{-s}, \quad \text{Re } s > 0,$$

denote the Dirichlet L-function associated with the character $\chi(n) = \left(\frac{n}{3}\right)$, where $\left(\frac{n}{3}\right)$ denotes the Legendre symbol, i.e., $\chi(n) = \pm 1$, if $n \equiv \pm 1 \pmod 3$, and $\chi(n) = 0$, if $n \equiv 0 \pmod 3$. For each integer $n \geq 2$, let

$$a_n = \frac{4\Gamma(n)\zeta(n)L(n+1,\chi)}{(2\pi/\sqrt{3})^{2n+1}}.$$

Then, for $x > 0$,

$$\frac{(3x)^{1/3}}{1} - \frac{1}{1+e^x} - \frac{1}{1+e^{2x}} - \frac{1}{1+e^{3x}} - \cdots = \frac{\Gamma\left(\frac{1}{3}\right)}{\Gamma\left(\frac{2}{3}\right)} e^{G(x)}, \tag{5.20}$$

where, as $x \to 0+$,

$$G(x) \sim \sum_{n=1}^{\infty} a_{2n} x^{2n}.$$

In particular,

$$a_2 = \frac{1}{108}, \qquad a_4 = \frac{1}{4320}, \qquad a_6 = \frac{1}{38880}.$$

Both of Ramanujan's asymptotic formulas follow as special cases from a general theorem recently proved by Berndt and Sohn [20]. For what continued fractions can one obtain such elegant asymptotic formulas? Can one find asymptotic formulas as $q = e^{-x}$ approaches other points on the unit circle $|q| = 1$? The convergence and divergence of (5.3)–(5.10) on the unit circle is not well understood. I. Schur [53] and Ramanujan [9, p. 35, Entry 12, Theorem 12.1] have examined the convergence of $R(q)$ when q is a primitive root of unity. Zhang [69] has examined (5.3), (5.4), and (5.10) when q is a root of unity, but the other cited continued fractions have yet to be examined. We have no knowledge about the convergence or divergence at other points on $|q| = 1$ for any of the continued fractions (5.3)–(5.10).

6. Elliptic Integrals

On pages 51–53 in his lost notebook [51], Ramanujan recorded several identities involving integrals of eta-functions and incomplete elliptic integrals of the first kind. We offer here one typical example. Recall that $f(-q)$ is defined in (2.6). Let

$$v := v(q) := q\frac{f^3(-q)f^3(-q^{15})}{f^3(-q^3)f^3(-q^5)}. \tag{6.1}$$

Then

$$\int_0^q f(-t)f(-t^3)f(-t^5)f(-t^{15})dt \tag{6.2}$$

$$= \frac{1}{5}\int_{2\tan^{-1}\left(\frac{1}{\sqrt{5}}\sqrt{\frac{1-11v-v^2}{1+v-v^2}}\right)}^{2\tan^{-1}(1/\sqrt{5})} \frac{d\varphi}{\sqrt{1-\frac{9}{25}\sin^2\varphi}}.$$

The reader will immediately realize that these are rather uncommon integrals.

S. Raghavan and S. S. Rangachari [46] proved all of these formulas, but almost all of their proofs use results with which Ramanujan would have been unfamiliar, in particular, the theory of modular forms, which was evidently not known to Ramanujan. For example, for four identities, including (6.2), Raghavan and Rangachari appealed to differential equations satisfied by certain quotients of eta-functions, such as (6.1). The requisite differential equation for (6.2) is given by

$$\frac{dv}{dq} = f(-q)f(-q^3)f(-q^5)f(-q^{15})\sqrt{1-10v-13v^2+10v^3+v^4},$$

where v is given by (6.1). In an effort to better discern Ramanujan's methods and to better understand the origins of identities like (6.2), Berndt, Chan, and Huang [14] devised proofs independent of the theory of modular forms and other ideas with which Ramanujan would have been unfamiliar. Particularly troublesome are the aforementioned four differential equations for quotients of eta-functions for which the authors devised proofs depending on identities for Eisenstein series found in Chapter 21 of Ramanujan's second notebook and eta-function identities such as (4.12). A better understanding and more systematic derivation of these nonlinear differential equations likely will play a key role in our understanding of integral identities like (6.2).

References

1. N. H. Abel, *Recherches sur les fonctions elliptiques*, J. Reine Angew. Math. **2** (1828), 160–190; *Oeuvres Completes de Niels Henrik Abel*, Vol. 1, Gröndahl, Oslo, 1881, pp. 380–382.

2. G. E. Andrews, *Ramanujan's "lost" notebook. III. The Rogers–Ramanujan continued fraction*, Adv. Math. **41** (1981), 186–208.

3. G. E. Andrews, R. A. Askey, and R. Roy, *Special Functions*, Cambridge University Press, Cambridge, 1999.

4. G. E. Andrews and B. C. Berndt, *Ramanujan's Lost Notebook, Part I*, Springer–Verlag, to appear.

5. R. Apéry, *Interpolation de fractions continues et irrationalite de certaines constantes*, Bull. Sect. des Sci., t. III, Bibliothéque Nationale, Paris, 1981, 37–63.

6. B. C. Berndt, *Ramanujan's Notebooks, Part II*, Springer–Verlag, New York, 1989.

7. B. C. Berndt, *Ramanujan's Notebooks, Part III*, Springer–Verlag, New York, 1991.

8. B. C. Berndt, *Ramanujan's Notebooks, Part IV*, Springer–Verlag, New York, 1994.

9. B. C. Berndt, *Ramanujan's Notebooks, Part V*, Springer–Verlag, New York, 1998.

10. B. C. Berndt, *Modular equations in Ramanujan's lost notebook*, in: "Number Theory", R. P. Bambah, V. C. Dumir, and R. Hans–Gill, eds., Hindustan Book Co., Delhi, 1999, pp. 55–74.

11. B. C. Berndt, S. Bhargava, and F. G. Garvan, *Ramanujan's theories of elliptic functions to alternative bases*, Trans. Amer. Math. Soc. **347** (1995), 4163–4244.

12. B. C. Berndt and H. H. Chan, *Ramanujan's explicit values for the classical theta-function*, Mathematika **42** (1995), 278–294.

13. B. C. Berndt and H. H. Chan, *Eisenstein series and approximations to π*, Illinois J. Math. **45** (2001), to appear.

14. B. C. Berndt, H. H. Chan, and S.-S. Huang, *Incomplete elliptic integrals in Ramanujan's lost notebook*, in: "q–Series From a Contemporary Perspective", M. E. H. Ismail and D. W. Stanton, eds., Amer. Math. Soc., Providence, 2000, pp. 79–126.

15. B. C. Berndt, H. H. Chan, S.-Y. Kang, and L.-C. Zhang, *A certain quotient of eta-functions found in Ramanujan's lost notebook*, Pacific J. Math., to appear.

16. B. C. Berndt, H. H. Chan, and W.-C. Liaw, *On Ramanujan's quartic theory of elliptic functions*, J. Number Theory **88** (2001), 129–156.

17. B. C. Berndt, H. H. Chan, and L.-C. Zhang, *Explicit evaluations of the Rogers–Ramanujan continued fraction*, J. Reine Angew. Math. **480** (1996), 141–159.

18. B. C. Berndt, H. H. Chan, and L.-C. Zhang, *Ramanujan's class invariants, Kronecker's limit formula, and modular equations*, Trans. Amer. Math. Soc. **349** (1997), 2125–2173.

19. B. C. Berndt, S.-S. Huang, J. Sohn, and S. H. Son, *Some theorems on the Rogers–Ramanujan continued fraction in Ramanujan's lost notebook*, Trans. Amer. Math. Soc. **352** (2000), 2157–2177.

20. B. C. Berndt and J. Sohn, *Asymptotic formulas for two continued fractions in Ramanujan's lost notebook*, submitted for publication.

21. J. M. and P. B. Borwein, *Pi and the AGM*, Wiley, New York, 1987.

22. J. M. and P. B. Borwein, *A cubic counterpart of Jacobi's identity and the AGM*,

Trans. Amer. Math. Soc. **323** (1991), 691–701.

23. D. Bowman and G. Choi, *The Rogers–Ramanujan q-difference equations of arbitrary order*, submitted for publication.

24. H. H. Chan, *On Ramanujan's cubic continued fraction*, Acta Arith. **73** (1995), 343–355.

25. H. H. Chan, *On Ramanujan's cubic transformation formula for* $_2F_1(1/3, 2/3; 1; z)$, Math. Proc. Cambridge Philos. Soc. **124** (1998), 193–204.

26. H. H. Chan, A. Gee, and V. Tan, *Cubic singular moduli, Ramanujan's class invariant* λ_n *and the explicit Shimura reciprocity law*, submitted for publication.

27. H. H. Chan and S.-S. Huang, *On the Ramanujan–Gordon–Göllnitz continued fraction*, The Ramanujan J. **1** (1997), 75–90.

28. H. H. Chan and W.-C. Liaw, *Cubic modular equations and new Ramanujan-type series for* $1/\pi$, Pacific J. Math. **192** (2000), 219–238.

29. H. H. Chan, W.-C. Liaw, and V. Tan, *Ramanujan's class invariant* λ_n *and a new class of series for* $1/\pi$, J. London Math. Soc., to appear.

30. H. H. Chan and Y. L. Ong, *On Eisenstein series and* $\sum_{m,n=-\infty}^{\infty} q^{m^2+mn+2n^2}$, Proc. Amer. Math. Soc. **127** (1999), 1735–1744.

31. D. A. Cox, *Primes of the Form* $x^2 + ny^2$, Wiley, New York, 1989.

32. F. J. Dyson, *A walk through Ramanujan's garden*, in: "Ramanujan Revisited", G. E. Andrews, R. A. Askey, B. C. Berndt, K. G. Ramanathan, and R. A. Rankin, eds., Academic Press, Boston, 1988, pp. 7–28.

33. L. Euler, *De fractionibus continuis observationes*, in: Opera Omnia, Ser. I, Vol. 14, B. G. Teubner, Lipsiae, 1925, pp. 291–349.

34. F. G. Garvan, *Ramanujan's theories of elliptic functions to alternative bases–A symbolic excursion*, J. Symbolic Comput. **20** (1995), 517–536.

35. E. Goursat, *Sur l'equation différentielle linéaire qui admet pour intégrale la série hypergéométrique*, Ann. Sci. École Norm. Sup. (2) **10** (1881), 3–142.

36. D. P. Gupta and D. R. Masson, *Watson's basic analogue of Ramanujan's Entry 40 and its generalizations*, SIAM J. Math. Anal. **25** (1994), 429–440.

37. F. Hajir and F. Rodriguez Villegas, *Explicit elliptic units, I*, Duke Math. J. **90** (1997), 495–521.

38. S.-S. Huang, *Ramanujan's evaluations of Rogers–Ramanujan type continued fractions at primitive roots of unity*, Acta Arith. **80** (1997), 49–60.

39. J. G. Huard, P. Kaplan, and K. S. Williams, *The Chowla–Selberg formula for genera*, Acta Arith. **73** (1995), 271–301.

40. G. S. Joyce and I. J. Zucker, *Special values of the hypergeometric series*, Math. Proc. Cambridge Philos. Soc. **109** (1991), 257–261.

41. S.-Y. Kang, *Some theorems on the Rogers–Ramanujan continued fraction and associated theta function identities in Ramanujan's lost notebook*, The Ramanujan J. **3** (1999), 91–111.

42. S.-Y. Kang, *Ramanujan's formulas for the explicit evaluation of the Rogers–Ramanujan continued fraction and theta-functions*, Acta Arith. **90** (1999), 49–68.

43. D. R. Masson, *Some continued fractions of Ramanujan and Meixner–Pollaczek polynomials*, Canad. Math. Bull. **32** (1989), 177–181.

44. D. R. Masson, *Wilson polynomials and some continued fractions of Ramanujan*, Rocky Mt. J. Math. **21** (1991), 489–499.

45. D. R. Masson, *A generalization of Ramanujan's best theorem on continued fractions*, C. R. Math. Rep. Acad. Sci. Canada **13** (1991), 167–172.

46. S. Raghavan and S. S. Rangachari, *On Ramanujan's elliptic integrals and modular identities*, in: "Number Theory", Oxford University Press, Bombay, 1989, pp. 119–149.

47. K. G. Ramanathan, *On some theorems stated by Ramanujan*, in: "Number Theory and Related Topics", Tata Institute of Fundamental Research Studies in Mathematics, Oxford University Press, Bombay, 1989, pp. 151–160.

48. S. Ramanujan, *Modular equations and approximations to* π, Quart. J. Math. **45** (1914), 350–372.
49. S. Ramanujan, *Collected Papers*, Cambridge University Press, Cambridge, 1927; reprinted by Chelsea, New York, 1962; reprinted by the American Mathematical Society, Providence, RI, 2000.
50. S. Ramanujan, *Notebooks* (2 volumes), Tata Institute of Fundamental Research, Bombay, 1957.
51. S. Ramanujan, *The Lost Notebook and Other Unpublished Papers*, Narosa, New Delhi, 1988.
52. L. J. Rogers, *Second memoir on the expansion of certain infinite products*, Proc. London Math. Soc. **25** (1894), 318–343.
53. I. Schur, *Ein Beitrag zur additiven Zahlentheorie und zur Theorie der Kettenbrüche*, Sitz. Preus. Akad. Wiss., Phys.–Math. Kl. (1917), 302–321.
54. A. Selberg, *Über einige arithmetische Identitäten* Avh. Norske Vid.–Akad. Oslo I. Mat.–Naturv. Kl, No. 8, (1936), 3–23.
55. A. Selberg, *Collected Papers*, Vol. I, Springer–Verlag, Berlin, 1989.
56. A. Selberg and S. Chowla, *On Epstein's zeta-function*, J. Reine Angew. Math. **227** (1967), 86–110.
57. L.-C. Shen, *On an identity of Ramanujan based on the hypergeometric series* $_2F_1\left(\frac{1}{3}, \frac{2}{3}; \frac{1}{2}; x\right)$, J. Number Theory **69** (1998), 125–134.
58. S. H. Son, *Some theta function identities related to the Rogers–Ramanujan continued fraction*, Proc. Amer. Math. Soc. **126** (1998), 2895–2902.
59. A. van der Poorten and K. S. Williams, *Values of the Dedekind eta function at quadratic irrationalities*, Canad. J. Math. **51** (1999), 176–224.
60. K. R. Vasuki and M. S. Mahadeva Naika, *Some evaluations of the Rogers–Ramanujan continued fractions*, preprint.
61. G. N. Watson, *Theorems stated by Ramanujan (VII): Theorems on continued fractions*, J. London Math. Soc. **4** (1929), 39–48.
62. G. N. Watson, *Theorems stated by Ramanujan (IX): Two continued fractions*, J. London Math. Soc. **4** (1929), 231–237.
63. G. N. Watson, *Ramanujan's continued fraction*, Proc. Cambridge Philos. Soc. **31** (1935), 7–17.
64. H. Weber, *Lehrbuch der Algebra*, Bd. 3, Chelsea, New York, 1961.
65. E. T. Whittaker and G. N. Watson, *A Course of Modern Analysis*, fourth ed., University Press, Cambridge, 1966.
66. J. Yi, *Evaluations of the Rogers–Ramanujan continued fraction* $R(q)$ *by modular equations*, Acta Arith. **97** (2001), 103–127.
67. J. Yi, *Modular equations for the Rogers–Ramanujan continued fraction and the Dedekind eta-function*, submitted for publication.
68. J. Yi, The Construction and Applications of Modular Equations, Ph.D. Thesis, University of Illinois at Urbana–Champaign, Urbana, 2001.
69. L.-C. Zhang, *q-difference equations and Ramanujan–Selberg continued fractions*, Acta Arith. **57** (1991), 307–355.
70. L.-C. Zhang, *Ramanujan's continued fractions for products of gamma functions*, J. Math. Anal. Applics. **174** (1993), 22–52.
71. I. J. Zucker, *The evaluation in terms of* Γ-*functions of the periods of elliptic curves admitting complex multiplication*, Math. Proc. Cambridge Philos. Soc. **82** (1977), 111–118.

ORTHOGONAL RATIONAL FUNCTIONS AND CONTINUED FRACTIONS

ADHEMAR BULTHEEL
Department of Computer Science
Katholieke Universiteit Leuven
Celestijnenlaan 200A
B-3001 Leuven Belgium

PABLO GONZALEZ-VERA
Department of Mathematical Analysis
University La Laguna
Tenerife, 38271 Spain

ERIK HENDRIKSEN
Universiteit van Amsterdam
KdV Institute for Mathematics
Pl. Muidergracht 24, 1018 TV Amsterdam
The Netherlands

AND

OLAV NJÅSTAD
Department of Mathematical Sciences
Norwegian University of Science and Technology
Trondheim, N-7491 Norway

Abstract. A class of continued fractions is discussed that generalize the real J-fractions, and which have the same relationship to orthogonal rational functions, multipoint Padé approximants and rational moment problems as real J-fractions have to orthogonal polynomials, one point Padé approximants and classical moment problems.

The setting is as follows. A sequence $\{\alpha_n\}$ of interpolation points in $\hat{\mathbb{R}} \setminus \{0\}$ is given. The space $\mathcal{L}$ consists of all rational functions of the form $p(z)/\omega_n(z)$ for some n where $p(z)$ is a polynomial of degree at most n and $\omega_n(z) = (1 - z/\alpha_1)(1 - z/\alpha_2)\cdots(1 - z/\alpha_n)$. A positive linear functional M on the product space $\mathcal{L} \cdot \mathcal{L}$ is given and defines an inner product on $\mathcal{L}$. A positive measure μ on $\mathbb{R}$ is a solution of the moment problem on $\mathcal{L}$ (on

J. Bustoz et al. (eds.), Special Functions 2000, 87–109.

$\mathcal{L} \cdot \mathcal{L}$) if $\int_{-\infty}^{\infty} R(t) d\mu(t) = M[R]$ for all R in $\mathcal{L}$ (for all R in $\mathcal{L} \cdot \mathcal{L}$). The inner product on $\mathcal{L}$ defines an orthogonal sequence $\{\varphi_n(z)\}$ of functions associated with the basis $\{z^n/\omega_n(z)\}$. The sequence $\{\varphi_n(z)\}$ satisfies a three-term recurrence relation associated with a continued fraction. The approximants $\sigma_n(z)/\varphi_n(z)$ of that continued fraction are multipoint Padé approximants to the Stieltjes transform $S(z,\mu) = \int_{-\infty}^{\infty}(z-t)^{-1}d\mu(t)$ of solutions μ of the moment problem on $\mathcal{L} \cdot \mathcal{L}$. As in the case of J-fractions, a theory of nested disks can be developed.

1. Introduction

It is well known that there is a close relationship between orthogonal polynomials, special continued fractions, Padé approximants and the classical moment problems. This relationship can very briefly be described as follows.

Let a positive linear functional M on the space Π of all polynomials be given. (Equivalently: A positive definite sequence $\{s_n\}_{n=0}^{\infty}$ may be given, and a positive linear functional is then defined by setting $M[t^n] = s_n$, $n = 0, 1, 2, \ldots$.) A solution of the (Hamburger) moment problem for M is a positive measure μ on $\mathbb{R}$ with infinite support such that $\int_{\mathbb{R}} P(t)d\mu(t) = M[P]$ for all $P \in \Pi$. The functional M gives rise to an inner product $\langle \cdot, \cdot \rangle$ on Π through the formula $\langle P, Q \rangle = M[P \cdot \overline{Q}]$. By orthogonalization of the (ordered) basis $\{1, t, t^2, \ldots\}$ an orthogonal sequence $\{P_n\}$ is obtained. The sequence $\{P_n\}$ satisfies a three-term recurrence relation of a special form, and any polynomial sequence satisfying such a recurrence formula is a sequence of orthogonal polynomials determined by a positive linear functional on Π. Associated polynomials $\{Q_n\}$ are defined through the formula $Q_n(z) = M[(P_n(t) - P_n(z))/(t-z)]$. The quotients $Q_n(z)/P_n(z)$ are the approximants of a continued fraction of the type real J-fraction. These approximants are Padé approximants to the Stieltjes transform $S(z,\mu) = \int_{\mathbb{R}}(z-t)^{-1}d\mu(t)$ for any solution μ of the moment problem for M which means that $Q_n(z)/P_n(z) - S(z,\mu) = O(z^{-(2n+1)})$ at $z = \infty$.

The continued fraction determines for each $z \in \mathbb{C}\backslash\mathbb{R}$ a sequence of linear fractional transformations which map the upper half-plane (including the real line) onto a nested sequence of (closed) disks (for a precise definition see Section 8). The intersections $\Delta_{\infty}(z)$ of these disks have an invariance property: $\Delta_{\infty}(z)$ is either a single point for every z (the limit point case) or a proper closed disk for every z (the limit circle case). The set $\{w \in \mathbb{C} : w = S(z,\mu)$ for some solution μ of the moment problem$\}$ equals $\Delta_{\infty}(z)$, and the moment problem has a unique solution if and only if the limit point case occurs.

For details of this classical theory we refer to [1, 13, 16, 24, 30, 35, 37, 40, 41].

The aim of this paper is to present the basic theory of a rational extension of the polynomial situation sketched above. The interpolation (Padé approximation) at $z = \infty$ is replaced by interpolation at arbitrary points on the extended real line, the orthogonal polynomials become orthogonal rational functions, a continued fraction generalizing the real J-fraction is associated with the orthogonal functions, the approximants of the continued fraction are multipoint Padé approximants to Stieltjes transforms at certain tables determined by the interpolation points. Furthermore the continued fraction again determines sequences of nested disks, the intersections again have the invariance property, and the intersections are related to Stieltjes transforms of measures solving two in general different moment problems for the functional determining the orthogonal functions.

For details concerning this rational theory which will be presented here, we refer especially to [9], and also to [3, 4, 5, 6, 7, 8, 17, 18, 26, 31, 34]. In these publications are also treated problems where the interpolation points are situated outside the real axis. Such problems are not considered in this paper. Also we do not treat explicitly the (equivalent) situation of orthogonality on the unit circle, with interpolation points on or within the circle.

Other approaches to a study of orthogonality and biorthogonality of rational functions with applications to special functions have been studied by various authors, in particular Ismail, Masson, Rachmanov, Spiridonov, and Zhedanov, and can be found e.g., in [15, 20, 36, 38, 42]. It can be shown that by a special choice of the poles in [20] and of the poles used in this paper, the orthogonal rational functions coincide. See Sections 5 and 6.

It should also be mentioned that orthogonality of rational functions is closely related to orthogonality of polynomials with respect to varying weights, see especially work by Lopez, Stahl, Totik and others [14, 27, 28, 29, 39].

A theory of orthogonal rational functions was initiated by Djrbashian in 1969, see the survey paper in [12] and independently by Bultheel, Dewilde and Dym, see [2, 11]. A study of orthogonal Laurent polynomials (i.e., interpolation points at $z = 0$ and $z = \infty$) and corresponding continued fractions was initiated by Jones and Thron, see [25]. For special developments in this direction we refer to the survey article [21], and also to the papers [19, 22, 23, 33].

2. Spaces of Rational Functions

Let $\{\alpha_n\}_{n=1}^{\infty}$ be a sequence of not necessarily distinct points on $\hat{\mathbb{R}} \setminus \{0\}$ where $\hat{\mathbb{R}}$ denotes the extended real line $\mathbb{R} \cup \{\infty\}$. (For technical reasons, it is convenient to select a distinguished fixed point in $\hat{\mathbb{R}}$ which is different from all the α_n, and there is no loss of generality in placing this point at the origin.) We shall consistently use the notation $\beta_n = \alpha_n^{-1}$.

We define

$$u_0 = 1, \quad u_n(z) = \prod_{k=1}^{n} \frac{z}{1 - \beta_k z}, \quad n = 1, 2, \ldots$$

and set

$$\omega_0 = 1, \quad \omega_n(z) = \prod_{k=1}^{n} (1 - \beta_k z), \quad n = 1, 2, \ldots .$$

Thus we may write

$$u_n(z) = \frac{z^n}{\omega_n(z)}, \quad n = 1, 2, \ldots .$$

The linear spaces $\mathcal{L}_n$ and $\mathcal{L}$ are defined by

$$\mathcal{L}_n = \mathrm{span}\{u_0, u_1, \ldots, u_n\}, \quad n = 1, 2, \ldots$$

(the space is over the complex scalars) and

$$\mathcal{L} = \bigcup_{n=0}^{\infty} \mathcal{L}_n.$$

A function f belongs to $\mathcal{L}_n$ if and only of it is of the form

$$f(z) = \frac{p(z)}{\omega_n(z)} \tag{1}$$

where p is a polynomial of degree at most n.

If $\alpha_k = \infty$ for all k, then $\omega_n(z) = 1$, $u_n(z) = z^n$ for all n, and $\mathcal{L}_n$ is the space Π_n of all polynomials of degree at most n, $\mathcal{L}$ is the space Π of all polynomials.

Instead of $\{u_0, u_1, \ldots, u_n, \ldots\}$, we could work with other simple bases $\{v_0, v_1, \ldots, v_n, \ldots\}$ for $\mathcal{L}$ with the property that $\mathcal{L}_n = \mathrm{span}\{v_0, v_1, \ldots, v_n\}$ for all n. For example, if all the points α_k are finite, then $\{1, 1/\omega_1, \ldots, 1/\omega_n, \ldots\}$ is such a basis, and if all the points α_k are finite and distinct, then $\{1, 1/(1 - \beta_1 z), \ldots, 1/(1 - \beta_n z), \ldots\}$ is such a basis.

Every function f in $\mathcal{L}_n$ has a unique representation $f = \sum_{k=0}^n \lambda_k u_k$. We call λ_n the leading coefficient of f (with respect to the basis $\{u_0, u_1, \ldots, u_n\}$). When $\lambda_n = 1$, the rational function f is said to be *monic*.

We shall also deal with the spaces $\mathcal{L}_m \cdot \mathcal{L}_n$ consisting of all functions $h = f \cdot g$, with $f \in \mathcal{L}_m$ and $g \in \mathcal{L}_n$. It follows from (1) and the factorization theorem for polynomials that a function h belongs to $\mathcal{L}_m \cdot \mathcal{L}_n$ if and only if it is of the form

$$h(z) = \frac{r(z)}{\omega_m(z)\omega_n(z)},$$

where r is a polynomial of degree at most $m + n$. Thus $\mathcal{L}_m \cdot \mathcal{L}_n$ is a linear space of dimension $m + n + 1$. We use the notation $\mathcal{L} \cdot \mathcal{L}$ for the space $\{h = f \cdot g : f \in \mathcal{L}, g \in \mathcal{L}\} = \bigcup_{m,n=0}^\infty \mathcal{L}_m \cdot \mathcal{L}_n$.

Trivially $\mathcal{L} \subset \mathcal{L} \cdot \mathcal{L}$. If all the points in the sequence $\{\alpha_k\}$ are repeated an infinite number of times, then we have $\mathcal{L} \cdot \mathcal{L} = \mathcal{L}$. This is e.g., the case in the cyclic situation, where the sequence $\{\alpha_k\}$ consists of a finite number of points cyclically repeated. In particular, when $\alpha_k = \infty$ for all n (the polynomial situation), we have $\mathcal{L} \cdot \mathcal{L} = \mathcal{L} = \Pi$.

3. Moment Problems

The *substar transform* h_* of the function h is defined by

$$h_*(z) = \overline{h(\bar{z})}.$$

We note that $h_* \in \mathcal{L} \cdot \mathcal{L}$ if $h \in \mathcal{L} \cdot \mathcal{L}$, and that $h_* = h$ if h belongs to $\mathcal{L} \cdot \mathcal{L}$ and if $h = fg$ with $f, g \in \mathcal{L}$ having real coefficients with respect to a basis $\{v_0, v_1, \ldots, v_n, \ldots\}$ for $\mathcal{L}$ such that $v_{k*} = v_k$ for all k, so in particular if we use the basis $\{u_0, u_1, \ldots, u_n, \ldots\}$.

A linear functional M on $\mathcal{L} \cdot \mathcal{L}$ is said to be *positive* if it satisfies

$$M[h_*] = \overline{M[h]}, \quad \text{for all } h \in \mathcal{L} \cdot \mathcal{L} \tag{2}$$

and

$$M[f \cdot f_*] > 0, \quad \text{for all } f \in \mathcal{L}, \, f \neq 0.$$

It follows from (2) that $M[h]$ is real when $h = fg$ with $f, g \in \mathcal{L}$ having real coefficients with respect to the basis $\{u_0, u_1, \ldots, u_n, \ldots\}$. For simplicity we shall in the following assume M to be normalized such that $M[1] = 1$.

A probability measure μ on $\mathbb{R}$ with infinite support is said to solve the *moment problem on $\mathcal{L}$* if

$$M[f] = \int_{\mathbb{R}} f(t)d\mu(t), \quad \text{for all } f \in \mathcal{L} \tag{3}$$

and to solve the *moment problem on* $\mathcal{L} \cdot \mathcal{L}$ if

$$M[h] = \int_{\mathbb{R}} h(t) \mathrm{d}\mu(t), \quad \text{for all } h \in \mathcal{L} \cdot \mathcal{L}. \tag{4}$$

A measure which solves the moment problem on $\mathcal{L} \cdot \mathcal{L}$ clearly also solves the moment problem on $\mathcal{L}$. In the polynomial situation ($\alpha_k = 0$ for all k) and more generally in the cyclic situation, the moment problems are equivalent.

The moment problem on $\mathcal{L}$ is always solvable when the linear functional M is positive (see [9], and [7] for the equivalent problem on the unit circle). Clearly it is sufficient for μ to be a solution of the moment problem on $\mathcal{L}$ or on $\mathcal{L} \cdot \mathcal{L}$ that (3) or (4) is satisfied for some generating system for $\mathcal{L}$ or $\mathcal{L} \cdot \mathcal{L}$. In particular, (3) is equivalent to

$$M[u_n] = \int_{\mathbb{R}} u_n(t) \mathrm{d}\mu(t), \quad n = 0, 1, 2, \ldots$$

and (4) is equivalent to

$$M[u_m \cdot u_n] = \int_{\mathbb{R}} u_m(t) u_n(t) \mathrm{d}\mu(t), \quad m, n = 0, 1, 2, \ldots .$$

The constants $M[u_m]$ and $M[u_m \cdot u_n]$ are called *moments*. The moments are real numbers since $(u_n)_* = u_n$.

4. Orthogonal Rational Functions

Let M be a positive linear functional on $\mathcal{L} \cdot \mathcal{L}$. An inner product $\langle \cdot, \cdot \rangle$ on $\mathcal{L}$ can then be defined by the formula

$$\langle f, g \rangle = M[f \cdot g_*], \quad \text{for } f, g \in \mathcal{L}.$$

Let $\{\varphi_n\}_{n=0}^{\infty}$ be the orthonormal basis of $\mathcal{L}$ associated with the basis $\{u_n\}$ and with positive leading coefficients. Thus $\mathcal{L}_n = \text{span}\{\varphi_0, \varphi_1, \ldots, \varphi_n\}$ for all n, $\langle \varphi_j, \varphi_k \rangle = \delta_{j,k}$ for all j, k, and $\varphi_n = \lambda_n u_n + \lambda_{n-1} u_{n-1} + \cdots + \lambda_0 u_0$ with $\lambda_n > 0$. The function φ_n may be written in the form

$$\varphi_n(z) = \frac{p_n(z)}{\omega_n(z)}$$

where p_n is a polynomial of degree at most n.

For each n we may define the linear functional M_n and the (in general not positive definite) inner product $\langle \cdot, \cdot \rangle_n$ on the space Π of polynomials by

$$M_n[p] = M[p(z)/(1 - \beta_n z) \omega_{n-1}(z)^2]$$

and

$$\langle p, q \rangle_n = M_n[p \cdot q_*].$$

Since φ_n is orthogonal to all functions of the form $z^m/\omega_{n-1}(z)$ for $m = 0, 1, \ldots, n-1$, we see that $M_n[p_n(z) \cdot z^m] = 0$ for $m = 0, 1, \ldots, n-1$. Thus the polynomial sequence $\{p_n\}$ is orthogonal with respect to the sequence of varying (in general not positive definite) inner products $\{\langle \cdot, \cdot \rangle_n\}$. For treatment of orthogonality with respect to varying measures, we refer to [28, 29, 39].

We note that by construction we always have $p_n(\alpha_n) \neq 0$. We call φ_n *regular* if $p_n(\alpha_{n-1}) \neq 0$, *completely regular* if $p_n(\alpha_k) \neq 0$ for $k = 1, 2, \ldots, n-1$. (When $\alpha_{n-1} = \infty$, regularity means $\deg p_n = n$.) We say that the sequence $\{\varphi_n\}$ is regular (completely regular) if all φ_n are regular (completely regular).

A regular sequence $\{\varphi_n\}$ satisfies a three-term recurrence relation. For a proof of the following theorem see [9] and [17] (the last for the situation that a finite number of finite interpolation points are cyclically repeated). See also the treatment of the equivalent situation concerning orthogonality on the unit circle in [3].

THEOREM 4.1 *Assume that the sequence $\{\varphi_n\}$ is regular. Then $\{\varphi_n\}$ satisfies a three-term recurrence relation of the form*

$$\varphi_n(z) = \left(E_n \frac{z}{1 - \beta_n z} + B_n \frac{1 - \beta_{n-1} z}{1 - \beta_n z} \right) \varphi_{n-1}(z) - \frac{E_n}{E_{n-1}} \frac{1 - \beta_{n-2} z}{1 - \beta_n z} \varphi_{n-2}(z), \tag{5}$$

for $n = 1, 2, \ldots$ where B_n and E_n are real constants,

$$E_n \neq 0, \quad \text{for all } n \tag{6}$$

and by convention

$$\beta_0 = 0, \quad \beta_{-1} = 0, \quad (i.e., \ \alpha_0 = \infty, \quad \alpha_{-1} = \infty), \quad \varphi_{-1} = 0. \tag{7}$$

Recall that $\varphi_0 = 1$. Also note that E_0 can be arbitrarily chosen, since $\varphi_{-1} = 0$.

We define a new orthogonal sequence $\{\Phi_n\}_{n=0}^\infty$ by the normalization

$$\Phi_0 = \varphi_0, \quad \Phi_n(z) = (E_1 E_2 \cdots E_n)^{-1} \varphi_n(z), \quad n = 1, 2, \ldots.$$

The recurrence relation of Theorem 4.1 can then be written

$$\Phi_n(z) = \left(\frac{z}{1 - \beta_n z} + \frac{B_n}{E_n} \frac{1 - \beta_{n-1} z}{1 - \beta_n z} \right) \Phi_{n-1}(z) - \frac{1}{E_{n-1}^2} \frac{1 - \beta_{n-2} z}{1 - \beta_n z} \Phi_{n-2}(z),$$

for $n = 1, 2, \ldots$ with initial values

$$\Phi_0 = 1, \quad \Phi_{-1} = 0, \quad \text{and} \quad \alpha_0 = \infty, \quad \alpha_{-1} = \infty.$$

In the polynomial case ($\alpha_k = \infty$ for all k) this recurrence relation takes the form

$$\Phi_n(z) = \left(z + \frac{B_n}{E_n} \right) \Phi_{n-1}(z) - \frac{1}{E_{n-1}^2} \Phi_{n-2}(z), \quad n = 1, 2, \ldots . \quad (8)$$

We observe that the polynomials Φ_n are monic and that (8) is the classical recurrence relation for monic orthogonal polynomials.

In general, we may write $\Phi_n(z) = P_n(z)/\omega_n(z)$, where $P_0 = p_0$; $P_n(z) = (E_1 E_2 \cdots E_n)^{-1} p_n(z)$ for $n = 1, 2, \ldots$. The polynomial sequence $\{P_n\}$ satisfies the recurrence relation

$$P_n(z) = \left(z + \frac{B_n}{E_n}(1 - \beta_{n-1}z) \right) P_{n-1}(z) - \frac{(1 - \beta_{n-2}z)(1 - \beta_{n-1}z)}{E_{n-1}^2} P_{n-2}(z) \quad (9)$$

for $n = 1, 2, \ldots$ with initial values

$$P_0 = 1, \quad P_{-1} = 0, \quad \alpha_0 = \alpha_{-1} = \infty. \quad (10)$$

A proof of the following *Favard-type theorem* can be found in [9]. See also [6] where the equivalent unit circle situation is treated.

THEOREM 4.2 *Let sequences $\{B_n\}_{n=1}^{\infty}$ and $\{E_n\}_{n=0}^{\infty}$ of real constants be given such that (6) is satisfied, and let the functions $\{\varphi_n\}$ be defined recursively by $\varphi_0 = 1$ and by (5), (7) for $n = 1, 2, \ldots$. Then $\varphi_n \in \mathcal{L}_n \setminus \mathcal{L}_{n-1}$ for $n = 1, 2, \ldots$, φ_n is regular for all n, and there exists a positive linear functional M on $\mathcal{L} \cdot \mathcal{L}$ such that the sequence $\{\varphi_n\}$ is orthonormal with respect to the associated inner product.*

The situation here is more complicated than the classical (polynomial) situation. This is partly so because for a given recursion in $\mathcal{L}$, we need to define a functional M on $\mathcal{L} \cdot \mathcal{L}$ with respect to which the sequence $\{\varphi_n\}$ is to be orthonormal.

REMARK 4.1 To obtain the recurrence relations of this section, we need not to assume the points α_k to lie on the real axis and the functional M need not be assumed positive. A sufficient condition is the regularity of the functional in the sense that $M[f \cdot g] \neq 0$ for $f \in \mathcal{L}$, $f \not\equiv 0$. The regular orthogonal functions must then be defined in terms of the Euclidean inner product $\langle f, g \rangle = M[f \cdot g]$. The recurrence coefficients may of course be (non-real) complex numbers.

5. R_I-Recursion

We shall in this and the next section briefly review the basic facts concerning the three-term recurrence relation introduced by Ismail and Masson [20], see also [38, 42], and point out some connections with our theory of orthogonal rational functions.

The R_I-*recursion* of Ismail and Masson arises in the following way: Let $\{a_n\}_{n=2}^\infty$, $\{c_n\}_{n=1}^\infty$, $\{\lambda_n\}_{n=1}^\infty$ be sequences of complex numbers. A system $\{P_n\}_{n=0}^\infty$ of monic polynomials is generated by the recursion

$$P_n(z) = (z - c_n)P_{n-1}(z) - \lambda_n(z - a_n)P_{n-2}(z), \quad n = 1, 2, \ldots$$

$$(11)$$

$$P_{-1}(z) = 0, \quad P_0(z) = 1.$$

It is assumed that

$$\lambda_n \neq 0, \quad P_n(a_{n+1}) \neq 0, \quad n = 1, 2, \ldots . \tag{12}$$

The rational functions R_n are defined by

$$R_{-1}(z) = 0, \quad R_0(z) = 1, \quad R_n(z) = \frac{P_n(z)}{\prod_{k=2}^{n+1}(z - a_k)}, \quad n = 1, 2, \ldots .$$

These functions satisfy the three-term recursion

$$R_n(z) = \frac{z - c_n}{z - a_{n+1}} R_{n-1}(z) - \frac{\lambda_n}{z - a_{n+1}} R_{n-2}(z), \quad n = 1, 2, \ldots .$$

A Favard type theorem states that there exists a linear functional L defined on the span of the functions $\{z^k R_n(z) : k = 0, \ldots, n; n = 0, 1, 2, \ldots\}$ such that the orthogonality property

$$L[z^k R_n(z)] = 0, \quad k = 0, 1, \ldots, n - 1; \; n = 1, 2, \ldots \tag{13}$$

holds and such that

$$L[z^n R_n(z)] = \lambda_1 \lambda_2 \cdots \lambda_{n+1}.$$

In the special case

$$a_{2m} = a_{2m+1} \quad m = 1, 2, \ldots ,$$

the recurrence (11) may be written in the form

$$P_{2m}(z) = (z - c_{2m})P_{2m-1}(z) - \lambda_{2m}(z - a_{2m})P_{2m-2}(z), \quad m = 1, 2, \ldots$$

$$(14)$$

$$P_{2m+1}(z) = (z - c_{2m+1})P_{2m}(z) - \lambda_{2m+1}(z - a_{2m})P_{2m-1}(z), \; m = 0, 1, 2, \ldots \tag{15}$$

The functions R_n have the form

$$R_{2m}(z) = \frac{P_{2m}(z)}{\prod_{k=1}^{m}(z - a_{2k})^2}, \quad m = 1, 2, \ldots \tag{16}$$

$$R_{2m+1}(z) = \frac{P_{2m+1}(z)}{(z - a_{2m+2})\prod_{k=1}^{m}(z - a_{2k})^2}, \quad m = 0, 1, 2, \ldots \tag{17}$$

Now we consider the situation in Section 4 for regular φ_n where we have

$$\alpha_{2m} = \infty, \quad \alpha_{2m+1} \neq \infty, \quad m = 1, 2, \ldots. \tag{18}$$

The orthogonal rational functions Φ_n may then be written in the form

$$\Phi_{2m}(z) = \frac{C_{2m}P_{2m}(z)}{\prod_{k=0}^{m-1}(z - \alpha_{2k+1})}, \quad m = 1, 2, \ldots$$

$$\Phi_{2m+1}(z) = \frac{C_{2m+1}P_{2m+1}(z)}{\prod_{k=0}^{m}(z - \alpha_{2k+1})}, \quad m = 0, 1, 2, \ldots$$

where C_n are constants and P_n are the numerator polynomials determined by (9)-(10). We observe the following: We may write for $n = 1, 2, \ldots$

$$z + B_n E_n^{-1}(1 - \beta_{n-1}z) = (1 - B_n E_n^{-1}\beta_{n-1})\left(z + \frac{B_n E_n^{-1}}{1 - B_n E_n^{-1}\beta_{n-1}}\right). \tag{19}$$

It follows from (19), (9) and the regularity of φ_n (which is implied by (9)) that $1 - B_n E_n^{-1}\beta_{n-1} \neq 0$. Hence (9) may be written as

$$P_n(z) = \rho_n(z - c_n)P_{n-1}(z) - E_{n-1}^{-2}(1 - \beta_{n-2}z)(1 - \beta_{n-1}z)P_{n-2}(z),$$

for $n = 1, 2, \ldots$, where

$$\rho_n = 1 - B_n E_n^{-1}\beta_{n-1} \neq 0, \quad c_n = -\frac{B_n E_n^{-1}}{1 - B_n E_n^{-1}\beta_{n-1}}. \tag{20}$$

Since in our situation $\beta_{2m} = 0$, $\beta_{2m+1} \neq 0$, this gives

$$P_{2m}(z) = \rho_{2m}(z - c_{2m})P_{2m-1}(z) - \pi_{2m}(z - \alpha_{2m-1})P_{2m-2}(z), \tag{21}$$

for $m = 1, 2, \ldots$ while for $m = 0, 1, 2, \ldots$ we have

$$P_{2m+1}(z) = \rho_{2m+1}(z - c_{2m+1})P_{2m}(z) - \pi_{2m+1}(z - \alpha_{2m-1})P_{2m-1}(z),$$
$$(22)$$

where $\pi_n \neq 0$, ρ_n and c_n given by (20).

Conversely, if $\{P_n\}$ satisfies a recurrence relation of the form (21)-(22) with $\rho_n \neq 0$, $\pi_n \neq 0$, then $\{P_n\}$ also satisfies a recurrence relation of the form (9) with $\beta_{2m} = 0$, i.e., $\alpha_{2m} = \infty$.

We now define a_n by

$$a_{2m} = a_{2m+1} = \alpha_{2m-1}, \quad m = 1, 2, \ldots. \tag{23}$$

Note that $P_{2m}(a_{2m+1}) = P_{2m}(\alpha_{2m-1}) \neq 0$ by regularity and $P_{2m+1}(a_{2m+2}) = P_{2m+1}(\alpha_{2m+1}) \neq 0$ by construction, so that (12) is satisfied. With the definition (23) we see that apart from a normalization factor the formulas (21)-(22) coincide in form with the formulas (14)-(15), and the initial conditions are identical. Thus the polynomials P_n of Section 4 are in this situation (apart from a normalization factor) polynomials obtained by a suitable R_I-recursion.

Next consider the recurrence relation (14)-(15) and assume that $a_n \neq 0$ for all n. We define α_n by

$$\alpha_{2m} = \infty, \quad \alpha_{2m-1} = a_{2m} = a_{2m+1}, \quad m = 1, 2, \ldots. \tag{24}$$

With this definition, formulas (14)-(15) may be written in the form (21)-(22) with $\rho_n = 1$. Again the initial conditions are identical. Hence we conclude that the polynomials P_n determined by the R_I-recurrence in this situation are polynomials obtained by a recurrence (9)-(10).

Moreover, when (18), (23) (or equivalently (24)) are satisfied, formulas (16)-(17) may be written as

$$R_{2m}(z) = A_{2m} \frac{\Phi_{2m}(z)}{\prod_{j=0}^{m-1}(z - \alpha_{2j+1})}, \quad m = 1, 2, \ldots$$

$$R_{2m+1}(z) = A_{2m+1} \frac{\Phi_{2m+1}(z)}{\prod_{j=0}^{m-1}(z - \alpha_{2j+1})}, \quad m = 0, 1, 2, \ldots$$

where A_n are constants. The orthogonality (13) may be written

$$L\left[\frac{z^k}{\prod_{j=0}^{m-1}(z - \alpha_{2j+1})}\Phi_{2m}(z)\right] = 0, \quad \begin{matrix} k = 0, 1, 2, \ldots, 2m - 1 \\ m = 1, 2, \ldots \end{matrix}$$

$$L\left[\frac{z^k}{\prod_{j=0}^{m-1}(z - \alpha_{2j+1})}\Phi_{2m+1}(z)\right] = 0, \quad \begin{matrix} k = 0, 1, 2, \ldots, 2m \\ m = 0, 1, \ldots \end{matrix}$$

When all the α_n (or a_n) are real and all the E_n^2 (or λ_n) are positive, this is the orthogonality of Section 4.

We may roughly sum up the main results of this section as follows: *The special case of the R_I-recursion where $a_{2m} = a_{2m+1} \neq 0$ for all m is equivalent (up to a normalizing factor) to the special case of Section 4 where $\alpha_{2m} = \infty$, $\alpha_{2m-1} \neq \infty$ for all m. The correspondence is established by the equality*

$$a_{2m} = a_{2m+1} = \alpha_{2m-1}.$$

The equivalence means that the numerator polynomials of Φ_n and of R_n for suitable parameters are equal. Moreover, the orthogonality satisfied by $\{\Phi_n\}$ coincides with the orthogonality (derived through the Favard theorem) belonging to the R_I-recurrence when all the α_n (or a_n) are real and all E_n^2 (or λ_n) are positive.

6. R_{II}-Recursion

The R_{II}-*recursion* of Ismail and Masson has the following setting: Let $\{a_n\}_{n=2}^\infty$, $\{b_n\}_{n=2}^\infty$, $\{c_n\}_{n=1}^\infty$, $\{\lambda_n\}_{n=1}^\infty$ be sequences of complex numbers. A system $\{P_n\}_{n=0}^\infty$ of polynomials is generated by the recursion

$$P_n(z) = (z - c_n)P_{n-1}(z) - \lambda_n(z - a_n)(z - b_n)P_{n-2}(z), \quad n = 1, 2, \ldots \tag{25}$$

$$P_{-1}(z) = 0, \quad P_0(z) = 1.$$

(Note that a_1 and b_1 in the formulas can be freely chosen.) It is assumed that

$$\lambda_{n+1} \neq 0, \quad P_n(a_{n+1}) \neq 0, \quad P_n(b_{n+1}) \neq 0, \quad \text{for } n = 1, 2, \ldots \tag{26}$$

The rational functions S_n are defined by

$$S_{-1}(z) = 0, \quad S_0(z) = 1, \quad \text{and} \quad S_n(z) = \frac{P_n(z)}{\prod_{k=1}^n (z - a_{k+1})(z - b_{k+1})}, \tag{27}$$

for $n = 1, 2, \ldots$. These functions satisfy the three-term recurrence relation

$$S_n(z) = \frac{z - c_n}{(z - a_{n+1})(z - b_{n+1})} S_{n-1}(z) - \frac{\lambda_n}{(z - a_{n+1})(z - b_{n+1})} S_{n-2}(z),$$

for $n = 1, 2, \ldots$. A Favard type theorem states that there exists a linear functional L defined on the span of the functions $\{z^k S_n(z) : k = 0, 1, \ldots, n; n = 0, 1, 2, \ldots\}$ such that the orthogonality property

$$L[z^k S_n(z)] = 0, \quad \text{for } k = 0, 1, \ldots, n-1; n = 1, 2, \ldots \tag{28}$$

holds.

To get a simple and direct connection between R_{II}-recursion and the recursion of Section 4, we replace the initial equations in (25) (for $n = 1$ and $n = 2$) by

$$P_2(z) = (z - c_2)P_1(z) - \lambda_2(z - a_2)P_0(z)$$

$$P_1(z) = (z - c_1)P_0(z) - \lambda_1(z - a_1)P_{-1}(z).$$

Here no b_2 is involved. The formulas can be interpreted as setting $b_2 = \infty$ in (25), cf. the use of the factors $(1 - z/\alpha_k)$ in Sections 2-4. We shall call this recursion *modified R_{II}-recursion*.

The rational functions in (27) are now replaced by the functions

$$S_1(z) = \frac{P_1(z)}{z - a_2}, \quad \text{and} \quad S_n(z) = \frac{P_n(z)}{(z - a_2) \prod_{k=2}^{n}(z - a_{k+1})(z - b_{k+1})}, \tag{29}$$

for $n = 2, 3, \dots$. The linear functional L is now defined on the span of the functions $\{z^k S_n(z) : k = 0, 1, \dots, n; n = 0, 1, 2, \dots\}$ as before (with the new meaning of S_n for $n = 1, 2, \dots$), and the orthogonality condition can be expressed by (28) as before. (Note that the degree of the denominator of S_n is now $2n - 1$, while the degree of the numerator of $z^k S_n(z)$ is at most $k + n$ as before.)

In the special case

$$b_n = a_{n-1}, \quad k = 3, 4, \dots \tag{30}$$

the modified R_{II}-recursion may be written in the form

$$P_n(z) = (z - c_n)P_{n-1}(z) - \lambda_n(z - a_n)(z - a_{n-1})P_{n-2}(z), \quad k = 3, 4, \dots \tag{31}$$

$$P_2(z) = (z - c_2)P_1(z) - \lambda_2(z - a_2)P_0(z) \tag{32}$$

$$P_1(z) = (z - c_1)P_0(z) - \lambda_1(z - a_1)P_{-1}(z) \tag{33}$$

$$P_{-1}(z) = 0, \quad P_0(z) = 1. \tag{34}$$

The functions S_n of (29) now have the form

$$S_1(z) = \frac{P_1(z)}{z - a_2}, \quad S_n(z) = \frac{P_n(z)}{(z - a_{n+1}) \prod_{k=1}^{n-1}(z - a_{k+1})^2}, \quad k = 2, 3, 4, \dots. \tag{35}$$

Now consider the situation in Section 4 where we have

$$\alpha_n \neq \infty, \quad \text{for all } n.$$

The orthogonal rational functions Φ_n may then be written in the form

$$\Phi_n(z) = \frac{C_n P_n(z)}{\prod_{k=1}^n (z - \alpha_k)}, \quad k = 1, 2, \dots, \tag{36}$$

where C_n are constants and P_n are the numerator polynomials defined by (9)-(10).

It follows from (19) (recall that the Φ_n are regular) that (9) may equivalently be written in the form

$$P_n(z) = \rho_n(z - c_n)P_{n-1}(z) - \pi_n(z - \alpha_{n-1})(z - \alpha_{n-2})P_{n-2}(z), \quad k = 3, 4, \dots \tag{37}$$

$$P_2(z) = \rho_2(z - c_2)P_1(z) - \pi_2(z)(z - \alpha_1)P_0(z) \tag{38}$$

$$P_1(z) = \rho_1(z - c_1)P_0(z) - \pi_1(z)(z - \alpha_1)P_{-1}(z) \tag{39}$$

where $\pi_n \neq 0$ and c_n given as in (20). We now define a_n by

$$a_n = \alpha_{n-1}, \quad n = 2, 3, \dots . \tag{40}$$

Note that $P_n(a_{n+1}) = P_n(\alpha_n) \neq 0$ by construction and $P_n(b_{n+1}) = P_n(a_n) = P_n(\alpha_{n-a}) \neq 0$ by regularity, so that (26) is satisfied. With the definition (40) we see that apart from a normalization factor, the formulas (37)-(39) coincide in form with the formulas (31)-(33), and the initial conditions are identical. Thus the polynomials P_n of Section 4 are in this situation (apart from a normalization factor) polynomials obtained by a suitable modified R_{II}-recursion.

Next consider the recurrence relation (31)-(34) and assume $a_n \neq 0$ for all n. We define α_n by

$$\alpha_n = a_{n+1} = b_{n+2}, \quad n = 1, 2, 3, \dots . \tag{41}$$

With this definition, formulas (31)-(34) may be written in the form (37)-(39) with $\rho_n = 1$. Again the initial conditions are identical. Hence we conclude that the polynomials P_n determined by the modified R_{II}-recursion in this situation are polynomials obtained by a recursion (9)-(10).

Moreover, when (30), (40) (or equivalently (41)) is satisfied, formula (36) may be written as

$$S_n(z) = A_n \frac{\Phi_n(z)}{\prod_{j=1}^{n-1}(z - \alpha_j)}, \quad n = 1, 2, 3, \dots$$

where A_n are constants (recall formula (35)). The orthogonality (28) may then be written

$$L\left[\frac{z^k}{\prod_{j=1}^{n-1}(z-\alpha_j)}\Phi_n(z)\right] = 0, \quad k = 0, 1, 2, \ldots, n-1; n = 1, 2, 3, \ldots.$$

When all the α_n (or a_n) are real and all E_n^2 (or λ_n) are positive, this is the orthogonality of Section 4.

We may roughly sum up the main results of this section in the following way:

The special case of the modified R_{II}-recursion where $b_n = a_{n-1} \neq 0$ for all n is equivalent (up to a normalization factor) to the special case of Section 4 where $\alpha_n \neq \infty$ for all n. The correspondence is established by the equality

$$a_n = b_{n+1} = \alpha_{n-1}.$$

The equivalence means that the numerator polynomials of Φ_n and of S_n for suitable parameters are equal. Moreover, the orthogonality satisfied by $\{\Phi_n\}$ coincides with the orthogonality (derived through the Favard theorem) belonging to the modified R_{II}-recurrence when all the α_n (or a_n) are real and all the E_n^2 (or λ_n) are positive.

7. Continued Fractions

We now return to the setting of Sections 2-4. We define the *associated orthogonal functions* σ_n and Σ_n by

$$\sigma_n(z) = M\left[\frac{\varphi_n(t) - \varphi_n(z)}{t-z}\right], \quad n = 0, 1, 2, \ldots$$

$$\Sigma_n(z) = M\left[\frac{\Phi_n(t) - \Phi_n(z)}{t-z}\right], \quad n = 0, 1, 2, \ldots,$$

(where M operates on its argument as a function of t). Note that we may write

$$\sigma_n(z) = \frac{q_n(z)}{\omega_n(z)}$$

where q_n is a polynomial of degree at most $n-1$. Clearly

$$\sigma_0 = \Sigma_0 = 0, \quad \Sigma_n(z) = (E_1 E_2 \cdots E_n)^{-1}\sigma_n(z), \quad n = 1, 2, \ldots.$$

$$(42)$$

In [9] are considered *functions of the second kind* defined by

$$\psi_0(z) = iz, \quad \psi_n(z) = -iM\left[\frac{1+tz}{t-z}(\varphi_n(t) - \varphi_n(z))\right], \quad n = 1, 2, \ldots.$$

The functions σ_n and ψ_n are related through the formula

$$\psi_n(z) = -i(1 + z^2)\sigma_n(z) + iz\varphi_n(z), \quad n = 0, 1, 2, \ldots . \tag{43}$$

From [9], (42) and (43) we find that the sequence $\{\Sigma_n\}$ satisfies the recurrence relation

$$\Sigma_n(z) = \left(\frac{z}{1 - \beta_n z} + \frac{B_n}{E_n}\frac{1 - \beta_{n-1}z}{1 - \beta_n z}\right)\Sigma_{n-1}(z) - \frac{1}{E_{n-1}^2}\frac{1 - \beta_{n-2}z}{1 - \beta_n z}\Sigma_{n-2}(z),$$

for $n = 1, 2, \ldots$ where B_n and E_n for $n = 1, 2, \ldots$ are the coefficients appearing in Theorem 4.1, and initial values are given by

$$\Sigma_0 = 0, \quad \Sigma_{-1} = 1, \quad E_0 = -1, \quad \text{and} \quad \alpha_0 = \infty, \quad \alpha_{-1} = \infty.$$

(Note that a similar recurrence is obtained in Proposition 2.1 of [10].) From this follows

THEOREM 7.1 *Assume that the sequence $\{\varphi_n\}$ is regular. Then $\Sigma_n(z)$ and $\Phi_n(z)$ are the canonical numerators and denominators of a continued fraction* $\mathrm{K}_{n=1}^{\infty}\frac{a_n(z)}{b_n(z)}$, *where*

$$a_n(z) = -\frac{1}{E_{n-1}^2}\frac{1 - \beta_{n-2}z}{1 - \beta_n z}, \quad n = 1, 2, \ldots , \tag{44}$$

$$b_n(z) = \frac{z}{1 - \beta_n z} + \frac{B_n}{E_n}\frac{1 - \beta_{n-1}z}{1 - \beta_n z}, \quad n = 1, 2, \ldots . \tag{45}$$

We shall call a continued fraction of this form a *Multipoint Padé continued fraction* or an *MP-fraction*. A motivation for this terminology is given in Theorem 7.2. Clearly $\sigma_n(z)$ and $\varphi_n(z)$ are the canonical numerators and denominators of a continued fraction $\mathrm{K}_{n=1}^{\infty}\frac{c_n(z)}{d_n(z)}$ equivalent to $\mathrm{K}_{n=1}^{\infty}\frac{a_n(z)}{b_n(z)}$, with elements

$$c_n(z) = -\frac{E_n}{E_{n-1}}\frac{1 - \beta_{n-2}z}{1 - \beta_n z}, \quad n = 1, 2, \ldots ,$$

$$d_n(z) = E_n\frac{z}{1 - \beta_n z} + B_n\frac{1 - \beta_{n-1}z}{1 - \beta_n z}, \quad n = 1, 2, \ldots .$$

We also note that in the polynomial situation ($\alpha_k = \infty$ for all k) we have

$$a_n(z) = -E_{n-1}^{-2}, \quad b_n(z) = z + E_n^{-1}B_n, \quad n = 1, 2, \ldots ,$$

and the MP-fraction thus becomes the real *J*-fraction associated with the classical Hamburger moment problem.

We shall here define the *Stieltjes transform* $S(z, \mu)$ of a finite measure μ on $\mathbb{R}$ by the formula

$$S(z, \mu) = \int_{\mathbb{R}} \frac{d\mu(t)}{z - t}.$$

In [9] the *Nevanlinna transform*

$$\Omega_\mu(z) = i \int_{\mathbb{R}} \frac{1 + tz}{z - t} d\mu(t)$$

is worked with. The two transforms are related through the formula

$$\Omega_\mu(z) = i(1 + z^2) S(z, \mu) - iz.$$

We shall consider interpolation tables of the form $\{\infty, \alpha_1, \alpha_1, \ldots, \alpha_{n-1}, \alpha_{n-1}, \alpha_n\}$. We denote by $\alpha^\#$ the multiplicity of α as an entry in the table in question. By the limit at a point α of the table, we shall mean the one-sided limit in the upper half-plane along the normal to the real axis at the point α, when $\alpha \neq \infty$ and along the imaginary axis when $\alpha = \infty$.

The following result can be deduced from [8, 9].

THEOREM 7.2 *Let the positive linear functional M on $\mathcal{L} \cdot \mathcal{L}$ give rise to a completely regular sequence $\{\varphi_n\}$ and let μ be an arbitrary solution of the moment problem on $\mathcal{L} \cdot \mathcal{L}$. Then the approximants $\sigma_n(z)/\varphi_n(z) = \Sigma_n(z)/\Phi_n(z)$ of the MP-fraction determined by M are $[n - 1/n]$ multipoint Padé approximants to $S(z, \mu)$ at the table $\{\infty, \alpha_1, \alpha_1, \ldots, \alpha_{n-1}, \alpha_{n-1}, \alpha_n\}$ in the following sense:*

$$\lim_{z \to \alpha} \left[\frac{\sigma_n(z)}{\varphi_n(z)} - S(z, \mu) \right]^{(k)} = 0, \quad for \ k = 0, 1, \ldots, \alpha^\# - 1, \qquad (46)$$

where $\alpha \in \{\alpha_1, \alpha_1, \ldots, \alpha_{n-1}, \alpha_{n-1}, \alpha_n\}$, $\alpha \neq \infty$ and

$$\lim_{z \to \infty} \left[\frac{\sigma_n(z)}{\varphi_n(z)} - S(z, \mu) \right] z^{\infty^\#} = 0. \qquad (47)$$

In the polynomial case ($\alpha_n = \infty$ for all n) the interpolation property (46)-(47) of Theorem 7.2 takes the form

$$\lim_{z \to \infty} \left[\frac{\sigma_n(z)}{\varphi_n(z)} - S(z, \mu) \right] z^{2n} = 0.$$

In this case the slightly more general property holds:

$$\frac{\sigma_n(z)}{\varphi_n(z)} - S(z, \mu) = O(1/z^{2n+1}) \quad \text{as } z \to \infty.$$

See e.g., [1]. In the rational case when $\mathcal{L} \cdot \mathcal{L} = \mathcal{L}$, a similar strengthening of the limiting properties (46)-(47) is true at all the interpolation points. (See [4].)

8. Modified Approximants and Nested Disks

In all of this section, we assume that the sequence $\{\varphi_n\}$ determined by the functional M is regular.

We define the linear fractional transformations $w \to t_n(z, w)$ and $w \to T_n(z, w)$ by

$$t_0(z, w) = w, \quad t_n(z, w) = \frac{a_n(z)}{b_n(z) + w}, \quad \text{for } n = 1, 2, \ldots,$$

$$T_0(z, w) = w, \quad T_n(z, w) = T_{n-1}(z, t_n(z, w)), \quad \text{for } n = 1, 2, \ldots,$$

where $a_n(z)$ and $b_n(z)$ are as defined in (44)-(45). (Recall that by convention $\alpha_0 = \alpha_{-1} = \infty$.) Then $T_n(z, 0) = \Sigma_n(z)/\Phi_n(z) = \sigma_n(z)/\varphi_n(z)$ and in general

$$T_n(z, w) = \frac{\Sigma_n(z) + w\Sigma_{n-1}(z)}{\Phi_n(z) + w\Phi_{n-1}(z)} = \frac{\sigma_n(z) + wE_n\sigma_{n-1}(z)}{\varphi_n(z) + wE_n\varphi_{n-1}(z)}, \quad n = 0, 1, 2, \ldots.$$

See e.g., [24, 30, 35]. We shall call the expressions $T_n(z, w)$ *modified approximants* of the MP-fraction $\mathrm{K}_{n=1}^{\infty} \frac{a_n(z)}{b_n(z)}$ when w is of the form $w = [(1 - \beta_{n-1}z)/(1 - \beta_n z)]\tau$, $\tau \in \hat{\mathbb{R}}$. Thus modified approximants $R_n(z, \tau)$ are the expressions

$$R_n(z, \tau) = T_n\left(z, \tau\frac{1 - \beta_{n-1}z}{1 - \beta_n z}\right) = \frac{\Sigma_n(z) + \tau(1 - \beta_{n-1}z)(1 - \beta_n z)^{-1}\Sigma_{n-1}(z)}{\Phi_n(z) + \tau(1 - \beta_{n-1}z)(1 - \beta_n z)^{-1}\Phi_{n-1}(z)},$$

with $\tau \in \hat{\mathbb{R}}$. We note that in the polynomial case the concept of modified approximants reduces to the traditional concept of modified approximants to continued fractions.

We shall consider mapping properties of the transformations $w \to t_n(z, w)$ and $w \to T_n(z, w)$. Let z be a fixed point in the upper half-plane $\mathbb{U} = \{z : \mathrm{Im}\, z > 0\}$. We set $\theta_n = \arg(1 - \beta_n z)$ (note that then $\theta_0 = 0$) and define

$$\Omega_0 = \overline{\mathbb{U}} = \{w \in \mathbb{C} : \mathrm{Im}\, w \geq 0\}$$

$$\Omega_n = \{w \in \mathbb{C} : \theta_{n-1} - \theta_n \leq \arg w \leq \pi + \theta_{n-1} - \theta_n\}, \quad \text{for } n = 1, 2, \ldots.$$

With $\mathrm{Im}\, z > 0$ we have $\theta_n < \arg z < \theta_n + \pi$ for all n. It follows that $W = z(1 - \beta_n z)^{-1} + B_n E_n^{-1}(1 - \beta_{n-1}z)(1 - \beta_n z)^{-1} + w \in \Omega_n$ when $w \in \Omega_n$.

Hence $-E_n^{-2}(1-\beta_{n-2}z)(1-\beta_n z)^{-1}W^{-1} \in \Omega_{n-1}$ when $w \in \Omega_n$. Thus $w \in \Omega_n$ implies $a_n(z)/[b_n(z)+w] \in \Omega_{n-1}$. This means that for $n = 1, 2, \ldots$ we have

$$t_n(z, \Omega_n) \subset \Omega_{n-1}. \tag{48}$$

(Cf. the analogous argument in [18].)

For each z in the open upper half-plane, $\mathbb{U}$, we set $\Delta_n(z) = T_n(z, \Omega_n)$. Clearly $\Delta_n(z)$ is a closed half-plane or a closed disk or a closed exterior of a disk. It follows from the definition that $b_1(z)+w$ vanishes at a point outside Ω_1 and hence $t_1(z, \Omega_1)$ is a bounded set. Formula (48) gives $t_1(z, \Omega_1) \subset \Omega_0 = \overline{\mathbb{U}}$. Again by using (48) we conclude that $\Delta_n(z) = T_n(z, \Omega_n) = T_{n-1}(z, t_n(z, \Omega_n)) \subset T_{n-1}(z, \Omega_{n-1}) = \Delta_{n-1}(z)$ for $n = 2, 3, \ldots$. Thus $\Delta_n(z)$ must be a closed disk contained in Ω_0. Consequently the following result holds:

THEOREM 8.1 *For $n = 1, 2, \ldots$ we have:*

A. $\Delta_n(z)$ *is a closed disk contained in the closed upper half-plane* $\overline{\mathbb{U}}$,

B. $\Delta_n(z) \subset \Delta_{n-1}(z)$.

The transformation $\tau \to \tau(1 - \beta_{n-1}z)/(1 - \beta_n z)$ maps $\overline{\mathbb{U}}$ onto Ω_n. Hence we find that the modified approximants $R_n(z, \tau)$ map the closed upper half-plane $\overline{\mathbb{U}}$ onto $\Delta_n(z)$ and the extended line $\hat{\mathbb{R}}$ onto the circle periphery $\partial \Delta_n(z)$.

By solving the equation $w = R_n(z, \tau)$ with respect to τ and setting $\text{Im } \tau = 0$ we find that the equation for the circle $\partial \Delta_n(z)$ may be written

$$(1 - \beta_n z)(1 - \beta_{n-1}\overline{z})\Gamma_n(z, w)\overline{\Gamma_n(z, w)}$$
$$-(1 - \beta_n\overline{z})(1 - \beta_{n-1}z)\overline{\Gamma_{n-1}(z, w)}\Gamma_{n-1}(z, w) = 0, \tag{49}$$

where

$$\Gamma_n(z, w) = \sigma_n(z) - w\varphi_n(z).$$

(Note that here w has a different meaning from that in the first part of this section.)

In [9] is proved a general Green's formula involving arbitrary solutions of the recurrence relation of the continued fraction. By replacing w by z, and $x_n(z)$ and $y_n(z)$ by $\sigma_n(z) - w\varphi_n(z)$ in Theorem 11.4.1 of [9], we get

$$\frac{1}{E_n(z - \overline{z})}[(1 - \beta_n z)(1 - \beta_{n-1}\overline{z})\Gamma_n(z, w)\overline{\Gamma_{n-1}(z, w)}$$
$$-(1 - \beta_n\overline{z})(1 - \beta_{n-1}z)\overline{\Gamma_n(z, w)}\Gamma_{n-1}(z, w)] \tag{50}$$
$$= \sum_{k=0}^{n-1} |\Gamma_k(z, w)|^2 - \frac{w - \overline{w}}{z - \overline{z}}.$$

Combining (49) and (50) we find that the circle $\partial\Delta_n(z)$ is given by

$$\sum_{k=0}^{n-1} |\sigma_k(z) - w\varphi_k(z)|^2 = \frac{w - \overline{w}}{z - \overline{z}}$$

and the disk $\Delta_n(z)$ is given by

$$\sum_{k=0}^{n-1} |\sigma_k(z) - w\varphi_k(z)|^2 \le \frac{w - \overline{w}}{z - \overline{z}}.$$

(Note that the last term in Theorem 11.4.1 of [9] has wrong sign.)

Since the sequence $\{\Delta_n(z)\}_n$ is nested, i.e., $\Delta_{n+1}(z) \subset \Delta_n(z)$ for all n, the intersection $\Delta_\infty(z) = \bigcap_{n=1}^\infty \Delta_n(z)$ is either a single point or a proper closed disk. The following invariance result follows from [9]. See also [7] for the equivalent unit circle problem and [32] for the cyclic situation.

THEOREM 8.2 *The intersection $\Delta_\infty(z)$ consists either of a single point for every $z \in \mathbb{U}$ (the limit point case) or is a proper closed disk for every $z \in \mathbb{U}$ (the limit circle case).*

The MP-fraction $K_{n=1}^\infty \frac{a_n(z)}{b_n(z)}$ is said to *converge completely* for z if $T_n(z, \tau(1 - \beta_{n-1}z)/(1 - \beta_n z))$ converges uniformly for $\tau \in \hat{\mathbb{R}}$ to a value independent of τ. (In the polynomial situation this definition coincides with the usual definition of complete convergence of real J-fractions, see e.g., [16].) Thus complete convergence occurs exactly in the limit point case, and the continued fraction converges completely for all or no z in $\mathbb{U}$.

9. Convergence and Uniqueness

Also in this section we shall always assume that the sequence $\{\varphi_n\}$ is regular.

We introduce the notation $\Sigma(z, \mathcal{L})$ and $\Sigma(z, \mathcal{L} \cdot \mathcal{L})$ for the set of Stieltjes transforms at z of all solutions of the moment problem on $\mathcal{L}$ and on $\mathcal{L} \cdot \mathcal{L}$ respectively:

$$\Sigma(z, \mathcal{L}) = \{w \in \mathbb{C} : w = S(z, \mu) \text{ for some solution } \mu \\ \text{of the moment problem on } \mathcal{L}\}$$

$$\Sigma(z, \mathcal{L} \cdot \mathcal{L}) = \{w \in \mathbb{C} : w = S(z, \mu) \text{ for some solution } \mu \\ \text{of the moment problem on } \mathcal{L} \cdot \mathcal{L}\}$$

Let n be fixed. By using quadrature formulas determined by the modified approximants $R_n(z, \tau)$ it can be shown (see [9] and [3, 7] for the circle and [32] for the cyclic situation on the real line) that except for

n values of the real parameter τ, there exists a discrete measure $\mu_n(\cdot, \tau)$ on $\mathbb{R}$ with the following properties: $\mu_n(\cdot, \tau)$ solves the truncated moment problem on $\mathcal{L}_{n-1} \cdot \mathcal{L}_{n-1}$, and $R_n(z, \tau) = \int_{\mathbb{R}} (z - t)^{-1} d\mu_n(t, \tau)$. A limiting argument involving convergent subsequences of sequences $\{\mu_n(\cdot, \tau_n)\}$ leads for every point w on the boundary circle $\partial\Delta_\infty(z)$ to a measure μ_w which solves the moment problem on $\mathcal{L}$ and such that $S(z, \mu_w) = w$. Thus $\partial\Delta_\infty(z) \subset \Sigma(z, \mathcal{L})$. (Note that although $\mu_n(\cdot, \tau)$ solves the truncated moment problem on $\mathcal{L}_{n-1} \cdot \mathcal{L}_{n-1}$, we can in general not conclude that the measures μ_w solve the moment problem on $\mathcal{L} \cdot \mathcal{L}$, only on $\mathcal{L}$.) The set $\Sigma(z, \mathcal{L})$ is easily seen to be convex, and hence $\Delta_\infty(z) \subset \Sigma(z, \mathcal{L})$.

A Hilbert space argument using Bessel's inequality analogous to the proof in the classical situation (see e.g., [1]) shows that $\Sigma(z, \mathcal{L} \cdot \mathcal{L}) \subset \Delta_\infty(z)$.

Thus the following result holds.

THEOREM 9.1 *For every $z \in \mathbb{U}$ we have*

$$\Sigma(z, \mathcal{L} \cdot \mathcal{L}) \subset \Delta_\infty(z) \subset \Sigma(z, \mathcal{L}).$$

We call a moment problem *determinate* if it has exactly one solution, *indeterminate* if it has more than one solution.

All the convergent subsequences of sequences $\{\mu_n(\cdot, \tau_n)\}$ have the same solution μ as limit if and only if the continued fraction is completely convergent.

The above considerations and the results of Section 8 lead to the following theorem.

THEOREM 9.2   A. *If the moment problem on $\mathcal{L}$ is determinate with solution μ, then the MP-fraction is completely convergent to $S(z, \mu)$ for all $z \in \mathbb{U}$.*

B. *If the MP-fraction is completely convergent for some $z \in \mathbb{U}$, then it is completely convergent for every $z \in \mathbb{U}$ with limit $S(z, \mu)$, where μ is a certain solution of the moment problem on $\mathcal{L}$.*

C. *If the moment problem on $\mathcal{L} \cdot \mathcal{L}$ is solvable and the MP-fraction is completely convergent for some $z \in \mathbb{U}$, then the moment problem on $\mathcal{L} \cdot \mathcal{L}$ is determinate and the MP-fraction converges completely for every $z \in \mathbb{U}$ to $S(z, \mu)$ where μ is the unique solution of the moment problem on $\mathcal{L} \cdot \mathcal{L}$.*

Acknowledgements. The work of the first author is partially supported by the Belgian Program on Interuniversity Poles of Attraction, initiated by the Belgian State, Prime Minister's Office for Science, Technology and Culture. The scientific responsibility rests with the authors. The work of the second author was supported by the Scientific Research Project of the Spanish D.G.E.S. under contract PB96-1029.

References

1. N. Akhiezer, *The classical moment problem and some related questions in analysis*, Hafner, New York, 1965.
2. A. Bultheel and P. Dewilde: 1979, *Orthogonal functions related to the Nevanlinna-Pick problem*, in: "*Proc. 4th Int. Conf. on Math. Theory of Networks and Systems at Delft*", P. Dewilde, ed., North-Hollywood, pp. 207–212.
3. A. Bultheel, P. González-Vera, E. Hendriksen, and O. Njåstad, *Orthogonal rational functions with poles on the unit circle*. J. Math. Anal. Appl. **182** (1994), 221–243.
4. A. Bultheel, P. González-Vera, E. Hendriksen, and O. Njåstad, *Orthogonality and boundary interpolation*. in: "*Nonlinear Numerical Methods and Rational Approximation II*, A. Cuyt, ed., 1994, pp. 37–48.
5. A. Bultheel, P. González-Vera, E. Hendriksen, and O. Njåstad, *Recurrence relations for orthogonal functions*, in: "*Continued Fractions and Orthogonal Functions*", S. Cooper and W. Thron, eds., Vol. 154 of Lecture Notes in Pure and Appl. Math., 1994, pp. 24–46.
6. A. Bultheel, P. González-Vera, E. Hendriksen, and O. Njåstad, *A Favard theorem for rational functions with poles on the unit circle*, East J. Approx. **3** (1997), 21–37.
7. A. Bultheel, P. González-Vera, E. Hendriksen, and O. Njåstad, *A rational moment problem on the unit circle*, Methods Appl. Anal. **4** (1997), 283–310.
8. A. Bultheel, P. González-Vera, E. Hendriksen, and O. Njåstad, *Interpolation of Nevanlinna functions by rationals with poles on the real line*, in: "*Orthogonal Functions, Moment Theory and Continued Fractions: Theory and Applications*", W. Jones and A. Ranga, eds., Vol. 199 of Lecture Notes in Pure and Applied Mathematics, 1998, pp. 101–110.
9. A. Bultheel, P. González-Vera, E. Hendriksen, and O. Njåstad, *Orthogonal rational functions*, Vol. 5 of Cambridge Monographs on Applied and Computational Mathematics, Cambridge University Press, 1999.
10. A. Bultheel, P. González-Vera, E. Hendriksen, and O. Njåstad, *A rational Stieltjes problem*, Appl. Math. Comput. (2001).
11. P. Dewilde and H. Dym, *Schur recursions, error formulas, and convergence of rational estimators for stationary stochastic sequences*, IEEE Trans. Inf. Th. **IT-27** (1981), 446–461.
12. M. Djrbashian, *A survey on the theory of orthogonal systems and some open problems*, in: textsl"Orthogonal polynomials: Theory and practice", P. Nevai, ed., Vol. 294 of *Series C: Mathematical and Physical Sciences*, Boston, 1990, pp. 135–146.
13. G. Freud, *Orthogonal polynomials*, Pergamon Press, Oxford, 1971.
14. A. Gonchar and G. López, *On Markov's theorem for multipoint Padé approximants for functions of Stieltjes type*, Math. USSR-Sb. **105** (1978), 512–524. English translation, Math. USSR-Sb **34** (1978), 449–459.
15. D. Gupta and D. Masson, *Continued fractions and orthogonality*, Trans. Amer. Math. Soc. **350** (1998), 769–808.
16. H. Hamburger, *Ueber eine Erweiterung des Stieltjesschen Moment Problems I*, Math. Ann. **81** (1920), 235–319.
17. E. Hendriksen and O. Njåstad, *A Favard theorem for rational functions*, J. Math. Anal. Appl. **142** (1989), 508–520.
18. E. Hendriksen and O. Njåstad, *Positive multipoint Padé continued fractions*, Proc. Edinburgh Math. Soc. **32** (1989), 261–269.
19. E. Hendriksen and H. van Rossum, *Orthogonal Laurent polynomials*, Proc. of the Kon. Nederl. Akad. Wetensch, Proceedings A **89** (1986), 17–36.
20. M. Ismail and D. Masson, *Generalized orthogonality and continued fractions*, J. Approx. Theory **83** (1995), 1–40.
21. W. Jones and O. Njåstad, *Orthogonal Laurent polynomials and strong moment theory*, J. Comput. Appl. Math. **105** (1999), 51–91.
22. W. Jones, O. Njåstad and W. Thron, *Continued fractions and strong Hamburger*

moment problems, Proc. London Math. Soc., Ser III **47** (1983), 363–384.

23. W. Jones, O. Njåstad, and W. Thron, *Orthogonal Laurent polynomials and the strong Hamburger moment problem*, J. Math. Anal. Appl. **98** (1984), 528–554.

24. W. Jones and W. Thron, *Continued Fractions. Analytic Theory and Applications*, Addison-Wesley, Reading, Mass., 1980.

25. W. Jones and W. Thron, *Orthogonal Laurent polynomials and Gaussian quadrature*, in: "Quantum mechanics in mathematics, chemistry and physics", K. Gustafson and W. Reinhardt, eds., New York, 1984, pp. 449–455.

26. X. Li, 1999, *Regularity of orthogonal rational functions with poles on the unit circle*, J. Comput. Appl. Math. **105** (1999), 371–383.

27. G. L. López-Lagomasino, *Conditions for convergence of multipoint Padé approximants for functions of Stieltjes type*, Math. USSR-Sb. **35** (1979), 363–376.

28. G. L. López-Lagomasino, *On the asymptotics of the ratio of orthogonal polynomials and convergence of multipoint Padé approximants*, Math. USSR-Sb. **56** (1985), 207–219.

29. G. L. López-Lagomasino, *Asymptotics of polynomials orthogonal with respect to varying measures*, Constr. Approx. **5** (1989), 199–219.

30. L. Lorentzen and H. Waadeland, *Continued fractions with applications*, Vol. 3 of Studies in Computational Mathematics, North-Holland, 1992.

31. O. Njåstad, *An extended Hamburger moment problem*, Proc. Edinburgh Math. Soc. **28** (1985), 167–183.

32. O. Njåstad, 1987, *Unique solvability of an extended Hamburger moment problem*, J. Math. Anal. Appl. **124** (1987), 502–519.

33. O. Njåstad, *Solutions of the strong Hamburger moment problem*, J. Math. Anal. Appl. **197** (1996), 227–248.

34. O. Njåstad and W. Thron, *Unique solvability of the strong Hamburger moment problem*, J. Austral. Math. Soc. (Series A) **40** (1986), 5–19.

35. O. Perron, *Die Lehre von den Kettenbrüchen*, Teubner, 1977.

36. M. Rahman and S. Suslov, *Classical biothogonal rational functions*, in: "Methods of Approximation Theory in Complex Analysis and Mathematical Physics", Vol. 1550 of Lecture Notes in Math. (1993), pp. 131–146.

37. J. Shohat and J. Tamarkin, *The problem of moments*, Vol. 1 of Math. Surveys, Amer. Math. Soc., Providence, R.I., 1943.

38. V. Spiridonov and A. Zhedanov, *Spectral transformation chains and some new biorthogonal rational functions*, Comm. Math. Phys. **210** (2000), 49–83.

39. H. Stahl and V. Totik, *General orthogonal polynomials*, Encyclopedia of Mathematics and its Applications, Cambridge University Press, 1992.

40. T. Stieltjes, *Recherches sur les fractions continues*, Ann. Fac. Sci. Toulouse **8** (1894), J.1–122, 9:A.1–47. English transl.: Oeuvres Complèts, Collected Papers, Vol. 2, Springer-Verlag, 1993, pp. 609–745.

41. G. Szegő, 1975, *Orthogonal polynomials*, Vol. 33 of Amer. Math. Soc. Colloq. Publ., 4th edition, Amer. Math. Soc., Providence, Rhode Island, 1975. First edition, 1939.

42. A. Zhedanov, *Biorthogonal rational functions and generalized eigenvalue problem*, J. Approx. Theory **101** (1999), 303–329.

ORTHOGONAL POLYNOMIALS AND REFLECTION GROUPS

CHARLES F. DUNKL
Department of Mathematics
University of Virginia
Charlottesville, VA 22904-4137 USA

1. Orthogonal Polynomials of Classical Type of Several Variables

Classical orthogonal polynomials of one variable are studied by means of differential equations, standard definite integrals like the gamma and beta functions, and other tools such as Rodrigues' relations. In this survey article we show how these ideas can be extended to several variables. We start with some simple examples which serve to motivate the use of reflection groups as symmetry groups of the weight functions. It is necessary to generalize differential operators to differential-difference operators (parametrized commutative algebras associated to reflection groups). Proofs are omitted in this survey, but references are given to the research literature as well as to the monograph by Yuan Xu and the author [11]. Orthogonal polynomials of several variables are studied for their applicability in topics such as harmonic analysis on homogeneous spaces, solutions of partial differential equations, quantum-mechanical wave functions, eigenvalues of random symmetric matrices, algebraic combinatorics, approximation and numerical cubature.

1.1. NOTATION AND EXAMPLES

We list some notations for indexing and variables:

$\mathbb{N} = \{1, 2, 3, \dots\}$, the set of natural numbers; $\mathbb{N}_0 = \{0, 1, 2, 3, \dots\}$;

the Pochhammer symbol (shifted factorial) is $(a)_n = \prod_{j=1}^{n} (a + j - 1)$ (and $(a)_0 = 1$);

$\alpha = (\alpha_1, \alpha_2, \dots, \alpha_d) \in \mathbb{N}_0^d$; α is called a multi-index or composition;

$\alpha! = \alpha_1! \dots \alpha_d!, |\alpha| = \sum_{i=1}^{d} \alpha_i$; the cardinality of a set E is denoted by $\#E$.

J. Bustoz et al. (eds.), Special Functions 2000, 111–128.

For $x = (x_1, x_2, \ldots, x_d) \in \mathbb{R}^d$, $||x|| = \left(\sum_{i=1}^{d} x_i^2 \right)^{1/2}$, and the unit ball is $B^d = \left\{ x \in \mathbb{R}^d : ||x|| < 1 \right\}$. For $\alpha \in \mathbb{N}_0^d$ the monomial $x^\alpha = \prod_{i=1}^{d} x_i^{\alpha_i}$ has degree $|\alpha|$; the space of polynomials of degree $\leq n$ is $\Pi_n^d = \text{span}\{x^\alpha : \alpha \in \mathbb{N}_0^d, |\alpha| \leq n\}$, (the field of coefficients could be $\mathbb{R}$ or $\mathbb{Q}(\kappa)$, formal parameters κ). The space of all polynomials is denoted by $\Pi^d = \cup_{n=0}^{\infty} \Pi_n^d$. The space of homogeneous polynomials of degree n is $\mathcal{P}_n^d = \text{span}\{x^\alpha : \alpha \in \mathbb{N}_0^d, |\alpha| = n\}$. Also $\partial_i = \frac{\partial}{\partial x_i}$ for $1 \leq i \leq d$. The *beta function* is $B(u, v) = \Gamma(u) \Gamma(v) / \Gamma(u + v)$. One of the useful orderings on compositions is the dominance order (see Definition 1.7 for a modification of it).

Definition 1. For $\alpha, \beta \in \mathbb{N}_0^d$ say α dominates β (denoted $\alpha \succeq \beta$) when $\sum_{i=1}^{j} \alpha_i \geq \sum_{i=1}^{j} \beta_i$ for $1 \leq j \leq d$. Write $\alpha \succ \beta$ to mean $\alpha \succeq \beta$ and $\alpha \neq \beta$.

Let μ be a finite positive Baire measure on $\mathbb{R}^d$ such that $\int_{\mathbb{R}^d} |x^\alpha| \, d\mu(x) < \infty$ for all $\alpha \in \mathbb{N}_0^d$, so that polynomials are in $L^2(\mu)$, (and we use $\langle f, g \rangle = \int_{\mathbb{R}^d} fg \, d\mu$ for the inner product).

Definition 2. For $n \in \mathbb{N}_0$, the space of orthogonal polynomials of degree n is $\mathcal{V}_n^d = \{p \in \Pi_n^d : \langle p, x^\alpha \rangle = 0 \text{ whenever } |\alpha| < n\}$ (the orthogonal complement of Π_{n-1}^d in Π_n^d).

We will use the term *classical type* for systems of orthogonal polynomials where each $\mathcal{V}_n^d$ is the space of eigenfunctions of a differential (or differential-difference) operator. In addition it is desirable to construct an orthogonal basis for $\mathcal{V}_n^d$ as simultaneous eigenfunctions of commuting self-adjoint operators.

Usually, μ is supported on a "solid" subset (= closure of its interior) or on the sphere $S^{d-1} = \{x : ||x|| = 1\}$; and the dimensions of $\mathcal{V}_n^d$ are $\#\left\{ \alpha \in \mathbb{N}_0^d : |\alpha| = n \right\} = \binom{n+d-1}{n}$ or $\#\left\{ \alpha \in \mathbb{N}_0^d : |\alpha| = n, \alpha_d = 0 \text{ or } 1 \right\} = \binom{n+d-2}{n} \frac{2n+d-2}{n+d-2}$ respectively.

1.1.1. *The classical orthogonal polynomials of one variable*
1. Hermite (weight is $e^{-t^2} dt$ on $\mathbb{R}$):

$$H_n(t) = \sum_{j \leq n/2} \frac{n!}{(n-2j)! \, j!} (-1)^j (2t)^{n-2j};$$

2. Laguerre (weight is $t^\kappa e^{-t} dt$ on $\mathbb{R}_+$):

$$L_n^\kappa(t) = \frac{(\kappa+1)_n}{n!} \sum_{j=0}^{n} \frac{(-n)_j}{(\kappa+1)_j} \frac{t^j}{j!};$$

3. Jacobi (weight $(1-t)^\alpha (1+t)^\beta dt$, on $-1 \leq t \leq 1$) :

$$P_n^{(\alpha,\beta)}(t) = \frac{(\alpha+1)_n}{n!} {}_2F_1 \left(\begin{array}{c} -n, n+\alpha+\beta+1 \\ \alpha+1 \end{array} ; \frac{1-t}{2} \right);$$

4. Gegenbauer (ultraspherical) (weight $(1-t^2)^{\lambda-\frac{1}{2}} dt$, on $-1 \leq t \leq 1$) :

$$C_n^\lambda(t) = \frac{(\lambda)_n 2^n}{n!} t^n {}_2F_1 \left(\begin{array}{c} -\frac{n}{2}, \frac{1-n}{2} \\ 1-n-\lambda \end{array} ; \frac{1}{t^2} \right).$$

The parameters are restricted by $\kappa, \alpha, \beta > -1, \lambda > -\frac{1}{2}$ and $n \in \mathbb{N}_0$.

1.2. POLYNOMIALS ON THE BALL

For any $\lambda > -\frac{1}{2}$ consider the measure $(1-||x||^2)^{\lambda-\frac{1}{2}} dx$ on the ball B^d (§2.3.2 [11]); then each $p \in \mathcal{V}_n^d$ satisfies

$$\Delta p - \sum_{j=1}^d \partial_j x_j ((2\lambda-1) + \sum_{i=1}^d x_i \partial_i) p = -(n+d)(n+2\lambda-1)p,$$

or equivalently,

$$\Delta p - \left(\sum_{i=1}^d x_i \partial_i \right)^2 p - (2\lambda+d-1) \left(\sum_{i=1}^d x_i \partial_i \right) p = -n(n+d+2\lambda-1)p,$$

where the Laplacian $\Delta = \sum_i \partial_i^2$.

1.2.1. *The disc*

There are two bases of orthogonal polynomials for the disc (the ball with $d = 2$); the first is related to polar or complex coordinates, and the second to Cartesian coordinates:

For complex/polar coordinates: let $z = x_1 + ix_2 = re^{i\theta}$ ($i = \sqrt{-1}, 0 \leq r \leq 1, -\pi < \theta \leq \pi$), and let $m \geq n$; then

$$\begin{aligned} p_{mn}(z, \bar{z}) &= z^{m-n} P_n^{(\lambda-1/2, m-n)}(2z\bar{z}-1) \\ &= e^{i(m-n)\theta} r^{m-n} P_n^{(\lambda-1/2, m-n)}(2r^2-1), \\ p_{nm}(z, \bar{z}) &= p_{mn}(\bar{z}, z); \end{aligned}$$

are orthogonal polynomials (called the *disc polynomials* (§2.4.3 [11])) of degree $m+n$.

For the Cartesian coordinates the orthogonal polynomials of degree n are given by (§2.4.2 [11]): for $0 \leq k \leq n$,

$$p_{n,k}(x_1, x_2) = C_{n-k}^{k+\lambda+\frac{1}{2}}(x_1)(1 - x_1^2)^{\frac{k}{2}} C_k^\lambda \left(\frac{x_2}{\sqrt{1 - x_1^2}} \right).$$

1.2.2. *Biorthogonal polynomials*

We define (§2.3 [11]) biorthogonal polynomials for the ball, $\{U_\alpha(x),$ $V_\alpha(x) : \alpha \in \mathbb{N}_0^d\}$; by means of generating functions

$$\sum_{\alpha \in \mathbb{N}_0^d} t^\alpha V_\alpha(x) = (1 - 2\langle t, x \rangle + \|t\|^2)^{-\lambda - (d-1)/2}$$

and

$$\sum_{\alpha \in \mathbb{N}_0^d} t^\alpha U_\alpha(x) = \left((1 - \langle t, x \rangle)^2 + \|t\|^2 \left(1 - \|x\|^2 \right) \right)^{-\lambda}$$

where $t \in \mathbb{R}^d$ and the standard inner product $\langle t, x \rangle = \sum_i t_i x_i$; then

$$w_\lambda \int_{B^d} U_\alpha(x) V_\beta(x) \left(1 - \|x\|^2 \right)^{\lambda - \frac{1}{2}} dx = \delta_{\alpha\beta} \frac{(2\lambda)_{|\alpha|}}{\alpha!} \frac{\lambda + \frac{d-1}{2}}{|\alpha| + \lambda + \frac{d-1}{2}},$$

where the normalizing constant is $w_\lambda = \pi^{-d/2} \Gamma(\lambda + \frac{d+1}{2}) / \Gamma\left(\lambda + \frac{1}{2}\right)$.

1.3. OTHER EXAMPLES

Example 3. For $\alpha \in \mathbb{N}_0^d$ the multi-Hermite polynomial is $H_\alpha(x) = \prod_{i=1}^d \times H_{\alpha_i}(x_i)$, orthogonal for $e^{-\|x\|^2} dx$ on $\mathbb{R}^d$; and is a solution of $(\Delta - 2\sum_{i=1}^d \times x_i \partial_i) H_\alpha = -2|\alpha| H_\alpha$.

Example 4. For $\alpha \in \mathbb{N}_0^d$ the multi-Laguerre polynomial is $L_\alpha^{(\kappa)}(x) = \prod_{i=1}^d L_{\alpha_i}^{\kappa_i}(x_i)$, weight $\prod_i x_i^{\kappa_i} \exp(-\sum_i x_i) dx$ on $\mathbb{R}_+^d = \left\{ x \in \mathbb{R}^d : x_i > 0 \right\}$, $\kappa_i > -1$; note that under the transformation $x_i = y_i^2, 1 \leq i \leq d$, the Laguerre polynomials can be considered as the polynomials even in each y_i, orthogonal for the weight $\prod_i |y_i|^{2\kappa_i + 1} e^{-\|y\|^2} dy$ on $\mathbb{R}^d$. These polynomials are invariant under the group of sign-changes (called $\mathbb{Z}_2^d$).

This group is an example of a reflection group. Another such is the symmetric group S_d acting on $\mathbb{R}^d$ by permuting the coordinates. The natural weight function associated with S_d is $\prod_{1 \leq i < j \leq d} |x_i - x_j|^{2\kappa}$, some parameter $\kappa > 0$. There are two classical one-variable results in this flavor (§1.3.1 [11]). Let μ be a positive measure on an interval $[a, b] \subset \mathbb{R}$ (possibly infinite) with

$\int_a^b |x|^n d\mu(x) < \infty$ for all $n \geq 0$. The moments are $c_n = \int_a^b x^n d\mu(x)$; put $d_n = \det(c_{i+j-2})_{i,j=1}^n$ and denote the orthonormal polynomials in $L^2(\mu)$ by p_n. Then for $n \geq 1$

$$\int_{[a,b]^n} \prod_{1 \leq i < j \leq n} (x_i - x_j)^2 d\mu(x_1) \dots d\mu(x_n) = n! d_n,$$

$$\int_{[a,b]^n} \prod_{i=1}^n (x - x_i) \prod_{1 \leq i < j \leq n} (x_i - x_j)^2 d\mu(x_1) \dots d\mu(x_n) = n! (d_n d_{n+1})^{1/2} p_n(x).$$

1.4. REFLECTION GROUPS AND ROOT SYSTEMS

We proceed to the general framework for the systems that were described. A subset R of $\mathbb{R}^d \setminus \{0\}$ is called a (reduced) *root system* if $u, v \in R$ implies (i) $u\sigma_v \in R$ and (ii) $\mathbb{R}u \cap R = \{\pm u\}$; where σ_v is the reflection along v (fixing the hyperplane $v^\perp$), defined by

$$x\sigma_v = x - 2(\langle x, v \rangle / \|v\|^2)v.$$

This is an orthogonal transformation and $\sigma_v^2 = 1$. The root system can be written as $R_+ \cup (-R_+)$; fix some vector $z \in \mathbb{R}^d$ such that $\langle z, v \rangle \neq 0$ for all $v \in R$ then put $R_+ = \{v \in R : \langle z, v \rangle > 0\}$. The group $W(R)$ generated by $\{\sigma_v : v \in R_+\}$ is a finite subgroup of $O(d)$, the orthogonal group on $\mathbb{R}^d$. Thus $\langle xw, yw \rangle = \langle x, y \rangle$ for $w \in W(R)$ and $x, y \in \mathbb{R}^d$. We will be concerned with certain weight functions invariant under the action of $W(R)$:

Definition 5. A *multiplicity function* κ is a function on R with the property that $\kappa_u = \kappa_v$ whenever $u, v \in R$ and there exists $w \in W(R)$ such that $v = uw$ (equivalently, σ_u, σ_v are in the same conjugacy class).

Given κ, let

$$h(x) = \prod_{v \in R_+} |\langle x, v \rangle|^{\kappa_v}.$$

This is a positively homogeneous (degree $\gamma = \sum_{v \in R_+} \kappa_v$) and $W(R)$-invariant function. The alternating polynomial is

$$a(x) = \prod_{v \in R_+} \langle x, v \rangle,$$

and is the minimal polynomial with the property $a(x\sigma_v) = -a(x)$ for each $v \in R$. Three examples of root systems: let ε_i denote the ith unit basis vector $(0, \dots, 1, 0, \dots) \in \mathbb{R}^d$;

1. the group of sign-changes ($\mathbb{Z}_2^d$ or A_1^d), $R_+ = \{\varepsilon_i : 1 \leq i \leq d\}$;

2. the symmetric group $(S_d$ or $A_{d-1})$, $R_+ = \{\varepsilon_i - \varepsilon_j : 1 \le i < j \le d\}$;
3. the hyperoctahedral group $(W_d$ or $B_d)$, $R_+ = \{\varepsilon_i \pm \varepsilon_j : 1 \le i < j \le d\} \cup \{\varepsilon_i\}$.

Extend the class of differential operators to include difference (reflection) operators: given a root system R and a multiplicity function κ, define the first-order operator $(1 \le i \le d, p \in \Pi^d)$

$$\mathcal{D}_i p(x) = \frac{\partial}{\partial x_i} p(x) + \sum_{v \in R_+} \kappa_v \frac{p(x) - p(x\sigma_v)}{\langle x, v \rangle} v_i;$$

then $\mathcal{D}_i$ maps $\mathcal{P}_n^d$ to $\mathcal{P}_{n-1}^d$ for each $n \ge 1$. The important aspect of these operators is that they commute $(\mathcal{D}_i\mathcal{D}_j = \mathcal{D}_j\mathcal{D}_i)$ (Dunkl [5], or see Ch.4 [11]). There are several connections to inner product structures. First there are analogues of spherical harmonics (called h-harmonics); let $\Delta_h = \sum_{i=1}^d \mathcal{D}_i^2$; also let $d\omega$ denote the normalized rotation-invariant surface measure on the unit sphere S^{d-1}. The orthogonal polynomials of degree n $(\mathcal{V}_n^d)$ for $h^2 d\omega$ are exactly the homogenous h-harmonic polynomials $\mathcal{H}_n^d = \{p \in \mathcal{P}_n^d : \Delta_h p = 0\}$ (Dunkl [4]). When the group is $\mathbb{Z}_2^d$, the multiplicity function is a d-tuple (κ_i) and the weight function is $h(x) = \prod_i |x_i|^{\kappa_i}$; the invariant polynomials are those even in each x_i and can be considered as polynomials on the simplex $T^{d-1} = \{y : \sum_{i=1}^d y_i = 1, \text{ each } y_i \ge 0\}$ for $y_i = x_i^2$. The measure $h^2 d\omega$ maps to $\prod_{i=1}^d y_i^{\kappa_i - \frac{1}{2}} dy_1 \ldots dy_{d-1}$ (a Dirichlet measure).

Adjoin an extra dimension, that is, form $R' = R \cup \{\varepsilon_{d+1}\}$ so that $W(R') = W(R) \times \mathbb{Z}_2$ (direct product). Then $W(R)$ acts on the first d coordinates, and we can write $h'(x) = h(x_1, \ldots, x_d)|x_{d+1}|^\lambda$; then the polynomials in $\mathcal{H}_n^{d+1}$ and even in x_{d+1} correspond to orthogonal polynomials on the ball $B^d = \{(x_1, \ldots, x_d) : \sum_{i=1}^d x_i^2 \le 1\}$ with measure $h(x)^2 \left(1 - \sum_{i=1}^d x_i^2\right)^{\lambda - \frac{1}{2}} dx_1 \ldots dx_d$. This allows a transfer of the equation $\Delta_{h'} p = 0$ to the new variables (and characterizing $\mathcal{V}_n^d$ for the respective measures). First we state the explicit formula:

$$\Delta_h p(x) = \Delta p(x) + \sum_{v \in R_+} \kappa_v \left(\frac{2 \langle v, \nabla p(x) \rangle}{\langle v, x \rangle} - ||v||^2 \frac{p(x) - p(x\sigma_v)}{\langle v, x \rangle^2} \right);$$

where $\nabla p(x) = (\partial_i p)_{i=1}^d$, the gradient. Transferring this to the B^d-situation (§6.1.1 [11]) we obtain the equation

$$\left(\Delta_h - \langle x, \nabla \rangle^2 - (2\gamma + 2\lambda + d - 1)\langle x, \nabla \rangle\right) p = -n(n + 2\gamma + 2\lambda + d - 1) p,$$

satisfied by the elements of $\mathcal{V}_n^d$ (recall $\gamma = \sum_v \kappa_v$, the degree of h).

Observe that Δ_h is a differential operator when restricted to invariant polynomials ($p(xw) = p(x)$ for all $w \in W(R)$). It is desirable to find a commuting set of self-adjoint operators on $\mathcal{H}_n$ in order to determine an orthogonal basis. One building block: for $p \in \mathcal{H}_n$, the adjoint of $\mathcal{D}_i$ is

$$\mathcal{D}_i^* p(x) = (d + 2n + \gamma)(x_i p(x) - (d + 2n + 2\gamma - 2)^{-1} ||x||^2 \mathcal{D}_i p(x)).$$

Clearly the operators $\mathcal{D}_i^* \mathcal{D}_i$ and $(\mathcal{D}_i^* \mathcal{D}_j - \mathcal{D}_j^* \mathcal{D}_i)^2$ are self-adjoint, but the commutation relations are generally intractable (except for the abelian case $\mathbb{Z}_2^d$).

Another fundamental inner product structure (the "pairing" norm) is defined as follows: for $p, q \in \Pi^d$ define the linear operator $p(\mathcal{D}) = p(\mathcal{D}_1, \ldots, \mathcal{D}_d)$ (there is no ambiguity because of the commutativity) then let

$$\langle p, q \rangle_h = p(\mathcal{D})q(x)|_{x=0};$$

with the properties: $\langle p, q \rangle_h = \langle q, p \rangle_h$, $\langle wp, wq \rangle_h = \langle p, q \rangle_h$ (for $w \in W(R)$, $wp(x) = p(xw)$), $\langle p, q \rangle_h = 0$ if p, q are homogeneous of different degrees, and $\mathcal{D}_i x_i$ is self-adjoint. The inner product is positive-definite when $\kappa_v \geq 0$. There is a key relation to a Gaussian type measure (Dunkl [6]):

$$\langle p, q \rangle_h = c_\kappa \int_{\mathbb{R}^d} \left(e^{-\Delta_h/2} p\right) \left(e^{-\Delta_h/2} q\right) h^2 e^{-||x||^2/2} dx$$

(the normalizing constant c_κ, explicitly known by the Macdonald-Mehta-Selberg integral makes $\langle 1, 1 \rangle_h = 1$). Note that this especially useful when p, q are h-harmonic. Later we discuss the application of this formula to produce Hermite-type polynomials. The self-adjoint operators can be transformed:

$$e^{-\Delta_h/2} \mathcal{D}_i x_i e^{\Delta_h/2} = \mathcal{D}_i x_i - \mathcal{D}_i^2$$

(self-adjoint for $L^2(\mathbb{R}^d, h(x)^2 e^{-||x||^2/2} dx)$).

1.5. HARMONICS FOR DIRECT PRODUCTS

Suppose that W_1, W_2 act on $(x_1, \ldots, x_m)$ and $(x_{m+1}, \ldots, x_d)$ respectively; that their respective sets of positive roots are $R_{1,+}$ and $R_{2,+}$; then let $t_1 = \sum_{i=1}^m x_i^2$, $t_2 = \sum_{i=m+1}^d x_i^2$ and $\lambda_1 = \sum_{v \in R_{1,+}} k_v$, $\lambda_2 = \sum_{v \in R_{2,+}} k_v$.

Proposition 6. *Suppose $p_1(x_1, \ldots, x_m)$ and $p_2(x_{m+1}, \ldots, x_d)$ are harmonic, that is, $\sum_{i=1}^m \mathcal{D}_i^2 p_1 = 0$ and $\sum_{i=m+1}^d \mathcal{D}_i^2 p_2 = 0$. Then $\Delta_h(f(t_1, t_2) \times p_1 p_2) = 0$ if and only if f satisfies the differential equation $(t_1(\frac{\partial}{\partial t_1})^2 + t_2(\frac{\partial}{\partial t_2})^2 + (\lambda_1 + \deg(p_1) + \frac{m}{2})\frac{\partial}{\partial t_1} + (\lambda_2 + \deg(p_2) + \frac{d-m}{2})\frac{\partial}{\partial t_2})f(t_1, t_2) = 0$.*

The solutions for f are $(t_1 + t_2)^n P_n^{(\alpha,\beta)} \left(\frac{t_1 - t_2}{t_1 + t_2}\right)$ for $n = 0, 1, 2, 3, \ldots$, where $\alpha = \frac{d-m}{2} + \lambda_2 + \deg(p_2) - 1$ and $\beta := \frac{m}{2} + \lambda_1 + \deg(p_1) - 1$ and $P_n^{(\alpha,\beta)}$ denotes the Jacobi polynomial (Dunkl [9]). For ordinary spherical harmonics, this method was used by Braaksma and Meulenbeld [3].

1.6. HARMONICS IN THE ABELIAN CASE

This subsection concerns Dirichlet measure on the simplex and a related measure on the ball. Note that $\mathcal{D}_i p(x) = \partial_i p(x) + \kappa_i (p(x) - p(\ldots, -x_i, \ldots)) / x_i$ (for $1 \le i \le d$; so there are d arbitrary parameters κ_i). The direct product method gives a construction for an orthogonal basis for $\mathcal{H}_n^d$. It can be shown that $\dim \mathcal{H}_n^d = \dim \mathcal{P}_n^d - \dim \mathcal{P}_{n-2}^d = \#\{\alpha \in \mathbb{N}_0^d : |\alpha| = n, \alpha_d = 0 \text{ or } 1\}$. For each such α there is a harmonic homogeneous polynomial $p(\alpha; x)$, of degree $|\alpha|$, constructed inductively as follows: let $\alpha_i = 2a_i + \epsilon_i$, for each i, with $a_i \in \mathbb{N}_0$ and $\epsilon_i = 0$ or 1 (and $a_d = 0$); $p((0, \ldots, \epsilon_d); x) = x_d^{\epsilon_d}$, and when $\alpha_1 = \alpha_2 = \ldots = \alpha_{d-m-2} = 0$ for some m, let

$$\alpha' = (0, \ldots, 0, \alpha_{d-m}, \ldots, \alpha_d),$$

$$\gamma = \frac{m+1}{2} + \sum_{i=d-m}^{d} (\kappa_i + a_i) - 1,$$

$$\delta = \kappa_{d-m-1} + \epsilon_{d-m-1} - \frac{1}{2},$$

$$t_1 = x_{d-m-1}^2, \quad t_2 = \sum_{i=d-m}^{d} x_i^2;$$

then

$$p(\alpha; x) = (t_1 + t_2)^{a_{d-m-1}} P_{a_{d-m-1}}^{(\gamma,\delta)} \left(\frac{t_1 - t_2}{t_1 + t_2}\right) (x_{d-m-1})^{\epsilon_{d-m-1}} p(\alpha'; x).$$

The induction runs over $m = 0, 1, \ldots, d - 2$. To get polynomials on the simplex (in the variables $y_i = x_i^2$), indexed by $\beta \in \mathbb{N}_0^{d-1}$, put $\alpha_i = 2\beta_i, \alpha_d = 0$ and apply the construction.

This is the construction of commuting self-adjoint operators for $\mathcal{H}_n^d$ associated with the group $\mathbb{Z}_2^d$: using a rescaled version of $\mathcal{D}_i^*$ defined above let

$$\theta = d - 4 + 2 \sum_{v \in R_+} \kappa_v + 2 \sum_{i=1}^{d} x_i \partial_i,$$

$$S_i = \theta x_i \mathcal{D}_i - \|x\|^2 \mathcal{D}_i^2, \quad (1 \le i \le d)$$

$$\mathcal{R}_{ij} = (x_i \mathcal{D}_j - x_j \mathcal{D}_i)^2, \quad (i \ne j),$$

$$\mathcal{U}_i = \mathcal{S}_i + \sum_{1 \le j < i} \mathcal{R}_{ij}, \ (1 \le i \le d).$$

Then θ is constant on each $\mathcal{P}_n^d$. Because $\mathcal{S}_i$ and $\mathcal{R}_{ij}$ are (scalar on each $\mathcal{P}_n^d$) multiples of $\mathcal{D}_i^* \mathcal{D}_i$ and $(\mathcal{D}_i^* \mathcal{D}_j - \mathcal{D}_j^* \mathcal{D}_i)^2$ they are self-adjoint (on $L^2(S^{d-1}, h^2 d\omega)$). It can be shown that the operators $\mathcal{U}_i$ commute pairwise, are triangular with respect to the dominance order on monomials, and have the $p(\alpha; x)$ as eigenfunctions (Dunkl [9]). The purely differential operators (that is, applied to polynomials even in each x_i) were first studied by Kalnins, Miller, and Tratnik [13].

1.7. NONSYMMETRIC JACK POLYNOMIALS AND THE GROUP S_d

The group S_d acts on $\mathbb{R}^d$ by permuting the coordinates. The multiplicity function is constant (one class of reflections, only) with value κ. For $i \ne j$ the transposition (reflection with root $\varepsilon_i - \varepsilon_j$) is denoted by (i, j), so that $x(i, j) = (\ldots, \overset{i}{x_j}, \ldots, \overset{j}{x_i}, \ldots)$ and $(i, j)p(x) = p(x(i, j))$. The differential-difference operators are:

$$\mathcal{D}_i = \partial_i + \kappa \sum_{j \ne i} \frac{1 - (i, j)}{x_i - x_j}$$

for $1 \le i \le d$, and

$$\Delta_h = \Delta + 2\kappa \sum_{i < j} \left\{ \frac{\partial_i - \partial_j}{x_i - x_j} - \frac{1 - (i, j)}{(x_i - x_j)^2} \right\}.$$

The weight function is $h(x)^2 = \prod_{i<j} |x_i - x_j|^{2\kappa}$. (The commutator of two linear operators is $[A, B] = AB - BA$.) From the fact $[\mathcal{D}_i x_i, \mathcal{D}_j x_j] = \kappa (\mathcal{D}_i x_i - \mathcal{D}_j x_j)(i, j)$ it can be shown that the operators

$$\mathcal{U}_i = \mathcal{D}_i x_i + \kappa - \kappa \sum_{j=1}^{i-1} (j, i)$$

commute pairwise and are self-adjoint (in fact for several inner products) for the pairing $\langle \cdot, \cdot \rangle_h$. They have a triangular property for a modification of the dominance order.

A composition $\lambda \in \mathbb{N}_0^d$ with $\lambda_1 \ge \lambda_2 \ge \ldots \ge \lambda_d$ is called a *partition* (with $\le d$ parts). The action of S_d on compositions $\alpha \to w\alpha$ is chosen so that $w(x^\alpha) = (xw)^\alpha = x^{w\alpha}$; thus $(xw)_i = x_{w(i)}$ for $1 \le i \le d$; and $(w\alpha)_i = \alpha_{w^{-1}(i)}$ for $\alpha \in \mathbb{N}_0^d$. For any $\alpha \in \mathbb{N}_0^d$ there is a unique partition α^+ so that $\alpha^+ = w\alpha$ for some $w \in S^d$.

Definition 7. The partial ordering $\rhd$ is given by: $\alpha \rhd \beta$ means that $|\alpha| = |\beta|$ and either $\alpha^+ \succ \beta^+$ or $\alpha^+ = \beta^+$ and $\alpha \succ \beta$.

For $\alpha \in \mathbb{N}_0^d$ and $1 \leq i \leq d$ let

$$\xi_i(\alpha) = \kappa\,(d - \#\{j : \alpha_j > \alpha_i\} - \#\{j : j < i; \alpha_j = \alpha_i\}) + \alpha_i + 1.$$

For a partition λ, the formula specializes to $\xi_i(\lambda) = \kappa\,(d - i + 1) + \lambda_i + 1$.

Then $\mathcal{U}_i x^\alpha = \xi_i(\alpha) x^\alpha + \sum_{\alpha \rhd \beta} B_{\beta\alpha}^{(i)} x^\beta$ (matrix of coefficients from $\mathbb{Q}[\kappa]$; note the triangularity) (Knop and Sahi [14]). The simultaneous eigenfunctions of $\{\mathcal{U}_i\}$ are called nonsymmetric Jack polynomials, indexed by $\mathbb{N}_0^d$. There is a useful basis defined by

$$\sum_{\alpha \in \mathbb{N}_0^d} p_\alpha(x)\, z^\alpha = \prod_{i=1}^d \left((1 - x_i z_i)^{-1} \prod_{j=1}^d (1 - x_j z_i)^{-\kappa} \right).$$

In this basis the triangularity of $\{\mathcal{U}_i\}$ is in the opposite direction $\lhd$. The eigenfunction with eigenvalues $\xi_i(\alpha)$ (for $\mathcal{U}_i$) which is monic in p_α is denoted by $\zeta_\alpha(x)$. The coefficients in the expansion of ζ_α in the p-basis are independent of the number of trailing zeros of α (Dunkl [8]). Thus

$$\zeta_\alpha = p_\alpha + \sum_{\beta \rhd \alpha} A_{\beta\alpha} p_\beta,$$

where $A_{\beta\alpha} \neq 0$ implies $\beta \rhd \alpha$, $A_{\beta\alpha} \in \mathbb{Q}(\kappa)$, and if $\beta' = (\beta_1, \dots, \beta_d, 0)$, $\alpha' = (\alpha_1, \dots, \alpha_d, 0)$ then $A_{\beta'\alpha'} = A_{\beta\alpha}$ (the independence on trailing zeros). For any partition λ, the space $\text{span}\{\zeta_\alpha : \alpha^+ = \lambda\}$ is closed under the actions of $w \in S_d$ and each $\mathcal{D}_i x_i$. Each ζ_α can be constructed from ζ_λ by a sequence of adjacent transpositions $(i, i+1)$. Indeed suppose $\alpha_i > \alpha_{i+1}$ for some i, and $\sigma = (i, i+1)$ then

$$\zeta_{\sigma\alpha} = \sigma\zeta_\alpha - \frac{\kappa}{\xi_i(\alpha) - \xi_{i+1}(\alpha)} \zeta_\alpha.$$

Note that $\alpha^+ \unrhd \alpha \rhd \sigma\alpha$. From this one can obtain recursions for $\langle \zeta_\alpha, \zeta_\alpha \rangle_h$ and $\zeta_\alpha(1, 1, \dots, 1)$ in terms of the corresponding values for ζ_λ (where $\lambda = \alpha^+$).

The computation of these values uses a degree-changing formula: fix a partition λ with $\lambda_m > \lambda_{m+1} = 0$ for some $m \leq d$, then

$$\mathcal{D}_m \zeta_\lambda = (\kappa\,(d - m + 1) + \lambda_m)\, \theta_m^{-1} \zeta_{\widetilde{\lambda}}$$

where $\theta_m = (1, 2)(2, 3) \dots (m - 1, m)$ (a cyclic shift) and $\widetilde{\lambda} = (\lambda_m - 1, \lambda_1, \lambda_2, \dots, \lambda_{m-1}, 0, \dots)$. To express the norms: for $\alpha \in \mathbb{N}_0^d$ and $\varepsilon = \pm$,

$$\mathcal{E}_\varepsilon(\alpha) = \prod \left\{ 1 + \frac{\varepsilon\kappa}{\xi_j(\alpha) - \xi_i(\alpha)} : i < j \text{ and } \alpha_i < \alpha_j \right\};$$

for a partition λ (with $\lambda_m > \lambda_{m+1} = 0$), variable t, the generalized Pochhammer symbol is

$$(t)_\lambda = \prod_{i=1}^{d} (t - (i-1)\kappa)_{\lambda_i}$$

and the hook-length product (from [14]) is

$$h(\lambda, t) = \prod_{i=1}^{m} \prod_{j=1}^{\lambda_i} (\lambda_i - j + t + \kappa \#\{k : k > i; j \leq \lambda_k\}).$$

Then

$$\zeta_\alpha(1, 1, \ldots, 1) = \mathcal{E}_-(\alpha) \frac{(d\kappa + 1)_{\alpha^+}}{h(\alpha^+, 1)}$$

$$\langle \zeta_\alpha, \zeta_\alpha \rangle_h = \mathcal{E}_+(\alpha) \mathcal{E}_-(\alpha) (d\kappa + 1)_{\alpha^+} \frac{h(\alpha^+, \kappa + 1)}{h(\alpha^+, 1)}.$$

The polynomials $e^{-\Delta_h/2}\zeta_\alpha$ are the corresponding *Hermite-type* polynomials. See Ch. 8 [11].

1.8. THE HYPEROCTAHEDRAL GROUPS W_d (TYPE B)

We use κ, κ' for the values of the multiplicity function; with κ assigned to the class of roots $\{\pm \varepsilon_i \pm \varepsilon_j : 1 \leq i < j \leq d\}$ and κ' assigned to the class $\{\pm \varepsilon_i : 1 \leq i \leq d\}$. For $i \neq j$ the roots $\varepsilon_i - \varepsilon_j$, $\varepsilon_i + \varepsilon_j$ correspond to the reflections σ_{ij}, τ_{ij} respectively, where (for $x \in \mathbb{R}^d$)

$$x\sigma_{ij} = (x_1, \ldots, \overset{i}{x_j}, \ldots, \overset{j}{x_i}, \ldots, x_d),$$

$$x\tau_{ij} = (x_1, \ldots, -\overset{i}{x_j}, \ldots, -\overset{j}{x_i}, \ldots, x_d).$$

For $1 \leq i \leq d$ the root ε_i corresponds to the reflection (sign-change)

$$x\sigma_i = (x_1, \ldots, -\overset{i}{x_i}, \ldots, x_d).$$

The first-order operators are

$$\mathcal{D}_i = \partial_i + \kappa' \frac{1 - \sigma_i}{x_i} + \kappa \sum_{j \neq i} \left\{ \frac{1 - \sigma_{ij}}{x_i - x_j} + \frac{1 - \tau_{ij}}{x_i + x_j} \right\}.$$

With respect to the pairing norm there are commuting self-adjoint operators

$$\mathcal{U}_i = \mathcal{D}_i x_i - \kappa \sum_{j=1}^{i-1} (\sigma_{ij} + \tau_{ij}).$$

The simultaneous eigenfunctions are expressed in terms of the nonsymmetric Jack polynomials, using coordinates $y_i = x_i^2$. Fix $\beta \in \mathbb{N}_0^d$ and m with $0 \leq m \leq d$; let $x_E = \prod_{i=1}^{m} x_i, \eta = \sum_{i=1}^{m} \varepsilon_i$ and $\alpha = 2\beta + \eta$; then the eigenfunction labeled by α (any composition with the odd components coming before the even components) is $x_E \zeta_\beta(y)$. The pairing norm is

$$
\langle x_E \zeta_\beta(y), x_E \zeta_\beta(y) \rangle_h = 2^{|\alpha|} (d\kappa + 1)_{\beta+} \left((d-1)\kappa + \kappa' + \frac{1}{2} \right)_{(\beta+\eta)+}
$$
$$
\times \quad \mathcal{E}_+(\beta) \mathcal{E}_-(\beta) \frac{h(\beta^+, \kappa+1)}{h(\beta^+, 1)}.
$$

The remaining eigenfunctions (with the odd parts in arbitrary places) are constructed as follows (continuing the same notation): let $w \in S_d$ with the property that $w(i) < w(j)$ whenever $1 \leq i < j \leq m$ or $m+1 \leq i < j \leq d$; then $w x_E \zeta_\beta(y)$ is an eigenfunction of each $\mathcal{U}_i$ (and the same norm, since w is an isometry), (the type B polynomials were constructed in Dunkl [8], also see (§9.3 [11])).

The corresponding Hermite polynomials (for the weight function $\prod_{i=1}^{d} |x_i|^{2\kappa'} \prod_{i<j} \left| x_i^2 - x_j^2 \right|^{2\kappa} e^{-\|x\|^2/2}$ on $\mathbb{R}^d$) can be written "explicitly" using the generalized binomial coefficients of Baker and Forrester [2], but not much is known about them (see §9.4 [11]). The problem of constructing h-harmonic polynomials of type B is open.

2. Boundedness and Summability Problems for Expansions

This section concerns a sufficient condition for the density of polynomials in $L^2(\mu)$; the generalized Fourier transform and orthogonal polynomials associated with reflection groups, and the Cesàro and Poisson summation processes for orthogonal expansions. The Fourier transform involves the intertwining operator and the analogue of the exponential function for reflection groups.

As in Section 1, let μ be a finite positive Baire measure on $\mathbb{R}^d$ such that $\int_{\mathbb{R}^d} |x^\alpha| \, d\mu(x) < \infty$ for all $\alpha \in \mathbb{N}_0^d$, and $\mathcal{V}_n^d$ is the orthogonal complement of Π_{n-1}^d in Π_n^d (for $n \in \mathbb{N}_0$). Suppose $\{p_{n,i} : 1 \leq i \leq \dim \mathcal{V}_n^d\}$ is an orthonormal basis for $\mathcal{V}_n^d$ then let $P_n(x, y) = \sum_i p_{n,i}(x) p_{n,i}(y)$. For $f \in L^2(\mu)$ the orthogonal projection onto $\mathcal{V}_n^d$ is given by $\pi_n f(x) = \int_{\mathbb{R}^d} f(y) P_n(x, y) d\mu(y)$ (thus $P_n(x, y)$ is independent of the choice of basis). We will consider the problem of recovering f from the projections $(\pi_n f)_{n=0}^{\infty}$. When the support of μ is compact the polynomials are of course dense in $L^2(\mu)$. Otherwise, there is an elegant sufficient condition due to Hamburger (for one variable) which applies to our examples.

Theorem 8. *Suppose $\int_{\mathbb{R}^d} e^{c\|x\|} d\mu(x) < \infty$ for some $c > 0$. Then the polynomials Π^d are dense in $L^2(\mu)$.*

The idea of the proof is to take the Fourier-Stieltjes transform of the measure $f d\mu$ for some $f \in L^2(\mu)$; by the boundedness condition the transform is analytic on a strip and orthogonality to all polynomials would imply the transform is zero in a neighborhood of $0 \in \mathbb{C}^d$ (see §3.1 [11]).

The condition applies to the measures $h(x)^2 e^{-\|x\|^2/2} dx$ on $\mathbb{R}^d$ discussed in Section 1.

A motivating classical one-variable result for the Fourier transform is:

$$(2\pi)^{-1/2} \int_{\mathbb{R}} H_n(x) e^{-x^2/2} e^{-ixy} dx = (-i)^n H_n(y) e^{-y^2/2}$$

(which proves the Plancherel theorem) where H_n is the Hermite polynomial.

Suppose R is a root system with group $W = W(R)$ and κ is a multiplicity function on R (and notation as in Section 1). The analogue (denoted by $K(x, y)$) of the exponential function $\exp(\langle x, y \rangle)$ $(x, y \in \mathbb{R}^d)$ should have certain properties: entire in x and y, $K(y, x) = K(x, y)$, $K(xw, yw) = K(x, y)$ for $w \in W$; and the key property $D_i^{(x)} K(x, y) = y_i K(x, y)$ for each i, (where the superscript (x) indicates the action variable). The construction follows from the *intertwining operator* V, defined as follows: V is a linear map on polynomials such that $V\mathcal{P}_n^d \subseteq \mathcal{P}_n^d$ (preserves homogeneity) for each n, $V1 = 1$, and $\mathcal{D}_i V p(x) = V(\partial_i p)(x)$ for each i and $p \in \Pi^d$.

There is an inductive construction for V (on $1, \mathcal{P}_1^d, \mathcal{P}_2^d, \ldots$) using a group algebra method which works for any $\kappa > 0$ (Dunkl [6], or §4.6 [11])). In fact V^{-1} exists for all κ but fails to be one-to-one for certain "singular" values of κ. (If κ has just one value, as for S_d, then the singularities are negative rational numbers, for the general situation, see Dunkl, de Jeu, and Opdam [10]). The case $d = 1, W = \mathbb{Z}_2, \kappa > 0$ is explicit:

$$Vp(x) = c_\kappa \int_{-1}^1 p(xt)(1 - t)^{\kappa-1}(1 + t)^\kappa dt$$

(where $c_\kappa = B\left(\kappa, \frac{1}{2}\right)^{-1}$). The operator $\mathcal{D}$ is given by

$$\mathcal{D}p(x) = p'(x) + \kappa \frac{p(x) - p(-x)}{x}, x \in \mathbb{R}.$$

Then $Vx^{2n} = \dfrac{(\frac{1}{2})_n}{(\kappa+\frac{1}{2})_n} x^{2n}, Vx^{2n+1} = \dfrac{(\frac{1}{2})_{n+1}}{(\kappa+\frac{1}{2})_{n+1}} x^{2n+1}$ which shows that the singularities of V are at $-\frac{1}{2}, -\frac{3}{2}, -\frac{5}{2}, \ldots$ (for this particular group).

To construct $K(x,y)$ let

$$K_n(x,y) = V^{(x)}\left(\frac{\langle x,y\rangle^n}{n!}\right)$$

for $n \in \mathbb{N}_0$. It can be shown (using the van der Corput-Schaake inequality) that $|K_n(x,y)| \leq \max_{w \in W} |\langle xw, y\rangle|^n /n!$. Thus the series $K(x,y) = \sum_{n=0}^{\infty} \times K_n(x,y)$ is uniformly convergent on compact subsets and is an entire function (thus $|K(x,y)| \leq \exp(\max_{w \in W} |\langle xw, y\rangle|)$). The property $K_n(x,y) = K_n(y,x)$ is not obvious (its proof uses the fact that w and w^{-1} are in the same conjugacy class in W for each w).

2.1. THE GENERALIZED FOURIER TRANSFORM

For a suitable function f on $\mathbb{R}^d$, let

$$\widehat{f}(y) = c_\kappa \int_{\mathbb{R}^d} f(x) K(x, -iy) h(x)^2 dx$$

(with the same normalizing constant as before: $c_\kappa \int_{\mathbb{R}^d} h(x)^2 e^{-\|x\|^2/2} dx = 1$). The transform is isometric (Dunkl [7], or §5.7 [11]) and can be considered as the spectral resolution of the operators $\mathcal{D}_i$ since $(\mathcal{D}_j f)\widehat{}(y) = iy_j \widehat{f}(y)$ for each j (and appropriate f). The h-harmonic polynomials can be used to describe an orthogonal basis for $L^2(\mathbb{R}^d, h(x)^2 dx)$.

Theorem 9. *Suppose* $p \in \mathcal{H}_n$ *(recall* $\Delta_h p = \sum_{i=1}^d \mathcal{D}_i^2 p = 0$*) and* $m, n \in \mathbb{N}_0$ *then let* $f(x) = p(x) L_m^A(\|x\|^2) e^{-\|x\|^2/2}$ *where the index* $A = n + \gamma + \frac{d}{2} - 1$ *and* $\gamma = \sum_{v \in R_+} \kappa_v$. *Then* $\widehat{f}(y) = (-i)^{n+2m} f(y), y \in \mathbb{R}^d$.

This shows that the transform is an isometry on $L^2(\mathbb{R}^d, h(x)^2 dx)$; just note that if $\{p_{n,\alpha} \in \mathcal{H}_n : n \in \mathbb{N}_0$ and $\alpha \in \mathbb{N}_0^d$ with $\alpha_d = 0$ or $1\}$ is an orthogonal basis for $L^2(S^{d-1}, h^2 d\omega)$ then $\{p_{n,\alpha}(x) L_m^{(n+\gamma-1+d/2)}(\|x\|^2) e^{-\|x\|^2/2} : m, n \in \mathbb{N}_0\}$ is an orthogonal basis for $L^2(\mathbb{R}^d, h(x)^2 dx)$, (this is proven by means of spherical polar coordinates and Hamburger's theorem).

De Jeu [12] showed that $|K(x, -iy)| \leq (\#W(R))^{1/2}$ for $x, y \in \mathbb{R}^d$ which shows that the domain of the transform includes $L^1(\mathbb{R}^d, h(x)^2 dx)$. A stronger result was then proven by Rösler [17]: the positivity of V (that is, for each $x \in \mathbb{R}^d$ the functional $f \to Vf(x)$ is given by a positive measure supported by the closed convex hull of $\{xw : w \in W\}$); and for each $y \in \mathbb{R}^d$ the function $x \to K(x, -iy)$ is of positive type, thus $|K(x, -iy)| \leq K(0, -iy) = 1$.

2.1.1. *The symmetric group case*

Since $K(x, y)$ is the reproducing kernel for the pairing norm, we can express $K_n(x, y)$ in terms of the nonsymmetric Jack polynomials (see §8.9 [11]):

$$K_n(x, y) = \sum_{|\alpha|=n} \frac{h(\alpha^+, 1)}{h(\alpha^+, \kappa+1)(d\kappa+1)_{\alpha^+} \mathcal{E}_+(\alpha) \mathcal{E}_-(\alpha)} \zeta_\alpha(x) \zeta_\alpha(y).$$

2.2. SERIES OF HARMONICS

For a series of functions $\sum_{n=0}^\infty f_n(x)$, ordinary convergence refers to the sequence of partial sums $s_n(x) = \sum_{i=0}^n f_i(x)$, Poisson (Abel) summability is used when the function $P_r(f)(x) = \sum_{n=0}^\infty f_n(x)r^n$ is analytic for $|r| < 1$ for each x and means that $\lim_{r\to 1_-} P_r(f)(x)$ exists. For a parameter $\delta > 0$ define the Cesàro (C, δ)-means by $s_n^\delta = \sum_{i=0}^n \frac{(-n)_i}{(-n-\delta)_i} f_i$ (also $s_n^\delta = \frac{\delta}{n+\delta} \sum_{i=0}^n \frac{(-n)_i}{(1-n-\delta)_i} s_i$, so when $\delta = 1$ we obtain the familiar $s_n^1 = \frac{1}{n+1} \sum_{i=0}^n s_i$). A key fact is that if $\lim_{n\to\infty} s_n^\delta = s$ for some δ then $\lim_{n\to\infty} s_n^\tau = s$ for each $\tau > \delta$. We will apply these methods to the series $\sum_{n=0}^\infty \pi_n f(x)$ where π_n is the orthogonal projection onto $\mathcal{V}_n^d$ (actually, $\mathcal{H}_n^d$ for most of the examples). It turns out that the main results can be derived from Gegenbauer series.

Fix $\lambda > 0$ and the measure $w_\lambda(x)dx = B\left(\lambda + \frac{1}{2}, \frac{1}{2}\right)^{-1}(1-x^2)^{\lambda-\frac{1}{2}}dx$ on $[-1, 1]$ and for $f \in L^1([-1, 1], w_\lambda(x)dx), n \in \mathbb{N}_0$ let

$$a_n = \int_{-1}^1 f(x)C_n^\lambda(x)w_\lambda(x)dx.$$

The Gegenbauer series of f is

$$f(x) \approx \sum_{n=0}^\infty a_n \frac{n+\lambda}{\lambda} \frac{n!}{(2\lambda)_n} C_n^\lambda(x).$$

We ask which summability methods give uniform convergence if f is continuous, or L^1-convergence. Because of the product formula (explained below for spherical harmonics) the summability behavior is determined by the series $\sum_{n=0}^\infty \frac{n+\lambda}{\lambda} C_n^\lambda(x)$. Poisson summability follows immediately from the generating function

$$\sum_{n=0}^\infty \frac{n+\lambda}{\lambda} C_n^\lambda(x)r^n = \frac{(1-r^2)}{(1-2xr+r^2)^{\lambda+1}}.$$

For (C, δ) summability consider the polynomials $\sum\limits_{k=0}^{n} \frac{(-n)_k}{(-n-\delta)_k} \frac{k+\lambda}{\lambda} C_k^\lambda(x)$; by a result of Kogbetliantz [15](an elegant proof by completely monotone functions can be found in Askey [1]) they are positive for $\delta \geq 2\lambda + 1$; in fact

$$\sum_{k=0}^{n} \frac{(-n)_k}{(-n-2\lambda-1)_k} \frac{k+\lambda}{\lambda} C_k^\lambda(x) > 0 \text{ for } -1 < x \leq 1.$$

For the sphere S^{d-1} the polynomial $\|x\|^n C_n^\lambda(\langle x, y\rangle / \|x\|)$ is harmonic for any $y \in S^{d-1}$ (a zonal harmonic) with $\lambda = \frac{d}{2} - 1$. The reproducing kernel for $\mathcal{H}_n^d$ is $\frac{n+\lambda}{\lambda} C_n^\lambda(\langle x, y\rangle)$ (for $x, y \in S^{d-1}$). Suppose $f(x) \approx \sum\limits_{n=0}^{\infty} f_n(x)$ is a Laplace series (each $f_n \in \mathcal{H}_n^d$) then (formally)

$$\sum_{n=0}^{\infty} b_n f_n(x) = \int_{S^{d-1}} f(y) \sum_{n=0}^{\infty} b_n \frac{n+\lambda}{\lambda} C_n^\lambda(\langle x, y\rangle) d\omega(y).$$

This explains why it suffices to consider one-variable series for summability; for a fixed z let

$$f(x) = \sum_{n=0}^{\infty} a_n \frac{n+\lambda}{\lambda} \frac{n!}{(2\lambda)_n} C_n^\lambda(\langle x, z\rangle).$$

We return to the general reflection group case. The normalizing constant c_κ' satisfies $c_\kappa' \int_{S^{d-1}} h^2 d\omega = 1$; for a fixed $y \in \mathbb{R}^d$ with $\|y\| < 1$ let

$$P(x, y) = V^{(x)} \left(\frac{1 - \|y\|^2}{(1 - 2\langle x, y\rangle + \|y\|^2)^{\gamma + d/2}} \right)$$

(recall $\gamma = \sum\limits_{v \in R_+} \kappa_v$). Then for any $f \in \mathcal{H}_n$ we have

$$f(y) = c_\kappa' \int_{S^{d-1}} f(x) P(x, y) h(x)^2 d\omega(x), (\|y\| < 1).$$

Since this is independent of the degree (and $P(x, y) > 0$, [6]) the Poisson summability for expansions with respect to $\sum_{n=0}^{\infty} \oplus \mathcal{H}_n^d$ is proven. Now expand the formula for $P(x, y)$ to get the homogeneous expansion in y (where $\lambda = \gamma + \frac{d}{2} - 1$):

$$\sum_{n=0}^{\infty} P_n(x, y) = \sum_{n=0}^{\infty} \frac{n+\lambda}{\lambda} \|y\|^n V^{(x)} \left(C_n^\lambda \left(\frac{\langle x, y\rangle}{\|y\|} \right) \right).$$

From this we obtain the kernel for the (C,δ)-means

$$S_n^\delta(x,y) = V^{(x)} \sum_{k=0}^n \frac{(-n)_k}{(-n-\delta)_k} \frac{k+\lambda}{\lambda} \|y\|^k C_k^\lambda \left(\frac{\langle x,y \rangle}{\|y\|} \right)$$

(the (C,δ)-means for λ-Gegenbauer series, combined with the Poisson kernel, valid for $\|x\| = 1$). The positivity of V (due to Rösler [17]) together with the Kogbetliantz result now show that the (C,δ)-means are positive operators when $\delta \geq 2\gamma + d - 1$. There are obvious implications for the measures on the simplex and the ball discussed before. The convergence is uniform for continuous functions and in the norm for L^p for $1 \leq p < \infty$. If one relaxes the positivity condition on the summation kernels and requires only uniform boundedness, then $\delta > \gamma + \frac{d-1}{2}$ suffices (see §7.2 [11]).

Li and Xu [16] have some results about necessary conditions for δ: (example) consider the measure $\prod_{i=1}^d |x_i|^{2\kappa_i} d\omega(x)$; the (C,δ)-means converge uniformly for every continuous function on S^{d-1} if and only if $\delta > \frac{d}{2} - 1 + \sum_{i=1}^d \kappa_i - \min_{1 \leq i \leq d} \kappa_i$.

2.3. THE REPRODUCING KERNEL FOR THE ABELIAN CASE

Consider $h(x) = \prod_{i=1}^d |x_i|^{\kappa_i} |x_{d+1}|^\lambda$ and consider the polynomials even in x_{d+1} as polynomials on the ball B^d (replace x_{d+1}^2 by $r^2 - \sum_{i=1}^d x_i^2$ and restrict $r = 1$) so the measure $h^2 d\omega$ transforms to

$$\prod_{i=1}^d |x_i|^{2\kappa_i} (1 - \sum_{i=1}^d x_i^2)^{\lambda - \frac{1}{2}} dx_1 \ldots dx_d.$$

The $\mathbb{Z}_2$-intertwining operator can be used separately in each coordinate (and is specialized to functions even in x_{d+1}). Let $\mu = \frac{d+1}{2} - 1 + \sum_{i=1}^d \kappa_i + \lambda$, then the reproducing kernel for $\mathcal{V}_n^d$ is

$$P_n(x,y) = \frac{n+\mu}{\mu} c \int_{[-1,1]^{d+1}} C_n^\mu \left(\sum_{i=1}^d x_i y_i t_i + t_{d+1} ((1 - \|x\|^2)(1 - \|y\|^2))^{1/2} \right)$$

$$\times \prod_{i=1}^d ((1 - t_i)^{\kappa_i - 1}(1 + t_i)^{\kappa_i})(1 - t_{d+1}^2)^{\lambda - 1} dt;$$

(for $x, y \in B^d$; the normalizing constant is $c = \left(B\left(\lambda, \frac{1}{2}\right) \prod_{i=1}^d B\left(\kappa_i, \frac{1}{2}\right) \right)^{-1}$, so that $P_0 = 1$). See (§6.1 [11]).

2.4. PROBLEMS

These are some of the open problems: finding explicit integral formulae for the intertwining operator V for types A and B; construction of orthogonal bases for harmonics for nonabelian $W(R)$; finding orthogonal polynomials (with respect to the pairing norm) for the icosahedral group H_3.

Acknowledgement. During the preparation of this article, the author was partially supported by NSF grant # DMS-9970389.

References

1. R. Askey, *Orthogonal Polynomials and Special Functions*, Reg. Conf. Series in Appl. Math. **21**, SIAM, Philadelphia, 1975.
2. T. Baker and P. Forrester, *Nonsymmetric Jack polynomials and integral kernels*, Duke Math. J. **95** (1998), 1–50.
3. B. L. Braaksma and B. Meulenbeld, *Jacobi polynomials as spherical harmonics*, Indag. Math. **30** (1968), 384–389.
4. C. F. Dunkl, *Reflection groups and orthogonal polynomials on the sphere*, Math. Z. **197** (1988), 33–60.
5. C. F. Dunkl, *Differential-difference operators associated to reflection groups*, Trans. Amer. Math. Soc. **311** (1989), 167–183.
6. C. F. Dunkl, *Integral kernels with reflection group invariance*, Canadian J. Math. **43** (1991), 1213–1227.
7. C. F. Dunkl, *Hankel transforms associated to finite reflection groups*, Contemp. Math. **138** (1992), Amer. Math. Soc. Providence RI, pp. 123–138.
8. C. F. Dunkl, *Orthogonal polynomials of types A and B and related Calogero models*, Commun. Math. Phys. **197** (1998), 451–487.
9. C. F. Dunkl, *Computing with differential-difference operators*, J. Symb. Comp. **28** (1999), 819–826.
10. C. F. Dunkl, M. de Jeu, and E. Opdam, *Singular polynomials for finite reflection groups*, Trans. Amer. Math. Soc. **346** (1994), 237–256.
11. C. F. Dunkl and Y. Xu, *Orthogonal Polynomials of Several Variables*, Encyclopedia of Mathematics and its Applications series, Cambridge University Press, 2001.
12. M. de Jeu, *The Dunkl transform*, Invent. Math. **113** (1993), 147–162.
13. E. G. Kalnins, W. Miller, Jr., and M. V. Tratnik, *Families of orthogonal and biorthogonal polynomials on the N-sphere*, SIAM J. Math. Analysis **22** (1991), 272–294.
14. F. Knop and S. Sahi, *A recursion and a combinatorial formula for Jack polynomials*, Invent. Math. **128** (1997), no. 1, 9–22.
15. E. Kogbetliantz, *Recherches sur la summabilité des séries ultrasphériques par la méthode des moyennes arithmetiques*, J. Math. Pures Appl. (9) **3** (1924), 107–187.
16. Zh.-K. Li and Y. Xu, *Summability of orthogonal expansions on spheres, balls and simplices*, to appear.
17. M. Rösler, *Positivity of Dunkl's intertwining operator*, Duke Math. J. **98** (1999), 445–463.

THE BISPECTRAL PROBLEM: AN OVERVIEW

F. ALBERTO GRUNBAUM

Department of Mathematics
University of California
Berkeley, CA 94720 USA

Abstract. I describe the problem in its original form, talk about possible variants of it, describe some results, and then indicate the role of the bispectral problem as a discovery tool in unknown territory.

1. Introduction

There are many places in mathematics where one can see a useful bridge between the old and the new. One of those bridges has to do with the relation of Special Functions with Group Representation Theory. Besides the intrinsic beauty involved here, there is an element of "efficient packaging" for material that had been developed in previous centuries to address different problems. In the sixties and seventies we witnessed another such "gluing phenomenon" centered around equations connected with names like Korteweg-deVries, nonlinear Schroedinger, and Toda. The resulting development that could be loosely called Soliton Mathematics gives a nice picture of several parts of mathematics coming together for their mutual benefit.

The bispectral problem, which I describe below, can be seen to have connections with both of these areas of rich mathematical encounters.

The selection of material here is intended to go in a few pages from the original formulation to some open problems of personal interest. There is no attempt to give a fair and balanced picture of all the work in this area. In particular, I do not touch on the very important work in [2], [3], [4], [5], [7], [8], [39], [53], [54]. For a guide to the literature the reader should consult [33] and the many references in that monograph.

J. Bustoz et al. (eds.), Special Functions 2000, 129–140.

2. The Original Problem

The problem as posed and solved in [14] is as follows:
    Find all nontrivial instances where a function $\varphi(x, k)$ satisfies

$$L\left(x, \frac{d}{dx}\right) \varphi(x, k) \equiv (-D^2 + V(x))\varphi(x, k) = k\varphi(x, k)$$

as well as

$$B\left(k, \frac{d}{dk}\right) \varphi(x, k) \equiv \left(\sum_{i=0}^{M} b_i(k) \left(\frac{d}{dk}\right)^i\right) \varphi(x, k) = \Theta(x)\varphi(x, k).$$

All the functions $V(x), b_i(k), \Theta(x)$ are, in principle, arbitrary except for smoothness assumptions. Notice that here M is arbitrary (finite). The complete solution is given as follows:

Theorem 2.1 [14] *If $M = 2$, then $V(x)$ is (except for translation) either c/x^2 or ax, i.e., we have a Bessel or an Airy case. If $M > 2$, there are two families of solutions*

  a) *L is obtained from $L_0 = -D^2$ by a finite number of Darboux transformations ($L = AA^* \rightarrow \tilde{L} = A^*A$). In this case V is a rational solution of KdV.*
  b) *L is obtained from $L_0 = -D^2 + \frac{1}{4x^2}$ after a finite number of rational Darboux transformations.*

It was later observed in [42] that in the second case we are dealing with rational solutions of the Virasoro or master symmetries of KdV.
    In the first case the space of common solutions has dimension one, in the second it has dimension two. We refer to these as the rank one and rank two situations.
    Observe that the "trivial cases" when $M = 2$ are self-dual in the sense that since the eigenfunctions $f(x, k)$ are functions either of the product xk or of the sum $x + k$, one gets B by replacing k for x in L.
    I remark that there are at least *three* very different issues that one may want to consider in extending the problem to other settings.

  1. The spatial variable x and/or the spectral variable k can be allowed to run in a higher-dimensional space, in which case it is natural to replace L and B by a commuting set of operators L_i and B_i, respectively.
  2. The spatial and/or spectral variables can be allowed to range over either a continuum or a discrete set, maybe some appropriate lattice in either one or higher dimensions.
  3. The eigenfunctions $\varphi(x, k)$ could be scalar or either vector or matrix valued. The coefficients in L and B can also be scalar or matrix valued.

In this lecture I will describe two areas of recent work that deal with some extensions in the directions 2 and 3 described above. It should be clear that the problem posed by Krall [41] quite a while back fits nicely under variant 2 described above. At the "basic or trivial" level the problem posed by Bochner [10] is a precedent in this program too. It is important, however, to see that in these older cases the insistence on polynomial solutions has certain restrictive consequences.

There are many useful ways to modify the original formulation of the bispectral problem. A very interesting development in this direction can be seen in the lecture by Spiridonov and Zhedanov in this volume, [49]. There are also interesting connections with the work of P. Terwilliger, see [50] as well as his presentation at this meeting. There are also some connections to the methods in [18].

For interesting relations to the Yang-Baxter as well as Knizhnik-Zamolodchikov equations, see [17]. See also [48], [51].

There is a lot of very deep work already available on what one may call the basic or self dual instance of the bispectral property for several variables. See for instance [15], [12], [34], [44], [52]. These efforts have started to produce new classes of functions that could find novel uses in many areas of mathematical physics and hopefully even in some engineering applications. In an attempt to make these functions more widely known I have recently, see [21], produced some plots of some of these functions. Much more could be done in the direction of making these functions part of some symbolic-numerical package.

3. The Associated Polynomials from a Different Viewpoint

For a very nice description of work in this area, look at the paper by Mizan Rahman in this volume, [46].

In [22], [23] one finds a treatment of the following problem.

Find all nontrivial instances of a sequence $f_n(z)$ of functions, $n \in Z$, $z \in \mathbb{C}$, such that

$$(Lf)_n(z) = z f_n(z)$$

and

$$B f_n(z) = \lambda_n f_n(z).$$

Here L is a doubly infinite tridiagonal matrix independent of z and B is a second-order differential operator independent of n. We see that we are replacing the continuous variable x by the discrete variable n, and restricting M to be equal to two. The continuous variable k is renamed z.

This can be seen as the original problem of Bochner [10] *without the restriction* of the usual boundary condition $f_{-1}(z) = 0$ which leads to polynomials. In that case we can have at most two degrees of freedom (the case of Jacobi polynomials, parametrized by α, β), while now we will see that there is a three-parameter family of solutions.

The solution of the problem above goes as follows.

Pick a, b, c in generic position.

Pick any solution $w(z)$ of Gauss's equation

$$z(1-z)w'' + (c - (a+b+1)z)w' - abw = 0.$$

Put $f_0(z) = w(z)$ and $f_1(z)$ given by

$$f_1(z) = \frac{1}{(b-a)\sqrt{a_1}}\left(z(1-z)w'(z) - a\left(z + \frac{b-c}{a-b+1}\right)w(z)\right).$$

Define now $f_n(z)$ for $n \in Z$, $n \neq 0, 1$, by means of

$$\sqrt{a_n}f_{n-1}(z) + b_{n+1}f_n(z) + \sqrt{a_{n+1}}f_{n+1}(z) = zf_n(z)$$

where

$$c = \beta + 1, \quad b = -t, \quad a = \alpha + \beta + 1 + t$$

and

$$a_n = \tilde{a}_{n+t}, \quad b_n = \tilde{b}_{n+t}.$$

Here we have put

$$\tilde{a}_n = \frac{n(n+\alpha)(n+\beta)(n+\alpha+\beta)}{(2n+\alpha+\beta-1)(2n+\alpha+\beta)^2(2n+\alpha+\beta+1)}$$

$$\tilde{b}_n = \frac{1}{2}\left(1 + \frac{\beta^2 - \alpha^2}{(2n+\alpha+\beta-2)(2n+\alpha+\beta)}\right).$$

Furthermore, B is, after an appropriate conjugation, exactly the Gauss operator, and $\lambda_n = (a+n)(b+n)$.

In other words, the solution to the discrete-continuous bispectral problem with $M = 2$ is given by functions obtained by the translation $n \to n+t$ in the doubly infinite Jacobi matrix. From the point of view of "soliton mathematics", translation is the basic case, followed in a natural hierarchy by the KdV flows or the Toda flows. It is thus nice to see translations showing up here in the "trivial" case $M = 2$.

The case of $M > 2$ is still not completely settled. In the case of rank one there is very nice recent work in [31], [32]. In the case of rank two one can consult [30].

In [22] one also finds a complete solution to the problem of determining the "spectral matrix orthogonality measure" corresponding to the 2×2 matrix of polynomials

$$P_n(z) = \begin{pmatrix} q_{-n-1}(z) & p_{-n-1}(z) \\ q_n(z) & p_n(z) \end{pmatrix} \quad n = 0, 1, 2, \ldots$$

Here, p_n and q_n denote the two families of polynomials obtained by solving

$$Lf = zf$$

with initial conditions

$$p_{-1} = 0, p_0 = 1; \quad q_{-1} = 1, \quad q_0 = 0.$$

The actual determination of the entries of the matrix $\sigma(z)$ satisfying

$$\int_R P_n(z)\sigma(z)P_m^T(z)dz = \delta_{nm}I$$

is a real tour de force comparable to that in the very influential paper by J. Wimp [55] in the case of the associated Jacobi polynomials.

In the same paper [22] one sees a possible advantage in using as a basis for the space of solutions of $Lf = zf$, one that is *different* from the pair $(p_n(z), q_n(z))$ above.

Consider the functions $(u_n(z), v_n(z))$ given by

$$\begin{aligned}
u_n(z) &= k_n \alpha_n \, {}_2F_1(a+n, b-n, c; z) \\
v_n(z) &= k_n \beta_n z^{1-c} \, {}_2F_1(a-c+1+n, b-c+1-n, 2-c; z)
\end{aligned}$$

with

$$\begin{aligned}
\alpha_n &= (-1)^n (a)_n \frac{(c-b)_n}{(a-b)_n} \\
\beta_n &= (-1)^n \frac{(a+1-c)_n, (1-b)_n}{(a-b)_n}, \quad k_0 = 1, \quad \frac{k_{n-1}}{k_n} = \sqrt{a_n}.
\end{aligned}$$

These ones are a pair of *bispectral solutions*, i.e., not only do they satisfy the three-term recursion relation implied by L, but they also satisfy

$$Bu_n(z) = \lambda_n u_n(z) \text{ and } Bv_n(z) = \lambda_n v_n(z)$$

with B as described above. In terms of this new basis, we put, as above,

$$U_n(z) = \begin{pmatrix} u_{-n-1}(z) & v_{-n-1}(z) \\ u_n(z) & v_n(z) \end{pmatrix}$$

and obtain

$$\int U_n(z)R(z)U_m^T(z)dz = \delta_{nm}I$$

with

$$R(z) = \begin{pmatrix} 1 & 0 \\ 0 & e_1 \end{pmatrix} e_2\rho(z)$$

and

$$\rho(z) = z^{c-1}(1 - z)^{a+b-c}$$

with e_1, e_2 a pair of constants. In [26] there is an analogous treatment in the case when B is replaced by a q-difference second-order operator. In that paper, the role of the Gauss equation is taken over by what we have called the Gauss-Askey-Wilson equation.

Finally, as a very special but important case of the results in [26], I describe the problem of finding bispectral "trivial" pairs on the purely discrete and bi-infinite case.

Consider, as in [19], the problem of determining the situations where there is a pair of bi-infinite tridiagonal matrices L, B, a family of either column or row vectors obtained from the matrix $F_{i,j}$ by fixing one or the other index, and some infinite sequence of scalars t_i such that

$$(LF)_{i,j} = jF_{i,j}$$

and

$$(BF)_{i,j} = t_i F_{i,j}.$$

In [19] one sees that this leads in a natural way to consider a possible discrete analog of Gauss's hypergeometric equation, namely

$$
\begin{aligned}
&4w_{j+1} - 4[2(j - b_1)^2 u - (1 + 3u)r(j - b_1) + 2\epsilon - 4u\delta]w_j \\
&+[4u^2(j - b_1)^4 - 4u((3r + 2)u + r)(j - b_1)^3 \\
&+(8u\epsilon - 16u^2\delta + ((3r + 1)u + r)((3r + 5)u + r) - 1)(j - b_1)^2 \\
&+(8u((3r + 2)u + r)\delta - 4((3r + 2)u + r)\epsilon \\
&-((3r + 1)u + r - 1)((3r + 1)u + r + 1))(j - b_1) \\
&+2r(3u + 1)(\epsilon - 2u\delta) + 4u^2\delta(4\delta - 1) \\
&-2u\epsilon(8\delta + 1) + 2(2\epsilon^2 + \delta)]w_{j-1} = 0.
\end{aligned}
$$

Here

$$
\begin{aligned}
\epsilon &= a_0 - a_1 \\
\delta &= a_0 + a_1.
\end{aligned}
$$

The form of the equation for w_j is as follows

$$
\begin{aligned}
(Bw)(j) &= 4w_{j+1} + (aj^2 + bj + c)w_j \\
&+ (ej^4 + fj^3 + gj^2 + hj + i)w_{j-1} = 0.
\end{aligned}
$$

The parameters a, b, c, e, f, g, h, i are not all independent. They are given by polynomials in the free parameters u, a_0, a_1, b_1, b_2 as seen above.

Generically, the solution is not given by an hypergeometric sequence in the sense of [45]. This is reminiscent of the situation in [37], [36].

4. A Matrix Valued Version of the Bispectral Property

As mentioned earlier, there are at least two directions in which one can extend naturally the problem posed in [14]. One could go from one to several variables, or one could go from the scalar valued case to the matrix or operator valued one. The first direction is explored, for instance, in [6], [11]. In the second direction I give below some very brief indication of very recent work in [29] leading (to the best of my knowledge) to the first explicit examples of (nonscalar) matrix valued spherical functions on symmetric spaces. From our point of view these can be seen as giving rise to families of "classical matrix valued orthogonal polynomials" in the original sense of Bochner. The subject of matrix valued orthogonal polynomials has been discussed for instance in [13], [16]. For related work in this area, one can see [40] and [35].

If G and K are a locally compact unimodular group and a compact subgroup of it, respectively, and π an irreducible representation of K acting on V, one says that $\Phi : G \to \text{End}(V)$ is a spherical function of type π if $\Phi(e) = I$, Φ is continuous and

$$\Phi(x)\Phi(y) = \int_K \chi_\pi(k^{-1})\Phi(xky)dk$$

for all x, y in G. dk denotes Haar measure on K.

When $G = SU(3)$, $K = U(2) \cong S(U(2) \times U(1))$ the complex projective plane CP_2 is the corresponding homogeneous space G/K. In this case the representations of K are parametrized by a pair of integers (n, l), $n \in Z$, $l \in Z_{\geq 0}$ and $\dim V = l + 1$. It is convenient to introduce a matrix valued function Φ_π and to define $H = \Phi\Phi_\pi^{-1}$.

It is a profound fact, due to a string of workers going from E. Cartan to Harish-Chandra that for symmetric spaces of rank one (as in this case), the spherical functions corresponding to the trivial representation with $\chi_\pi(k^{-1}) = 1$ are always expressible in terms of Gauss' $_2F_1$ function. In fact in this case if $n = l = 0$ we get to deal with $_2F_1\left(\begin{array}{c} -w, w+2 \\ 1 \end{array}; t\right)$, $w = 0, 1, 2, \ldots$.

We see here that spherical functions depend on a continuous parameter t and a discrete parameter $w = 0, 1, \ldots$. In fact in the classical case we are dealing with Jacobi polynomials.

Among the results in [29] we mention the following small sample.

a) The entries of the $(l + 1) \times (l + 1)$ matrices $H_{n,l}$ can be expressed explicitly in terms of generalized hypergeometric functions

$$_{p+1}F_p \left(\begin{array}{c} a_1 \ldots a_{p+1} \\ b_1 \ldots b_p \end{array} ; t \right), \ 0 \leq p \leq l+1.$$

In particular, if $l = 0$ we get to deal only with Gauss's $_2F_1$.

b) For each (n, l) there are $(l+1)$ families of spherical functions each one of them given (in an appropriate basis) by a diagonal matrix (multiplied by the fixed matrix valued function Φ_π, which happens to be diagonal). One can then form for each (n, l) and $(l+1) \times (l+1)$ matrix whose rows are given by the entries in the diagonals of H_{nl}. Denote by $\Psi(w, t)$ the corresponding (full) matrix. We then have the following *matrix valued bispectral property.* The matrices $\Psi(w, t)$, $0 \leq t \leq 1$, $w = 0, 1, 2, \ldots$ satisfy

$$D\Psi^+(w, t) = (\Lambda\Psi)^+(w, t)$$

$$A_w \Psi(w - 1, t) + B_w \Psi(w, t) + C_w \Psi(w + 1, t) = t\Psi(w, t).$$

D is a second order differential operator independent of w, Λ is a diagonal matrix which depends on w. The matrices A_w, B_w, C_w are independent of t.

5. Heat Kernel Expansions on the Integers

The subject of heat kernel expansions on Riemannian manifolds with or without boundaries cuts across a number of branches of mathematics and serves as an interesting playground for a whole array of interactions with physics.

It suffices to mention, for instance, the work of M. Kac [38] on the issue of recovering the shape of a drum from its pure tones, as well as the suggestion by H. P. McKean and I. Singer [43] that one should be able to find a "heat equation proof" of the index theorem. For a more updated account, see [47], [9].

In the simpler case of a potential for the whole real line, one has that the fundamental solution of the equation

$$u_t = u_{xx} + V(x)u, \ u(x, 0) = \delta_y(x)$$

admits an asymptotic expansion valid for *small t and x close to y,* in the form

$$u(x, y, t) \sim \frac{e^{\frac{-(x-y)^2}{4t}}}{\sqrt{t}} \left(1 + \sum_{n=1}^{\infty} H_n(x, y)t \right).$$

When V is taken to be a potential such that $L = -(d/dx)^2 - V$ belongs to a rank one bispectral ring, then something remarkable happens, namely this expansion gives rise to an exact formula consisting of a finite number of terms and valid *for all x, y as well as all t*. For a few examples of this, one can see my contribution in [33] as well as references in [1].

If one replaces the real line by the integers, one can consider anew the issue of these expansions when one replaces the second difference by an appropriate perturbation of it. I have not seen such treatment in the literature; and, in fact, I am proposing to use some of the first results of the bispectral search to anticipate what kind of expansions might hold in general.

When L is a tridiagonal matrix, the corresponding rank-one bispectral rings have recently been determined, see [31], [32]; and it appears (as I discuss below) that some interesting "finite number of terms phenomenon" is present here too.

From my point of view it would be very nice to use this "bispectral technology" as a tool for the discovery of the general form for a heat kernel expansion on the integers.

If one considers the fundamental solution of

$$u_t(n, t) = u(n + 1, t) - 2u(n, t) + u(n - 1, t)$$

with $u(n, 0) = \delta(n - m)$, one obtains, in terms of the Bessel function of imaginary argument $I_n(t)$, the well-known result

$$u(n, t) = e^{-2t} I_{n-m}(2t).$$

When one applies repeatedly the Darboux process to the matrix L_0, corresponding to the second difference given above, the fundamental solution for the resulting problem

$$u_t = Lu$$

can be seen to be given by a *finite* sum of terms of the form

$$e^{-2t} I_r(2t) t^p,$$

and one extra term.

The sum extends over a few integer values of r and nonnegative integer values of p. The crucial property here is that if $\psi(n, x)$ denotes the wave function of the appropriate maximal rank 1 ring of differential operators, then the product $\psi(n, x)\psi(m, x^{-1})$ can be seen to be given by a constant scalar multiple of x^{n-m} multiplied by a series in x of the form

$$1 + \sum_{j=1}^{\infty} x^j \alpha_j$$

where each coefficient α_j is a *polynomial* in j of a bounded degree. For instance, in the case of three applications of the Darboux process the appropriate τ function is

$$\tau = \det \begin{pmatrix} S_1(n,t) & S_3(n+1,t) & S_5(n+2,t) \\ S_1(n+1,t) & S_3(n+2,t) & S_5(n+3,t) \\ S_1(n+2,t) & S_3(n+3,t) & S_5(n+4,t) \end{pmatrix}$$

with $S_j(n;t)$ the appropriately shifted Schur polynomials introduced in [31], and one can see that the α_j are polynomials (in j) of degree 5.

To conclude the finiteness property mentioned above one needs to resort also to certain identities satisfied by the Bessel functions of imaginary argument. In the case of one Darboux step (when the degree of α_j is one) the identity in question is

$$\sum_{n=p+1}^{\infty} n I_n(z) = \frac{z}{2}(I_p(z) + I_{p+1}(z)).$$

For more details, see [20].

Acknowledgements. The author was supported in part by NSF Grant # DMS94-00097 and by AFOSR under Contract FDF49620-96-1-0127.

References

1. I. G. Avramidi and R. Schimming, *Heat kernel coefficients for the matrix Schrödinger operator*, J. Math. Physics **36** (1995), 5042–5054.
2. B. Bakalov, E. Horozov, and M. Yakimov, *Bispectral algebras of commuting ordinary differential operators*, Comm. Math. Phys. **190** (1997), 331–373.
3. B. Bakalov, E. Horozov, and M. Yakimov, *Highest weight modules over the $W_{1+\infty}$ algebra and the bispectral problem*, Duke Math. J. **93**, 41–72.
4. B. Bakalov, E. Horozov, and M. Yakimov, *General methods for constructing bispectral operators*, Phys. Lett. A **222**(1996), 59–66.
5. Y. Berest, *Huygens principle and the bispectral problem*, in: "The Bispectral Problem", CRM, Harnard and Kasman, eds., Proceedings, Vol. 14 (1998), 11–30.
6. Y. Berest and A. Kasman, *D-modules and Darboux transformations*, Lett. in Math. Physics **43** (1998), 279–294.
7. Y. Berest and A. Veselov, *Huygens principle and integrability*, Uspekhi Math. Nauk. **49** (1994), Russian Math. Surveys **49** (1994), 5–77.
8. Y. Berest and G. Wilson, *Classification of rings of differential operators on affine curves*, Intern. Math. Research Notes **2** (1999), 105–109.
9. N. Berline, E. Getzler, and M. Vergue, *Heat kernels and Dirac operators*, Grundlehren der Mathematischen Wissenchaften, Vol 298, Springer, Berlin, Heidelberg (1992).
10. Bochner, *Über Sturm-Liouvillesche Polynomsysteme*, Math. Z. **29** (1929), 730–736.
11. O. A. Chalykh, M. V. Feigin, and A. P. Veselov, *Multidimensional Baker–Akhiezer functions and Huygens' principle*, Comm. Math. Phys. **206** (1999), 533–566.
12. M. F. E. de Jeu, *The Dunkl transform*, Inventiones Math. **113** (1993), 147–162.
13. Ph. Delsarte, Y. Genin, and V. Kamp, *Orthogonal polynomial matrices on the unit circle*, IEEE Transact. Circuits & Systems **25** (1978), 149–160.

14. H. Duistermaat and F. A. Grünbaum, *Differential equations in the spectral parameter*, Comm. Math. Phys. **103** (1986), 177–240.
15. Ch. Dunkl, *Hankel transforms associated to finite reflection groups, hypergeometric functions in domains of positivity, Jack polynomials and applications* (Tampa, Florida, 1991), Contemporary Mathematics **138** (1992), 123–138.
16. A. J. Duran and W. Van Assche, *Orthogonal matrix polynomials and higher order recurrences*, Linear Algebra Appl. **219** (1995), 261–280.
17. G. Felder, Y. Markov, V. Tarasov, and A. Varchenko, *Differential equations compatible with KZ equations*, to appear.
18. Ya. I. Granovskii, A. S. Zhedanov, and I. M. Lutsenko, *Mutual integrability, quadratic algebras, and dynamical symmetry*, Ann. Phys. **217** (1992), 1–20.
19. F. A. Grünbaum, *Discrete models of the harmonic oscillator and a discrete analog of Gauss' hypergeometric equation*, to appear in The Ramanujan Journal.
20. F. A. Grünbaum and P. Iliev, *Heat kernel expansions on the integers*, in preparation.
21. F. A. Grünbaum, *Some plots of Bessel functions of two variables*, to appear in Proceedings of the Nato Workshop on Harmonic Analysis, Ciocco, Italy, 2000.
22. F. A. Grünbaum and L. Haine, *Associated polynomials, spectral matrices and the bispectral problem*, Meth. Appl. Anal. **6**(2000), 209–224.
23. F. A. Grünbaum and L. Haine, *A theorem of Bochner, revisited*, in: "Algebraic Aspects of Integrable Systems: In Memory of Irene Dorfman", A. S. Fokas and I. M. Gelfand, eds., Progr. Nonlinear Differential Equations, Birkhäuser, Boston, 1996, pp. 143–172.
24. F. A. Grünbaum and L. Haine, *Bispectral Darboux transformations: An extension of the Krall polynomials*, IMRN **8** (Internat. Math Res. Notices) (1997), 359–392.
25. F. A. Grünbaum and L. Haine, *The q-version of a Theorem of Bochner*, J. Computational Appl. Math. **68** (1996), 103–114.
26. F. A. Grünbaum and L. Haine, *Some functions that generalize the Askey-Wilson polynomials*, Commun. Math. Phys. **184** (1997), 173–202.
27. F. A. Grünbaum and L. Haine, *On a q-analog of Gauss' equation and some q-Riccati equations*, in: "Special Functions, q-Series and Related Topics", M. E. H. Ismail, D. R. Masson and M. Rahman, eds., Fields Institute Commun., Vol. 14, Amer. Math. Soc, Providence, 1997 ,pp. 77–81.
28. F. A. Grünbaum, L. Haine, and E. Horozov, E., *Some functions that generalize the Krall–Laguerre polynomials*, J. Compl. Appl. Math. **106** (1999), 271–297.
29. F. A. Grünbaum, I. Pacharoni, and J. A. Tirao, *Matrix valued spherical functions associated to the complex projective plane*, submitted for publication.
30. F. A. Grünbaum and M. Yakimov, *Discrete bispectral Darboux transformations from Jacobi polynomials*, in preparation.
31. L. Haine and P. Iliev, *The bispectral property of a q-deformation of the Schur polynomials and the q-KdV hierarchy*, J. Phys. A: Math. Gen. **30** (1997), 7217–7227.
32. L. Haine and P. Iliev, *Commutative rings of difference operators of an adelic flag manifold*, Internat. Math. Res. Notices **6** (2000), 281–323.
33. J. Harnard and A. Kasman, *The bispectral problem*, CRM Montreal Proceedings, Vol 14 (1998).
34. G. Heckman, *Integrable systems and reflection groups*, Unpuglished lecture notes, Fall School on Hamiltonian Geometry, Woudschoten, 1992.
35. G. Heckman and H. Schlicktkrull, *Harmonic analysis and special functions on symmetric spaces*, Academic Press, San Diego, 1994. Series title: Perspective in Mathematics **16**.
36. M. E. H. Ismail and D. Masson, *Some continued fractions related to elliptic functions*, Contemporary Mathematics **236** (1999), 149–166.
37. M. E. H. Ismail and G. Valent, *On a family of orthogonal polynomials related to elliptic functions*, Ill. J. Math., **42** (1998), 294–312.
38. M. Kac, *Can one hear the shape of a drum?*, Amer. Math. Month. **73** (1966), 1–23.
39. A. Kasman and M. Rothstein, *Bispectral Darboux transformations: the generalized*

Airy case, Physics D **102** (1997), 159–176.

40. T. Koorwinder, *Matrix elements of irreducible representations of $SU(2) \times SU(2)$ and vector valued polynomials*, SIAM J. Math. Anal. **16** (1985), 602–613.

41. H. L. Krall, *Certain differential equations for Tchebycheff polynomials*, Duke Math. J. **4** (1938), 705–718.

42. F. Magri and J. Zubelli, *Differential equations in the spectral parameter, Darboux transformations and a hierarchy of master symmetries for KdV*, Comm. Math. Phys. **141** (1991) , 329–351.

43. J. P. McKean and I. Singer, *Curvature and the eigenvalues of the Laplacian*, J. Differential Geometry **1** (1967), 43–69.

44. E. Opdam, *Dunkl operators, Bessel functions and the discriminant of a finite Coxeter group*, Compositio Math. **85** (1993), 333–373.

45. M. Petkovsek, H. Wilf, and D. Zeilberger, $A = B$, A. K. Peters, Wellesley, 1996.

46. M. Rahman, *The associated classical orthogonal polynomials*, this volume.

47. S. Rosenberg, *The Laplacian on a Riemannian Manifold*, London Math. Society, Student Texts **31**, Cambridge, UK, 1997.

48. S. Ruijsenaars, *Special functions defined by analytic difference equations*, this volume.

49. V. Spiridonov and A. Zhedanov, *Generalized eigenvalue problem and a new family of rational functions biorthogonal on elliptic grids*, this volume.

50. P. Terwilliger, *Two linear transformations each tridiagonal with respect to an eigenbasis of the other*, to appear.

51. J. F. van Diejen and V. P. Spiridonov, *An elliptic Macdonald-Morris conjecture and multiple modular hypergeometric sums*, preprint, May 2000.

52. J. F. van Diejen and L. Vinet, *The quantum dynamics of the compactified trigonometric Ruijsenaars-Schneider Model*, Comm. Math. Phys. **197** (1998), 33–74.

53. G. Wilson, *Bispectral commutative ordinary differential operators*, J. reine angew Math. **442** (1993), 177–204.

54. G. Wilson, *Collisions of Calogero-Moser particles and an adelic Grassmanian*, Inventiones Math. **133** (1998), 1–41.

55. J. Wimp, *Explicit formulas for the associated Jacobi polynomials and some applications*, Canad. J. Math. **39** (1987), 983–1000.

THE BOCHNER-KRALL PROBLEM: SOME NEW PERSPECTIVES

LUC HAINE
Department of Mathematics
Université Catholique de Louvain
Chemin du Cyclotron 2
1348 Louvain-la-Neuve Belgium

Abstract. The Bochner-Krall problem asks for the classification of all families of orthogonal polynomials that are also eigenfunctions of a differential operator. We survey some recent work on this problem. An important issue is to extend the problem by allowing for a doubly infinite three-term recursion relation, instead of a semi-infinite one. In this way, the problem gets connected with the associated classical orthogonal polynomials, the Darboux factorization method and the Burchnall-Chaundy theory of commutative rings of difference operators.

1. Introduction

We denote by L a doubly infinite tridiagonal matrix of the form

$$L = a(n)E^{-1} + b(n)I + E, \qquad (1.1)$$

where I is the identity operator and E, E^{-1} denote respectively the forward and the backward shift operators acting on an infinite vector $h = (\dots, h_{-1}, h_0, h_1, \dots)^T$ as follows

$$(Eh)_n = h_{n+1}, \quad (E^{-1}h)_n = h_{n-1}, \quad n \in \mathbb{Z}. \qquad (1.2)$$

The aim of this article is to survey some recent results on the following problem:

EXTENDED BOCHNER-KRALL PROBLEM. *Find all doubly infinite tridiagonal matrices L, for which some family of eigenfunctions $f(z) = (\dots, f_{-1}(z), f_0(z), f_1(z), \dots)$ satisfying*

$$a(n)f_{n-1}(z) + b(n)f_n(z) + f_{n+1}(z) = zf_n(z), \qquad (1.3)$$

J. Bustoz et al. (eds.), Special Functions 2000, 141–178.

is also a family of eigenfunctions of a differential operator in the spectral parameter z

$$\left\{ \sum_{i=0}^{K} c_i(z) \frac{d^i}{dz^i} \right\} f_n(z) = \lambda(n) f_n(z). \qquad (1.4)$$

We always assume that that all $a(n) \neq 0$ and that the eigenvalues $\lambda(n)$ are distinct. A doubly infinite problem where these assumptions are violated is usually rather degenerate and uninteresting. We write (1.3) and (1.4) as

$$L f = z f, \qquad (1.5)$$

$$B f = \Lambda f, \qquad (1.6)$$

where B denotes a differential operator in z and Λ is the diagonal matrix with entries $\lambda(n)$. If we impose the additional conditions $f_{-1}(z) = 0, f_0(z) = 1$, then the $f_n(z), n \geq 0$, are polynomials of degree n, and the problem reduces to the so-called Bochner-Krall problem of classifying all families of orthogonal polynomials which are eigenfunctions of a differential operator. If this operator is of order two, Bochner (1929) [9] proved that (with similar non-degeneracy assumptions as those stated above) the only solutions are provided by the Hermite, the Laguerre, the Jacobi and the (lesser known) Bessel polynomials. Following Koornwinder [40], I shall refer to these polynomials as to the *very classical orthogonal polynomials*, to distinguish them from the more general orthogonal polynomials composing the Askey tableau. The general question was asked by H.L. Krall (1938) [42]. He proved that the operator has to be of even order and in [43], he solved the problem completely for the case of an operator of order four. I refer the reader to the work of A. M. Krall [41] and L. L. Littlejohn and collaborators [15], [46], for a thorough account on the subject. Though no complete results on the higher order solutions are known, a conjecture has emerged thanks to some of the recent work on the problem. The motivation for relaxing the special conditions leading to the orthogonal polynomial solutions, comes from the analogy with the Duistermaat-Grünbaum problem which we now state:

DUISTERMAAT-GRÜNBAUM PROBLEM. *To determine all potentials $V(x)$ for which the corresponding Schrödinger operators possess a family of eigenfunctions $\Psi(x, z)$ satisfying*

$$L\Psi(x, z) \equiv \left\{ \frac{d^2}{dx^2} + V(x) \right\} \Psi(x, z) = z\Psi(x, z), \qquad (1.7)$$

*which is also a family of eigenfunctions of a differential operator in the
spectral variable z, for an eigenvalue Λ which is a function of x*

$$B\Psi(x,z) \equiv \left\{ \sum_{i=1}^{K} c_i(z)\frac{\mathrm{d}^i}{\mathrm{d}z^i} \right\} \Psi(x,z) = \Lambda(x)\Psi(x,z). \tag{1.8}$$

This problem as well as the previous one are now part of a big family of problems known as "bispectral problems," see the contribution of
F. A. Grünbaum [18] to these proceedings for an overview of the subject.
One can look for purely discretized versions of the problem, q-versions,
versions involving higher order than three-term recursion relations, etc.,
see [21], [22], [24], [25], [29], [30], [32], [52]. A purely discrete version of the
problem involving only finite tridiagonal matrices, is related to the so-called
theory of Leonard pairs and association schemes, see the survey article [57]
of P. Terwilliger for many exciting results in this direction. There is even
quite a bit of recent activity about bispectral problems in several variables, notably in connection with the so-called Huygens' principle, see [7],
[11]. The (extended) Bochner-Krall problem can be thought of as a partially discretized version of the Duistermaat-Grünbaum problem, with the
Schrödinger operator replaced by a doubly infinite tridiagonal matrix.

One of the key findings of Duistermaat and Grünbaum [14] is that all
the solutions of the problem they posed can be obtained from a few *basic solutions* by iteration of the so-called Darboux factorization method. I
remind the reader that the method consists in factorizing the Schrödinger
operator as a product of two first-order operators

$$L = \left(\frac{\mathrm{d}}{\mathrm{d}x} + \frac{\phi'(x)}{\phi(x)} \right)\left(\frac{\mathrm{d}}{\mathrm{d}x} - \frac{\phi'(x)}{\phi(x)} \right),$$

with $\phi(x)$ an *arbitrary* function in the kernel of L, and to producing a new
operator $\hat{L}$, by exchanging the order of the factors

$$\hat{L} = \left(\frac{\mathrm{d}}{\mathrm{d}x} - \frac{\phi'(x)}{\phi(x)} \right)\left(\frac{\mathrm{d}}{\mathrm{d}x} + \frac{\phi'(x)}{\phi(x)} \right) \equiv \frac{\mathrm{d}^2}{\mathrm{d}x^2} + \hat{V}(x),$$

$$\text{with } \hat{V}(x) = V(x) + 2\frac{\mathrm{d}^2}{\mathrm{d}x^2}\log\phi(x).$$

Because of the freedom in the choice of $\phi(x)$, *one free parameter* appears
in $\hat{V}(x)$. This method has been reinvented several times in the context of
differential operators, see [12], [33], [55]. Darboux himself (see [12], Livre IV,
Chap. IX, No. 408) credits the method to Moutard. The modern theory of
solitons resurrected the method, with fundamental contributions by Adler

and Moser [1], Deift [13], Flaschka and McLaughlin [16], Matveev [47], [48] and Matveev and Salle [49], [50]. It seems however not to be widely known how this method did appear as a systematic tool in the study of bispectral problems. It is worth reminding the story, which is relevant for my purpose.

Assuming that the operator B is of order K, a fundamental identity in [14] expresses the compatibility between the pair of equations (1.7) and (1.8), namely:

$$(\text{ad } L)^{K+1}(\Lambda) \equiv [L, [L, \ldots , [L, \Lambda] \ldots]] = 0. \qquad (1.9)$$

Solving this equation for $K = 1$ gives $V = $ constant (which after a translation in z can always be put to be zero), as the only solution. For $K = 2$, one finds that the only solutions modulo translations and scalings in x and z by some constants, are provided by the Airy, $V(x) = x$, and the Bessel, $V(x) = c/x^2$, potentials. $K = 3$ doesn't bring anything new, but $K = 4$ leads to two families of potentials, one of which Duistermaat and Grünbaum [14] recognized as the first of the families of rational solutions of the Korteweg-de Vries (KdV) equation, vanishing as $x \to \infty$. The fact that the families of rational solutions of the KdV equation were shown by Adler and Moser [1] to be obtainable via the Darboux factorization method, starting at $V(x) = 0$, hinted at the connection between bispectrality and the method of the Darboux transformation. The second new family of bispectral potentials that was discovered in [14], was shown in that paper to be obtainable by the same method starting from some instances of the Bessel potentials. This beautiful discovery has revolutionized the subject, though it seems fair to say that a direct relation between (1.9) and the Darboux factorization method still resists any transparent explanation.

Coming back to the extended Bochner-Krall problem (1.5), (1.6), the same compatibility argument as the one used in [14] leads to (1.9), where this equation must now be interpreted as a matrix equation to be satisfied by the tridiagonal matrix L and the diagonal matrix Λ. It is already instructive to solve this equation for $K = 1$. A quick computation shows that it amounts to the equations

$$\lambda(n+1) - 2\lambda(n) + \lambda(n-1) = 0, \quad [\lambda(n) - \lambda(n-1)][b(n) - b(n-1)] = 0,$$
$$[\lambda(n) - \lambda(n-1)]a(n) - [\lambda(n-1) - \lambda(n-2)]a(n-1) = 0.$$

One sees immediately that, under the non-degeneracy assumption that all $a(n) \neq 0$ and all $\lambda(n)$ be distinct, $\lambda(n) = \gamma n + \delta$, $b(n) = b$ and $a(n) = a$, is the only solution of these equations, with a, b, γ, δ arbitrary constants, $a \neq 0$ and $\gamma \neq 0$. Thus there are *no* solutions within the context of *semi-infinite* matrices, and up to translation and scaling, the only solution within the realm of *doubly infinite* matrices, is provided by the *discrete second-order derivative operator*

$$L_0 = E^{-1} - 2I + E,$$

with E, E^{-1} as in (1.2). This is reminiscent of the role played by the rational solutions of the KdV equation in the Duistermaat-Grünbaum problem. In collaboration with P. Iliev [30], [31], we have shown that iteration of the (matrix) Darboux transformation at both ends of the continuous spectrum of this operator (which lies on the interval $[-4, 0]$), leads to solutions of the *extended* Bochner-Krall problem. When suitably interpreted, the free parameters in these solutions coming from the iteration of the Darboux transformation, correspond to a deformation of L_0 according to the flows of the (doubly infinite) Toda lattice hierarchy. This is one other famous integrable hierarchy of equations which can be thought of as a discretization of the KdV hierarchy, see [58]. These results are reported in Section 6.

Solving the matrix version of equation (1.9) for $K = 2$ is already much more involved than for its differential version, for which the only solutions are easily seen to be given by the Airy and the Bessel potentials. This work was done in [20] (see also [28]) in collaboration with F. A. Grünbaum, where the fundamental role played by *Gauss' hypergeometric equation* was discovered. In Section 2, I present an improved review of this result, showing that the heart of the matter amounts to evaluating a sum with hypergeometric term, see (2.15) and Remark 2.2. With the non-degeneracy assumptions stated above, the only possible tridiagonal matrices L are given by what we propose to call the *associated very classical matrices*. The entries of these matrices are obtained by making an *arbitrary* shift in the coefficients of the recursion relations satisfied by any one of the four families of very classical orthogonal polynomials, extending them over *all* integers, that is

$$a(n) = \tilde{a}(n + \varepsilon), \quad b(n) = \tilde{b}(n + \varepsilon), \quad n \in \mathbb{Z}, \qquad (1.10)$$

where $\tilde{a}(n)$, $n \geq 1$, and $\tilde{b}(n)$, $n \geq 0$, are the coefficients of the three-term recursion relations satisfied by the monic Jacobi, Bessel, Laguerre and Hermite polynomials. The same coefficients are precisely those that define the recursion relations satisfied by the *associated polynomials* when the conditions $f_{-1}(z) = 0$, $f_0(z) = 1$, are imposed. However, it is important to realize that, as soon as $\varepsilon \neq 0$, the functions $f_n(z)$ that solve (1.3) and (1.4) with a differential operator B of order two, are *not* given by the associated polynomials, but rather by a *two-dimensional* space of contiguous hypergeometric (or confluent hypergeometric) functions; see Theorem 2.1, formulas (2.4) and (2.5) for the precise definition of these functions, as well as Subsection 2.2 for explicit formulas.

The history and the recent developments on the subject of the associated classical orthogonal polynomials is beautifully told in the contribution of Rahman [54] to these proceedings. Some pioneering contributions belong to Askey and Ismail [3], Askey and Wimp [4], and Wimp [62], where the orthogonality measures for several families of associated very classical orthogonal polynomials were explicitly computed. In the remarkable paper of Wimp [62], the orthogonality measure is computed for the associated Jacobi polynomials. The formula is quite involved! In [23], in collaboration with F. A. Grünbaum, we use Wimp's results to compute the two by two spectral matrix measure, corresponding to the doubly infinite tridiagonal matrix whose entries are formed with the coefficients of the recursion relation satisfied by the associated Jacobi polynomials, extended over all integers. We were pleased to find that, when expressed in the basis of bispectral functions given by the $f_n(z)$ above, the spectral matrix measure becomes diagonal and quite simple, restoring the simplicity of the familiar orthogonality relations satisfied by the Jacobi polynomials, see especially formula (2.64). These results are reviewed in Subsection 2.3.

In Section 3, I explain the so-called method of bispectral Darboux transformations, which gives some sufficient conditions (most probably also necessary) for the Darboux factorization method to preserve bispectrality, see Theorem 3.1. This part reviews a joint work with F. A. Grünbaum and E. Horozov [25], [26]. It takes some of its inspiration from previous works by Bakalov, Horozov and Yakimov [5], [6], as well as Kasman and Rothstein [35], on the purely continuous version of the bispectral problem. The method is illustrated in Section 4 on an example (which is presented here for the first time), leading to the so-called Krall-Jacobi polynomials and some of their generalizations. A similar result was obtained previously by J. Koekoek and R. Koekoek [37] and Zhedanov [63], by different means. This example is complicated enough so that it illustrates quite clearly some of the main difficulties that can occur in applying Theorem 3.1. In Section 5, I list the results which have been obtained by using the method. Finally, Section 6 reviews my joint work with P. Iliev [30], [31], about the discrete analogues of the rational solutions of the Korteweg-de Vries equation, already alluded to above. These solutions are built in terms of some rational curves, with only cusplike singularities. This connects our subject with the so-called Burchnall-Chaundy theory [10] of commutative rings of difference operators. In the purely differential version of the bispectral problem, this line of work was initiated by G. Wilson [60], [61].

In analogy with the result of Duistermaat and Grünbaum [14], it is natural to expect that all the solutions to the extended Bochner-Krall problem are obtained by iteration of the Darboux transformation, starting from some of the basic solutions in Section 2. A conjecture to this effect is formulated at the end of the paper.

2. A New Road to the Associated Very Classical Orthogonal Polynomials

2.1. SOLVING $(ad\,L)^3(\Lambda) = 0$: THE BASIC SOLUTIONS

Theorem 2.1 ([20], [28]) *Assume that all $a(n)$ in (1.3) are nonzero and that all $\lambda(n)$ in (1.4) are distinct. Then the only solutions of (1.3) and (1.4), with B an operator of order 2, are described as follows:*

i) The entries $\lambda(n), b(n)$ and $a(n)$ of the matrices Λ and L are defined by the equations below with α, β, γ, δ, λ, a, b seven arbitrary free parameters:

$$\lambda(n) = \frac{n(n+1)}{2}\gamma + n\delta + \lambda, \tag{2.1}$$

$$[\lambda(n) - \lambda(n+1)][\lambda(n-1) - \lambda(n)]b(n) =$$
$$[\lambda(0) - \lambda(1)][\lambda(-1) - \lambda(0)]b + [\lambda(0) - \lambda(n)]\beta, \tag{2.2}$$

$$[\lambda(n+1) - \lambda(n-1)][\lambda(n) - \lambda(n-2)]a(n) =$$
$$[\lambda(1) - \lambda(-1)][\lambda(0) - \lambda(-2)]a + [\lambda(n) - \lambda(n+1)]^2 b^2(n)$$
$$- [\lambda(0) - \lambda(1)]^2 b^2 + [\lambda(-1) + \lambda(0) - \lambda(n-1) - \lambda(n)]\alpha$$
$$+ \Big\{[\lambda(0) - \lambda(1)]b - [\lambda(n) - \lambda(n+1)]b(n)\Big\}\beta. \tag{2.3}$$

ii) The functions $f_n(z), n \in \mathbb{Z}$, are defined as the solutions of the doubly infinite three-term recursion relation determined in i), with initial conditions $f_0(z)$ and $f_1(z)$ given by

$$f_0(z) = w(z), \tag{2.4}$$

$$f_1(z) = (z - b)w(z) + \frac{H(z)}{\gamma + 2\delta}\,w'(z), \text{ with } H(z) = \alpha + \beta z + \gamma z^2, \tag{2.5}$$

where $w(z)$ in (2.4) denotes an arbitrary solution of the equation

$$Bw(z) = 0, \tag{2.6}$$

$$B = \frac{H(z)}{2}\frac{d^2}{dz^2} + (\gamma + \delta)(z - b)\frac{d}{dz} - \frac{a(\gamma - 2\delta)(\gamma + 2\delta)}{2H(z)}. \tag{2.7}$$

The functions $f_n(z)$ so defined, satisfy

$$Bf_n(z) = [\lambda(n) - \lambda(0)]f_n(z), \quad \forall n \in \mathbb{Z}, \tag{2.8}$$

with B the second-order differential operator defined in (2.7).

Proof. <u>Step i)</u> The best way to solve the equation

$$(adL)^3(\Lambda) = 0, \tag{2.9}$$

is to notice that this equation can be "integrated", i.e.

$$(adL)^3(\Lambda) = 0 \Leftrightarrow [L, [L, \Lambda]] = \alpha I + \beta L + \gamma L^2, \tag{2.10}$$

with α, β and γ three integrating constants. Equating the diagonals of (2.9) to zero, starting with the upper one, we obtain at the $(n, n+3)$-th entry the equation:

$$\lambda(n+2) - 3\lambda(n+1) + 3\lambda(n) - \lambda(n-1) = 0,$$

which can be written as

$$[\lambda(n+2) - 2\lambda(n+1) + \lambda(n)] - [\lambda(n+1) - 2\lambda(n) + \lambda(n-1)] = 0,$$

or equivalently

$$\lambda(n+1) - 2\lambda(n) + \lambda(n-1) = \gamma, \tag{2.11}$$

with γ some constant independent of n. This last equation can be further integrated twice, giving

$$\lambda(n) = \frac{n(n+1)}{2}\gamma + n\delta + \lambda,$$

as announced in (2.1).

The $(n, n+2)$-th entry of (2.9) gives

$$[\lambda(n+1) - \lambda(n+2)]b(n+1) + [\lambda(n+1) - \lambda(n-1)]b(n)$$
$$+ [\lambda(n-2) - \lambda(n-1)]b(n-1) = 0,$$

which can be rewritten as

$$\left\{ [\lambda(n+1) - \lambda(n+2)]b(n+1) + [\lambda(n) - \lambda(n-1)]b(n) \right\}$$
$$- \left\{ [\lambda(n) - \lambda(n+1)]b(n) + [\lambda(n-1) - \lambda(n-2)]b(n-1) \right\} = 0,$$

i.e.

$$[\lambda(n) - \lambda(n+1)]b(n) + [\lambda(n-1) - \lambda(n-2)]b(n-1) = \beta, \tag{2.12}$$

with β some constant independent of n. Multiplying this last equation by $\lambda(n-1) - \lambda(n)$ and summing from $n = 1$ till n, we get a telescoping sum leading to

$$[\lambda(n-1) - \lambda(n)][\lambda(n) - \lambda(n+1)]b(n)$$
$$+ [\lambda(0) - \lambda(1)][\lambda(0) - \lambda(-1)]b(0) = [\lambda(0) - \lambda(n)]\beta,$$

which is precisely (2.2), with $b = b(0)$.

Finally the $(n, n+1)$-th entry of (2.9) gives

$$[\lambda(n) - \lambda(n+2)]a(n+1) + [\lambda(n+1) + \lambda(n) - \lambda(n-1) - \lambda(n-2)]a(n)$$
$$+ [\lambda(n-3) - \lambda(n-1)]a(n-1) + [\lambda(n) - \lambda(n-1)][b(n) - b(n-1)]^2 = 0,$$

which using (2.11) and (2.12) becomes

$$\left\{ [\lambda(n) - \lambda(n+2)]a(n+1) + [\lambda(n) - \lambda(n-2)]a(n) - \gamma b^2(n) - \beta b(n) \right\}$$
$$- \left\{ [\lambda(n-1) - \lambda(n+1)]a(n) \right.$$
$$+ [\lambda(n-1) - \lambda(n-3)]a(n-1) - \gamma b^2(n-1) - \beta b(n-1) \Big\} = 0,$$

or equivalently

$$[\lambda(n-1) - \lambda(n+1)]a(n) + [\lambda(n-1) - \lambda(n-3)]a(n-1)$$
$$- \gamma b^2(n-1) - \beta b(n-1) = \alpha, \quad (2.13)$$

with α some other constant independent of n. Multiplying this equation by $\lambda(n-2) - \lambda(n)$ and summing from $n = 1$ till n we obtain

$$[\lambda(n-2) - \lambda(n)][\lambda(n-1) - \lambda(n+1)]a(n) + [\lambda(-1) - \lambda(1)][\lambda(0) - \lambda(-2)]a(0)$$
$$= [\lambda(-1) + \lambda(0) - \lambda(n-1) - \lambda(n)]\alpha$$
$$+ \sum_{i=0}^{n-1} [\lambda(i-1) - \lambda(i+1)][\beta + \gamma b(i)]b(i). \quad (2.14)$$

Using (2.11) and (2.12), one shows by induction on n that

$$\sum_{i=0}^{n-1} [\lambda(i-1) - \lambda(i+1)][\beta + \gamma b(i)]b(i)$$
$$= \left\{ [\lambda(n) - \lambda(n+1)]^2 b^2(n) - [\lambda(0) - \lambda(1)]^2 b^2 \right\}$$
$$+ \beta \left\{ [\lambda(0) - \lambda(1)]b - [\lambda(n) - \lambda(n+1)]b(n) \right\}, \quad (2.15)$$

and thus (2.14) is identical with (2.3), with $a = a(0)$. The diagonal entries of (2.9) are automatically identically zero and the equations obtained by looking below the main diagonal are proportional to those obtained before. It is easy to check that (2.11), (2.12) and (2.13) are equivalent to (2.10).

Step ii) We observe that (2.10) is precisely the compatibility condition between the pair of equations

$$Lf = zf, \tag{2.16}$$
$$Af = Mf, \tag{2.17}$$

with $M = [L, \Lambda]$ and

$$A = H(z)\frac{d}{dz}, \text{ with } H(z) = \alpha + \beta z + \gamma z^2. \tag{2.18}$$

We determine the most general family of functions $f_n(z)$ which satisfy both equations, and we show that these functions are automatically the eigenfunctions of a second-order differential operator. We proceed as follows. We write

$$(Mf(z))_n = M_{n-1,n}f_{n-1}(z) + M_{n,n}f_n(z) + M_{n,n+1}f_{n+1}(z).$$

We can always modify the operator $M = [L, \Lambda]$ into

$$M = [L, \Lambda] + xI + yL, \tag{2.19}$$

with x, y arbitrary constants. Then

$$M_{n,n-1} = [\lambda(n-1) - \lambda(n) + y]a(n),$$
$$M_{n,n} = x + yb(n), \quad M_{n,n+1} = \lambda(n+1) - \lambda(n) + y,$$

and, using the three-term recursion relation (2.16), we can rewrite (2.17), (2.18) either as

$$H(z)f_n'(z) = [\lambda(n-1) - \lambda(n+1)]a(n)f_{n-1}(z)$$
$$+ \Big\{[\lambda(n+1) - \lambda(n) + y]z + [\lambda(n) - \lambda(n+1)]b(n) + x\Big\}f_n(z), \tag{2.20}$$

or

$$H(z)f_{n-1}'(z) = [\lambda(n) - \lambda(n-2)]f_n(z)$$
$$+ \Big\{[\lambda(n-2) - \lambda(n-1) + y]z +$$
$$[\lambda(n-1) - \lambda(n-2)]b(n-1) + x\Big\}f_{n-1}(z). \tag{2.21}$$

We differentiate (2.20) once and replace in it $f'_{n-1}(z)$ by its expression obtained from (2.21). Replacing then $f_{n-1}(z)$ in that equation by the expression obtained from (2.20), we finally get a second-order differential equation for $f_n(z)$. Using (2.1), (2.2) and (2.3), this equation remarkably simplifies, reducing to

$$Bf_n(z) = [\lambda(n) - \lambda(0)]f_n(z),$$

with

$$B = \frac{H(z)}{2}\frac{\mathrm{d}^2}{\mathrm{d}z^2} - (x + yz)\frac{\mathrm{d}}{\mathrm{d}z}$$
$$+ \frac{1}{2H(z)}\Big\{(y - \delta)(y + \gamma + \delta)z^2 + 2(x(y + \gamma) + b\delta(\gamma + \delta))z$$
$$+ (x - b(\gamma + \delta))(x + b(\gamma + \delta) + \beta) - \alpha(y + \gamma + \delta) - a(\gamma - 2\delta)(\gamma + 2\delta)\Big\}.$$

Since x and y in (2.19) are arbitrary constants, we can pick for instance

$$x = b(\gamma + \delta), \quad y = -(\gamma + \delta), \tag{2.22}$$

which leads to (2.8) with B as in (2.7). Putting $n = 0$ in this equation gives equation (2.6) for $f_0(z)$ as specified in (2.4), and putting $n = 1$ in (2.21), with x and y as in (2.22), gives (2.5) determining $f_1(z)$. All the other f_n's are then uniquely determined by the three-term recursion relation or, equivalently, by (2.20) and (2.21), since (2.16) and (2.17) are compatible. This completes the proof of the theorem.   □

Remark 2.2 As mentioned in the introduction, the heart of the matter in the above proof lies in the end in our ability of evaluating the sum (2.15) in "closed form." I leave it to the reader to check that this sum can in fact be evaluated using the famous Gosper's algorithm [17] for indefinite hypergeometric summation. I refer the reader to [2] or [53] for a nice account of Gosper's algorithm. This seems to indicate that much more work could be done in the direction of using the methods developed by Petkovšek, Wilf and Zeilberger (see [53]) for solving bispectral problems.

We can always translate and rescale the $\lambda(n)$'s by an arbitrary constant. Similarly, we can translate the $b(n)$'s by an arbitrary constant and rescale z. So, in fact, we are left with $7 - 4 = 3$ free parameters in the general solution described in part i) of Theorem 2.1. For generic values of α, β and γ, the polynomial $H(z)$ in (2.5) has two distinct roots. Thus, the second-order homogeneous equation (2.6) has three regular singular points, and can be converted to *Gauss' hypergeometric equation*. Below, we show that the three effective free parameters in the solution can be (generically) identified with the standard three parameters of Gauss' equation. We also discuss all the possible confluent forms. We show that the only matrices L that can occur are precisely those defining the recursion relations satisfied by the associated very classical orthogonal polynomials, extended over *all* integers, as announced in (1.10). There are also some interesting limiting solutions, see Remark 2.3. We call these matrices the *associated very classical matrices*. However, as soon as the parameter $a \neq 0$, the functions $f_n(z)$ defined in part ii) of Theorem 2.1 are not polynomials in general. It is only when $a = 0$ that we can get polynomials, by picking $w(z) = 1$ as a solution of (2.6). These polynomials coincide then with the very classical orthogonal polynomials.

2.2. THE ASSOCIATED VERY CLASSICAL MATRICES

2.2.1. *The associated Jacobi matrices*
This case corresponds to the generic situation, when $H(z)$ in (2.5) has two distinct roots. Then (2.6) can be reduced to Gauss' hypergeometric equation

$$z(z-1)w''(z) + ((a+b+1)z - c)w'(z) + abw(z) = 0.$$
$$(2.23)$$

The entries of the doubly infinite tridiagonal matrices L are given by

$$a(n) = \frac{(n+a-1)(n-b)(n-c+a)(n+c-b-1)}{(2n-b+a-2)(2n-b+a-1)^2(2n-b+a)},$$

$$b(n) = \frac{1}{2}\left(\frac{(b+a-1)(2c-b-a-1)}{(2n-b+a-1)(2n-b+a+1)} + 1\right). \qquad (2.24)$$

We notice that by putting

$$a = \alpha + \beta + 1 + \varepsilon, \quad b = -\varepsilon, \quad c = \beta + 1, \qquad (2.25)$$

we obtain that

$$a(n) = \tilde{a}(n+\varepsilon), \quad b(n) = \tilde{b}(n+\varepsilon), \qquad (2.26)$$

with $\tilde{a}(n)$ and $\tilde{b}(n)$ the coefficients of the three-term recursion relation satisfied by the monic Jacobi polynomials shifted to the interval $[0,1]$:

$$\tilde{a}(n) = \frac{n(n+\alpha)(n+\beta)(n+\alpha+\beta)}{(2n+\alpha+\beta-1)(2n+\alpha+\beta)^2(2n+\alpha+\beta+1)},$$

$$\tilde{b}(n) = \frac{1}{2}\left(\frac{\beta^2-\alpha^2}{(2n+\alpha+\beta)(2n+\alpha+\beta+2)} + 1\right). \qquad (2.27)$$

For generic values of α, β and ε, the general solution of the doubly infinite three-term recursion relation with initial conditions

$$f_0(z) = w(z),$$

$$f_1(z) = \left\{ \frac{\varepsilon + \alpha + \beta + 1}{2\varepsilon + \alpha + \beta + 1} z - \frac{(\varepsilon + \beta + 1)(\varepsilon + \alpha + \beta + 1)}{(2\varepsilon + \alpha + \beta + 1)(2\varepsilon + \alpha + \beta + 2)} \right\} w(z)$$

$$+ \frac{z(z-1)}{2\varepsilon + \alpha + \beta + 1} w'(z), \tag{2.28}$$

with $w(z)$ an arbitrary solution of (2.23), is given by

$$f_n(z) = c_1 u_n(z) + c_2 v_n(z), \tag{2.29}$$

with

$$u_n(z) = (-1)^n \frac{(\alpha + \beta + 1 + \varepsilon)_n (\beta + 1 + \varepsilon)_n}{(\alpha + \beta + 1 + 2\varepsilon)_{2n}}$$

$$\times {}_2F_1(\alpha + \beta + 1 + n + \varepsilon, -n - \varepsilon; \beta + 1; z), \tag{2.30}$$

$$v_n(z) = (-1)^n \frac{(1 + \varepsilon)_n (\alpha + 1 + \varepsilon)_n}{(\alpha + \beta + 1 + 2\varepsilon)_{2n}} z^{-\beta}$$

$$\times {}_2F_1(\alpha + 1 + n + \varepsilon, -\beta - n - \varepsilon; 1 - \beta; z). \tag{2.31}$$

Here, as well as in the rest of the paper, ${}_2F_1(a, b; c; z)$ denotes Gauss' hypergeometric series, and $(a)_n = \Gamma(a+n)/\Gamma(a)$ is the standard shifted factorial. The $f_n(z)$ satisfy

$$z f_n(z) = (Lf(z))_n, \quad A f_n(z) = (Mf(z))_n, \quad B f_n(z) = \lambda(n) f_n(z),$$

with L as in (2.26), (2.27) and

$$A = z(z-1)\frac{d}{dz}, \quad B = z(z-1)\frac{d^2}{dz^2} + ((\alpha + \beta + 2)z - (\beta + 1))\frac{d}{dz}, \tag{2.32}$$

$$M_{n-1,n} = -a(n)(n + \varepsilon + \alpha + \beta + 1),$$

$$M_{n,n} = \frac{(\beta - \alpha)(n + \varepsilon)(n + \varepsilon + \alpha + \beta + 1)}{(2n + 2\varepsilon + \alpha + \beta)(2n + 2\varepsilon + \alpha + \beta + 2)},$$

$$M_{n,n+1} = n + \varepsilon, \tag{2.33}$$

$$\lambda(n) = (n + \varepsilon)(n + \varepsilon + \alpha + \beta + 1). \tag{2.34}$$

When $\varepsilon = 0$, the choice $w(z) = 1$ in (2.28) leads to the functions $u_n(z)$ above, which agree then with the monic Jacobi polynomials.

Remark 2.3 As pointed out to me by Mourad Ismail, the case $\alpha + \beta = 0$ should be treated with some care when $\varepsilon = 0$. If we first take the limit $\varepsilon \to 0$ and then let $\alpha + \beta \to 0$, we are led to the usual Jacobi polynomials with parameters $\alpha, -\alpha$. But, if we first let $\alpha + \beta \to 0$ in (2.26) and then take the limit $\varepsilon \to 0$, we obtain a doubly infinite solution to the extended Bochner-Krall problem with all $a(n) \neq 0$ (for generic α). The limiting coefficients $a(n), n \geq 1$ and $b(n), n \geq 0$, coincide then with the coefficients defining the recursion relation for the exceptional Jacobi polynomials introduced by Ismail and Masson in [34]. However, the common solution to (1.5) and (1.6) (for the doubly infinite problem) is not provided by these polynomials.

2.2.2. *The associated Bessel matrices*
This situation corresponds to the case when $H(z)$ in (2.5) is still of degree 2 ($\gamma \neq 0$), but its two roots coincide. Then equation (2.6) can be brought to the form

$$z^2 w''(z) + (az + 1)w'(z) - \varepsilon(\varepsilon + a - 1)w(z) = 0. \tag{2.35}$$

The entries of the matrix L are given by

$$a(n) = \tilde{a}(n + \varepsilon), \quad b(n) = \tilde{b}(n + \varepsilon), \tag{2.36}$$

with $\tilde{a}(n)$ and $\tilde{b}(n)$ the coefficients of the three-term recursion relation satisfied by the Bessel polynomials

$$\tilde{a}(n) = -\frac{n(n + a - 2)}{(2n + a - 3)(2n + a - 2)^2(2n + a - 1)},$$
$$\tilde{b}(n) = \frac{2 - a}{(2n + a - 2)(2n + a)}. \tag{2.37}$$

Any solution $f_n(z), n \in \mathbb{Z}$, of the doubly infinite three-term recursion relation with initial condition

$$f_0(z) = w(z),$$
$$f_1(z) = \left\{ \frac{\varepsilon + a - 1}{(2\varepsilon + a - 1)(2\varepsilon + a)} + \frac{\varepsilon + a - 1}{2\varepsilon + a - 1}z \right\}w(z)$$
$$+ \frac{z^2}{2\varepsilon + a - 1}w'(z), \tag{2.38}$$

with $w(z)$ an arbitrary solution of (2.35) satisfies

$$z f_n(z) = (Lf(z))_n, \quad A f_n(z) = (Mf(z))_n, \quad B f_n(z) = \lambda(n) f_n(z),$$

with L as in (2.36), (2.37) and

$$A = z^2 \frac{d}{dz}, \quad B = z^2 \frac{d^2}{dz^2} + (az+1)\frac{d}{dz}, \qquad (2.39)$$

$$M_{n,n-1} = -a(n)(n+\varepsilon+a-1),$$

$$M_{n,n} = -\frac{2(n+\varepsilon)(n+\varepsilon+a-1)}{(2n+2\varepsilon+a-2)(2n+2\varepsilon+a)}, \quad M_{n,n+1} = n+\varepsilon, \qquad (2.40)$$

$$\lambda(n) = (n+\varepsilon)(n+\varepsilon+a-1). \qquad (2.41)$$

When $\varepsilon = 0$, the choice $w(z) = 1$ in (2.38) leads to the monic Bessel polynomials.

2.2.3. *The associated Laguerre matrices*

This case arises when $\gamma = 0$ and $\beta \neq 0$, for $H(z)$ in (2.5). Equation (2.6) can then be reduced to the confluent hypergeometric equation

$$zw''(z) + (\alpha + 1 - z)w'(z) + \varepsilon w(z) = 0. \qquad (2.42)$$

The entries of the corresponding tridiagonal matrices L are given by

$$a(n) = \tilde{a}(n+\varepsilon), \quad b(n) = \tilde{b}(n+\varepsilon), \qquad (2.43)$$

with $\tilde{a}(n)$ and $\tilde{b}(n)$ the coefficients of the three-term recursion relation satisfied by the monic Laguerre polynomials

$$\tilde{a}(n) = n(n+\alpha), \quad \tilde{b}(n) = 2n+\alpha+1. \qquad (2.44)$$

For generic values of α and ε, the general solution $f_n(z), n \in \mathbb{Z}$, of the doubly infinite three-term recursion relation with initial conditions

$$f_0(z) = w(z), \quad f_1(z) = \{z - (\varepsilon + \alpha + 1)\}w(z) - zw'(z), \qquad (2.45)$$

with $w(z)$ an arbitrary solution of (2.42), is given by

$$f_n(z) = c_1 u_n(z) + c_2 v_n(z), \qquad (2.46)$$

with

$$u_n(z) = (-1)^n(\alpha+1+\varepsilon)_n \, {}_1F_1(-n-\varepsilon; \alpha+1; z), \qquad (2.47)$$

$$v_n(z) = (-1)^n(1+\varepsilon)_n \, z^{-\alpha} \, {}_1F_1(-\alpha-n-\varepsilon; 1-\alpha; z). \qquad (2.48)$$

The $f_n(z)$ satisfy

$$z f_n(z) = (Lf(z))_n, \quad A f_n(z) = (Mf(z))_n, \quad B f_n(z) = \lambda(n) f_n(z),$$

with L defined by (2.43) and (2.44) and

$$A = z\frac{d}{dz}, \quad B = -z\frac{d^2}{dz^2} + (z - \alpha - 1)\frac{d}{dz}, \tag{2.49}$$

$$M_{n,n-1} = a(n), \quad M_{n,n} = n + \varepsilon, \quad M_{n,n+1} = 0, \quad \lambda(n) = n + \varepsilon. \tag{2.50}$$

When $\varepsilon = 0$, the special choice $w(z) = 1$ in (2.45) leads to the functions $u_n(z)$ above, which agree then with the monic Laguerre polynomials.

2.2.4. *The associated Hermite matrix*
This case is reached when $\beta = \gamma = 0$, $\alpha \neq 0$, for $H(z)$ in (2.5). Then (2.6) can be reduced to

$$w''(z) - 2zw'(z) + 2\varepsilon w(z) = 0. \tag{2.51}$$

The entries of the matrix L are again given by

$$a(n) = \tilde{a}(n + \varepsilon), \quad b(n) = \tilde{b}(n + \varepsilon), \tag{2.52}$$

where now $\tilde{a}(n)$ and $\tilde{b}(n)$ are the coefficients of the three-term recursion relation satisfied by the monic Hermite polynomials

$$\tilde{a}(n) = \frac{n}{2}, \quad \tilde{b}(n) = 0. \tag{2.53}$$

For a generic value of ε, the general solution $f_n(z), n \in \mathbb{Z}$, of the doubly infinite three-term recursion with initial conditions

$$f_0(z) = w(z), \quad f_1(z) = zw(z) - \frac{1}{2}w'(z), \tag{2.54}$$

with $w(z)$ an arbitrary solution of (2.51), is given by

$$f_n(z) = c_1 u_n(z) + c_2 v_n(z), \tag{2.55}$$

with

$$u_n(z) = \frac{1}{(\sqrt{2})^{n+\varepsilon}} e^{z^2/2} D_{n+\varepsilon}(\sqrt{2}z), \tag{2.56}$$

$$v_n(z) = \frac{(-1)^n}{(\sqrt{2})^{n+\varepsilon}} e^{z^2/2} D_{n+\varepsilon}(-\sqrt{2}z), \tag{2.57}$$

where $D_\nu(z)$ denote the Weber-Hermite functions (also known as the parabolic cylinder functions). The $f_n(z)$ satisfy

$$z f_n(z) = (Lf(z))_n, \quad A f_n(z) = (Mf(z))_n, \quad B f_n(z) = \lambda(n) f_n(z),$$

with the matrix L as in (2.52), (2.53) and

$$A = \frac{1}{2}\frac{d}{dz}, \quad B = -\frac{1}{2}\frac{d^2}{dz^2} + z\frac{d}{dz}, \tag{2.58}$$

$$M_{n,n-1} = a(n), \quad M_{n,n} = 0, \quad M_{n,n+1} = 0, \quad \lambda(n) = n + \varepsilon. \tag{2.59}$$

When $\varepsilon = 0$, the choice $w(z) = 1$ in (2.54) leads to the functions $u_n(z)$ above, which agree then with the monic Hermite polynomials.

2.3. THE SPECTRAL MATRIX

Coming back to the associated Jacobi matrix in (2.24), one can show that as long as $0 < c < 2$ and $-1 < c - a - b < 1$, all entries $a(n) > 0$. In this case, when put in symmetric form, the operator

$$(Lh)_n = \sqrt{a(n)} h_{n-1} + b(n) h_n + \sqrt{a(n+1)} h_{n+1},$$

defines a bounded self-adjoint operator from $l_2 \to l_2$, with $l_2 = \{h = (h_n)_{n \in \mathbb{Z}} : h_n \in \mathbb{C}, \sum_{n=-\infty}^{\infty} |h_n|^2 < \infty\}$. Then, by the standard spectral theory [8], there is a uniquely determined spectral 2×2 matrix measure $d\Sigma(z)$ such that the following orthogonality relations hold

$$\int_{-\infty}^{\infty} P_n(z) \, d\Sigma(z) \, P_m(z)^T = \delta_{nm} I, \quad n, m \geq 0. \tag{2.60}$$

Here $P_n(z), n \geq 0$, denote the following 2×2 matrices

$$P_n(z) = \begin{pmatrix} q_{-n-1}(z) & p_{-n-1}(z) \\ q_n(z) & p_n(z) \end{pmatrix}, \tag{2.61}$$

with $p_n(z)$ and $q_n(z)$, $n \in \mathbb{Z}$, the polynomial solutions of the three-term recursion relation with initial conditions

$$p_{-1}(z) = 0, \ p_0(z) = 1 \quad \text{and} \quad q_{-1}(z) = 1, \ q_0(z) = 0.$$

We introduce the functions:

$$G_1 = {}_2F_1(a, b; c; z), \quad G_2 = {}_2F_1(a - c + 1, b - c + 1; 2 - c; z),$$
$$G_3 = {}_2F_1(a - 1, b + 1; c; z), \quad G_4 = {}_2F_1(a - c, b - c + 2; 2 - c; z).$$

Theorem 2.4 ([23]) *Assume that* $0 < c < 2$ *and* $-1 < c-a-b < 1$. *Then, the spectral matrix of the associated Jacobi matrix* (2.24) *is given by*

$$d\Sigma(z) = 0, \quad \text{for } z \notin [0,1],$$
$$= \Sigma'(z)dz, \quad \text{for } z \in [0,1],$$

with the entries $\Sigma'_{ij}, \Sigma'_{12} = \Sigma'_{21},$ *of* $\Sigma'(z)$ *given by*

$$\Sigma'_{11} = \frac{L}{\gamma}\Big(G_3^2 - \nu^2\mu K^2 z^{2(1-c)} G_4^2\Big)\rho(z),$$

$$\Sigma'_{12} = -L\Big(G_1 G_3 - \nu\mu K^2 z^{2(1-c)} G_2 G_4\Big)\rho(z),$$

$$\Sigma'_{22} = \gamma L\Big(G_1^2 - \mu K^2 z^{2(1-c)} G_2^2\Big)\rho(z), \tag{2.62}$$

where

$$\rho(z) = z^{c-1}(1-z)^{a+b-c},$$

and

$$\mu = \frac{\sin(\pi b)\sin(\pi(c-a))}{\sin(\pi a)\sin(\pi(c-b))},$$

$$\nu = \frac{(a-1)(c-b-1)}{b(c-a)}, \quad \gamma = \frac{(a-b)_{-2}}{\sqrt{a(0)}(a)_{-1}(c-b)_{-1}},$$

$$K = \frac{\Gamma(1-a)\Gamma(1-b)\Gamma(c-1)\sin(\pi a)}{\Gamma(c-a)\Gamma(c-b)\Gamma(1-c)\sin(\pi(c-a))},$$

$$L = \frac{b(a-c)\sin(\pi(1-c))}{\pi\sqrt{a(0)}(a-b-1)(1-c)(1-\mu)K}.$$

The proof of Theorem 2.4 uses heavily the results of [62], where the orthogonality measure for the associated Jacobi polynomials is explicitly computed. The big surprise however comes when expressing the orthogonality relations (2.60) in a basis of bispectral functions. Let

$$U_n(z) = \begin{pmatrix} u_{-n-1}(z) & v_{-n-1}(z) \\ u_n(z) & v_n(z) \end{pmatrix},$$

with $u_n(z)$ and $v_n(z)$ defined as in (2.30), (2.31). Since $\{u_n(z), v_n(z)\}$ and $\{q_n(z), p_n(z)\}$ both satisfy the three-term recursion relation defined by the associated Jacobi matrix, there is as 2×2 matrix $S(z)$ (z dependent, but n independent), such that

$$P_n(z) = U_n(z)S(z),$$

with $P_n(z)$ as in (2.61). Since $P_0 = I$, we have that

$$S(z) = U_0^{-1}(z) = \frac{1}{\nu G_1 G_4 - G_2 G_3} \begin{pmatrix} \gamma G_2 & \nu G_4 \\ -\gamma z^{c-1} G_1 & -z^{c-1} G_3 \end{pmatrix}. \tag{2.63}$$

Using (2.62) and (2.63), one computes that

$$P_n(z) d\Sigma(z) P_m(z)^T = U_n(z) S(z) d\Sigma(z) S(z)^T U_m(z)^T,$$

with

$$S(z) d\Sigma(z) S(z)^T = \begin{pmatrix} 1 & 0 \\ 0 & -\mu K^2 \end{pmatrix} \gamma L \rho(z) dz.$$

This leads finally to the following result, see [23] for details. Define

$$\tilde{u}_n(z) = \sqrt{\gamma L} u_n(z), \quad \tilde{v}_n(z) = \sqrt{-\mu K^2 \gamma L} v_n(z).$$

Theorem 2.5 ([23])

$$\int_0^1 \{\tilde{u}_n(z)\tilde{u}_m(z) + \tilde{v}_n(z)\tilde{v}_m(z)\} z^\beta (1-z)^\alpha dz = \delta_{nm}, \quad \forall n, m \in \mathbb{Z}, \tag{2.64}$$

with α and β defined in terms of a, b, c as in (2.25).

3. Bispectral Darboux Transformations

We denote by

$$\mathcal{B} = \langle L, \Lambda \rangle, \tag{3.1}$$

the subalgebra of the algebra of finite band bi-infinite matrices generated by L and Λ appearing in (1.5), (1.6). Similarly

$$\mathcal{B}' = \langle z, B \rangle, \tag{3.2}$$

will denote the subalgebra of the algebra of differential operators generated by the operator of multiplication by z and the operator B appearing in these same equations. Formulas (1.5) and (1.6) serve to define an *anti-isomorphism*

$$b : \mathcal{B} \to \mathcal{B}' \tag{3.3}$$

between these two algebras, i.e. it is given on the generators by

$$b(L) = z \quad \text{and} \quad b(\Lambda) = B. \tag{3.4}$$

More precisely, any monomial $L^i \Lambda^j \in \mathcal{B}$, $i, j \geq 0$, acting on the space of bispectral functions $\{f_n\}_{n \in \mathbb{Z}}$, gives

$$L^i \Lambda^j f = B^j z^i f \quad \text{i.e.} \quad b(L^i \Lambda^j) = b(\Lambda^j)b(L^i) = B^j z^i. \qquad (3.5)$$

We shall also need the commutative subalgebras (the algebras of "functions") of $\mathcal{B}$ and $\mathcal{B}'$ defined by

$$\mathcal{K} = \langle \Lambda \rangle \quad \text{and} \quad \mathcal{K}' = \langle z \rangle. \qquad (3.6)$$

Their images by b and b^{-1} will be denoted by $\mathcal{A}'$ and $\mathcal{A}$ respectively and provide obvious bispectral operators. We denote with a bar the fields of quotients of $\mathcal{K}, \mathcal{K}', \mathcal{A}$ and $\mathcal{A}'$. Obviously, b extends to isomorphisms $\bar{\mathcal{K}} \to \bar{\mathcal{A}}'$ and $\bar{\mathcal{A}} \to \bar{\mathcal{K}}'$. The next theorem is the main tool that we need to produce out of some given bispectral equations, new non-trivial bispectral operators by mean of the so-called Darboux factorization method.

Theorem 3.1 ([25], [26]) *Let $\mathcal{L} \in \mathcal{A}$ be a constant coefficient polynomial in L, which factorizes "rationally" as*

$$\mathcal{L} = QP, \qquad (3.7)$$

in such a way that

$$Q = SV^{-1}, \quad P = \Theta^{-1}R, \qquad (3.8)$$

with $R, S \in \mathcal{B}$, and $\Theta V \in \mathcal{K}$. Then the Darboux transform of $\mathcal{L}$ given by

$$\hat{\mathcal{L}} = PQ, \qquad (3.9)$$

is a bispectral operator. More precisely, defining $\mu \equiv b(\mathcal{L}) \in \mathcal{K}'$ and $\hat{f} \equiv Pf$, with f satisfying (1.5) and (1.6), we have

$$\hat{\mathcal{L}}\hat{f} = \mu \hat{f}, \qquad (3.10)$$
$$\hat{B}\hat{f} = \Theta V \hat{f}, \qquad (3.11)$$

with

$$\hat{B} = b(R)b(S)\mu^{-1}. \qquad (3.12)$$

Proof. Equation (3.10) follows immediately from the definitions. Let

$$P_b \equiv b(R) \quad \text{and} \quad Q_b \equiv b(S)\mu^{-1}. \qquad (3.13)$$

Clearly, using the anti-isomorphism introduced in (3.3), (3.4), (3.5),

$$\hat{f} = Pf = \Theta^{-1}P_b f. \qquad (3.14)$$

Since B has no zero divisors, (3.7), (3.8) imply

$$\Theta V = R\mathcal{L}^{-1}S, \quad \mathcal{L}^{-1} \in \bar{A}.$$

Applying the anti-isomorphism b, we obtain

$$b(\Theta V) = b(S)\mu^{-1}b(R). \tag{3.15}$$

From (3.13),(3.14) and (3.15), we have

$$\Theta V f = Q_b P_b f = \Theta Q_b \hat{f},$$

and thus, using (3.7) and this last relation, we deduce that

$$f = \mu^{-1}Q\hat{f} = V^{-1}Q_b\hat{f}.$$

This equation combined with (3.13) and (3.14) gives (3.11), which completes the proof of the theorem. □

Conjecture 3.2 In general, Theorem 3.1 does not produce a bispectral operator $\hat{B}$ of the lowest possible order. In all the applications of the theorem that have been worked out so far (see [25], [26]), the following fact seems to be true:

> *ring of bispectral operators $\hat{B}$ =*
> *ring of polynomials $p(\lambda(n))$ (in $\lambda(n)$) such that*
> $$p(\lambda(n)) - p(\lambda(n-1)) \text{ is divisible by } \theta(n), \tag{3.16}$$

with $\theta(n)$ denoting the (diagonal) entries of the matrix Θ in (3.8), $(\Theta h)_n = \theta(n)h_n$. The corresponding bispectral operators $\hat{B}$ are obtained by solving the following equation for $\hat{B}$

$$\hat{B}P_b = P_b p(B), \tag{3.17}$$

with P_b as in (3.13). For the particular choice (3.12) given in Theorem 3.1, one has $\hat{B} = P_b Q_b$ and $p(B) \equiv b(\Theta V) = Q_b P_b$. I don't know how to prove the conjecture in general. See however [29] and [32] for some indications, as well as Remark 4.3 for an illustration.

4. A Detailed Example: The Krall-Jacobi Polynomials with Weight Function $z^\beta(1-z)^\alpha + \gamma\delta(z)$ on $[0,1]$, $\alpha > -1$, $\beta \in \mathbb{N}$

In this section, we denote by $L_{\alpha,\beta,\varepsilon}$, the doubly infinite associated Jacobi matrix, with entries $a_n^{\alpha,\beta,\varepsilon}, b_n^{\alpha,\beta,\varepsilon}$ given as in (2.26) and (2.27), to emphasize the values of the parameters $\alpha, \beta, \varepsilon$ that we are considering. Similarly, we denote by $A, B_{\alpha,\beta}$, the differential operators defined in (2.32), and by $M_{\alpha,\beta,\varepsilon}, \Lambda_{\alpha,\beta,\varepsilon}$, the matrices with entries given respectively in (2.33) and (2.34). $\mathcal{B}_{\alpha,\beta,\varepsilon} = \langle L_{\alpha,\beta,\varepsilon}, \Lambda_{\alpha,\beta,\varepsilon} \rangle$ and $\mathcal{B}'_{\alpha,\beta} = \langle z, B_{\alpha,\beta} \rangle$ denote the corresponding algebras, as introduced in (3.1) and (3.2). Notice that A is independent of $\alpha, \beta, \varepsilon$, and $B_{\alpha,\beta}$ does not depend on ε. Because of (2.19), we have that $M_{\alpha,\beta,\varepsilon} \in \mathcal{B}_{\alpha,\beta,\varepsilon}$, and thus also $A_{\alpha,\beta} \in \mathcal{B}'_{\alpha,\beta}$.

We start with $L_{\alpha,\beta+1,\varepsilon}$ and we try to factorize it as

$$L_{\alpha,\beta+1,\varepsilon} = QP, \tag{4.1}$$

with Q an upper triangular matrix and P a lower triangular matrix. The most general factorization is given by

$$(Ph)_n = h_n - \frac{\phi_n}{\phi_{n-1}} h_{n-1}, \tag{4.2}$$

$$(Qh)_n = -a_n^{\alpha,\beta+1,\varepsilon} \frac{\phi_{n-1}}{\phi_n} h_n + h_{n+1}, \tag{4.3}$$

with ϕ an arbitrary element in the kernel of $L_{\alpha,\beta+1,\varepsilon}$. Thus there is *one free parameter* in the factorization. The reason for performing an upper-lower instead of a lower-upper factorization, is that in the case of a semi-infinite tridiagonal matrix, only the upper-lower factorization contains a free parameter, see [19]. Denoting the functions in (2.30) and (2.31) by $u_n^{\alpha,\beta,\varepsilon}(z)$ and $v_n^{\alpha,\beta,\varepsilon}(z)$, since these functions belong to the kernel of $L_{\alpha,\beta,\varepsilon} - zI$, we easily determine that a basis for the kernel of $L_{\alpha,\beta+1,\varepsilon}$ is given by

$$u_n = (-1)^n \frac{(\alpha + \beta + 2 + \varepsilon)_n (\beta + 2 + \varepsilon)_n}{(\alpha + \beta + 2 + 2\varepsilon)_{2n}},$$

$$v_n = (-1)^n \frac{(\alpha + 1 + \varepsilon)_n (1 + \varepsilon)_n}{(\alpha + \beta + 2 + 2\varepsilon)_{2n}}, \tag{4.4}$$

and thus the most general element in the kernel has the form

$$\phi_n = v_n + \gamma u_n, \tag{4.5}$$

with γ a free parameter. The Darboux transformation $\hat{L}$ of $L_{\alpha,\beta+1,\varepsilon}$ is obtained by exchanging the order of the factors in (4.1)

$$\hat{L} = PQ. \tag{4.6}$$

Explicitly, the entries $\hat{a}_n$ and $\hat{b}_n$ of $\hat{L}$ are given by

$$\hat{a}_n = a_{n-1}^{\alpha,\beta+1,\varepsilon} \frac{\phi_{n-2}\phi_n}{\phi_{n-1}^2}, \quad \hat{b}_n = b_n^{\alpha,\beta+1,\varepsilon} + \frac{\phi_{n+1}}{\phi_n} - \frac{\phi_n}{\phi_{n-1}}.$$

Assume now that $\beta + 1$ is a *positive integer*, i.e. $\beta = 0, 1, 2, \ldots$, then (4.5), with u_n and v_n as in (4.4), can be rewritten as

$$\phi_n = (-1)^n \frac{(\alpha + 1 + \varepsilon)_n (1 + \varepsilon)_n}{(\alpha + \beta + 2 + 2\varepsilon)_{2n}} \theta(n+1), \tag{4.7}$$

$$\theta(n) = 1 + \rho(n + \varepsilon)_{\beta+1} (\alpha + n + \varepsilon)_{\beta+1},$$

$$\rho = \frac{\gamma}{(1 + \varepsilon)_{\beta+1} (\alpha + 1 + \varepsilon)_{\beta+1}}. \tag{4.8}$$

It is easy to convince oneself that it is not possible to generate the algebra $\mathcal{B}_{\alpha,\beta,\varepsilon}$ with matrices containing at most two diagonals. Since the matrices P and Q involved in (4.1) contain only two diagonals, it looks a little bit puzzling how we could possibly fit this example within the framework of Theorem 3.1! The key out of this difficulty is to introduce the four matrices C, D, F, G acting as follows:

$$(Ch)_n = h_n + c(n)h_{n-1}, \tag{4.9}$$

$$(Dh)_n = h_{n+1} + d(n)h_n, \tag{4.10}$$

$$(Fh)_n = (\varepsilon + n)h_n - c(n)(\alpha + \beta + n + \varepsilon + 1)h_{n-1}, \tag{4.11}$$

$$(Gh)_n = (\alpha + \beta + n + \varepsilon + 2)h_{n+1} - (n + \varepsilon)d(n)h_n, \tag{4.12}$$

with

$$c(n) = \frac{(n + \varepsilon)(\alpha + n + \varepsilon)}{(\alpha + \beta + 2n + 2\varepsilon)(\alpha + \beta + 2n + 2\varepsilon + 1)}, \tag{4.13}$$

$$d(n) = \frac{(\beta + n + \varepsilon + 1)(\alpha + \beta + n + \varepsilon + 1)}{(\alpha + \beta + 2n + 2\varepsilon + 1)(\alpha + \beta + 2n + 2\varepsilon + 2)}. \tag{4.14}$$

Denoting by $f^{\alpha,\beta,\varepsilon}(z) \equiv \left(f_n^{\alpha,\beta,\varepsilon}(z) \right)_{n \in \mathbb{Z}}$ the vector formed with the functions $u_n^{\alpha,\beta,\varepsilon}(z)$ or $v_n^{\alpha,\beta,\varepsilon}(z)$ in (2.30) or (2.31), one easily checks using some of the *contiguous relations* satisfied by the hypergeometric function, that the following equations hold:

$$f^{\alpha,\beta,\varepsilon}(z) = k_{\alpha,\beta,\varepsilon} \, C f^{\alpha,\beta+1,\varepsilon}(z), \tag{4.15}$$

$$z f^{\alpha,\beta+1,\varepsilon}(z) = k_{\alpha,\beta,\varepsilon}^{-1} \, D f^{\alpha,\beta,\varepsilon}(z), \tag{4.16}$$

$$(z - 1)\frac{d}{dz} f^{\alpha,\beta,\varepsilon}(z) = k_{\alpha,\beta,\varepsilon} \, F f^{\alpha,\beta+1,\varepsilon}(z), \tag{4.17}$$

$$\left\{ z(z-1)\frac{d}{dz} + (\alpha + \beta + 2)z - \beta \right\} f^{\alpha,\beta+1,\varepsilon}(z) = k_{\alpha,\beta,\varepsilon}^{-1} \, G f^{\alpha,\beta,\varepsilon}(z). \tag{4.18}$$

In the above relations, $k_{\alpha,\beta,\varepsilon}$ denotes some constant depending on the parameters $\alpha, \beta, \varepsilon$. This constant is different whether we substitute in them the functions $u_n^{\alpha,\beta,\varepsilon}(z)$ or the functions $v_n^{\alpha,\beta,\varepsilon}(z)$, but this plays no role in what follows. From (4.15) and (4.16), we get that

$$L_{\alpha,\beta,\varepsilon} = CD \quad \text{and} \quad L_{\alpha,\beta+1,\varepsilon} = DC, \tag{4.19}$$

and thus, using (4.1) and (4.6), we obtain the two equations

$$L_{\alpha,\beta,\varepsilon}^2 = C(DC)C = CL_{\alpha,\beta+1,\varepsilon}D = (CQ)(PD), \tag{4.20}$$

and

$$\hat{L}^2 = P(QP)Q = PL_{\alpha,\beta+1,\varepsilon}Q = (PD)(CQ), \tag{4.21}$$

which exhibit $\hat{L}^2$ as a Darboux transformation of $L_{\alpha,\beta,\varepsilon}^2$.

The trick now is to show that Theorem 3.1 is relevant to the factorization (4.20), i.e. this factorization can be recast as prescribed in (3.7) and (3.8), within the algebra $\mathcal{B}_{\alpha,\beta,\varepsilon}$. From (2.32), (4.16) and (4.17), we deduce that $M_{\alpha,\beta,\varepsilon}$ in (2.33) satisfies

$$M_{\alpha,\beta,\varepsilon} = FD, \tag{4.22}$$

and, from (4.15) and (4.18), we find that

$$CG = M_{\alpha,\beta,\varepsilon} + (\alpha + \beta + 2)L_{\alpha,\beta,\varepsilon} - \beta I. \tag{4.23}$$

We need also to introduce the diagonal matrices Θ and V, acting as follows

$$(\Theta h)_n = \theta(n)h_n, \quad (Vh)_n = \theta(n+1)h_n, \tag{4.24}$$

with $\theta(n)$ defined as in (4.8).

Lemma 4.1 *With the notations introduced above, we have*

$$L_{\alpha,\beta+1,\varepsilon} = QP \equiv (UV^{-1})(\Theta^{-1}T), \tag{4.25}$$

with

$$T = C + \rho\, W(\Lambda_{\alpha,\beta,\varepsilon})\big\{\Lambda_{\alpha,\beta,\varepsilon}C - (\beta+1)F\big\}, \tag{4.26}$$

$$U = D + \rho\,\big\{D\Lambda_{\alpha,\beta,\varepsilon} + (\beta+1)G\big\}W(\Lambda_{\alpha,\beta,\varepsilon}), \tag{4.27}$$

and $W(\Lambda_{\alpha,\beta,\varepsilon})$ the following polynomial of degree β in $\Lambda_{\alpha,\beta,\varepsilon}$:

$$W(\Lambda_{\alpha,\beta,\varepsilon}) = \prod_{j=1}^{\beta}\Big(\Lambda_{\alpha,\beta,\varepsilon} + j(\alpha+\beta+1-j)I\Big). \tag{4.28}$$

Proof. The key is to observe that $\theta(n)$ in (4.8) can be rewritten in terms of the entries $\lambda(n) = (n+\varepsilon)(n+\varepsilon+\alpha+\beta+1)$ of the diagonal matrix $\Lambda_{\alpha,\beta,\varepsilon}$ (2.34), as follows:

$$\theta(n) = 1 + \rho(n+\varepsilon)(n+\varepsilon+\alpha)\prod_{j=1}^{\beta}(n+\varepsilon+j)(n+\varepsilon+\alpha+\beta+1-j),$$

$$= 1 + \rho(n+\varepsilon)(n+\varepsilon+\alpha)\prod_{j=1}^{\beta}\Big(\lambda(n)+j(\alpha+\beta+1-j)\Big),$$

$$\equiv 1 + \rho(n+\varepsilon)(n+\varepsilon+\alpha)W(\lambda(n)). \tag{4.29}$$

From (4.2), (4.7), (4.9), (4.13) and (4.29) we get that $P = \Theta^{-1}T$, with Θ as in (4.24) and T given by

$$
\begin{aligned}
(Th)_n &= \theta(n)h_n + c(n)\theta(n+1)h_{n-1}, \\
&= (Ch)_n + \rho(n+\varepsilon)(n+\varepsilon+\alpha)W(\lambda(n))h_n \\
&\quad + \rho c(n)(n+\varepsilon+1)(n+\varepsilon+1+\alpha)W(\lambda(n+1))h_{n-1}, \\
&= (Ch)_n + \rho(n+\varepsilon)(n+\varepsilon+\alpha)W(\lambda(n))h_n \\
&\quad + \rho c(n)W(\lambda(n))(n+\varepsilon+\beta+1)(n+\varepsilon+\alpha+\beta+1)h_{n-1}, \\
&= (Ch)_n + \rho W(\lambda(n))\Big(\lambda(n)-(\beta+1)(n+\varepsilon)\Big)h_n \\
&\quad + \rho c(n)W(\lambda(n))\Big(\lambda(n)+(\beta+1)(n+\varepsilon+\alpha+\beta+1)\Big)h_{n-1}.
\end{aligned}
$$

Using the definition (4.9) and (4.11) of the matrices C and F, this last equation is immediately seen to agree with (4.26).

Similarly, using now (2.26), (2.27), (4.3), (4.7), (4.10), (4.14) and (4.29), we obtain that $Q = UV^{-1}$, with V as in (4.24) and U given by

$$
\begin{aligned}
(Uh)_n &= d(n)\theta(n)h_n + \theta(n+2)h_{n+1}, \\
&= (Dh)_n + \rho d(n)\Big(\lambda(n)-(\beta+1)(n+\varepsilon)\Big)W(\lambda(n))h_n \\
&\quad + \rho\Big(\lambda(n+1)+(\beta+1)(\alpha+\beta+n+\varepsilon+2)\Big)W(\lambda(n+1))h_{n+1},
\end{aligned}
$$

which, using (4.10) and (4.12), agrees with (4.27), establishing the lemma. $\square$

Theorem 4.2 Let $\alpha > -1$ and let $\beta+1$ be a positive integer. The Darboux transform $\hat{L} = PQ$ of $L_{\alpha,\beta+1,\varepsilon}$, with P and Q as in (4.2) and (4.3), is again bispectral. Explicitly, defining

$$\hat{f}_n = (Pf^{\alpha,\beta+1,\varepsilon})_n, \tag{4.30}$$

with $f_n^{\alpha,\beta+1,\varepsilon}(z) = c_1 u_n^{\alpha,\beta+1,\varepsilon}(z) + c_2 v_n^{\alpha,\beta+1,\varepsilon}(z)$, c_1, c_2 *arbitrary constants,*
we have

$$\hat{L}\hat{f}_n = z\hat{f}_n, \quad \hat{B}\hat{f}_n = \theta(n)\theta(n+1)\hat{f}_n,$$

with

$$\hat{B} = b(R)b(S)z^{-2}, \tag{4.31}$$

$$b(R) = z + \rho\Big\{z\, B_{\alpha,\beta} - (\beta+1)A\Big\}W(B_{\alpha,\beta}), \tag{4.32}$$

$$b(S) = z + \rho W(B_{\alpha,\beta})\Big\{B_{\alpha,\beta}\, z + (\beta+1)(A + (\alpha+\beta+2)z - \beta)\Big\},$$

where $A, B_{\alpha,\beta}, \theta(n)$ *and* W *are defined as in* (2.32), (4.8) *and* (4.28).

Proof. The proof amounts to a straightforward application of Theorem
3.1, with $\mathcal{L} = L_{\alpha,\beta,\varepsilon}^2$, factorized as in (4.20), and with the role of P and
Q in Theorem 3.1 played here by PD and CQ respectively. From (4.19),
(4.22), (4.23), (4.25), (4.26) and (4.27) we obtain that

$$PD = \Theta^{-1}R, \quad \text{with}$$

$$R = L_{\alpha,\beta,\varepsilon} + \rho W(\Lambda_{\alpha,\beta,\varepsilon})\Big\{\Lambda_{\alpha,\beta,\varepsilon}L_{\alpha,\beta,\varepsilon} - (\beta+1)M_{\alpha,\beta,\varepsilon}\Big\},$$

and

$$CQ = SV^{-1}, \quad \text{with}$$
$$S = L_{\alpha,\beta,\varepsilon} +$$

$$\rho\Big\{L_{\alpha,\beta,\varepsilon}\Lambda_{\alpha,\beta,\varepsilon} + (\beta+1)\big(M_{\alpha,\beta,\varepsilon} + (\alpha+\beta+2)L_{\alpha,\beta,\varepsilon} - \beta I\big)\Big\}W(\Lambda_{\alpha,\beta,\varepsilon}).$$

From (4.8), (4.24) and (4.29), we check that the product $\Theta V \in \mathcal{K}$, that is
it is a constant coefficient polynomial in $\Lambda_{\alpha,\beta,\varepsilon}$:

$$\Theta V = 1 + \rho W(\Lambda_{\alpha,\beta,\varepsilon})\Big\{2\Lambda_{\alpha,\beta,\varepsilon} + (\beta+1)(\alpha+\beta+1)I\Big\}$$
$$+ \rho^2 W^2(\Lambda_{\alpha,\beta,\varepsilon})\Lambda_{\alpha,\beta,\varepsilon}\Big\{\Lambda_{\alpha,\beta,\varepsilon} + \alpha(\beta+1)I\Big\}. \tag{4.33}$$

From the definition of the bispectral anti-isomorphism in (3.4), (3.5), since
$\mu \equiv b(L_{\alpha,\beta,\varepsilon}^2) = z^2$, we deduce all the claims, which completes the proof. $\square$

Remark 4.3 Defining

$$q(n) = \frac{\rho}{\beta+2}(n+\varepsilon)_{\beta+2}(n+\varepsilon+\alpha)_{\beta+2} + \lambda(n) + c, \tag{4.34}$$

with c an arbitrary constant, it is easy to check that

$$q(n) - q(n-1) = (\lambda(n) - \lambda(n-1))\theta(n), \qquad (4.35)$$

with $\theta(n)$ as in (4.8). In fact, $q(n)$ is a polynomial of degree $\beta + 2$ in $\lambda(n)$:

$$q(n) = \frac{\rho}{\beta+2} \prod_{j=1}^{\beta+2} \left\{ \lambda(n) + (j-1)(\alpha+\beta+2-j) \right\} + \lambda(n) + c \equiv p(\lambda(n)).$$

It is the polynomial of the smallest possible degree in $\lambda(n)$ such that $p(\lambda(n)) - p(\lambda(n-1))$ is divisible by $\theta(n)$. According to Conjecture 3.2, the bispectral operator $\hat{B}$ of the smallest possible order is defined by $\hat{B} P_b = P_b \, p(B_{\alpha,\beta})$, with $P_b = b(R)$ as in (4.32). Since $B_{\alpha,\beta}$ is of order 2 and $p(\lambda(n))$ is of degree $\beta + 2$ (in $\lambda(n)$), $\hat{B}$ is of order $2\beta + 4$, in agreement with the results of [37] and [63]. Computer experiments show that the above equation can be solved for $\hat{B}$, giving additional evidence for Conjecture 3.2. Since the bispectral operator in (4.31) is conjugated via P_b to the operator $b(\Theta V)$, by (4.33), this operator is of order $4\beta + 4$. It is of lowest possible order only when $\beta = 0$.

Remark 4.4 If we choose in (4.30) to take $f_n^{\alpha,\beta+1,\varepsilon}(z) = u_n^{\alpha,\beta+1,\varepsilon}(z)$, as in (2.30), at the limit $\varepsilon = 0$, the functions $u_n^{\alpha,\beta+1,0}(z), n \geq 0$, coincide with the Jacobi polynomials (with parameters $\alpha, \beta+1$). It is easy to show that, when $\varepsilon = 0$, the $\hat{f}_n$'s in (4.30) are then orthogonal polynomials on the interval $[0,1]$, with weight function given by

$$\frac{\Gamma(\alpha+\beta+2)}{\Gamma(\alpha+1)\Gamma(\beta+1)} z^\beta (1-z)^\alpha + \gamma \delta(z),$$

with $\delta(z)$ the delta function and γ the free parameter in (4.5). When $\beta = 0$, we obtain one of the original cases discovered by H.L. Krall, see [43].

5. Krall-Laguerre and Krall-Jacobi type Associated Matrices

5.1. KRALL-LAGUERRE TYPE ASSOCIATED MATRICES

Theorem 5.1 ([26], see also [19], [22], [36]) *Let α be a positive integer and let ε be arbitrary. Then, the doubly infinite tridiagonal matrix which is obtained after a chain of k elementary Darboux transformations, with $1 \leq k \leq \alpha$, starting from the associated Laguerre matrix $L_{\alpha,\varepsilon}$ as defined in (2.43) and (2.44)*

$$L_0 \equiv L_{\alpha,\varepsilon} = Q_0 P_0 \mapsto L_1 = P_0 Q_0 = Q_1 P_1 \mapsto \ldots \mapsto L_k = P_{k-1} Q_{k-1}, \qquad (5.1)$$

solves the extended Bochner-Krall problem.

I refer the reader to [26] for the detailed proof of the theorem. The entries of the matrix L_k which is obtained at the end of the chain (5.1), are (rational) functions of $n + \varepsilon$. Thus, again, this matrix can be seen as the doubly infinite tridiagonal matrix *associated* with some class of orthogonal polynomials. These polynomials form an orthogonal sequence of polynomials with a moment functional given by the weight distribution

$$z^{\alpha-k}e^{-z} + \sum_{i=1}^{k} \rho_i \, \delta^{(k-i)}(z),$$

with ρ_i, the k free parameters coming from the successive elementary Darboux transformations, and $\delta^{(i)}(z)$ the derivatives of the delta function. When $k = 1$ (corresponding to the case of one elementary Darboux transformation), the bispectral property of these polynomials was first established by J. Koekoek and R. Koekoek in [36]. They are eigenfuntions of a differential operator of (minimal order) $2(\alpha+1)$, generalizing one of the cases of order four found by H. L. Krall [43], when $\alpha = k = 1$. The cases involving the derivatives of the delta function don't seem to have appeared prior to [26].

5.2. KRALL-JACOBI TYPE ASSOCIATED MATRICES

Theorem 5.2 ([27], see also [19], [22], [37], [38], [63]) i) *Let β be a positive integer, let $\alpha > -1$ and let ε be arbitrary. Then, the doubly infinite tridiagonal matrix which is obtained after k elementary Darboux transformations starting from the associated Jacobi matrix $L_{\alpha,\beta,\varepsilon}$ as defined in (2.26), (2.27), with $1 \le k \le \beta$, solves the extended Bochner-Krall problem.*

ii) Let α and β be positive integers and let ε be arbitrary. Then, the doubly infinite tridiagonal matrix $L_{k,l}$ which is obtained from the associated Jacobi matrix $L_{\alpha,\beta,\varepsilon}$ at the end of the chain of elementary Darboux transformations

$$L_0 \equiv L_{\alpha,\beta,\varepsilon} = Q_0 P_0 \mapsto L_1 = P_0 Q_0 = Q_1 P_1 \mapsto \ldots \mapsto L_k = P_{k-1} Q_{k-1}$$
$$L_k - I = Q_k P_k \mapsto L_{k,1} - I = P_k Q_k = Q_{k+1} P_{k+1} \mapsto \ldots$$
$$\mapsto L_{k,l} - I = P_{k+l-1} Q_{k+l-1}, \quad (5.2)$$

with $1 \le k \le \beta$ and $1 \le l \le \alpha$, solves the extended Bochner-Krall problem.

For part i), the case $k = 1$ was considered in the previous section. The result was obtained previously by A. Zhedanov [63] (see also [56]) and J. Koekoek and R. Koekoek [37], when $\varepsilon = 0$. For part ii), the case $k = 1$ and $l = 1$ was considered by R. Koekoek [38] in a special case, and then by J. Koekoek and R. Koekoek [37] in the general case, again when $\varepsilon = 0$ and by using somewhat different methods. When this paper was completed, F. A. Grünbaum and M. Yakimov [27] informed me that they have obtained the result concerning the iteration of the Darboux transformation, by performing a suitable bispectral Darboux transformation of the operator

$$\mathcal{L} = L_{\alpha,\beta,\varepsilon}^{k} \left(L_{\alpha,\beta,\varepsilon} - I \right)^{l}.$$

A similar (simpler, but perhaps more unexpected) situation corresponding to $\alpha = \beta = -1/2$ (in which case the resulting matrix $L_{\alpha,\beta,\varepsilon}$ is independent of ε) was considered by L. Haine and P. Iliev [31] and is discussed in the next section. The doubly infinite tridiagonal matrix $L_{k,l}$ (5.2) which is obtained at the end of the process, is the *associated* matrix with a family of orthogonal polynomials. These polynomials form an orthogonal polynomial sequence for the moment functional given by the weight distribution

$$z^{\beta-k}(1-z)^{\alpha-l} + \sum_{i=1}^{k} \rho_i \, \delta^{(k-i)}(z) + \sum_{i=1}^{l} \sigma_i \, \delta^{(l-i)}(z-1),$$

with ρ_i, σ_i, the free parameters in the successive elementary Darboux transformations. They generalize polynomials introduced by Koornwinder in [39].

6. Rank 1 Solutions

This section reviews a joint work with P. Iliev [30], [31]. I discuss a class of solutions which is the exact analogue of the rational solutions to the KdV equation, in the context of the Duistermaat-Grünbaum problem. These solutions are built by iteration of the Darboux transformation starting at

$$L_0 = E^{-1} - 2I + E. \tag{6.1}$$

As explained in the introduction, L_0 is (up to translation and scaling) the only solution of the equation $(\text{ad } L)^2(\Lambda) = 0$. According to a principle suggested by the study of the Krall polynomials, in order to preserve bispectrality, we have to perform the successive Darboux transformations at the end points of the interval of orthogonality, i.e. at the end points of the continuous spectrum of our starting operator L_0, which in the case of (6.1) lies on the interval $[-4, 0]$. Notice that by a suitable rescaling of the spectral variable, the continuous spectrum of the operator L_0 can be brought back to $[0, 1]$, making this operator agree with the case $\alpha = \beta = -1/2$ of (2.26), (2.27), the result being independent of ε. Thus, we look for the explicit solution of the following chain of successive Darboux transformations

$$L_0 = P_0 Q_0 \mapsto L_1 = Q_0 P_0 = P_1 Q_1 \mapsto \ldots \mapsto L_k = Q_{k-1} P_{k-1}$$

$$L_k + 4I = P_k Q_k \mapsto L_{k,1} + 4I = Q_k P_k = P_{k+1} Q_{k+1} \mapsto \cdots$$
$$\mapsto L_{k,l} + 4I = Q_{k+l-1} P_{k+l-1}. \quad (6.2)$$

We have chosen to perform a sequence of lower-upper (with P lower and Q upper) factorizations (instead of upper-lower). The motivation for doing upper-lower factorizations in (5.1), (5.2), was that only those factorizations contain a free parameter in the limit to semi-infinite matrices. Here, our construction does not have any orthogonal polynomials limiting situation, and an explanation for our choice will emerge soon!

An efficient way to solve the problem is to perform a lower-upper factorization of the operator

$$\mathcal{L} \equiv L_0^k (L_0 + 4I)^l = (E^{-(k+l)} P) Q, \quad (6.3)$$

with P and Q monic positive (i.e. upper) operators of order $k+l$. Since E^{-1} is the backward shift operator, the first factor in (6.3) $E^{-(k+l)} P$ is a lower matrix (with 1's on the main diagonal). The kernel of Q must be specified by $k + l$ functions $\phi_1(n), \phi_2(n), \ldots, \phi_k(n)$ and $\psi_1(n), \psi_2(n), \ldots, \psi_l(n)$ satisfying

$$L_0 \phi_j(n) = \phi_{j-1}(n), \quad 1 \leq j \leq k, \quad (6.4)$$
$$(L_0 + 4I)\psi_j(n) = \psi_{j-1}(n), \quad 1 \leq j \leq l, \quad (6.5)$$

with the convention that $\phi_0(n) = \psi_0(n) = 0$. Indeed, these equations imply that $\mathrm{Ker}\, Q \subset \mathrm{Ker}\, Q L_0$, that is $Q L_0$ can be factorized to the right by Q

$$Q L_0 = L_{k,l} Q. \quad (6.6)$$

It is easy to check that the operator $L_{k,l}$ defined by this equation coincides with the operator obtained at the end of the chain (6.2).

We introduce the elementary *shifted* Schur polynomials by

$$(1 + z)^n \exp\left(\sum_{i=1}^{\infty} t_i z^i \right) = \sum_{j \in \mathbb{Z}} S_j(n; t) z^j, \quad n \in \mathbb{Z},$$

with the understanding that $S_j(n; t) = 0$, for $j < 0$. In terms of the standard elementary Schur polynomials $S_j(t)$, corresponding to $n = 0$, we have

$$S_j(n; t) = S_j\left(t_1 + n, t_2 - \frac{n}{2}, t_3 + \frac{n}{3}, \ldots \right).$$

It is an elementary exercise to prove the following

Lemma 6.1 ([31])

$$\phi_j(n) = S_{2j-1}(n+j-1;t), \tag{6.7}$$

$$\psi_j(n) = (-1)^{n-j}S_{2j-1}(n+j-1;s), \tag{6.8}$$

solve (6.4) *and* (6.5), *with* $t_1, t_2, \ldots$ *and* $s_1, s_2, \ldots$ *arbitrary parameters.*
Let $\Delta = E - I$ and $\Delta^* = E^{-1} - I$. We introduce the discrete Wronskian
with respect to the variable n, the so-called Casorati determinant:

$$Wr_\Delta(g_1(n),\ldots,g_M(n)) = \det(\Delta^{i-1}g_j(n))_{1\leq i,j\leq M}.$$

We define similarly Wr_{Δ^*}, replacing Δ by Δ^*. We also denote by P^* the
adjoint (i.e. the transpose) of P. It is then possible to solve explicitly the
factorization problem (6.3).

Lemma 6.2 ([31])

$$Qh(n) = \frac{Wr_\Delta(\phi_1(n),\ldots,\phi_k(n),\psi_1(n),\ldots,\psi_l(n),h(n))}{Wr_\Delta(\phi_1(n),\ldots,\phi_k(n),\psi_1(n),\ldots,\psi_l(n))}, \tag{6.9}$$

$$P^*h(n) = \frac{Wr_{\Delta^*}(\phi_1^*(n),\ldots,\phi_k^*(n),\psi_1^*(n),\ldots,\psi_l^*(n),h(n))}{Wr_{\Delta^*}(\phi_1^*(n),\ldots,\phi_k^*(n),\psi_1^*(n),\ldots,\psi_l^*(n))}, \tag{6.10}$$

with

$$\phi_j^*(n) = \phi_j(n+k+l), \quad \psi_j^*(n) = \psi_j(n+k+l).$$

We could proceed by fitting Lemma 6.2 within the framework of Theorem
3.1. We choose a different road which brings more information and connects
our subject with recent ideas originating in the work of G. Wilson [60], [61].

The main observation is that the discrete Wronskian which appears on
the denominator of (6.9) is (essentially) a *tau function* for the discrete
Kadomtsev-Petviashvili (KP) hierarchy. Precisely, we have to define this
tau function as depending on the infinitely many time variables $t_1, t_2, \ldots$,
by substituting for s_k in (6.8), $k = 1, 2, \ldots$, the formal triangular change
of variables

$$s_k = (-1)^k \sum_{i=k}^{\infty} \binom{i}{k}(-2)^{i-k}t_i. \tag{6.11}$$

The tau function is then defined to be

$$\tau(n;t_1,t_2,\ldots) = (-1)^{nl}\,Wr_\Delta(\phi_1(n),\ldots,\phi_k(n),\psi_1(n),\ldots\psi_l(n)), \tag{6.12}$$

with $\phi_j(n)$ and $\psi_j(n)$ as in (6.7), (6.8), with both functions depending now on the variables $t = (t_1, t_2, \dots)$, taking into account (6.11).

It is beyond the scope of this review to explain the origins of these formulas, and I can only refer the reader to [30] and [31] for a complete discussion. The punch line is that the functions $w(n; t, z)$ and $w^*(n; t, z)$ which are defined through the operators P and Q in (6.9), (6.10), by the formulas

$$w(n; t, z) = Q(E - I)^{-k}(E + I)^{-l}(1 + z)^n, \tag{6.13}$$

$$w^*(n; t, z) = P^*_{n-1}(E^{-1} - I)^{-k}(E^{-1} + I)^{-l}(1 + z)^{-n}, \tag{6.14}$$

are precisely the so-called *wave function* and *adjoint wave function* of the discrete KP hierarchy. By the notation in (6.14), we mean that we replace $\phi_j^*(n), \psi_j^*(n), h(n)$ in (6.10), respectively by $\phi_j^*(n-1), \psi_j^*(n-1), (1+z)^{-n}$. Introducing the notation $[z] = (z, z^2/2, z^3/3, \dots)$, $w(n; t, z)$ and $w^*(n; t, z)$ can be written as follows

$$w(n; t, z) = (1 + z)^n \frac{\tau(n; t - [z^{-1}])}{\tau(n; t)}, \quad w^*(n; t, z) = (1 + z)^{-n} \frac{\tau(n; t + [z^{-1}])}{\tau(n; t)},$$

with $\tau(n; t)$ as in (6.12). Moreover, these functions satisfy the so-called *bilinear identities*:

$$\mathrm{res}_{z=\infty} w(n; t, z) w^*(m; t', z)$$

$$\times \exp\left(\sum_{i=1}^{\infty} (t_i - t_i') z^i \right) dz = 0, \ \forall n \geq m, \ \forall t, t'. \tag{6.15}$$

The connection between the Darboux factorization (6.3) and the wave and adjoint wave functions of the discrete KP hierarchy explains why we chose to make a lower-upper factorization instead of an upper-lower one.

From (6.6), the function $w(n; t, z)$ in (6.13) is an eigenfunction of the operator $L_{k,l}$, with eigenvalue $z^2/(1+z)$. It is in fact the common eigenfunction of a *rank 1 commutative ring of difference operators*, all commuting with $L_{k,l}$. By definition, the rank is the greatest common divisor of the orders of all the operators in the ring. We have strong evidence that the solutions to the extended Bochner-Krall problem obtained in this section are the *only ones* leading to *rank 1* commutative rings of difference operators. At present, a complete proof of this assertion is still missing. It would automatically imply that the Krall polynomials and their generalizations described in Theorem 5.1 and Theorem 5.2, necessarily correspond to *rank 2* commutative rings of difference operators. See [31] for a more complete discussion of this issue.

A theory developed originally by Burchnall and Chaundy [10] in the 1920's, establishes a dictionary between rank 1 commutative rings of differential operators and irreducible affine curves, which complete by adding *one* nonsingular point at infinity. These rings are isomorphic to the rings of meromorphic functions on the corresponding curves, with poles only at the added point at infinity. The order of the pole coincides with the order of the corresponding differential operator. The theory was revived in the late 1970's in connection with the developments in soliton theory, see [44]. A similar theory was then developed for rank 1 commutative rings of difference operators, establishing a one-to-one correspondence between these rings and irreducible affine curves, which complete by adding *two* nonsingular points at infinity, see [45], [51] and [59]. Intuitively, the orders of the poles of the meromorphic functions with singularities only at the two points at infinity, tell us now how many diagonals do the corresponding difference operators have above and below their main diagonal.

In [31], we prove that $L_{k,l}$ in (6.2) is part of a rank 1 commutative ring of difference operators. The equation of the corresponding curve is

$$v^2 = u^{2k+1}(u+1)^{2l+1}. \tag{6.16}$$

This curve can be rationally parametrized as follows:

$$x \mapsto \left(u = \frac{(x-1)^2}{4x}, v = \frac{(x-1)^{2k+1}(x+1)^{2l+1}}{(4x)^{k+l+1}} \right).$$

The two points at infinity correspond to $x = 0$ and $x = \infty$. There are also two *cuspidal singular points* with coordinates $x = \pm 1$. None of this is an accident, see [30], [60] and [61] for explanations!

Theorem 6.3 ([30], [31]) *The tridiagonal matrix $\hat{L} \equiv L_{k,l}$ which is obtained at the end of the chain of elementary Darboux transformations (6.2) or, equivalently, which is determined by equation (6.6), solves the extended Bochner-Krall problem. More precisely, the rational functions in x*

$$p_n(x) \equiv w(n; t, x-1) = (x-1)^{-k}(x+1)^{-l}Q\, x^n, \tag{6.17}$$

with Q as in (6.9) solve

$$\hat{L}\, p_n(x) = (x - 2 + x^{-1})p_n(x), \tag{6.18}$$

$$\hat{B}p_n(x) = \hat{\lambda}(n)p_n(x), \tag{6.19}$$

for a (rank 1) commutative ring of differential operators $\hat{B}$ in x. Moreover

$$\frac{1}{2\pi i} \oint p_n(x)p_m(x^{-1})\frac{dx}{x} = \frac{\tau(n+1;t)}{\tau(n;t)}\delta_{nm}, \quad \forall n, m \in \mathbb{Z}, \tag{6.20}$$

where the integral can be taken along any simple closed curve surrounding the origin $x = 0$ and avoiding the points $x = \pm 1$, and $\tau(n;t)$ is as in (6.12).

Proof. I shall only give a sketch of the proof, emphasizing some of the central ideas. Equation (6.18) follows immediately from the definition (6.17). To prove (6.19), one shows that the function $w'(y, z)$ defined as

$$w'(y, z) = w(z; t, e^y - 1),$$

is the wave function of a rank 1 commutative ring of differential operators in the variable y. The curve C' associated with this ring turns out to be a rational curve with ordinary double points. Hence

$$\hat{B}w'(y, z) = \hat{\lambda}(z)w'(y, z),$$

for all polynomials $\hat{\lambda}(z)$ which define meromorphic functions on C', having poles at the unique point at infinity which completes the curve. This establishes (6.19), changing the variable y in $\hat{B}$ to $x = e^y$.

From the explicit formulas (6.9), (6.10), (6.13) and (6.14), one checks that

$$p_n(x^{-1}) = \frac{\tau(n+1;t)}{\tau(n;t)} \, x p_{n+1}^*(x), \text{ with } p_n^*(x) = w^*(n; t, x - 1),$$
$$(6.21)$$

and $\tau(n;t)$ as in (6.12). Thus, proving the orthogonality relations (6.20), amounts to showing that

$$\text{res } _{x=0} p_n(x) p_{m+1}^*(x) \mathrm{d}x = \delta_{nm} \text{ and res } _{x=\pm 1} p_n(x) p_{m+1}^*(x) \mathrm{d}x = 0.$$

The first family of equations is easily checked by direct computation for $n \geq m$. Using the bilinear identities (6.15) with $t' = t$, and (6.21), one deduces then that the equations also hold for $n < m$. The second family of equations follows from the fact that the expression $p_n(x) p_m^*(x) \mathrm{d}x$ is a regular differential on the curve C (6.16), for all n, m, and the residue of such a form at a cusp, i.e. in this case at the points $x = \pm 1$, is always zero. This completes the sketch of the proof. $\qquad\square$

Our last theorem says that, when suitably parametrized, the matrix $L_{k,l}$ which is obtained at the end of the chain of elementary Darboux transformations (6.2), provides rational (in both the discrete and the continuous variables) solutions to the famous (doubly infinite) Toda lattice hierarchy.

Theorem 6.4 ([30]) *With the further change of time variables*

$$t_i = t_i' + \sum_{j=i+1}^{\infty} (-1)^{i+j} \binom{2j-i-1}{j-1} t_j',$$

the matrix $\hat{L} = L_{k,l}$, *obtained at the end of the chain of Darboux transformations* (6.2), *satisfies the Toda lattice hierarchy*

$$\frac{\partial \hat{L}}{\partial t_i'} = [(\hat{L}^i)_+, \hat{L}], \quad i = 1, 2, \ldots,$$

where $(\hat{L}^i)_+$ *denotes the upper part (including the diagonal) of* $\hat{L}^i$.

I like to conclude this survey with a challenge! Find a nice (instructive) proof or disprove (or improve!) the following conjecture:

Conjecture 6.5 *Besides the basic solutions described in Section 2, the solutions described in Theorem 5.1 (Krall-Laguerre type solutions), Theorem 5.2 (Krall-Jacobi type solutions) and Theorem 6.3 (Rank 1 solutions) give the complete list of solutions to the extended Bochner-Krall problem.*

Acknowledgements. I thank Mourad Ismail for several suggestions which led me to improve an earlier version of this paper. I also would like to thank the organizers of the NATO ASI "Special Functions 2000: Current Perspective and Future Directions" in Tempe, Arizona for providing such a stimulating atmosphere during the conference.

References

1. M. Adler and J. Moser, *On a class of polynomials connected with the Korteweg-de Vries equation*, Comm. Math. Phys. **61** (1978), 1–30.
2. G. E. Andrews, R. Askey and R. Roy, *Special functions*, Encyclopedia of mathematics and its applications **71**, Cambridge University Press, 1999.
3. R. Askey and M. E. H. Ismail, *Recurrence relations, continued fractions and orthogonal polynomials*, Mem. Amer. Math. Soc., No. 300, 1984.
4. R. Askey and J. Wimp, *Associated Laguerre and Hermite polynomials*, Proc. Roy. Soc. Edinburgh Sect. A **96** (1984), 15–37.
5. B. Bakalov, E. Horozov and M. Yakimov, *General methods for constructing bispectral operators*, Phys. Lett. A **222** (1996), 59–66.
6. B. Bakalov, E. Horozov and M. Yakimov, *Bispectral algebras of commuting ordinary differential operators*, Comm. Math. Phys. **190** (1997), 331–373.
7. Y. Berest, *Huygens' principle and the bispectral problem*, in: "The Bispectral Problem", J. Harnad and A. Kasman, eds, CRM Proc. Lecture Notes **14**, Amer. Math. Soc., Providence, 1998, pp. 11-30.
8. Ju. M. Berezanskii, *Expansions in eigenfunctions of selfadjoint operators*, Transl. Math. Monographs **17**, Amer. Math. Soc., Providence, 1968.
9. S. Bochner, *Über Sturm-Liouvillesche Polynomsysteme*, Math. Z. **29** (1929), 730–736.
10. J. L. Burchnall and T. W. Chaundy, a) *Commutative ordinary differential operators*, Proc. London Math. Soc. **21** (1923), 420–440; b) *Commutative ordinary differential operators*, Proc. Royal Soc. London (A) **118** (1928), 557–583; c) *Commutative ordinary differential operators II. The identity* $P^n = Q^m$, Proc. Royal Soc. London (A) **134** (1932), 471–485.
11. O. A. Chalykh, M. V. Feigin and A. P. Veselov, *Multidimensional Baker-Akhiezer functions and Huygens' principle*, Comm. Math. Phys. **206** (1999), 533–566.
12. G. Darboux, *Leçons sur la théorie générale des surfaces*, 2e partie, Gauthier-Villars, Paris, 1889.

13. P. Deift, *Applications of a commutation formula*, Duke Math. J. **45** (1978), 267–310.
14. J. J. Duistermaat and F. A. Grünbaum, *Differential equations in the spectral parameter*, Comm. Math. Phys. **103** (1986), 177–240.
15. W. N. Everitt, K. H. Kwon, L. L. Littlejohn and R. Wellman, *Orthogonal polynomial solutions of linear ordinary differential equations*, J. Comput. Appl. Math., to appear.
16. H. Flaschka and D. McLaughlin, *Some comments on Bäcklund transformations, canonical transformations and the inverse scattering method*, in: "Bäcklund Transformations", Lecture Notes in Mathematics **515**, Springer, Berlin, Heidelberg, New York, 1976, pp. 253–295.
17. R. W. Gosper, Jr., *Decision procedure for indefinite hypergeometric summation*, Proc. Natl. Acad. Sci. USA, **75** (1978), 40–42.
18. F. A. Grünbaum, *The bispectral problem: an overview*, this volume.
19. F. A. Grünbaum and L. Haine, *Orthogonal polynomials satisfying differential equations: The role of the Darboux transformation*, in: "Symmetries and Integrability of Difference Equations", D. Levi, L. Vinet, and P. Winternitz, eds, CRM Proc. Lecture Notes **9**, Amer. Math. Soc., Providence, 1996, pp. 143–154.
20. F. A. Grünbaum and L. Haine, *A theorem of Bochner, revisited* in: "Algebraic Aspects of Integrable Systems", A. S. Fokas and I. M. Gelfand, eds, Progr. Nonlinear Differential Equations **26**, Birkhäuser, Boston, 1997, pp. 143–172.
21. F. A. Grünbaum and L. Haine, *Some functions that generalize the Askey-Wilson polynomials*, Comm. Math. Phys. **184** (1997), 173–202.
22. F. A. Grünbaum and L. Haine, *Bispectral Darboux transformations: An extension of the Krall polynomials*, Internat. Math. Res. Notices 1997 (No.8), 359–392.
23. F. A. Grünbaum and L. Haine, *Associated polynomials, spectral matrices and the bispectral problem*, Meth. Appl. Anal. **6** (2) (1999), 209–224.
24. F. A. Grünbaum and L. Haine, *On a q-analogue of the string equation and a generalization of the classical orthogonal polynomials*, in: "Algebraic methods and q-special functions", J. F. van Diejen and L. Vinet, eds, CRM Proc. Lecture Notes **22**, Amer. Math. Soc., Providence, 1999, pp. 171–181.
25. F. A. Grünbaum, L. Haine and E. Horozov, *On the Krall-Hermite and the Krall-Bessel polynomials*, Internat. Math. Res. Notices 1997 (No. 19), 953–966.
26. F. A. Grünbaum, L. Haine and E. Horozov, *Some functions that generalize the Krall-Laguerre polynomials*, J. Comput. Appl. Math. **106** (1999), 271–297.
27. F. A. Grünbaum and M. Yakimov, *Discrete bispectral Darboux transformations from Jacobi operators*, Pacific J. Math., to appear.
28. L. Haine, *Beyond the classical orthogonal polynomials*, in: "The Bispectral Problem", J. Harnad and A. Kasman, eds., CRM Proc. Lecture Notes **14**, Amer. Math. Soc., Providence, 1998, pp. 47–65.
29. L. Haine and P. Iliev, *The bispectral property of a q-deformation of the Schur polynomials and the q-KdV hierarchy*, J. Phys. A: Math. Gen. **30** (1997), 7217–7227.
30. L. Haine and P. Iliev, *Commutative rings of difference operators and an adelic flag manifold*, Internat. Math. Res. Notices 2000 (No.6), 281–323.
31. L. Haine and P. Iliev, *A rational analogue of the Krall polynomials*, J. Phys. A: Math. Gen. **34** (2001), 2445–2457.
32. P. Iliev, *q-KP hierarchy, bispectrality and Calogero-Moser systems*, J. Geom. Phys. **35** (2000), 157–182.
33. L. Infeld and T. Hull, *The factorization method*, Rev. Modern Phys. **23** (1951), 21–68.
34. M. E. H. Ismail and D. R. Masson, *Two families of orthogonal polynomials related to Jacobi polynomials*, Rocky Mountain J. Math. **21** (1991), 359–375.
35. A. Kasman and M. Rothstein, *Bispectral Darboux transformations: the generalized Airy case*, Physica D **102** (1997), 159–176.
36. J. Koekoek and R. Koekoek, *On a differential equation for Koornwinder's generalized Laguerre polynomials*, Proc. Amer. Math. Soc. **112** (4) (1991), 1045–1054.

37. J. Koekoek and R. Koekoek, *Differential equations for generalized Jacobi polynomials*, J. Comput. Appl. Math. **126** (2000), 1–31.

38. R. Koekoek, *Differential equations for symmetric generalized ultraspherical polynomials*, Trans. Amer. Math. Soc. **345** (1994), 47–72.

39. T. H. Koornwinder, *Orthogonal polynomials with weight function* $(1-x)^\alpha(1+x)^\beta + M\delta(x+1) + N\delta(x-1)$, Can. Math. Bull. **27** (1984), 205–214.

40. T. H. Koornwinder, *Compact quantum groups and q-special functions*, in: "Representations of Lie groups and quantum groups", V. Baldoni and M. A. Picardello, eds., Pitman Research Notes in Mathematics Series **311**, Longman Scientific & Technical, 1994, pp. 46–128.

41. A. M. Krall, *Orthogonal polynomials satisfying fourth order differential equations*, Proc. Roy. Soc. Edinburgh Sect. A **87** (1981), 271–288.

42. H. L. Krall, *Certain differential equations for the Tchebycheff polynomials*, Duke Math. J. **4** (1938), 705–718.

43. H. L. Krall, *On orthogonal polynomials satisfying a certain fourth order differential equation*, The Pennsylvania State College Studies, No.6, 1940.

44. I. M. Krichever, *Methods of algebraic geometry in the theory of non-linear equations*, Russ. Math. Surveys **32** (1977), 185–214.

45. I. M. Krichever, *Algebraic curves and non-linear difference equations*, Russian Math. Surveys **33** (1978), 255–256.

46. L. L. Littlejohn, *The Krall polynomials: A new class of orthogonal polynomials*, Quaestiones Mathematicae **5** (1982), 255–265.

47. V. B. Matveev, *Darboux transformation and explicit solutions of the Kadomtcev-Petviaschvily equation, depending on functional parameters*, Lett. Math. Phys. **3** (1979), 213–216.

48. V. B. Matveev, *Darboux transformation and the explicit solutions of differential-difference and difference-difference evolution equations I*, Lett. Math. Phys. **3** (1979), 217–222.

49. V. B. Matveev and M. A. Salle, *Differential-difference evolution equations II: Darboux transformation for the Toda lattice*, Lett. Math. Phys. **3** (1979), 425–429.

50. V. B. Matveev and M. A. Salle, *Darboux transformations and Solitons*, Springer Series in Nonlinear Dynamics, 1991.

51. D. Mumford, *An algebro-geometric construction of commuting operators and of solutions to the Toda lattice equation, Korteweg-de Vries equation and related nonlinear equations*, in: "Proceedings of International Symposium on Algebraic Geometry", (Kyoto Univ., Kyoto 1977), M. Nagata, ed., Kinokuniya Book-Store, Tokyo, 1978, pp. 115–153.

52. F. W. Nijhoff and O. A. Chalykh, *Bispectral rings of difference operators*, Russian Math. Surveys **54** (1999), 644–645.

53. M. Petkovšek, H. S. Wilf and D. Zeilberger, $A = B$, A. K. Peters, Wellesley MA, 1996.

54. M. Rahman, *The associated classical orthogonal polynomials*, this volume.

55. E. Schrödinger, *A method of determining quantum mechanical eigenvalues and eigenfunctions*, Proc. Roy. Irish Acad. Sect. A **46** (1940), 9–16.

56. V. Spiridonov, L. Vinet and A. Zhedanov, *Bispectrality and Darboux transformations in the theory of orthogonal polynomials*, in: "The Bispectral Problem", J. Harnad and A. Kasman, eds, CRM Proc. Lecture Notes **14**, Amer. Math. Soc., Providence, 1998, pp. 111–122.

57. P. Terwilliger, *Two relations that generalize the q-Serre relations and the Dolan-Grady relations*, in: "Physics and Combinatorics", A. N. Kirillov, A. Tsuchiya, and H. Umemura, eds., World Scientific, 2001, to appear.

58. M. Toda, *Theory of Nonlinear Lattices*, Springer Series in Solid-State Sciences **20**, Springer, Berlin, Heidelberg, New-York, 1981.

59. P. van Moerbeke and D. Mumford, *The spectrum of difference operators and alge-*

braic curves, Acta Math. **143** (1979), 93–154.

60.   G. Wilson, *Bispectral commutative ordinary differential operators*, J. Reine Angew. Math. **442** (1993), 177–204.
61.   G. Wilson, *Collisions of Calogero-Moser particles and an adelic Grassmannian*, with an appendix by I.G. Macdonald, Invent. Math. **133** (1998), 1–41.
62.   J. Wimp, *Explicit formulas for the associated Jacobi polynomials and some applications*, Canad. J. Math. **39** (1987), 983–1000.
63.   A. Zhedanov, *A method of constructing Krall's polynomials*, J. Comput. Appl. Math. **107** (1999), 1–20.

LECTURES ON q-ORTHOGONAL POLYNOMIALS

MOURAD E. H. ISMAIL
Department of Mathematics
University of South Florida
Tampa, FL 33620-5700 USA

Abstract. These notes form a brief introduction to the theory of Askey-Wilson polynomials and related topics. We have used a novel approach to the development of those parts needed from the theory of basic hypergeometric functions. One important difference between our approach to basic hypergeometric functions and other approaches, for example those of Andrews, Askey, and Roy [7], of Gasper and Rahman [18] or of Bailey [12] is our frequent use of the Askey-Wilson divided difference operators, the q-difference operator, and the identity theorem for analytic functions.

We apply the bootstrap method from [13] and an attachment procedure to derive orthogonality relations for the Askey-Wilson polynomials in two steps using the Al-Salam-Chihara polynomials as an intermediate step. The continuous q-ultraspherical polynomialsand continuous q-Hermite polynomials are studied from a different approach. As an application of connection coefficient formulas we give the recent proof and generalizations of the Rogers-Ramanujan identities from [17].

These notes are summarized from a forthcoming Cambridge University Press monograph.

1. Introduction

This section contains preliminary material applied in the remaining lectures.

Theorem 1.1 (Identity Theorem) *Let $f(z)$ and $g(z)$ be analytic in a domain Ω and assume that $f(z_n) = g(z_n)$ for a sequence $\{z_n\}$ converging to an interior point of Ω. Then $f(z) = g(z)$ at all points of Ω.*

J. Bustoz et al. (eds.), Special Functions 2000, 179–219.

In §2 we develop those parts of the theory of basic hypergeometric functions that we shall use in later sections. Sometimes studying orthogonal polynomials leads to other results in special functions. For example the Askey-Wilson polynomials of §5 lead directly to the Sears transformation, so the Sears transformation (2.43) is stated and proved in §2 but another proof is given in §5.

1.1. ORTHOGONAL POLYNOMIALS

Let $\{p_n(x)\}$ be a sequence of polynomials such that p_n is of precise degree n and the p_n's satisfy an orthogonality relation

$$\int_{-\infty}^{\infty} p_m(x)p_n(x)d\mu(x) = \zeta_n \delta_{m,n},   \tag{1.1}$$

with μ a positive measure having moments of all orders and finite total mass.

Theorem 1.2 *Let $\{p_n(x)\}$ be as above and satisfy*

$$p_0(x) = 1, \quad p_1(x) = (x - b_0)/a_0, \quad a_0 \neq 0.   \tag{1.2}$$

Then the p_n's satisfy a three term recurrence relation

$$xp_n(x) = a_n p_{n+1}(x) + b_n p_n(x) + c_n p_{n-1}(x), \; n > 0,   \tag{1.3}$$

and

$$\zeta_n = \zeta_0 \prod_{k=1}^{n} \frac{c_k}{a_{k-1}}.   \tag{1.4}$$

Proof. Since $p_n(x)$ is of degree n for all n then

$$xp_n(x) = \sum_{k=0}^{n+1} a_{n,k} p_k(x), \quad a_{n,k}\zeta_k = \int_{-\infty}^{\infty} p_n(x)xp_k(x)d\mu(x).   \tag{1.5}$$

If $k < n - 1$ then $xp_k(x)$ is of degree $k + 1 < n$ and the integral in (1.5) will vanish. Now (1.5) with $k = n \pm 1$ yields

$$\int_{-\infty}^{\infty} xp_n(x)p_{n-1}(x)d\mu(x) = c_n\zeta_{n-1}, \quad \int_{-\infty}^{\infty} xp_n(x)p_{n+1}(x)d\mu(x) = a_n\zeta_{n+1}$$

These two relationships give $\zeta_n = \zeta_{n-1}c_n/a_{n-1}$, which implies (1.4) and the proof is complete. $\qquad\square$

Theorem 1.3 (Spectral Theorem) *Let $\{p_n(x)\}$ satisfy (1.4)–(1.5) and the positivity condition*

$$a_{n-1}c_n > 0, \quad n > 0. \tag{1.6}$$

Then there exists a positive measure μ whose moments $\int t^n d\mu$ are finite for all $n = 0, 1, \ldots$ such that (1.1) holds. Furthermore if the sequences $\{a_n\}$, $\{b_n\}$, $\{c_n\}$, are bounded then μ is unique and has compact support.

In most books on orthogonal polynomials Theorem 1.3 is known as Favard's theorem but we shall call it the spectral theorem because it describes the spectral measure of the symmetric tridiagonal matrix operator

$$b_n\delta_{m,n} + a_m\delta_{m,n-1} + c_m\delta_{m-1,n}, \quad m, n = 0, 1, \ldots,$$

which is similar to the tridiagonal matrix associated with the recursion (1.5). For details the reader may consult Akhiezer [1].

1.2. THE BOOTSTRAP METHOD

We shall often make use of a procedure we shall call the "bootstrap method" where we may obtain new orthogonal functions from old ones. Assume that we know a generating function for a sequence of orthogonal polynomials $\{p_n(x)\}$ satisfying (1.3). Let such a generating function be

$$\sum_{n=0}^{\infty} p_n(x)t^n/c_n = G(x, t), \tag{1.7}$$

with $\{c_n\}$ a suitable numerical sequence of nonzero elements. Thus the orthogonality relation (1.1) is equivalent to

$$\int_{-\infty}^{\infty} G(x, t_1)G(x, t_2)d\mu(x) = \sum_{0}^{\infty} \zeta_n \frac{(t_1t_2)^n}{c_n^2}, \tag{1.8}$$

provided that we can justify the interchange of integration and sums.

The idea is to use $G(x, t_1)G(x, t_2)\,d\mu(x)$ as a new measure, the total mass of which is given by (1.8), and then look for a system of functions (preferably polynomials) orthogonal or biorthogonal with respect to this new measure. If such a system is found one can then repeat the process. It it clear that we cannot indefinitely continue this process. The functions involved will become too complicated at a certain level, and the process will then terminate.

1.3. CONTINUED FRACTIONS

A continued fraction is

$$\frac{A_0}{A_0 z + B_0 -} \; \frac{C_1}{A_1 z + B_1 -} \; \cdots . \tag{1.9}$$

The nth convergent is a rational function $P_n^*(z)/P_n(z)$. The polynomials $P_n(z)$ and $P_n^*(z)$ are solutions of the recurrence relation

$$y_{n+1}(z) = [A_n z + B_n] y_n(z) - C_n y_{n-1}(z), \quad n > 0, \tag{1.10}$$

with the initial values

$$P_0(z) := 1, \; P_1(z) := A_0 z + B_0, \quad P_0^*(z) := 0, \; P_1^*(z) := A_0. \tag{1.11}$$

It is assumed that $A_0 \neq 0$. The positivity condition (1.6) in the normalization (1.10)–(1.11) is

$$A_n A_{n-1} C_n > 0. \tag{1.12}$$

Theorem 1.4 (Markov) *If μ is unique and is supported on $E \subset \mathbf{R}$ then*

$$\lim_{n \to \infty} \frac{P_n^*(z)}{P_n(z)} = \int\limits_E \frac{d\mu(t)}{z - t}, \quad z \notin E, \tag{1.13}$$

and the limit is uniform on compact subsets of $\mathbf{R} \setminus E$.

Theorem 1.4 is very useful in determining orthogonality measures for orthogonal polynomials from the knowledge of the recurrence relation they satisfy. For a proof see [26]. If $\{B_n/A_n\}$ and $\{C_n/A_n A_{n-1}\}$ are bounded then μ is unique and has compact support. Theorem 1.4 has been used very effectively in the 1980's, see for example [10].

In this normalization the orthogonality relation is

$$\int\limits_E P_m(x) P_n(x) \, d\mu(x) = \zeta_n \delta_{m,n}, \quad \zeta_n = \frac{A_0}{A_n} \prod_{k=1}^n C_k. \tag{1.14}$$

The Perron-Stieltjes inversion formula [1]

$$F(z) = \int\limits_{-\infty}^{\infty} \frac{d\mu(t)}{z - t}, z \notin \mathcal{R}, \text{ iff}$$

$$\mu(t) - \mu(s) = \lim_{\epsilon \to 0^+} \int\limits_s^t \frac{F(x - i\epsilon) - F(x + i\epsilon)}{2\pi i} \, dx. \tag{1.15}$$

The above inversion formula enables us to recover μ from knowing its Stieltjes transform $F(z)$.

1.4. q-DIFFERENCES

A discrete analogue of the derivative is the q-difference operator

$$(D_q f)(x) = (D_{q,x} f)(x) = \frac{f(x) - f(qx)}{(1-q)x}. \tag{1.16}$$

It is clear that

$$D_{q,x} x^n = \frac{1-q^n}{1-q} x^{n-1}, \tag{1.17}$$

and for differentiable functions $(D_q f)(x) \to f'(x)$ as $q \to 1^-$.

Some of the arguments in this subject become more transparent if we keep in mind the concept of q-differentiation and q-integration. The reason is that we can relate the q-results to the case $q = 1$ of classical special functions.

For finite a and b the q-integral is

$$\int_0^a f(x) d_q x := \sum_{n=0}^{\infty} [aq^n - aq^{n+1}] f(aq^n), \tag{1.18}$$

$$\int_a^b f(x) d_q x := \int_0^b f(x) d_q x - \int_0^a f(x) d_q x. \tag{1.19}$$

It is clear from (1.18)–(1.19) that the q-integral is an infinite Riemann sum with the division points in a geometric progression. We would then expect $\int_a^b f(x) d_q x \to \int_a^b f(x) dx$ as $q \to 1$ for continuous functions. The q-integral over $[0, \infty)$ uses the division points $\{q^n : -\infty < n < \infty\}$ and is

$$\int_0^\infty f(x) d_q x := (1-q) \sum_{-\infty}^{\infty} q^n f(q^n). \tag{1.20}$$

2. Summation Theorems

We recall some standard notation. The q-shifted factorials and the multiple q-shifted factorials are

$$(a;q)_0 := 1, \quad (a;q)_n := \prod_{k=1}^{n} (1 - aq^{k-1}), \quad n = 1, 2, \ldots, \text{ or } \infty, \tag{2.1}$$

$$(a_1, a_2, \ldots, a_k; q)_n := \prod_{j=1}^{k} (a_j; q)_n, \qquad (2.2)$$

respectively. The q-binomial coefficient is

$$\begin{bmatrix} n \\ k \end{bmatrix}_q := \frac{(q; q)_n}{(q; q)_k (q; q)_{n-k}}. \qquad (2.3)$$

Unless we say otherwise we shall always assume that $0 < q < 1$. A basic hypergeometric series is

$$_r\phi_s \left(\begin{array}{c} a_1, \ldots, a_r \\ b_1, \ldots, b_s \end{array} \middle| q, z \right) = {_r\phi_s}(a_1, \ldots, a_r; b_1, \ldots, b_s; q, z)$$

$$= \sum_{n=0}^{\infty} \frac{(a_1, \ldots, a_r; q)_n}{(q, b_1, \ldots, b_s; q)_n} z^n (-q^{(n-1)/2})^{n(s+1-r)}. \qquad (2.4)$$

Note that $(q^{-k}; q)_n = 0$ for $n > k$. To avoid trivial singularities or indeterminacies in (2.4) we shall always assume, unless otherwise stated, that none of the denominator parameters $b_1, \ldots, b_s$ in (2.4) has the form q^{-k}, $k = 0, 1, \ldots$. If one of the numerator parameters is of the form q^{-k} then the sum on the right-hand side of (2.4) is a finite sum and we say that the series in (2.4) is **terminating**. A series that does not terminate is called **nonterminating**.

The radius of convergence of the series in (2.4) is 1, 0 or ∞ according as $r = s + 1$, $r > s + 1$ or $r < s + 1$, as can be seen from the ratio test.

It is clear that

$$\lim_{q \to 1^-} \frac{(q^a; q)_n}{(1 - q)^n} = (a)_n, \qquad (2.5)$$

hence

$$\lim_{q \to 1^-} {_r\phi_s} \left(\begin{array}{c} q^{a_1}, \ldots, q^{a_r} \\ q^{b_1}, \ldots, q^{b_s} \end{array} \middle| q, z(1-q)^{s+1-r} \right)$$

$$= {_rF_s} \left(\begin{array}{c} a_1, \ldots, a_r \\ b_1, \ldots, b_s \end{array} \middle| (-1)^{s+1-r} z \right), \quad r \le s + 1. \qquad (2.6)$$

There are two key operators used in our analysis of q functions. The first is the q-difference operator D_q defined in (1.17). The second is the Askey-Wilson operator $\mathcal{D}_q$, which will be defined in the next paragraph. The operator $\mathcal{D}_q$ was introduced in [11] and Magnus showed in [24] how it arizes naturally from studying difference operators on different grids.

Given a polynomial f we set $\check{f}(e^{i\theta}) := f(x)$, $x = \cos\theta$, that is

$$\check{f}(z) = f((z + 1/z)/2), \quad z = e^{i\theta}. \tag{2.7}$$

In other words we think of $f(\cos\theta)$ as a function of $e^{i\theta}$. In this notation the Askey-Wilson divided difference operator $\mathcal{D}_q$ is defined by

$$(\mathcal{D}_q f)(x) := \frac{\check{f}(q^{1/2}e^{i\theta}) - \check{f}(q^{-1/2}e^{i\theta})}{\check{e}(q^{1/2}e^{i\theta}) - \check{e}(q^{-1/2}e^{i\theta})}, \quad x = \cos\theta, \tag{2.8}$$

with

$$e(x) = x, \quad \text{so that} \quad \check{e}(z) = (z + 1/z)/2. \tag{2.9}$$

A calculation reduces (2.8) to

$$(\mathcal{D}_q f)(x) = \frac{\check{f}(q^{1/2}e^{i\theta}) - \check{f}(q^{-1/2}e^{i\theta})}{(q^{1/2} - q^{-1/2})\, i\, \sin\theta}, \quad x = \cos\theta. \tag{2.10}$$

It is important to note that $e^{i\theta}$ is defined as

$$e^{i\theta} = x + \sqrt{x^2 - 1},$$

and the branch of the square root is taken such that $\sqrt{x^2 - 1} \approx x$ as $x \to \infty$.

As an example let us apply $\mathcal{D}_q$ to a Chebyshev polynomial $T_n(x)$. Recall that the Chebyshev polynomials of the first kind and second kinds, $T_n(x)$ and $U_n(x)$, respectively are

$$T_n(\cos\theta) := \cos(n\theta), \quad U_n(\cos\theta) := \frac{\sin((n+1)\theta)}{\sin\theta}. \tag{2.11}$$

Both T_n and U_n have degree n. Thus $\check{T}_n(z) = (z^n + z^{-n})/2$. Hence

$$\mathcal{D}_q T_n(x) = \frac{q^{n/2} - q^{-n/2}}{q^{1/2} - q^{-1/2}} U_{n-1}(x). \tag{2.12}$$

Therefore

$$\lim_{q \to 1} (\mathcal{D}_q f)(x) = f'(x), \tag{2.13}$$

holds for $f = T_n$, hence for all polynomials, since $\{T_n(x)\}_0^\infty$ is a basis for the vector space of all polynomials and $\mathcal{D}_q$ is a linear operator.

We will use the Askey-Wilson operator to derive some of the summation theorems needed in our treatment. The q-gamma function is

$$\Gamma_q(z) := \frac{(q; q)_\infty}{(1-q)^{z-1}(q^z; q)_\infty}. \tag{2.14}$$

There are several proofs of

$$\lim_{q \to 1^-} \Gamma_q(z) = \Gamma(z), \tag{2.15}$$

see for example [7], [4]. Formula (2.15) is useful in formulating the $q \to 1$ limiting cases of q-series results.

2.1. EXPANSION THEOREMS

In the calculus of the Askey-Wilson operator the basis $\{\phi_n(x; a) : 0 \leq n < \infty\}$

$$\phi_n(x; a) := (ae^{i\theta}, ae^{-i\theta}; q)_n = \prod_{k=0}^{n-1} \left[1 - 2axq^k + a^2 q^{2k} \right], \tag{2.16}$$

plays the role played by the monomials $\{x^n : 0 \leq n < \infty\}$ in the differential and integral calculus.

Theorem 2.1 *We have*

$$\mathcal{D}_q(ae^{i\theta}, ae^{-i\theta}; q)_n = -\frac{2a(1 - q^n)}{1 - q} (aq^{1/2}e^{i\theta}, aq^{1/2}e^{-i\theta}; q)_{n-1}. \tag{2.17}$$

Proof. With $f(x) = (ae^{i\theta}, ae^{-i\theta}; q)_n$, $\check{f}(z)$ is $(az, a/z; q)_n$ and the rest is an easy calculation. $\square$

Theorem 2.1 shows that it is natural to use $\{(ae^{i\theta}, ae^{-i\theta}; q)_n : n = 0, 1, \dots\}$ as a basis for polynomials when we deal with the Askey-Wilson operator.

Theorem 2.2 (Expansion Theorem, ([20])) . *Let f be a polynomial of degree n, then*

$$f(x) = \sum_{k=0}^{n} f_k \, (ae^{i\theta}, ae^{-i\theta}; q)_k, \tag{2.18}$$

where

$$f_k = \frac{(q - 1)^k}{(2a)^k (q; q)_k} q^{-k(k-1)/4} (\mathcal{D}_q^k f)(x_k), \quad x_k := \frac{1}{2} \left(aq^{k/2} + q^{-k/2}/a \right). \tag{2.19}$$

Proof. It is clear that the expansion (2.18) exists, and (2.17) yields

$$\mathcal{D}_q^k(ae^{i\theta}, ae^{-i\theta}; q)_n|_{x=x_k}$$

$$= (2a)^k \frac{q^{(0+1+\cdots+k-1)/2}(q;q)_n}{(q-1)^k (q;q)_{n-k}} (aq^{k/2}e^{i\theta}, aq^{k/2}e^{-i\theta}; q)_{n-k}\big|_{e^{i\theta}=aq^{k/2}}$$

$$= \frac{(q;q)_k}{(q-1)^k} (2a)^k q^{k(k-1)/4} \delta_{k,n}. \quad (2.20)$$

The theorem now follows by applying $\mathcal{D}_q^j$ to both sides of (2.18) then setting $x = x_j$. □

We need some elementary properties of the q-shifted factorials. It is clear from (2.1) that

$$(a;q)_n = (a;q)_\infty / (aq^n; q)_\infty, \qquad n = 0, 1, \ldots.$$

This suggests the following definition for q-shifted factorials of negative order

$$(a;q)_n := (a;q)_\infty / (aq^n; q)_\infty = 1/(aq^n; q)_{-n}, \qquad n = -1, -2, \ldots.$$
$$(2.21)$$

It is easy to see that

$$(a;q)_m (aq^m; q)_n = (a;q)_{m+n}, \qquad m, n = 0, \pm 1, \pm 2, \ldots.$$
$$(2.22)$$

Some useful identities involving q-shifted factorials are

$$(aq^{-n}; q)_k = \frac{(a;q)_k (q/a; q)_n}{(q^{1-k}/a; q)_n} q^{-nk},$$
$$(aq^{-n}; q)_n = (q/a; q)_n (-a)^n q^{-n(n+1)/2}, \quad (2.23)$$

$$(a;q)_{n-k} = \frac{(a;q)_n (-a)^{-k}}{(q^{1-n}/a; q)_k} q^{\frac{1}{2}k(k+1)-nk},$$
$$\frac{(a;q)_{n-k}}{(b;q)_{n-k}} = \frac{(a;q)_n (q^{1-n}/b; q)_k}{(b;q)_n (q^{1-n}/a; q)_k} \left(\frac{b}{a}\right)^k, \quad (2.24)$$

$$(a;q^{-1})_n = (1/a; q)_n (-a)^n q^{-n(n-1)/2}. \quad (2.25)$$

The identities (2.23)–(2.25) follow from the definitions (2.1) and (2.22). A basic hypergeometric function (2.4) is called **balanced** if

$$r = s+1 \quad \text{and} \quad qa_1 a_2 \cdots a_{s+1} = b_1 b_2 \cdots b_s. \quad (2.26)$$

Theorem 2.3 (q-Pfaff-Saalschütz) *The sum of a terminating balanced* $_3\phi_2$ *is given by*

$$_3\phi_2\left(\begin{array}{c} q^{-n}, a, b \\ c, d \end{array}\bigg|\, q,\, q\right) = \frac{(d/a, d/b; q)_n}{(d, d/ab; q)_n}, \qquad \text{with } cd = abq^{1-n}.$$

$$(2.27)$$

Proof. Apply Theorem 2.2 to $f(\cos\theta) = (be^{i\theta}, be^{-i\theta}; q)_n$, and use (2.17)–(2.19) to obtain

$$
\begin{aligned}
f_k &= \frac{(q;q)_n\,(b/a)^k}{(q;q)_k(q;q)_{n-k}}(bq^{k/2}e^{i\theta}, bq^{k/2}e^{-i\theta}; q)_{n-k}\big|_{e^{i\theta}=aq^{k/2}} \\
&= \frac{(q;q)_n\,(b/a)^k}{(q;q)_k(q;q)_{n-k}}(abq^k, b/a; q)_{n-k}.
\end{aligned}
$$

Therefore (2.18) becomes

$$\frac{(be^{i\theta}, be^{-i\theta}; q)_n}{(q;q)_n} = \sum_{k=0}^{n} \frac{b^k(ae^{i\theta}, ae^{-i\theta}; q)_k\,(abq^k, b/a; q)_{n-k}}{a^k(q;q)_k(q;q)_{n-k}}.$$

Using (2.24) we rewrite the above equation in basic hypergeometric form and the result is equivalent to (2.27). $\qquad\square$

Theorem 2.4 *We have the q-analogue of the Chu-Vandermonde sum*

$$_2\phi_1(q^{-n}, a; c; q, q) = \frac{(c/a; q)_n}{(c; q)_n}\, a^n,$$

$$(2.28)$$

and the q-analogue of Gauss's theorem

$$_2\phi_1(a, b; c; q, c/ab) = \frac{(c/a, c/b; q)_\infty}{(c, c/ab; q)_\infty}, \qquad |c/ab| < 1.$$

$$(2.29)$$

Proof. Let $n \to \infty$ in (2.27). Taking the limit inside the sum is justified since $(a, b; q)_k/(q, c; q)_k$ is bounded. The result is (2.29). When $b = q^{-n}$ then (2.29) becomes

$$_2\phi_1(q^{-n}, a; c; q, cq^n/a) = \frac{(c/a; q)_n}{(c; q)_n}.$$

$$(2.30)$$

To prove (2.28) we express the left-hand side of (2.30) as a sum, over k say, replace k by $n - k$, then apply (2.24) and arrive at (2.28) after some simplifications and substitutions. One can also prove (2.28) by letting $c = \lambda b$ then let $b \to 0$ and note that the presence of the free parameter λ allows a and b to be independent. $\qquad\square$

When we replace a, b, c by q^a, q^b, q^c, respectively, in (2.29), then apply (2.5), (2.14) and (2.15), we see that (2.29) reduces to Gauss's theorem [12]

$$_2F_1(a,b;c;1) = \frac{\Gamma(c)\Gamma(c-a-b)}{\Gamma(c-a)\Gamma(c-b)}, \qquad \Re(c-a-b) > 0.$$
(2.31)

Remark 1 Our proof of Theorem 2.4 shows that the terminating q-Gauss sum (2.30) is equivalent to the terminating q-Chu-Vandermonde sum (2.28). It is not true however that the nonterminating versions of (2.30) and (2.28) are equivalent. The nonterminating version of (2.30) is (2.29) but the non-terminating version of (2.28) is [18]

$$\frac{(aq/c, bq/c; q)_\infty}{(q/c; q)_\infty} {}_2\phi_1\left(\begin{array}{cc} a, b \\ c \end{array}\middle|\, q,\, q\right)$$
$$+ \frac{(a, b; q)_\infty}{(c/q; q)_\infty} {}_2\phi_1\left(\begin{array}{cc} aq/c, bq/c \\ q^2/c \end{array}\middle|\, q,\, q\right) = (abq/c; q)_\infty.$$
(2.32)

Theorem 2.5 (q-binomial Theorem) *If $|z| < 1$ or $a = q^{-n}$ then*

$$_1\phi_0(a; -; q, z) = \frac{(az; q)_\infty}{(z; q)_\infty}.$$
(2.33)

Proof. Let $c = abz$ in (2.29) then let $b \to 0$. The result is (2.33). $\quad\square$

Note that as $q \to 1^-$ the left side of (2.33), with a replaced by q^a, tends to $\sum_{n=0}^{\infty} (a)_n z^n / n!$, hence, by the binomial theorem the right-hand side must tend to $(1 - z)^{-a}$.

Theorem 2.6 (Euler) *We have*

$$e_q(z) := \sum_{n=0}^{\infty} \frac{z^n}{(q; q)_n} = \frac{1}{(z; q)_\infty}, \qquad |z| < 1,$$
(2.34)

and

$$E_q(z) := \sum_{n=0}^{\infty} \frac{z^n}{(q; q)_n} q^{n(n-1)/2} = (-z; q)_\infty.$$
(2.35)

Proof. Formula (2.34) is the special case $a = 0$ of (2.33). To get (2.35), we replace z by $-z/a$ in (2.33) and let $a \to \infty$. This and (2.1) establish (2.35) and the proof is complete. $\quad\square$

The left-hand sides of (2.34) and (2.35) are q-analogues of the exponential function. It readily follows that $e_q((1-q)x) \to e^x$, and $E_q((1-q)x) \to e^x$ as $q \to 1^-$.

The terminating version of the q-binomial theorem is

$$_1\phi_0(q^{-n}; -; q, z) = (q^{-n}z; q)_n = (-z)^n q^{-n(n+1)/2}(q/z; q)_n,$$

(2.36)

which follows from (2.33). The above identity may be written as

$$(z; q)_n = \sum_{k=0}^{n} \begin{bmatrix} n \\ k \end{bmatrix}_q q^{\binom{k}{2}}(-z)^k.$$

(2.37)

2.2. BILATERAL SERIES

Recall that $(a; q)_n$ for $n < 0$ has been defined in (2.21). A bilateral basic hypergeometric function is

$$_m\psi_m \left(\begin{array}{c} a_1, \ldots, a_m \\ b_1, \ldots, b_m \end{array} \middle| q, z \right) = \sum_{-\infty}^{\infty} \frac{(a_1, \ldots, a_m; q)_n}{(b_1, \ldots, b_m; q)_n} z^n.$$

(2.38)

It is easy to see that the series in (2.31) converges if

$$\left| \frac{b_1 b_2 \cdots b_m}{a_1 a_2 \cdots a_m} \right| < |z| < 1.$$

(2.39)

Our next result is the Ramanujan $_1\psi_1$ sum.

Theorem 2.7 *The following holds for* $|b/a| < |z| < 1$

$$_1\psi_1(a; b; q, z) = \frac{(b/a, q, q/az, az; q)_\infty}{(b, b/az, q/a, z; q)_\infty}.$$

(2.40)

Proof. ([19]) Observe that both sides of (2.40) are analytic function of b for $|b| < |az|$ since, by (2.21), we have

$$_1\psi_1(a; b; q, z) = \sum_{n=0}^{\infty} \frac{(a; q)_n}{(b; q)_n} z^n + \sum_{n=1}^{\infty} \frac{(q/b; q)_n}{(q/a; q)_n} \left(\frac{b}{az} \right)^n.$$

Furthermore when $b = q^{m+1}$, m a positive integer, then

$$1/(b; q)_n = (q/b; q)_{-n} = 0$$

for $n < -m$, see (2.21). Therefore

$$_1\psi_1(a; q^{m+1}; q, z) = \sum_{n=-m}^{\infty} \frac{(a; q)_n}{(q^{m+1}; q)_n} z^n = z^{-m} \frac{(a; q)_{-m}}{(q^{m+1}; q)_{-m}} \sum_{n=0}^{\infty} \frac{(aq^{-m}; q)_n}{(q; q)_n} z^n$$

$$= z^{-m} \frac{(a; q)_{-m}}{(q^{m+1}; q)_{-m}} \frac{(azq^{-m}; q)_\infty}{(z; q)_\infty}$$

$$= \frac{z^{-m}(az; q)_\infty (q, azq^{-m}; q)_m}{(z; q)_\infty (aq^{-m}; q)_m}.$$

Using (2.22) and (2.23) we simplify the above formula to

$$_1\psi_1(a; b; q, z) = \frac{(q^{m+1}/a, q, q/az, az; q)_\infty}{(q^{m+1}, q^{m+1}/az, q/a, z; q)_\infty},$$

which is (2.40) with $b = q^{m+1}$. The identity theorem for analytic functions then establishes the theorem.   □

Another proof of Theorem 2.7 using functional equations is in [5]. For other proofs see the references in [5], [18], [25].

Theorem 2.8 (Jacobi Triple Product Identity) *We have*

$$\sum_{-\infty}^{\infty} q^{n^2} z^n = (q^2, -qz, -q/z; q^2)_\infty. \tag{2.41}$$

Proof. Formula (2.40) implies

$$\sum_{-\infty}^{\infty} q^{n^2} z^n = \lim_{c \to 0} {}_1\psi_1(-1/c; 0; q^2, qzc) = \lim_{c \to 0} \frac{(q^2, -qz, -q/z; q^2)_\infty}{(-q^2c, qcz; q^2)_\infty},$$

which is (2.41).   □

2.3. TRANSFORMATIONS

A very important transformation in the theory of basic hypergeometric functions is the Sears transformation, [18, (III.15)]. It can be stated as

$$_4\phi_3 \left(\begin{array}{c} q^{-n}, a, b, c \\ d, e, f \end{array} \bigg| q, q \right)$$
$$= \left(\frac{bc}{d} \right)^n \frac{(de/bc, df/bc; q)_n}{(e, f; q)_n} \, {}_4\phi_3 \left(\begin{array}{c} q^{-n}, a, d/b, d/c \\ d, de/bc, df/bc \end{array} \bigg| q, q \right), \tag{2.42}$$

where $abc = defq^{n-1}$. We feel that this transformation can be better motivated if expressed in terms of the Askey-Wilson polynomials

$$w_n(x; a, b, c, d|q) := {}_4\phi_3 \left(\begin{array}{c} q^{-n}, abcdq^{n-1}, ae^{i\theta}, ae^{-i\theta} \\ ab, ac, ad \end{array} \bigg| q, q \right). \tag{2.43}$$

Theorem 2.9 *We have*

$$w_n(x; a, b, c, d \mid q) = \frac{a^n (bc, bd; q)_n}{b^n (ac, ad; q)_n} w_n(x; b, a, c, d \mid q). \tag{2.44}$$

It is clear that (2.42) and (2.44) are equivalent.

Proof. Using (2.17) we see that

$$\mathcal{D}_q \omega_n(x; a, b, c, d|q) = -\frac{2aq(1 - q^{-n})(1 - abcdq^{n-1})}{(1 - q)(1 - ab)(1 - ac)(1 - ad)}$$
$$\times \omega_{n-1}(x; aq^{1/2}, bq^{1/2}, cq^{1/2}, dq^{1/2}|q). \quad (2.45)$$

On the other hand we can expand $\omega_n(x; a, b, c, d|q)$ as $\sum_{k=0}^{n} f_k \psi_k(x; b)$ and (2.19) and (2.45) yield

$$f_k = \frac{a^k q^k (q^{-n}, abcdq^{n-1}; q)_k}{b^k (q, ab, ac, ad; q)_k}$$
$$\times \omega_{n-k}((bq^{k/2} + q^{-k/2}/b)/2; aq^{k/2}, bq^{k/2}, cq^{k/2}, dq^{k/2} \mid q)$$
$$= \frac{a^k (q^{-n}, abcdq^{n-1}; q)_k}{b^k (q, ab, ac, ad; q)_k} {}_3\phi_2 \left(\begin{array}{c} q^{k-n}, abcdq^{n+k-1}, a/b \\ acq^k, adq^k \end{array} \middle| q, q \right).$$

Now (2.27) sums the ${}_3\phi_2$ function and we find

$$f_k = \frac{a^k (q^{-n}, abcdq^{n-1}; q)_k (q^{1-n}/bd; q)_{n-k} (bc; q)_n}{b^k (q, ab, bc, ad; q)_k (q^{1-n}/ad; q)_{n-k} (ac; q)_n}$$

and we obtain (2.44) after some manipulations. □

Theorem 2.10 *The following* ${}_3\phi_2$ *transformation holds*

$${}_3\phi_2 \left(\begin{array}{c} q^{-n}, a, b \\ c, d \end{array} \middle| q, q \right) = \frac{b^n (d/b; q)_n}{(d; q)_n} {}_3\phi_2 \left(\begin{array}{c} q^{-n}, b, c/a \\ c, q^{1-n}b/d \end{array} \middle| q, aq/d \right). \quad (2.46)$$

Proof. In (2.42) set $f = abcq^{1-n}/de$ then let $c \to 0$ so that $f \to 0$ while the remaining parameters remain constant. The result is

$${}_3\phi_2 \left(\begin{array}{c} q^{-n}, a, b \\ d, e \end{array} \middle| q, q \right)$$
$$= \frac{(-e)^n q^{n(n-1)/2} (aq^{1-n}/e; q)_n}{(e; q)_n} {}_3\phi_2 \left(\begin{array}{c} q^{-n}, a, d/b \\ d, q^{1-n}a/e \end{array} \middle| q, bq/e \right).$$

The result now follows from (2.27). □

An interesting application of (2.46) follows by letting b and d tend to ∞ in such a way that b/d remains bounded. Let $b = \lambda d$ and let $d \to \infty$ in (2.46). The result is

$${}_2\phi_1 \left(\begin{array}{c} q^{-n}, a \\ c \end{array} \middle| q, q\lambda \right) = (\lambda q^{1-n}; q)_n \sum_{j=0}^{n} \frac{(q^{-n}, c/a; q)_j q^{j(j-1)/2}}{(q, c, \lambda q^{1-n}; q)_j} (-\lambda aq)^j.$$

Now replace λ by λq^{n-1} and observe that the above identity becomes the special case $\gamma = q^n$ of

$$_2\phi_1 \left(\begin{array}{c} a, 1/\gamma \\ c \end{array} \bigg| \, q, \, \gamma\lambda \right) = \frac{(\lambda; q)_\infty}{(\lambda\gamma; q)_\infty} \sum_{j=0}^\infty \frac{(1/\gamma, c/a; q)_j q^{j(j-1)/2}}{(q, c, \lambda; q)_j} (-\lambda a\gamma)^j. \tag{2.47}$$

Since both sides of the relationship (2.47) are analytic functions of γ when $|\gamma| < 1$ and they are equal when $\gamma = q^n$ then they must be identical for all γ if $|\gamma| < 1$.

It is more convenient to write the identity (2.47) in the form

$$\sum_{n=0}^\infty \frac{(A, C/B; q)_n}{(q, C, Az; q)_n} q^{n(n-1)/2}(-Bz)^n = \frac{(z; q)_\infty}{(Az; q)_\infty} {}_2\phi_1(A, B; C; q, z), \tag{2.48}$$

or equivalently

$$_2\phi_2(A, C/B; C, Az; q, Bz) = \frac{(z; q)_\infty}{(Az; q)_\infty} {}_2\phi_1(A, B; C; q, z). \tag{2.49}$$

Observe that (2.48) or (2.49) is the q-analogue of the Pfaff-Kummer transformation, [12, 7]

$$_2F_1(a, b; c; z) = (1 - z)^{-a} {}_2F_1(a, c - b; c; z/(z - 1)), \tag{2.50}$$

which holds when $|z| < 1$ and $|z/(z - 1)| < 1$.

3. Continuous q-Hermite Polynomials

In this section we study the continuous q-Hermite polynomials. These polynomials and the continuous q-ultraspherical first appeared in Rogers's memoirs on expansions of infinite products, where he proved, among other things, the Rogers Ramanujan identities in 1893–95 [9]. Fejer generalized the Legendre polynomials to polynomials $\{p_n(x)\}$ having generating functions

$$\sum_{n=0}^\infty p_n(\cos\theta)t^n = \left| F(re^{i\theta}) \right|^2, \tag{3.1}$$

where $F(z) = \sum_0^\infty f_n z^n$ is analytic in a neighborhood of $z = 0$ and f_n is real for all n. The Legendre polynomials correspond to the case $F(z) = (1 - z)^{-1/2}$. Fejer proved that the zeros of the generalized Legendre polynomials share many of the properties of the zeros of the Legendre and ultraspherical polynomials. For an account of these results see [26]. Feldheim [16] and Lanzewizky [23] independently proved that the only orthogonal generalized Legendre polynomials are either the ultraspherical polynomials or the q-ultraspherical polynomials. The weight function for the q-ultraspherical polynomials was first found by Askey and Ismail [9], and Askey and Wilson [11] using different methods.

3.1. q-HERMITE POLYNOMIALS

The continuous q-Hermite polynomials $\{H_n(x|q)\}$ are generated by the recursion relation

$$2xH_n(x|q) = H_{n+1}(x|q) + (1 - q^n)H_{n-1}(x|q), \tag{3.2}$$

and the initial conditions

$$H_0(x|q) = 1, \quad H_1(x|q) = 2x. \tag{3.3}$$

To find a generating function for $\{H_n(x|q)\}$ we let

$$H(x,t) := \sum_{n=0}^{\infty} H_n(x|q)\frac{t^n}{(q;q)_n}. \tag{3.4}$$

Multiply (3.2) by $t^n/(q;q)_n$ and add for $n = 1, 2, \cdots$ and take into account the initial conditions (3.3) we obtain the functional equation

$$H(x,t) - H(x,qt) = 2xtH(x,t) - t^2H(x,t).$$

Therefore

$$H(\cos\theta, t) = \frac{H(\cos\theta, qt)}{(1 - te^{i\theta})(1 - te^{-i\theta})}. \tag{3.5}$$

This suggests iterating the functional equation (3.5) to get

$$H(\cos\theta, t) = \frac{H(\cos\theta, q^n t)}{(te^{i\theta}, te^{-i\theta}; q)_n}.$$

As $n \to \infty$, $H(x, q^n t) \to H(x, 0) = 1$. This motivates the next theorem.

Theorem 3.1 *The continuous q-Hermite polynomials have the generating function*

$$\sum_{n=0}^{\infty} H_n(\cos\theta|q)\frac{t^n}{(q;q)_n} = \frac{1}{(te^{i\theta}, te^{-i\theta}; q)_\infty}. \tag{3.6}$$

Proof. It is straightforward to see that the left-hand side of (3.6) satisfies the functional equation

$$(1 - 2xt + t^2)F(x,t) = F(x, qt). \tag{3.7}$$

Since the right-hand side of (3.6) is analytic in t in a neighborhood of $t = 0$ then it can be expanded in a power series in t and by substituting the expansion $F(x,t) = \sum_{n=0}^{\infty} f_n(x)t^n/(q;q)_n$ into (3.7) and equating coefficients of t^n, we find that the f_n's satisfy the three term recurrence relation (3.2) and agree with $H_n(x \mid q)$ when $n = 0$, $n = 1$. Thus $f_n = H_n(x \mid q)$ for all n. $\qquad \square$

A rigorous proof of Theorem 3.1 was given in some detail in order to show how to justify the formal argument leading to it. In future we will only give the formal proof and the interested reader can easily fill in the details.

Next we expand $1/(te^{\pm i\theta}; q)_\infty$ by (2.34), then multiply the resulting series. This gives

$$H_n(\cos\theta|q) = \sum_{k=0}^{n} \frac{(q;q)_n}{(q;q)_k(q;q)_{n-k}} e^{i(n-2k)\theta}$$

$$= \sum_{k=0}^{n} \begin{bmatrix} n \\ k \end{bmatrix}_q e^{i(n-2k)\theta}. \tag{3.8}$$

Since $H_n(x|q)$ is a real polynomial we find

$$H_n(\cos\theta|q) = \sum_{k=0}^{n} \frac{(q;q)_n}{(q;q)_k(q;q)_{n-k}} \cos(n-2k)\theta \tag{3.9}$$

Theorem 3.2 *The continuous q-Hermite polynomials have the following properties*

$$H_n(-x|q) = (-1)^n H_n(x \mid q), \tag{3.10}$$

and

$$max\{|H_n(x|q)| : -1 \le x \le 1\} = H_n(1|q) = (-1)^n H_n(-1|q), \tag{3.11}$$

and the maximum is attained only at $x = \pm 1$.

Proof. Replace θ by $\pi - \theta$ in (3.9) to get (3.10). The rest follows from the triangular inequality. $\qquad \square$

Clearly (3.9) implies that $|H_n(1 \mid q)| \le Cn$, where C is a constant not depending on n. Now (3.11) indicates that the series on the left-hand side of (3.6) converges uniformly in x for $x \in [-1, 1]$ for every fixed t, $|t| < 1$.

Theorem 3.3 *The continuous q-Hermite polynomials satisfy the orthogonality relation*

$$\int\limits_{-1}^{1} H_m(x|q)H_n(x|q)w(x|q)dx = \frac{2\pi(q;q)_n}{(q;q)_\infty}\delta_{m,n}, \qquad (3.12)$$

where

$$w(x|q) = \frac{(e^{2i\theta}, e^{-2i\theta}; q)_\infty}{\sqrt{1-x^2}}, \qquad x = \cos\theta, \; 0 \le \theta \le \pi. \qquad (3.13)$$

Theorem 3.3 is a special case of Theorem 4.1 which we shall prove later. It is worth noting that

$$H_n(x|q) = (2x)^n + \text{lower order terms}, \qquad (3.14)$$

which follows from (3.2) and (3.3).

Theorem 3.4 *The linearization of products of continuous q-Hermite polynomials is given by*

$$H_m(x\mid q)H_n(x\mid q) = \sum_{k=0}^{m\wedge n} \frac{(q;q)_m(q;q)_n}{(q;q)_k(q;q)_{m-k}(q;q)_{n-k}}H_{m+n-2k}(x\mid q), \qquad (3.15)$$

where

$$m \wedge n := \min\{m, n\}. \qquad (3.16)$$

Proof. We only sketch a proof. The relationship (3.14) indicates that the product $H_m(x|q)H_n(x|q)$ has the same parity as $H_{m+n}(x|q)$, hence there exists a sequence $\{a_{m,n,k} : 0 \le k \le m \wedge n\}$ such that

$$H_m(x|q)H_n(x|q) = \sum_{k=0}^{m\wedge n} a_{m,n,k}H_{m+n-2k}(x|q) \qquad (3.17)$$

and $a_{m,n,k}$ satisfies

$$a_{m,0,k} = a_{0,n,k} = \delta_{k,0}, \quad a_{m,n.0} = 1. \qquad (3.18)$$

Multiply (3.17) by $2x$ and use the three term recurrence relation (3.2) to establish the system of difference equations

$$a_{m+1,n,k+1} - a_{m,n,k+1} = (1 - q^{m+n-2k})a_{m,n,k} - (1 - q^m)a_{m-1,n,k}, \qquad (3.19)$$

subject to the initial conditions (3.18). The cases $k = 0$ and $k = 1$ lead to the pattern

$$a_{m,n,k} = \frac{(q;q)_m (q;q)_n}{(q;q)_{m-k}(q;q)_{n-k}(q;q)_k},$$

which can be proved induction. $\qquad\square$

Theorem 3.5 *The Poisson kernel of the H_n's is*

$$\sum_{n=0}^{\infty} \frac{H_n(\cos\theta|q)H_n(\cos\phi|q)}{(q;q)_n} t^n \qquad (3.20)$$

$$= \frac{(t^2;q)_\infty}{(te^{i(\theta+\phi)}, te^{i(\theta-\phi)}, te^{-i(\theta+\phi)}, te^{-i(\theta-\phi)};q)_\infty}.$$

Proof. Multiply (3.15) by $t_1^m t_2^n/(q;q)_m(q;q)_n$ and add for $m, n = 0, 1, \cdots$ then apply the generating function (3.6) to get

$$\frac{1}{(t_1 e^{i\theta}, t_1 e^{-i\theta}, t_2 e^{i\theta}, t_2 e^{-i\theta};q)_\infty}$$

$$= \sum_{m \geq k, n \geq k, k \geq 0} \frac{H_{m+n-2k}(\cos\theta|q)t_1^m t_2^n}{(q;q)_{m-k}(q;q)_{n-k}(q;q)_k}$$

$$= \sum_{k,m,n=0}^{\infty} \frac{H_{m+n}(\cos\theta \mid q)t_1^m t_2^n (t_1 t_2)^k}{(q;q)_m(q;q)_n(q;q)_k}$$

$$= \frac{1}{(t_1 t_2;q)_\infty} \sum_{m,n=0}^{\infty} \frac{H_{m+n}(\cos\theta \mid q)t_1^m t_2^n}{(q;q)_m(q;q)_n},$$

where we used (2.34). In the last sum replace $m + n$ by s then replace t_1 and t_2 by $t_1 e^{i\phi}$ and $t_1 e^{-i\phi}$, respectively. Therefore

$$\frac{(t_1^2;q)_\infty}{(t_1 e^{i(\theta+\phi)}, t_1 e^{i(\phi-\theta)}, t_1 e^{i(\theta-\phi)}, t_1 e^{-i(\theta+\phi)};q)_\infty}$$

$$= \sum_{s=0}^{\infty} \frac{H_s(\cos\theta|q)t_1^s}{(q;q)_s} \sum_{n=0}^{s} \frac{(q;q)_s e^{i(s-2n)\phi}}{(q;q)_n(q;q)_{s-n}}.$$

In view of (3.9) the n sum is $H_s(\cos\phi \mid q)$ and the proof is complete. $\qquad\square$

Theorem 3.6 *The inverse to (3.15) is*

$$\frac{H_{n+m}(x \mid q)}{(q;q)_m(q;q)_n} = \sum_{k=0}^{m \wedge n} \frac{(-1)^k q^{k(k-1)/2}}{(q;q)_k} \frac{H_{n-k}(x \mid q)}{(q;q)_{n-k}} \frac{H_{m-k}(x \mid q)}{(q;q)_{m-k}}. \qquad (3.21)$$

Proof. As in the proof of Theorem 3.5 we have

$$\frac{(t_1t_2;q)_\infty}{(t_1e^{i\theta}, t_1e^{-i\theta}, t_2e^{i\theta}, t_2e^{-i\theta};q)_\infty} = \sum_{m,n=0}^{\infty} \frac{H_{n+m}(x\mid q)}{(q;q)_m(q;q)_n} t_1^m t_2^n. \tag{3.22}$$

Now expand $(t_1t_2;q)_\infty$ by (2.35) and use (3.6) to expand the rest of the left-hand side of (3.22) then equate coefficients of $t_1^m t_2^n$. The result is (3.21).

The Askey-Wilson operator acts on $H_n(x\mid q)$ in a natural way. Indeed

$$\mathcal{D}_q H_n(x\mid q) = \frac{2(1-q^n)}{1-q} q^{(1-n)/2} H_{n-1}(x\mid q), \tag{3.23}$$

follows from applying $\mathcal{D}_q$ to (3.6) and equating coefficients of like powers of t.   □

4. Continuous q-Ultraspherical Polynomials

The continuous q-ultraspherical polynomials form a one parameter family which generalize the q-Hermite polynomials. They are defined by

$$C_n(\cos\theta; \beta|q) = \sum_{k=0}^{n} \frac{(\beta;q)_k(\beta;q)_{n-k}}{(q;q)_k(q;q)_{n-k}} e^{i(n-2k)\theta}. \tag{4.1}$$

It is clear that

$$\begin{aligned} C_n(x;0\mid q) &= H_n(x\mid q)/(q;q)_n, \\ C_n(-x;\beta\mid q) &= (-1)^n C_n(x;\beta\mid q). \end{aligned} \tag{4.2}$$

Although the C_n's are special cases of the Askey-Wilson polynomials of §5 we, nevertheless, give an independent proof of their orthogonality.

Theorem 4.1 *The orthogonality relation*

$$\int_{-1}^{1} C_m(x;\beta\mid q)C_n(x;\beta\mid q)w(x\mid\beta)dx$$

$$= \frac{2\pi(\beta, q\beta;q)_\infty}{(q, \beta^2;q)_\infty} \frac{(1-\beta)(\beta^2;q)_n}{(1-\beta q^n)(q;q)_n} \delta_{m,n}, \tag{4.3}$$

holds for $|\beta| < 1$, *with*

$$w(\cos\theta\mid\beta) = \frac{(e^{2i\theta}, e^{-2i\theta};q)_\infty}{(\beta e^{2i\theta}, \beta e^{-2i\theta};q)_\infty}(\sin\theta)^{-1}. \tag{4.4}$$

Proof. We shall first assume that $\beta \neq q^k$, $k = 1, 2, \ldots$, then extend the result by analytic continuation to include these values. Since the weight function is even, it follows that (4.3) holds trivially if $n - m$ is odd. Thus it suffices to evaluate

$$I_{m,n} := \int_0^\pi e^{i(n-2m)\theta} C_n(\cos\theta; \beta|q) \frac{(e^{2i\theta}, e^{-2i\theta}; q)_\infty}{(\beta e^{2i\theta}, \beta e^{-2i\theta}; q)_\infty} \, d\theta, \qquad (4.5)$$

for $0 \leq m \leq n$. From (4.4) and the $_1\psi_1$ sum (2.40) we find

$$\frac{1}{\pi} I_{m,n}$$

$$= \frac{1}{\pi} \sum_{k=0}^n \frac{(\beta; q)_k (\beta; q)_{n-k}}{(q; q)_k (q; q)_{n-k}} \int_0^\pi \frac{(e^{2i\theta}, qe^{-2i\theta}; q)_\infty}{(\beta e^{2i\theta}, \beta e^{-2i\theta}; q)_\infty} e^{(2n-2m-2k)i\theta} (1 - e^{-2i\theta}) \, d\theta$$

$$= \frac{(\beta, \beta q; q)_\infty}{(q, \beta^2; q)_\infty} \sum_{k=0}^n \frac{(\beta; q)_k (\beta; q)_{n-k}}{(q; q)_k (q; q)_{n-k}} \sum_{j=-\infty}^\infty \frac{\beta^j (1/\beta; q)_j}{(\beta; q)_j} [\delta_{j,k+m-n} - \delta_{j,k+m-n+1}]$$

$$= \frac{(\beta; q)_n (\beta, \beta q; q)_\infty}{(q; q)_n (q, \beta^2; q)_\infty} \left[\frac{\beta^{m-n}(1/\beta; q)_{m-n}}{(\beta; q)_{m-n}} {}_3\phi_2 \left(\begin{array}{c} q^{-n}, \beta, q^{m-n}/\beta \\ q^{1-n}/\beta, q^{m-n}\beta \end{array} \middle| q, q \right) \right.$$

$$\left. - \frac{\beta^{m-n+1}(1/\beta; q)_{m-n+1}}{(\beta; q)_{m-n+1}} {}_3\phi_2 \left(\begin{array}{c} q^{-n}, \beta, q^{m-n+1}/\beta \\ q^{1-n}/\beta, q^{m-n+1}\beta \end{array} \middle| q, q \right) \right].$$

Thus

$$I_{m,n} = \pi \frac{(\beta; q)_n (\beta, \beta q; q)_\infty}{(q; q)_n (q, \beta^2; q)_\infty} \left[\frac{\beta^{m-n}(1/\beta; q)_{m-n}(q^{m-n}, \beta^2; q)_n}{(\beta; q)_{m-n}(q^{m-n}\beta, \beta; q)_n} \right.$$

$$\left. - \frac{\beta^{m-n+1}(1/\beta; q)_{m-n+1}(q^{m-n+1}, \beta^2; q)_n}{(\beta; q)_{m-n+1}(q^{m-n+1}\beta, \beta; q)_n} \right]. \qquad (4.6)$$

We used (2.27) in the last step. The factor $(q^{m-n}; q)_n$ vanishes for $m \leq n$ and causes the first term in [ ] in (4.6) to vanish for $m \leq n$ while the factor $(q^{m-n+1}; q)_n$ annihilates the second term in [ ] for $m < n$. If $m = n$ then

$$I_{n,n} = \frac{\pi(\beta, q\beta; q)_\infty}{(q, \beta^2; q)_\infty} \frac{(\beta^2; q)_n}{(q\beta; q)_n}.$$

Thus (4.3) holds for $m < n$ and, if $m = n$ its left-hand side is equal to $2(\beta; q)_n I_{n,n}/(q; q)_n$. $\qquad \square$

It is straightforward to see that (4.1) implies the generating function

$$\sum_{n=0}^\infty C_n(\cos\theta; \beta \mid q) t^n = \frac{(t\beta e^{i\theta}, t\beta e^{-i\theta}; q)_\infty}{(te^{i\theta}, te^{-i\theta}; q)_\infty}. \qquad (4.7)$$

From (4.7) it follows that

$$\left[1 - 2xt + t^2\right] \sum_{n=0}^{\infty} C_n(x;\beta \mid q)t^n = \left[1 - 2x\beta t + \beta^2 t^2\right] \sum_{n=0}^{\infty} C_n(x;\beta \mid q)(qt)^n$$

which yields the recursion

$$2x(1 - \beta q^n)C_n(x;\beta \mid q) \tag{4.8}$$
$$= (1 - q^{n+1})C_{n+1}(x;\beta \mid q) + (1 - \beta^2 q^{n-1})C_{n-1}(x;\beta \mid q),$$

for $n > 0$. The initial values of the C_n's are

$$C_0(x;\beta \mid q) = 1, \quad C_1(x;\beta \mid q) = 2x(1 - \beta)/(1 - q). \tag{4.9}$$

The special cases $\beta = q$ of (4.1) or (4.7) give

$$C_n(x;q \mid q) = U_n(x). \tag{4.10}$$

On the other hand

$$\lim_{\beta \to 1} \frac{(1 - q^n)}{2(1 - \beta)}C_n(x;\beta \mid q) = T_n(x), \quad \lim_{q \to 1} C_n(x;q^\nu \mid q) = C_n^\nu(x), \tag{4.11}$$

$\{C_n^\nu(x)\}$ being the ultraspherical polynomials, [26].
It is clear from (4.1) that

$$\max\{C_n(x;\beta \mid q) : -1 \le x \le 1\} = C_n(1;\beta \mid q). \tag{4.12}$$

With $e^{i\theta} = \beta^{1/2}$ we find

$$C_n((\beta^{1/2} + \beta^{-1/2})/2;\beta \mid q) = \beta^{-n/2}\frac{(\beta^2;q)_n}{(q;q)_n}, \tag{4.13}$$

since the left-hand side of (4.1) becomes

$$\beta^{n/2}\frac{(\beta;q)_n}{(q;q)_n}\,{}_2\phi_1(q^{-n},\beta;q^{1-n}/\beta;q,q/\beta^2),$$

and can be summed by (2.30) and a simplification using (2.23) gives (4.13).
Furthermore

$$\max\{|C_n(x;\beta \mid q)| : |x| \le (\beta^{1/2} + \beta^{-1/2})/2, x \text{ real}\} \tag{4.14}$$
$$= \beta^{-n/2}\frac{(\beta^2;q)_n}{(q;q)_n}.$$

The value of $C_n(0; \beta \mid q)$ can be found from (4.8) to be

$$C_{2n+1}(0; \beta|q) = 0, \quad C_{2n}(0; \beta|q) = \frac{(-1)^n(\beta^2; q^2)_n}{(q^2; q^2)_n}. \tag{4.15}$$

In particular

$$H_{2n+1}(0 \mid q) = 0, \quad H_{2n}(0|q) = (-1)^n(q; q^2)_n. \tag{4.16}$$

An important limiting case of the C_n's is

$$\lim_{\beta \to \infty} \beta^{-n} C_n(x; \beta \mid q) = \frac{q^{n(n-1)/2}(-1)^n}{(q; q)_n} H_n(x \mid q^{-1}), \tag{4.17}$$

where

$$H_n(\cos \theta \mid q) = \sum_{k=0}^{n} \frac{(q; q)_n q^{k(k-n)}}{(q; q)_k (q; q)_{n-k}} e^{i(n-2k)\theta}.$$

Theorem 4.2 *The orthogonality relation (4.3) is equivalent to the evaluation of the integral*

$$\int_0^\pi \frac{(t_1\beta e^{i\theta}, t_1\beta e^{-i\theta}, t_2\beta e^{i\theta}, t_2\beta e^{-i\theta}, e^{2i\theta}, e^{-2i\theta}; q)_\infty}{(t_1 e^{i\theta}, t_1 e^{-i\theta}, t_2 e^{i\theta}, t_2 e^{-i\theta}, \beta e^{2i\theta}, \beta e^{-2i\theta}; q)_\infty} d\theta$$

$$= \frac{(\beta, q\beta; q)_\infty}{(q, \beta^2; q)_\infty} {}_2\phi_1(\beta^2, \beta; q\beta; q, t_1 t_2), \quad |t_1| < 1, |t_2| < 1. \tag{4.18}$$

Proof. The sequences $\{(\beta; q)_k\}$ and $\{(q; q)_k\}$ are bounded and their lower bounds are positive. Thus (4.1) implies $|C_n(x; \beta|q)| < Cn$, for a constant C which may depend on q. Now expand the integrand in (4.18) and interchange the summation and integration to establish the desired equivalence. $\square$

The analogue of (3.23) is

$$\mathcal{D}_q C_n(x; \beta|q) = \frac{2(1 - \beta)}{1 - q} q^{(1-n)/2} C_{n-1}(x; q\beta|q). \tag{4.19}$$

It can be proved similarly using the generating function (4.7).

It is worth noting that (4.1) is equivalent to

$$C_n(\cos \theta; \beta|q) = \frac{(\beta; q)_n e^{in\theta}}{(q; q)_n} {}_2\phi_1\left(\begin{array}{c} q^{-n}, \beta \\ q^{1-n}/\beta \end{array} \middle| q, \, qe^{-2i\theta}/\beta \right). \tag{4.20}$$

4.1. CONNECTION COEFFICIENTS

The connection coefficient formula for $\{C_n(x; \beta \mid q)\}$ is

$$C_n(x; \gamma \mid q) = \sum_{k=0}^{[n/2]} \frac{\beta^k (\gamma/\beta; q)_k (\gamma; q)_{n-k}}{(q; q)_k (q\beta; q)_{n-k}} \frac{(1 - \beta q^{n-2k})}{(1 - \beta)} C_{n-2k}(x; \beta \mid q),$$
$$\tag{4.21}$$

and is due to L. J. Rogers. Two important special cases are, cf. (4.2),

$$C_n(x; \gamma \mid q) = \sum_{k=0}^{[n/2]} \frac{(-\gamma)^k (\gamma; q)_{n-k}}{(q; q)_k} q^{\binom{k}{2}} \frac{H_{n-2k}(x \mid q)}{(q; q)_{n-2k}}, \tag{4.22}$$

$$\frac{H_n(x \mid q)}{(q; q)_n} = \sum_{k=0}^{[n/2]} \frac{\beta^k}{(q; q)_k (q\beta; q)_{n-k}} \frac{(1 - \beta q^{n-2k})}{(1 - \beta)} C_{n-2k}(x; \beta \mid q). \tag{4.23}$$

We first prove (4.21) for $\beta = q$ and general γ then use (4.19) to extend to $\beta = q^j$, then extend it for all β [8]. From (4.1) it is clear that

$$C_{n+1}(x; \gamma/q \mid q) - 2 \sum_{k=0}^{[n/2]} \frac{(\gamma/q; q)_k (\gamma/q; q)_{n+1-k}}{(q; q)_k (q; q)_{n+1-k}} T_{n+1-2k}(x) \tag{4.24}$$

is a constant in x for n odd and is zero for n even. Now apply $\mathcal{D}_q$ to (4.24) and take into account (4.19) and (2.12). The result is

$$C_n(x; \gamma \mid q) = \frac{(1 - q)}{(1 - \gamma/q)} q^{n/2} \sum_{k=0}^{[n/2]} \frac{(\gamma/q; q)_k (\gamma/q; q)_{n+1-k}}{(q; q)_k (q; q)_{n+1-k}}$$
$$\times \frac{q^{-k+(n+1)/2} - q^{k-(n+1)/2}}{q^{1/2} - q^{-1/2}} U_{n-2k}(x),$$

which is (4.21) with $\beta = q$. Apply $\mathcal{D}_q^j$ to (4.21) with $\beta = q$ and replace n by $n + j$. In view of (4.19) we get

$$(\gamma; q)_j q^{j(1-2n-j)/4} C_n(x; q^j \gamma \mid q) = \sum_{k=0}^{[n/2]} \frac{q^k (\gamma/q; q)_k (\gamma; q)_{n-k+j}}{(q; q)_k (q^2; q)_{n-k+j}} \frac{(1 - q^{j+n-2k+1})}{(1 - q)}$$
$$\times (q; q)_j q^{j(1-2n-j+4k)/4} C_{n-2k}(x; q^{j+1} \mid q).$$

The result now follows from

$$\frac{(q; q)_j (\gamma; q)_{n-k+j}}{(1 - q)(q^2; q)_{n-k+j} (\gamma; q)_j} = \frac{(\gamma q^j; q)_{n-k}}{(1 - q^{j+1})(q^{j+2}; q)_{n-k}},$$

and the fact that both sides of (4.21) are polynomials in β.

The limiting cases $\beta \to \infty$, and $\gamma \to \infty$ of (4.21) are the inverse relations

$$\frac{H_n(x \mid q)}{(q;q)_n} = \sum_{k=0}^{[n/2]} \frac{(-1)^k q^{k(3k-2n-1)/2}}{(q;q)_k (q;q)_{n-2k}} H_{n-2k}(x \mid q^{-1}), \tag{4.25}$$

$$H_n(x|q^{-1}) = \sum_{s=0}^{[n/2]} \frac{q^{-s(n-s)}(q;q)_n}{(q;q)_s(q;q)_{n-2s}} H_{n-2s}(x \mid q). \tag{4.26}$$

Furthermore the polynomials $\{H_n(x \mid q^{-1})\}$ have the generating function

$$\sum_{n=0}^{\infty} \frac{H_n(\cos\theta|q^{-1})}{(q;q)_n} q^{\binom{n}{2}}(-t)^n = (te^{i\theta}, te^{-i\theta}; q)_\infty. \tag{4.27}$$

In view of the orthogonality relation (4.3) the connection coefficient formula (4.21) is equivalent to the integral evaluation

$$\int_0^\pi \frac{(t\gamma e^{i\theta}, t\gamma e^{-i\theta}, e^{2i\theta}, e^{-2i\theta}; q)_\infty}{(te^{i\theta}, te^{-i\theta}, \beta e^{2i\theta}, \beta e^{-2i\theta}; q)_\infty} C_m(\cos\theta; \beta|q) d\theta$$

$$= \frac{(\beta, q\beta; q)_\infty (\gamma; q)_m \, t^m}{(q, \beta^2; q)_\infty (q\beta; q)_m} \, {}_2\phi_1(\gamma/\beta, \gamma q^m; q^{m+1}\beta; q, \beta t^2). \tag{4.28}$$

Since $C_n(x; q \mid q) = U_n(x)$ is independent of q we can then use (4.21) to establish the change of basis formula [14]

$$C_n(x; \gamma|q) = \sum_{k=0}^{[n/2]} \frac{q^k(\gamma/q; q)_k(\gamma; q)_{n-k}(1 - q^{n-2k+1})}{(q; q)_k(q^2; q)_{n-k}(1 - q)}$$

$$\times \sum_{j=0}^{[n/2]-k} \frac{(\beta)^j (p/\beta; q)_j (p; p)_{n-2k-j}}{(p; p)_j (p\beta; q)_{n-2k-j}} \tag{4.29}$$

$$\times \frac{(1 - \beta p^{n-2k-2j})}{(1 - \beta)} C_{n-2k-2j}(x; \beta \mid p).$$

$\square$

5. The Askey-Wilson Polynomials

In this section we shall build the theory of the Askey-Wilson polynomials through the bootstrap method and a method of attachment. This section closely follows [13]. We first derive the orthogonality relation of the Al-Salam-Chihara polynomials by starting from the q-Hermite polynomials. The special case $\beta = 0$ of (4.18) is

$$\int_0^\pi \frac{(e^{2i\theta}, e^{-2i\theta}; q)_\infty}{(t_1 e^{i\theta}, t_1 e^{-i\theta}, t_2 e^{i\theta}, t_2 e^{-i\theta}; q)_\infty} \, d\theta = \frac{2\pi}{(q, t_1 t_2; q)_\infty}, \quad |t_1|, |t_2| < 1. \tag{5.1}$$

The next step is to find polynomials $\{p_n(x; t_1, t_2 \mid q)\}$ orthogonal with respect to the weight function

$$w_1(x; t_1, t_2 | q) := \frac{(e^{2i\theta}, e^{-2i\theta}; q)_\infty}{(t_1 e^{i\theta}, t_1 e^{-i\theta}, t_2 e^{i\theta}, t_2 e^{-i\theta}; q)_\infty} \frac{1}{\sqrt{1 - x^2}}, \quad x = \cos\theta, \tag{5.2}$$

which is positive for $t_1, t_2 \in (-1, 1)$ and its total mass is given by (5.1). Here we follow a technique of attachment. Andrews and Askey [6], and Askey and Wilson [11] used this idea of attachment to prove the orthogonality of some polynomials. We will use it to discover the form of the polynomials orthogonal with respect to $w_1(x; t_1, t_2 \mid q)$ and later the form of the Askey-Wilson polynomials.

Write $\{p_n(x; t_1, t_2 \mid q))\}$ in the form

$$p_n(x; t_1, t_2 \mid q) = \sum_{k=0}^n \frac{(q^{-n}, t_1 e^{i\theta}, t_1 e^{-i\theta}; q)_k}{(q; q)_k} a_{n,k}, \tag{5.3}$$

then compute $a_{n,k}$ from the fact that the polynomial $p_n(x; t_1, t_2|q)$ is orthogonal to $\phi_j(\cos\theta; t_1)$, $j = 0, 1, \ldots, n-1$. As we saw in (2.16) $\phi_k(x; a)$ is a polynomial in x of degree k. The reason for choosing the bases $\{\phi_k(x; t_1)\}$ and $\{\phi_k(x; t_2)\}$ is that they attach nicely to the weight function in (5.2) in the following sense

$$(t_1 e^{i\theta}, t_1 e^{-i\theta}; q)_k (t_2 e^{i\theta}, t_2 e^{-i\theta}; q)_j w_1(x; t_1, t_2|q) = w_1(x; t_1 q^k, t_2 q^j \mid q).$$

Therefore (5.1) and (5.3) give

$$\int_{-1}^1 (t_2 e^{i\theta}, t_2 e^{-i\theta}; q)_j p_n(x; t_1, t_2 \mid q) w_1(x; t_1, t_2 \mid q) dx$$

$$= \sum_{k=0}^n \frac{(q^{-n}; q)_k}{(q; q)_k} a_{n,k} \int_0^\pi \frac{(e^{2i\theta}, e^{-2i\theta}; q)_\infty \, d\theta}{(t_1 q^k e^{i\theta}, t_1 q^k e^{-i\theta}, t_2 q^j e^{i\theta}, t_2 q^j e^{-i\theta}; q)_\infty}$$

$$= \frac{2\pi}{(q; q)_\infty} \sum_{k=0}^n \frac{(q^{-n}; q)_k a_{n,k}}{(q; q)_k (t_1 t_2 q^{k+j}; q)_\infty}$$

$$= \frac{2\pi}{(q, t_1 t_2 q^j; q)_\infty} \sum_{k=0}^n \frac{(q^{-n}, t_1 t_2 q^j; q)_k}{(q; q)_k} a_{n,k}.$$

The q-Chu-Vandermonde sum (2.28) suggests $a_{n,k} = q^k/(t_1 t_2; q)_k$. Therefore

$$\int_{-1}^1 (t_2 e^{i\theta}, t_2 e^{-i\theta}; q)_j p_n(x) w_1(x; t_1, t_2 \mid q) \, dx$$

$$= \frac{2\pi}{(q, t_1 t_2 q^j; q)_\infty} {}_2\phi_1(q^{-n}, t_1 t_2 q^j; t_1 t_2; q, q)$$

$$= \frac{2\pi (q^{-j}; q)_n}{(q, t_1 t_2 q^j; q)_\infty (t_1 t_2; q)_n} (t_1 t_2 q^j)^n.$$

It follows from (5.3) and (2.23) that the coefficient of x^n in $p_n(x; t_1, t_2 \mid q)$ is

$$(-2t_1)^n q^{n(n+1)/2} (q^{-n}; q)_n/(q, t_1 t_2; q)_n = (2t_1)^n/(t_1 t_2; q)_n.$$

This proves the following theorem.

Theorem 5.1 *The polynomials*

$$p_n(x; t_1, t_2 \mid q) = {}_3\phi_2 \left(\begin{array}{c} q^{-n}, t_1 e^{i\theta}, t_1 e^{-i\theta} \\ t_1 t_2, 0 \end{array} \middle| q, q \right). \tag{5.4}$$

satisfy the orthogonality relation

$$\int_{-1}^1 p_m(x; t_1, t_2 \mid q) p_n(x; t_1, t_2 \mid q) w_1(x; t_1, t_2 \mid q) dx$$

$$= \frac{2\pi (q; q)_n t_1^{2n}}{(q, t_1 t_2; q)_\infty (t_1 t_2; q)_n} \delta_{m,n}. \tag{5.5}$$

The polynomials in (5.4) were first identified by W. Al-Salam and T. Chihara [2]. Their weight function was given in [10] and [11] where there were called the Al-Salam-Chihara polynomials.

Observe that the orthogonality relation (5.5) and the uniqueness of the polynomials orthogonal with respect to a positive measure show that $t_1^{-n} p_n(x)$ is symmetric in t_1 and t_2. This gives the transformation formula

$${}_3\phi_2 \left(\begin{array}{c} q^{-n}, t_1 e^{i\theta}, t_1 e^{-i\theta} \\ t_1 t_2, 0 \end{array} \middle| q, q \right)$$

$$= (t_1/t_2)^n {}_3\phi_2 \left(\begin{array}{c} q^{-n}, t_2 e^{i\theta}, t_2 e^{-i\theta} \\ t_1 t_2, 0 \end{array} \middle| q, q \right). \tag{5.6}$$

The next step is to repeat the process with the Al-Salam-Chihara polynomials as our starting point, so we transform the representation (5.4) to a form more amenable to generating functions. Following [22] first write the $_3\phi_2$ as a sum over k then replace k by $n - k$ then apply (2.24) and obtain

$$p_n(x; t_1, t_2 \mid q) = \frac{(t_1 e^{i\theta}, t_1 e^{-i\theta}; q)_n}{(t_1 t_2; q)_n} q^{-n(n-1)/2} (-1)^n$$
$$\times \sum_{k=0}^{n} \frac{(-t_2/t_1)^k (q^{-n}, q^{1-n}/t_1 t_2; q)_k}{(q, q^{1-n} e^{i\theta}/t_1, q^{1-n} e^{-i\theta}/t_1; q)_k} q^{k(k+1)/2}. \quad (5.7)$$

Then apply the transformation (2.48) with

$$A = q^{-n}, \quad B = t_2 e^{i\theta}, \quad C = q^{1-n} e^{i\theta}/t_1, \quad z = q e^{-i\theta}/t_1$$

to get

$$p_n(x; t_1, t_2 | q) = \frac{(t_1 e^{-i\theta}; q)_n t_1^n e^{in\theta}}{(t_1 t_2; q)_n} {}_2\phi_1 \left(\begin{array}{c} q^{-n}, t_2 e^{i\theta} \\ q^{1-n} e^{i\theta}/t_1 \end{array} \middle| q, q e^{-i\theta}/t_1 \right). \quad (5.8)$$

Using (2.24) we express a multiple of p_n as a Cauchy product of two sequences. The result is

$$p_n(\cos\theta; t_1, t_2 | q) = \frac{(q; q)_n t_1^n}{(t_1 t_2; q)_n} \sum_{k=0}^{n} \frac{(t_2 e^{i\theta}; q)_k}{(q; q)_k}$$
$$\times e^{-ik\theta} \frac{(t_1 e^{-i\theta}; q)_{n-k}}{(q; q)_{n-k}} e^{i(n-k)\theta}. \quad (5.9)$$

This and the q-binomial theorem (2.33) establish the generating function

$$\sum_{n=0}^{\infty} \frac{(t_1 t_2; q)_n}{(q; q)_n} p_n(x; t_1, t_2 \mid q) (t/t_1)^n = \frac{(t t_1, t t_2; q)_\infty}{(t e^{-i\theta}, t e^{i\theta}; q)_\infty}. \quad (5.10)$$

It readily follows from (5.10) that the p_n's satisfy the three term recurrence relation

$$[2x - (t_1 + t_2)q^n] t_1 p_n(x; t_1, t_2 \mid q)$$
$$= (1 - t_1 t_2 q^n) p_{n+1}(x; t_1, t_2 \mid q) + t_1^2 (1 - q^n) p_{n-1}(x; t_1, t_2 \mid q). \quad (5.11)$$

Another consequence of (5.10) is

$$|p_n(x; t_1, t_2 \mid q)| \le |p_n(1; t_1, t_2 \mid q)| \le Cn|t_1|^n, \quad x \in [-1, 1], \quad (5.12)$$

for some constant C. Furthermore (5.10) yields

$$\mathcal{D}_q p_n(x; t_1, t_2 | q) = \frac{(1 - q^n) t_1 q^{n-1}}{(1 - t_1 t_2)(1 - q)} p_{n-1}(x; q^{1/2} t_1, q^{1/2} t_2 \mid q).$$
(5.13)

5.1. THE ASKEY-WILSON POLYNOMIALS

The orthogonality relation (5.5), the bound (5.12), and the generating function (5.10) imply the evaluation of the Askey-Wilson q-beta integral, [11], [18]

$$\int_0^\pi \frac{(e^{2i\theta}, e^{-2i\theta}; q)_\infty}{\prod_{j=1}^4 (t_j e^{i\theta}, t_j e^{-i\theta}; q)_\infty} \, d\theta = \frac{2\pi (t_1 t_2 t_3 t_4; q)_\infty}{(q; q)_\infty \prod_{1 \le j < k \le 4} (t_j t_k; q)_\infty}, \quad |t_1|, |t_2| < 1.$$
(5.14)

The polynomials orthogonal with respect to the weight function whose total mass is given by (5.14) are the Askey-Wilson polynomials. Their weight function is

$$w(x; t_1, t_2, t_3, t_4 | q) = \frac{(e^{2i\theta}, e^{-2i\theta}; q)_\infty}{\prod_{j=1}^4 (t_j e^{i\theta}, t_j e^{-i\theta}; q)_\infty} \frac{1}{\sqrt{1 - x^2}}, \quad x = \cos\theta.$$
(5.15)

To find an explicit representation we again set

$$p_n(x; t_1, t_2, t_3, t_4 | q) = \sum_{k=0}^n \frac{(q^{-n}, t_1 e^{i\theta}, t_1 e^{-i\theta}; q)_k}{(q; q)_k} a_{n,k},$$
(5.16)

with undetermined coefficients $\{a_{n,k}\}$. Therefore

$$\int_0^\pi (t_2 e^{i\theta}, t_2 e^{-i\theta}; q)_j p_n(\cos\theta; t_1, t_2, t_3, t_4 | q) w(\cos\theta; t_1, t_2, t_3, t_4 | q) \, \sin\theta \, d\theta$$

$$= \sum_{k=0}^n \frac{(q^{-n}; q)_k}{(q; q)_k} a_{n,k} \int_0^\pi w(\cos\theta; t_1 q^k, t_2 q^j, t_3, t_4 | q) \, \sin\theta \, d\theta$$

$$= \sum_{k=0}^n \frac{(q^{-n}; q)_k}{(q; q)_k} \frac{2\pi a_{n,k} (q^{j+k} t_1 t_2 t_3 t_4; q)_\infty}{(q; q)_\infty (q^{j+k} t_1 t_2, q^k t_1 t_3, q^k t_1 t_4, q^j t_2 t_3, q^j t_2 t_4, t_3 t_4; q)_\infty}$$

$$= \sum_{k=0}^n \frac{(q^{-n}; q)_k}{(q; q)_k} a_{n,k} \frac{(q^j t_1 t_2, t_1 t_3, t_1 t_4; q)_k}{(q^j t_1 t_2 t_3 t_4; q)_k}$$

$$\times \frac{2\pi (q^j t_1 t_2 t_3 t_4; q)_\infty}{(q, q^j t_1 t_2, t_1 t_3, t_1 t_4, q^j t_2 t_3, q^j t_2 t_4, t_3 t_4; q)_\infty}.$$

The choice

$$a_{n,k} = q^k \frac{(t_1 t_2 t_3 t_4 q^{n-1}; q)_k}{(t_1 t_2, t_1 t_3, t_1 t_4; q)_k} a_{n,0},$$

enables us to use the $_3\phi_2$ sum (2.27) and we have

$$\int_0^\pi (t_2 e^{i\theta}, t_2 e^{-i\theta}; q)_j p_n(\cos\theta; t_1, t_2, t_3, t_4|q) w(\cos\theta; t_1, t_2, t_3, t_4|q) \sin\theta \, d\theta$$

$$= \frac{2\pi (q^j t_1 t_2 t_3 t_4; q)_\infty a_{n,0}}{(q, q^j t_1 t_2, t_1 t_3, t_1 t_4, q^j t_2 t_3, q^j t_2 t_4, t_3 t_4; q)_\infty} \frac{(q^{-j}, q^{1-n}/t_3 t_4; q)_n}{(t_1 t_2, q^{1-j-n}/t_1 t_2 t_3 t_4; q)_n}.$$

For $j \leq n$ we use (2.24) to see that the right-hand side of the above equation is

$$\frac{2\pi (q^j t_1 t_2 t_3 t_4; q)_\infty a_{n,0}(-t_1 t_2)^n q^{n(n-1)/2} \delta_{j,n}}{(q, q^j t_1 t_2, t_1 t_3, t_1 t_4, q^j t_2 t_3, q^j t_2 t_4, t_3 t_4; q)_\infty} \frac{(q, t_3 t_4; q)_n}{(t_1 t_2, q^j t_1 t_2 t_3 t_4; q)_n}.$$

Since $(t_1 e^{i\theta}, t_1 e^{-i\theta}; q)_n - (t_1/t_2)^n (t_2 e^{i\theta}, t_2 e^{-i\theta}; q)_n$ is a polynomial in $\cos\theta$ of degree at most $n-1$ then

$$\int_0^\pi (t_1 e^{i\theta}, t_1 e^{-i\theta}; q)_j p_n(\cos\theta; t_1, t_2, t_3, t_4|q) w(\cos\theta; t_1, t_2, t_3, t_4|q) \sin\theta \, d\theta$$

$$= \frac{2\pi (q^j t_1 t_2 t_3 t_4; q)_\infty a_{n,0}}{(q, q^j t_1 t_2, t_1 t_3, t_1 t_4, q^j t_2 t_3, q^j t_2 t_4, t_3 t_4; q)_\infty}$$

$$\times \frac{(q, t_3 t_4; q)_n (-t_1^2)^n}{(t_1 t_2, q^j t_1 t_2 t_3 t_4; q)_n} q^{n(n-1)/2} \delta_{j,n}.$$

Hence if $m \leq n$ then

$$\int_{-1}^1 p_m(x; t_1, t_2, t_3, t_4|q) p_n(x; t_1, t_2, t_3, t_4|q) w(x; t_1, t_2, t_3, t_4) dx$$

$$= \frac{(q^{-m}, t_1 t_2 t_3 t_4 q^{m-1}; q)_m}{(q, t_1 t_2, t_1 t_3, t_1 t_4; q)_m} \frac{(q, t_3 t_4; q)_n}{(t_1 t_2, q^m t_1 t_2 t_3 t_4; q)_n}$$

$$\times \frac{2\pi (-1)^n (q^m t_1 t_2 t_3 t_4; q)_\infty a_{n,0}^2}{(q, q^m t_1 t_2, t_1 t_3, t_1 t_4, q^m t_2 t_3, q^m t_2 t_4, t_3 t_4; q)_\infty} (t_1)^{2n} q^{n(n+1)/2} \delta_{m,n}.$$

With the choice

$$a_{n,o} := t_1^{-n}(t_1 t_2, t_1 t_3, t_1 t_4; q)_n,$$

we have established the following result.

Theorem 5.2 *The Askey-Wilson polynomials satisfy the orthogonality relation*

$$\int\limits_{-1}^{1} p_m(x; t_1, t_2, t_3, t_4 \mid q) p_n(x; t_1, t_2, t_3, t_4 \mid q) \; w(x; t_1, t_2, t_3, t_4 \mid q) \; dx$$

$$= \frac{2\pi \, (t_1 t_2 t_3 t_4 q^{2n}; q)_\infty \, (t_1 t_2 t_3 t_4 q^{n-1}; q)_n}{(q^{n+1}; q)_\infty \displaystyle\prod_{1 \le j < k \le 4} (t_j t_k q^n; q)_\infty} \, \delta_{m,n}, \qquad (5.17)$$

for $\max\{|t_1|, |t_2|, |t_3|, |t_4|\} < 1$. *They also have the representation*

$$p_n(x; t_1, t_2, t_3, t_4 | q)$$
$$= t_1^{-n} (t_1 t_2, t_1 t_3, t_1 t_4; q)_n$$
$$\times \, {}_4\phi_3 \left(\begin{array}{c} q^{-n}, t_1 t_2 t_3 t_4 q^{n-1}, t_1 e^{i\theta}, t_1 e^{-i\theta} \\ t_1 t_2, \ t_1 t_3, \ t_1 t_4 \end{array} \; \middle| \; q, \, q \right). \qquad (5.18)$$

Observe that the weight function in (5.15) and the right-hand side of (5.17) are symmetric functions of t_1, t_2, t_3, t_4. The uniqueness of the polynomials orthogonal with respect to a positive measure indicates that the Askey-Wilson polynomials are symmetric in the four parameters t_1, t_2, t_3, t_4. This symmetry is the Sears transformation which we saw in §2, see (2.42) and Theorem 2.9.

Theorem 5.3 ([22]) *The Askey-Wilson polynomials have the generating function*

$$\sum_{n=0}^{\infty} \frac{p_n(\cos\theta; t_1, t_2, t_3, t_4 | q)}{(q, t_1 t_2, t_3 t_4; q)_n} \, t^n$$

$$= {}_2\phi_1 \left(\begin{array}{c} t_1 e^{i\theta}, t_2 e^{i\theta} \\ t_1 t_2 \end{array} \; \middle| \; q, \, te^{-i\theta} \right) {}_2\phi_1 \left(\begin{array}{c} t_3 e^{-i\theta}, t_4 e^{-i\theta} \\ t_3 t_4 \end{array} \; \middle| \; q, \, te^{i\theta} \right). \qquad (5.19)$$

Proof. Apply (2.42) with

$$a = t_1 e^{i\theta}, \; b = t_1 e^{-i\theta}, \; c = t_1 t_2 t_3 t_4 q^{n-1} \; d = t_1 t_2, \; e = t_1 t_3, \; f = t_1 t_4,$$

to obtain

$$p_n(x; t_1, t_2, t_3, t_4 \mid q) = (t_1 t_2, q^{1-n} e^{i\theta}/t_3, q^{1-n} e^{i\theta}/t_4; q)_n (t_3 t_4 q^{n-1} e^{-i\theta})^n$$

$$\times \, {}_4\phi_3 \left(\begin{array}{c} q^{-n}, t_1 e^{i\theta}, t_2 e^{i\theta}, q^{1-n}/t_3 t_4 \\ t_1 t_2, \ q^{1-n} e^{i\theta}/t_3, \ q^{1-n} e^{i\theta}/t_4 \end{array} \; \middle| \; q, \, q \right). \qquad (5.20)$$

Formula (2.23) leads to the simplification

$$(q^{1-n} e^{i\theta}/t_3, q^{1-n} e^{i\theta}/t_4; q)_n = e^{2in\theta} q^{-n(n-1)} (t_3 t_4)^{-n} (t_3 e^{-i\theta}, t_4 e^{-i\theta}; q)_n.$$

Let k be the summation index of the $_4\phi_3$ in (5.20) and apply (2.24) to get

$$\frac{(q^{-n}, q^{1-n}/t_3t_4; q)_k}{(q^{1-n}e^{i\theta}/t_3, q^{1-n}e^{i\theta}/t_4; q)_k} = \frac{(q, t_3t_4; q)_n}{(q, t_3t_4; q)_{n-k}} \frac{(t_3e^{-i\theta}, t_4e^{-i\theta}; q)_{n-k}}{(t_3e^{-i\theta}, t_4e^{-i\theta}; q)_n} (qe^{2i\theta})^{-k}.$$

Therefore

$$\frac{p_n(x; t_1, t_2, t_3, t_4|q)}{(q, t_1t_2, t_3t_4; q)_n} = \sum_{k=0}^{n} \frac{(t_1e^{i\theta}, t_2e^{i\theta}; q)_k}{(q, t_1t_2; q)_k} e^{-ik\theta}$$

$$\times \frac{(t_3e^{-i\theta}, t_4e^{-i\theta}; q)_{n-k}}{(q, t_3t_4; q)_{n-k}} e^{i(n-k)\theta}. \qquad (5.21)$$

It is clear that (5.21) implies (5.19). $\qquad\qquad\qquad\square$

We now write the orthogonality relation (5.17) in terms of the generating function (5.19). The result is the following integral representation for a $_6\phi_5$ function.

$$\int_0^\pi \prod_{j=5}^{6} {}_2\phi_1 \left(\begin{array}{c} t_1e^{i\theta}, t_2e^{i\theta} \\ t_1t_2 \end{array} \bigg| q, t_je^{-i\theta} \right) {}_2\phi_1 \left(\begin{array}{c} t_3e^{-i\theta}, t_4e^{-i\theta} \\ t_3t_4 \end{array} \bigg| q, t_je^{i\theta} \right)$$

$$\times \frac{(e^{2i\theta}, e^{-2i\theta}; q)_\infty}{\prod_{j=1}^{4}(t_je^{i\theta}, t_je^{-i\theta}; q)_\infty} d\theta$$

$$= \frac{2\pi (t_1t_2t_3t_4; q)_\infty}{(q; q)_\infty \prod_{1\le j<k\le 4}(t_jt_k; q)_\infty} \qquad (5.22)$$

$$\times {}_6\phi_5 \left(\begin{array}{c} \sqrt{t_1t_2t_3t_4/q}, -\sqrt{t_1t_2t_3t_4/q}, t_1t_3, t_1t_4, t_2t_3, t_2t_4 \\ \sqrt{t_1t_2t_3t_4q}, -\sqrt{t_1t_2t_3t_4q}, t_1t_2, t_3t_4, t_1t_2t_3t_4/q \end{array} \bigg| q, t_5t_6 \right),$$

valid for $\max\{|t_1|, |t_2|, |t_3|, |t_4|, |t_5|, |t_6|\} < 1$.

The Askey-Wilson polynomials are orthogonal polynomials, hence they satisfy a three term recurrence relation of the form

$$2xp_n(x) = A_np_{n+1}(x) + B_np_n(x) + C_np_{n-1}(x). \qquad (5.23)$$

Equating the coefficients of x^{n+1}, x^n and x^{n-1} on both sides of (5.23) we get

$$A_n = \frac{1 - t_1t_2t_3t_4q^{n-1}}{(1 - t_1t_2t_3t_4q^{2n-1})(1 - t_1t_2t_3t_4q^{2n})}, \qquad (5.24)$$

$$C_n = \frac{(1 - q^n) \prod_{1\le j<k\le 4}(1 - t_jt_kq^{n-1})}{(1 - t_1t_2t_3t_4q^{2n-2})(1 - t_1t_2t_3t_4q^{2n-1})}, \qquad (5.25)$$

$$B_n = t_1 + t_1^{-1} - A_n t_1^{-1} \prod_{j=2}^{4}(1 - t_1 t_j q^n)$$

$$- \frac{t_1 C_n}{\prod_{2 \le j < k \le 4}(1 - t_j t_k q^{n-1})}.$$

(5.26)

Remark 2 The continuous q-ultraspherical polynomials correspond to the case

$$t_1 = -t_2 = \sqrt{\beta}, \quad \text{and} \quad t_3 = -t_4 = \sqrt{q\beta}$$

as can be seen from comparing the weight functions (3.4) and (5.15). Thus $C_n(x; \beta \mid q)$ must be a constant multiple of an Askey-Wilson polynomial of degree n and the above parameters. Therefore

$$
{}_2\phi_1 \left(\begin{matrix} q^{-n}, \beta \\ q^{1-n}/\beta \end{matrix} \, \middle| \, q, \, \frac{qe^{-2i\theta}}{\beta} \right)
$$
$$
= \frac{(\beta^2; q)_n e^{-in\theta}}{\beta^{n/2}(\beta; q)_n} {}_4\phi_3 \left(\begin{matrix} q^{-n}, q^n\beta^2, \sqrt{\beta}e^{i\theta}, \sqrt{\beta}e^{-i\theta} \\ -\beta, \beta\sqrt{q}, -\beta\sqrt{q} \end{matrix} \, \middle| \, q, \, q \right).
$$

(5.27)

The constant multiple was computed from the fact that the leading coefficient in $C_n(x; \beta \mid q)$ is $2^n(\beta; q)_n/(q; q)_n$. When we replace β by q^b in (5.27) and let $q \to 1$ we obtain the quadratic transformation

$$
{}_2F_1 \left(\begin{matrix} -n, b \\ 1-n-b \end{matrix} \, \middle| \, q, \, x^2 \right)
$$
$$
= \frac{x^n (2b; q)_n}{(b; q)_n} {}_2F_1 \left(\begin{matrix} -n, n+2b \\ b+1/2 \end{matrix} \, \middle| \, -\frac{(1-x)^2}{4x} \right).
$$

(5.28)

The success in evaluating the Askey-Wilson integral (5.14) raises the question of evaluating the general integral

$$
I(t_1, t_2, \ldots, t_k) := \frac{(q; q)_\infty}{2\pi} \int_0^\pi \frac{(e^{2i\theta}, e^{-2i\theta}; q)_\infty}{\prod_{j=1}^{k}(t_j e^{i\theta}, t_j e^{-i\theta}; q)_\infty} \, d\theta.
$$

(5.29)

The evaluation of this integral is stated below but its known proof uses combinatorial ideas that are outside the scope of this work.

Theorem 5.4 ([21]) *Let*

$$
I(t_1, t_2, \ldots, t_k) = \sum_{n_1, \ldots n_k = 0}^{\infty} g(n_1, n_2, \ldots, n_k) \prod_{j=1}^{k} \frac{t_j^{n_j}}{(q; q)_{n_j}}.
$$

(5.30)

Then

$$g(n_1, n_2, \ldots, n_k) = \sum_{n_{ij}} \prod_{l=1}^{k} \begin{bmatrix} n_l \\ n_{l1}, \ldots n_{lk} \end{bmatrix}_q \prod_{1 \leq i < j \leq k} (q; q)_{n_{ij}} q^B, \tag{5.31}$$

where the summation is over all non-negative integral symmetric matrices
(n_{ij}) *such that* $n_{ii} = 0$ *and* $\sum_{i=1}^{k} n_{ij} = n_j$ *for* $1 \leq j \leq k$. *Furthermore*

$$B = \sum_{1 \leq i < j < m < l \leq k} n_{im} n_{jl}. \tag{5.32}$$

The q-multinomial coefficient in (5.31) is

$$\begin{bmatrix} n \\ n_1, \ldots n_k \end{bmatrix}_q := (q; q)_n / \prod_{j=1}^{k} (q; q)_{n_j}, \quad \sum_{j=1}^{k} n_j = n. \tag{5.33}$$

Observe that when $k = 4$ then $B = n_{13} n_{24}$ and

$$I(t_1, \ldots, t_4)$$
$$= \sum_{n_{ij}, 1 \leq i,j \leq 4} \frac{(t_1 t_2)^{n_{12}} (t_1 t_3)^{n_{13}} (t_1 t_4)^{n_{14}} (t_2 t_3)^{n_{23}} (t_2 t_4)^{n_{24}} (t_3 t_4)^{n_{34}}}{(q;q)_{n_{12}} (q;q)_{n_{13}} (q;q)_{n_{14}} (q;q)_{n_{23}} (q;q)_{n_{24}} (q;q)_{n_{34}}} q^{n_{13} n_{24}}.$$

The sums over n_{12}, n_{14}, n_{23}, n_{13}, and n_{34} are evaluable by (2.34) and we find

$$I(t_1, \ldots, t_4) = \frac{1}{(t_1 t_2, t_1 t_4, t_2 t_3, t_3 t_4; q)_\infty} \sum_{n_{24}=0}^{\infty} \frac{(t_2 t_4)^{n_{24}}}{(q;q)_{n_{24}} (t_1 t_3 q^{n_{24}}; q)_\infty}$$

$$= \frac{1}{(t_1 t_2, t_1 t_3, t_1 t_4, t_2 t_3, t_3 t_4; q)_\infty} \sum_{n_{24}=0}^{\infty} \frac{(t_1 t_3; q)_{n_{24}}}{(q;q)_{n_{24}}} (t_2 t_4)^{n_{24}},$$

and the evaluation of $I(t_1, t_2, t_3, t_4)$ follows from (2.33).

6. Applications

This section contains a proof of a generalization of the Rogers-Ramanujan identities. The Rogers-Ramanujan identities are

$$\sum_{n=0}^{\infty} \frac{q^{n^2}}{(q;q)_n} = \frac{1}{(q, q^4; q^5)_\infty}, \tag{6.1}$$

$$\sum_{n=0}^{\infty} \frac{q^{n^2+n}}{(q;q)_n} = \frac{1}{(q^2,q^3;q^5)_\infty}. \tag{6.2}$$

The right-hand sides of (6.1) and (6.2) are generating functions of the partitions of an integer m into parts $\equiv 1,4 \pmod 5$ and into parts $\equiv 2,3 \pmod 5$, respectively, as we shall indicate in the material following (6.11). MacMahon [3] interpreted the left-hand side of (6.1) as the generating function number of partitions of m into parts which differ by at least 2. He also showed that the left-hand side of (6.2) is the generating function for the number of partitions of m into parts which differ by at least 2, and each part is at least 2; [3]. The Rogers-Ramanujan identities and their generalizations play a central role in the theory of partition [4] and [3]. MacMahon's interpretations will be established later in this section but we first treat the analytic identities.

We now give a common generalization of the Rogers-Ramanujan identities.

Theorem 6.1 ([17]) *The following identity holds for* $m = 0, 1, \ldots$

$$\sum_{n=0}^{\infty} \frac{q^{n^2+mn}}{(q;q)_n} = \frac{1}{(q;q)_\infty} \sum_{s=0}^{m} \begin{bmatrix} m \\ s \end{bmatrix}_q q^{2s(s-m)} (q^5, q^{3+4s-2m}, q^{2-4s+2m}; q^5)_\infty. \tag{6.3}$$

Observe that the cases $m = 0, 1$ of (6.3) give (6.1) and (6.2), respectively.

Proof. Consider the integral

$$I_m(t) = \frac{(q;q)_\infty}{2\pi} \int_0^\pi H_m(\cos\theta|q) \, (te^{i\theta}, te^{-i\theta}, e^{2i\theta}, e^{-2i\theta}; q)_\infty \, d\theta. \tag{6.4}$$

Expand $(te^{i\theta}, te^{-i\theta}; q)_\infty$ in $\{H_j(x|q^{-1})\}$ by (4.26) then expand $H_j(x|q^{-1})$'s in terms of $H_j(x|q)$'s using (4.25), and apply the q-Hermite orthogonality. The details are

$I_m(t)$

$$= \frac{(q;q)_\infty}{2\pi} \sum_{n=0}^{\infty} \frac{(-t)^n}{(q;q)_n} q^{\binom{n}{2}} \int_0^\pi H_m(\cos\theta|q) H_n(\cos\theta|q^{-1})(e^{2i\theta}, e^{-2i\theta}; q)_\infty \, d\theta$$

$$= \sum_{n=0}^{\infty} \sum_{s=0}^{\lfloor n/2 \rfloor} \frac{(-t)^n}{(q;q)_n} q^{\binom{n}{2}} \frac{q^{s(s-n)}(q;q)_n}{(q;q)_s(q;q)_{n-2s}} (q;q)_m \delta_{m,n-2s}.$$

Hence $I_m(t)$ has the series representation

$$I_m(t) = (-t)^m q^{\binom{m}{2}} \sum_{n=0}^{\infty} \frac{q^{n^2-n}}{(q;q)_n} (t^2 q^m)^n. \tag{6.5}$$

We then choose $t = \sqrt{q}$, the fact that the integrand in $I_m(t)$ is an even function of θ, we integrate on $[-\pi, \pi]$, then apply (2.41) twice to obtain

$$I_m(\sqrt{q}) = \frac{(4\pi)^{-1}}{(q;q)_\infty} \sum_{k=0}^{m} \frac{(q;q)_m}{(q;q)_k(q;q)_{m-k}} \int_{-\pi}^{\pi} (q, \sqrt{q}e^{i\theta}, \sqrt{q}e^{-i\theta}; q)_\infty$$

$$\times (q, qe^{2i\theta}, e^{-2i\theta}; q)_\infty (1 - e^{2i\theta}) e^{i(m-2k)\theta} \, d\theta$$

$$= \sum_{k=0}^{m} \frac{(q;q)_m/(4\pi)}{(q;q)_\infty (q;q)_k (q;q)_{m-k}} \sum_{r,s=-\infty}^{\infty} (-1)^{r+s} q^{\binom{s+1}{2}+r^2/2}$$

$$\times \int_{-\pi}^{\pi} \left[e^{i(r+2s+m-2k)\theta} - e^{i(r+2s+m-2k+2)\theta} \right] d\theta$$

$$= \frac{(-1)^m}{2(q;q)_\infty} \sum_{k=0}^{m} \begin{bmatrix} m \\ k \end{bmatrix}_q \sum_{s=-\infty}^{\infty} (-1)^s q^{\binom{s+1}{2}}$$

$$\times \left[q^{(2s+m-2k)^2/2} - q^{(2s+m-2k+2)^2/2} \right].$$

Express the last sum as a difference of two sums. In the second sum replace s by $1 - s$ and k by $m - k$ then invoke the invariance of $\begin{bmatrix} m \\ k \end{bmatrix}_q$ under the replacement $k \to m - k$. The two sums become identical and we get

$$I_m(\sqrt{q}) = \frac{(-1)^m q^{m^2/2}}{(q;q)_\infty} \sum_{k=0}^{m} \begin{bmatrix} m \\ k \end{bmatrix}_q q^{2k(k-m)} \sum_{s=-\infty}^{\infty} (-1)^s q^{5s^2/2} q^{s(2m-4k+1/2)}$$

$$= \frac{(-1)^m q^{m^2/2}}{(q;q)_\infty} \sum_{k=0}^{m} \begin{bmatrix} m \\ k \end{bmatrix}_q q^{2k(k-m)} \frac{(q^5, q^{3+2m-4k}, q^{2-2m+4k}; q^5)_\infty}{(q;q)_\infty}.$$

The above evaluation of $I_m(\sqrt{q})$ and (6.5) establish the desired result. $\square$

Obviously the terms $4s - 2m \equiv 1 \pmod 5$ in (6.3) vanish. On the other hand if $4s - 2m \equiv 0, 4 \pmod 5$ in (6.3), the infinite products may be rewritten as a multiple of the Rogers-Ramanujan product $1/(q, q^4; q^5)_\infty$, while $4s - 2m \equiv 1, 3 \pmod 5$ lead to a multiple of $1/(q^2, q^3; q^5)_\infty$. A short calculation reveals that (6.3) is

$$\sum_{n=0}^{\infty} \frac{q^{n^2+mn}}{(q;q)_n} = \frac{(-1)^m q^{-\binom{m}{2}} a_m}{(q, q^4; q^5)_\infty} + \frac{(-1)^{m+1} q^{-\binom{m}{2}} b_m}{(q^2, q^3; q^5)_\infty} \tag{6.6}$$

where

$$a_m = \sum_\lambda (-1)^\lambda q^{\lambda(5\lambda-3)/2} \begin{bmatrix} m-1 \\ \lfloor \frac{m+1-5\lambda}{2} \rfloor \end{bmatrix}_q,$$

$$b_m = \sum_\lambda (-1)^\lambda q^{\lambda(5\lambda+1)/2} \begin{bmatrix} m-1 \\ \lfloor \frac{m+1-5\lambda}{2} \rfloor \end{bmatrix}_q. \qquad (6.7)$$

The polynomials a_m and b_m were considered by Schur in conjunction with his proof of the Rogers-Ramanujan identities. See [3] and [17] for details.

Theorem 6.2 ([17]) *The following identity holds*

$$\frac{(q^{3-2k}, q^{2+2k}; q^5)_\infty}{(q, q^2, q^3, q^4; q^5)_\infty} = \sum_{j=0}^{[k/2]} (-1)^j q^{2j(j-k)+j(j+1)/2} \begin{bmatrix} k-j \\ j \end{bmatrix}_q \sum_{s=0}^\infty \frac{q^{s^2+s(k-2j)}}{(q;q)_s} (6.8)$$

Observe that (6.8) provides an infinite family of extensions to both Rogers-Ramanujan identities. This is so since the cases $k = 0, 1$ of (6.8) yield (6.1) and (6.2) respectively. Furthermore the relationships (6.3) and (6.8) are inverse relations.

Proof. Define $J(k)$ by

$$J(k) := \frac{(q;q)_\infty}{2\pi} \int_0^\pi (\sqrt{q}e^{i\theta}, \sqrt{q}e^{-i\theta}, e^{2i\theta}, e^{-2i\theta}; q)_\infty U_k(\cos\theta) d\theta. \qquad (6.9)$$

Thus (2.11), (2.41) and the fact that the integrand in $J(k)$ is even yield

$J(k)$

$$= \frac{1}{4\pi(q;q)_\infty} \sum_{m,n=-\infty}^\infty q^{(m^2+n^2+n)/2}(-1)^{m+n} \int_{-\pi}^\pi e^{i(m-2n)\theta}[e^{ik\theta} - e^{-i(k+2)\theta}] d\theta$$

$$= \frac{1}{2(q;q)_\infty} \sum_{n=-\infty}^\infty q^{n(n+1)/2}(-1)^{n+k} \left[q^{(2n-k)^2} - q^{(2n+k+2)^2/2} \right].$$

In the second sum replace n by $-n-1$ to see that the two sums are equal and we get

$$J(k) = \frac{(-1)^k q^{k^2/2}}{(q;q)_\infty} \sum_{n=-\infty}^\infty (-1)^n q^{5n^2/2} q^{n(1-4k)/2}$$

$$= \frac{(-1)^k q^{k^2/2}}{(q;q)_\infty} (q^5, q^{3-2k}, q^{2+2k}; q^5)_\infty.$$

This leads to the evaluation

$$J(k) = (-1)^k q^{k^2/2} \frac{(q^{3-2k}, q^{2+2k}; q^5)_\infty}{(q, q^2, q^3, q^4; q^5)_\infty}. \tag{6.10}$$

We now evaluate the integral $J(k)$ in a different way. We have

$$J(k) = \frac{(q;q)_\infty}{2\pi} \sum_{n=0}^\infty \frac{(-1)^n q^{n^2/2}}{(q;q)_n} \int_0^\pi H_n(\cos\theta|q^{-1}) U_k(\cos\theta)(e^{2i\theta}, e^{-2i\theta}; q)_\infty d\theta$$

Since $U_k(x) = C_k(x; q \mid q)$ we use the connection coefficient formulas (4.25) and (4.22) with $\gamma = q$ to get

$$J(k) = \sum_{\infty > n \geq 2s \geq 0} \frac{(-1)^n q^{s(s-n)+n^2/2}}{(q;q)_s (q;q)_{n-2s}} \sum_{j=0}^{[k/2]} \frac{(q;q)_{k-j}(-1)^j}{(q;q)_j} q^{j(j+1)/2} \delta_{n-2s, k-2j}$$

Therefore

$$J(k) = \sum_{s \geq 0, 0 \leq 2j \leq k} \frac{(-1)^{k+j} q^{s^2+s(k-2j)+(k-2j)^2/2}}{(q;q)_s (q;q)_{k-2j}} \frac{(q;q)_{k-j}}{(q;q)_j} q^{j(j+1)/2}$$

$$= (-1)^k q^{k^2/2} \sum_{s=0}^\infty \sum_{j=0}^{[k/2]} \frac{(-1)^j q^{s^2+s(k-2j)+2j^2-2kj+j(j+1)/2}(q;q)_{k-j}}{(q;q)_s (q;q)_{k-2j}(q;q)_j}.$$

This completes the proof.　　　　　　　　　　　　　　　□

We now come to the number theoretic interpretations of the Rogers-Ramanujan identities. A **partition** of n, $n = 1, 2, \ldots$, is a finite sequence $(n_1, n_2, \ldots, n_k)$, with $n_i \leq n_{i+1}$, so that $n = \sum_{i=1}^k n_i$. For example $(1, 1, 1, 3)$, $(1, 2, 3)$ and $(1, 1, 2, 2)$ are partitions of 6. The number of parts in a partition $(n_1, n_2, \ldots, n_k)$ is k and its parts are $n_1, n_2, \ldots, n_k$.

Let $p(n)$ be the number of partitions of n, with $p(0) := 1$. Euler proved

$$\sum_{n=0}^\infty p(n) q^n = \frac{1}{(q;q)_\infty}. \tag{6.11}$$

Proof. Expand the infinite product in (6.11) as

$$\prod_{n=1}^\infty \left[\sum_{s=0}^\infty q^{ns} \right].$$

Thus the coefficient of q^m on the right-hand side of (6.11) is the number of ways of writing m as $\sum_j n_j s_j$. In other words s_j is the number of parts each of which is n_j and (6.11) follows.

From the idea behind Euler's theorem it follows that if

$$\sum_{n=0}^{\infty} p(n;1,4)q^n = 1/(q,q^4;q^5)_\infty, \quad \sum_{n=0}^{\infty} p(n;2,3)q^n = 1/(q^2,q^3;q^5)_\infty,$$

(6.12)

then $p(0;1,4) = p(0;2,3) = 1$ and $p(n;1,4)$ is the number of partitions of n into parts congruent to 1, or 4 modulo 5, while $p(n;2,3)$ is the number of partitions of n into parts congruent to 2, or 2 modulo 5. This provides a partition theoretic interpretation of the right-hand sides of (6.1) and (6.2). In order to interpret the left-hand sides of these identities we need to develop more machinery.

A partition $(n_1, n_2, \ldots, n_k)$ can be represented graphically by putting the parts of the partition on different levels and each part n_i is represented by n_i dots. The number of parts at a level is greater than or equal the number of parts at any level below it. For example the partition $(1, 1, 3, 4, 6, 8)$ will be represented graphically as

If we interchange rows and columns of a partition $(n_1, n_2, \ldots, n_k)$ we obtain its conjugate partition. For instance the partition conjugate to $(1, 1, 3, 4, 6, 8)$ is $(1, 1, 2, 2, 3, 4, 4, 6)$.

It is clear that $1/(q;q)_k$ is the generating function of partitions into parts each of which is at most k. Since conjugation is a bijection on partitions then $1/(q;q)_k$ is also the generating function of partitions into at most k-parts. Note that the identity $k^2 = \sum_{j=1}^{k} (2j-1)$ is a special partition, so for example 4^2 gives the partition

Therefore $q^{k^2}/(q;q)_k$ is the generating function of partitions of n into k parts differing by at least 2. Similarly $k^2 + k = \sum_{j=1}^{k} (2j)$ shows that $q^{k^2+k}/(q;q)_k$ is the generating function of partitions of n into k parts differing by at least 2, and each part is at least 2. $\square$

Theorem 6.3 *The number of partitions of n into parts congruent to 1 or 4 modulo 5 equals the number of partitions of n into parts which differ by at least 2. Furthermore the partitions of n into parts congruent to 2 or 3 modulo 5 are equinumerous as the partitions of n into parts which differ by at least 2 and the smallest part is at least 2.*

It is now clear that the left hand side of (6.3) is a generating function for the partitions of n into parts which differ by at least 2 and the smallest part is at least m.

Acknowledgments Thanks to Dick Askey for his valuable comments on these lectures and for correcting several minor slips. I am also grateful to the NATO ASI committee for giving me the opportunity to lecture on this material. My research was partially supported by NSF grant # DMS 99-70865.

References

1. N. I. Akhiezer, *The Classical Moment Problem and Some Related Questions in Analysis*, English translation, Oliver and Boyed, Edinburgh, 1965.
2. W. A. Al-Salam and T. S. Chihara, *Convolutions of orthogonal polynomials*, SIAM J. Math. Anal. **7** (1976), 16–28.
3. G. E. Andrews, *The Theory of Partitions*, Addison-Wesley, Reading, Massachusetts, 1976.
4. G. E. Andrews, *q-series: Their development and application in analysis, number theory, combinatorics, physics, and computer algebra*, CBMS Regional Conference Series, number 66, American Mathematical Society, Providence, R.I. 1986.
5. G. E. Andrews and R. A. Askey, *A simple proof of Ramanujan's summation $_1\psi_1$*, Aequationes Math. **18** (1978), 333–337.
6. G. E. Andrews and R. A. Askey, *Classical orthogonal polynomials*, in: "Polynomes Orthogonaux et Applications", C. Breziniski, et al., eds., Lecture Notes in Mathematics, Vol. 1171, Springer-Verlag, Berlin, 1984, pp. 36–63.
7. G. E. Andrews, R. A. Askey and R. Roy, *Special Functions*, Cambridge University Press, Cambridge, 1999.
8. R. A. Askey, *Divided difference operators and classical orthogonal polynomials*, Rocky Mountain J. Math. **19** (1989), 33–37.
9. R. A. Askey and M. E. H. Ismail, *A generalization of ultraspherical polynomials*, in: "Studies in Pure Mathematics", P. Erdös, ed., Birkhauser, Basel, 1983, pp. 55–78.
10. R. A. Askey and M. E. H. Ismail, *Recurrence relations, continued fractions and orthogonal polynomials*, Memoirs Amer. Math. Soc. Number **300** (1984).
11. R. A. Askey and J. A. Wilson, *Some basic hypergeometric orthogonal polynomials that generalize Jacobi polynomials*, Memoirs Amer. Math. Soc. Number **319** (1985).
12. W. N. Bailey, *Generalized Hypergeometric Series*, Cambridge University Press, Cambridge, 1935.
13. C. Berg and M. E. H. Ismail, *q-Hermite polynomials and classical orthogonal polynomials*, Canad. J. Math. **9** (1996), 43–63.
14. D. Bressoud, *On partitions, orthogonal polynomials and the expansion of certain infinite products*, Proc. London Math. Soc. **42** (1981), 478–500.
15. A. Erdelyi, W. Magnus, F. Oberhettinger and F. G. Tricomi, *Higher Transcendental Functions*, Vol. 2, McGraw-Hill, New York, 1953.
16. E. Feldheim, *Sur les polynomes généralisés de Legendre*, Izv. Akad. Nauk. SSSR Ser. Math. **5** (1941), 241–248.

17. K. Garrett, M. E. H. Ismail, and D. Stanton, *Variants of the Rogers-Ramanujan identities*, Advances in Appl. Math. **23** (1999), 274–299.
18. G. Gasper and M. Rahman, *Basic Hypergeometric Series*, Cambridge University Press, Cambridge, 1990.
19. M. E. H. Ismail, *A simple proof of Ramanujan's $_1\psi_1$ sum*, Proc. Amer. Math. Soc. **63** (1977), 185–186.
20. M. E. H. Ismail, *The Askey-Wilson operator and summation theorems*, in: "Mathematical Analysis, Wavelets, and Signal Processing,"*, M. E. H. Ismail, M. Z. Nashed, A. Zayed and A. Ghaleb, eds., Contemporary Mathematics, Vol. 190, Amer. Math. Soc., Providence, 1995, pp. 171–178.
21. M. E. H. Ismail, D. Stanton, and G. Viennot, *The combinatorics of the q-Hermite polynomials and the Askey-Wilson integral*, European J. Combinatorics **8** (1987), 379–392.
22. M. E. H. Ismail and J. Wilson, *Asymptotic and generating relations for the q-Jacobi and the $_4\phi_3$ polynomials*, J. Approx. Theory **36** (1982), 43–54.
23. I. L. Lanzewizky, *Über die orthogonalität der Fejer-Szegöschen polynome*, C. R. (Dokl.) Acad. Sci. URSS **31** (1941), 199–200.
24. A. P. Magnus, *Associated Askey-Wilson polynomials as Laguerre-Hahn orthogonal polynomials*, in: "Orthogonal Polynomials and Their Applications", M. Alfaro et al., eds., Lecture Notes in Mathematics, Vol. 1329, Springer-Verlag, Berlin, 1988, pp. 261–278.
25. M. Rahman and S. K. Suslov, *Unified approach to the summation and integration formulas for q-hypergeometric functions I*, J. Statistical Planning and Inference **54** (1996), 101–118.
26. G. Szegő, *Orthogonal Polynomials*, fourth edition, Amer. Math. Soc., Providence, 1975.

THE ASKEY-WILSON FUNCTION TRANSFORM SCHEME

ERIK KOELINK
Technische Universiteit Delft, Fac. ITS, Afd. TWA
Postbus 5031, 2600 GA Delft
The Netherlands

AND

JASPER V. STOKMAN
Universiteit van Amsterdam, KdV Institute for Mathematics
Pl. Muidergracht 24, 1018 TV Amsterdam
The Netherlands

Abstract. In this paper we present an addition to Askey's scheme of q-hypergeometric orthogonal polynomials involving classes of q-special functions which do not consist of polynomials only. The special functions are q-analogues of the Jacobi and Bessel function. The generalized orthogonality relations and the second order q-difference equations for these families are given. Limit transitions between these families are discussed. The quantum group theoretic interpretations are discussed shortly.

1. Introduction

The Askey-scheme of hypergeometric orthogonal polynomials is a scheme containing various known sets of orthogonal polynomials that can be written in terms of hypergeometric series, see e.g. Askey and Wilson [2], Koekoek and Swarttouw [15], Koornwinder [28]. A typical entry is the set of Jacobi polynomials defined by

$$R_n^{(\alpha,\beta)}(x) = {}_2F_1\left(\begin{matrix} -n, n+\alpha+\beta+1 \\ \alpha+1 \end{matrix}; \frac{1-x}{2}\right), \qquad n \in \mathbf{Z}_{\geq 0}, \qquad (1.1)$$

where we use the standard notation for hypergeometric series, see e.g. [7]. The Jacobi polynomials are orthogonal with respect to the beta distribution $(1-x)^\alpha(1+x)^\beta$ on the interval $[-1, 1]$. Moreover, we have the much bigger

J. Bustoz et al. (eds.), Special Functions 2000, 221–241.
© *2001 Kluwer Academic Publishers. Printed in the Netherlands.*

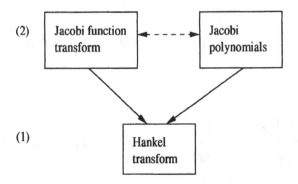

Figure 1.1. Jacobi function scheme.

q-analogue of the Askey-scheme having the Askey-Wilson polynomials and Racah polynomials at the top level with four degrees of freedom (apart from q), see e.g. [15].

The relation with schemes of non-polynomial special functions is less well-advertised, and here we discuss the q-analogue of the scheme displayed in Figure 1.1. The Jacobi function transform is an integral transform on $[0, \infty)$ in which the kernel is given by a Jacobi function

$$\phi_\lambda^{(\alpha,\beta)}(t) = {}_2F_1\left(\begin{matrix}\frac{1}{2}(\alpha+\beta+1-i\lambda), \frac{1}{2}(\alpha+\beta+1+i\lambda)\\ \alpha+1\end{matrix}; -\sinh^2 t\right) \quad (1.2)$$

for $|\sinh t| < 1$, which has a one-valued analytic continuation to $-\sinh^2 t \in \mathbb{C}\backslash[1,\infty)$, see Koornwinder's survey paper [25] for a nice introduction and references. We see that we may view the Jacobi function as an analytic continuation in its degree of the Jacobi polynomial; replace n in (1.1) by $\frac{1}{2}(i\lambda - \alpha - \beta - 1)$ and x by $\cosh 2t$ to get the Jacobi function of (1.2).

The Jacobi function transform is then given by

$$g(\lambda) = \int_0^\infty f(t)\phi_\lambda^{(\alpha,\beta)}(t)(2\sinh t)^{2\alpha+1}(2\cosh t)^{2\beta+1}\, dt, \quad (1.3)$$

$$f(t) = \int_0^\infty g(\lambda)\phi_\lambda^{(\alpha,\beta)}(t)\frac{|\Gamma(\frac{1}{2}(i\lambda+\alpha+\beta+1))\Gamma(\frac{1}{2}(i\lambda+\alpha-\beta+1))|^2}{4^{\alpha+\beta+1}|\Gamma(\alpha+1)\Gamma(i\lambda)|^2}\, d\lambda$$

for some suitable class of functions. Here we assume that $\alpha, \beta \in \mathbb{R}$ satisfy $|\beta| < \alpha + 1$, otherwise discrete mass points have to be added to the Plancherel measure, see [25, §2].

The Hankel transform is the integral transform on $[0, \infty)$ that has the Bessel function

$$J_\alpha(x) = \frac{(x/2)^\alpha}{\Gamma(\alpha+1)}\,{}_0F_1\left(\begin{matrix}-\\ \alpha+1\end{matrix}; -\frac{x^2}{4}\right) \quad (1.4)$$

as its kernel. For suitable functions and $\alpha > -1$ the Hankel transform pair is given by, see Watson [41, §14.3],

$$g(\lambda) = \int_0^\infty f(x)\, J_\alpha(x\lambda) x\, dx, \qquad f(x) = \int_0^\infty g(\lambda)\, J_\alpha(x\lambda) \lambda\, d\lambda. \qquad (1.5)$$

The Hankel transform can formally be obtained as a limit case from the orthogonality relations for the Jacobi polynomials using the limit

$$\lim_{N\to\infty} R_{n_N}^{(\alpha,\beta)}\left(1 - \frac{x^2}{2N^2}\right) = {}_0F_1(-;\alpha+1; -(x\lambda)^2/4), \quad n_N/N \to \lambda \text{ as } N \to \infty.$$
$$(1.6)$$

The Hankel transform can also be viewed as a limit case of the Jacobi function transform by use of the limit transition

$$\lim_{\epsilon\downarrow 0} \phi_{\lambda/\epsilon}^{(\alpha,\beta)}(t\epsilon) = 2^\alpha \Gamma(\alpha+1)(\lambda t)^{-\alpha} J_\alpha(\lambda t), \qquad (1.7)$$

see [25, (2.34)], and the limit transition (1.7) can be considered as a generalization of (1.6).

So Figure 1.1 gives an addition to the Askey-scheme of hypergeometric orthogonal polynomials by an analytic continuation in the spectral parameter (the dashed line) to the Jacobi functions, and by a limit transition to the Bessel functions. The Jacobi functions and polynomials have two degrees of freedom, namely α and β, and the Hankel transform has one degree of freedom, namely the order α of the Bessel function. We consider the Jacobi functions as the master functions, and the Jacobi polynomials and the Bessel functions as derivable functions. We discuss the group theoretic interpretation of Figure 1.1 in §7.1.

The purpose of this paper is to give an overview of results on q-analogues of the Jacobi and Bessel functions related to three q-analogues of the Jacobi polynomials from the q-analogue of Askey's scheme, namely the Askey-Wilson polynomials, the big q-Jacobi polynomials and the little q-Jacobi polynomials. We mainly focus on transforms that are analogues of the Jacobi function transform (1.3) and of the Hankel transform (1.5). There are also Fourier-Bessel type orthogonality relations for these kind of functions, and for this we refer to the work of Suslov and collaborators [5], [11], [37], [38] in the appropriate sections. Furthermore, Ruijsenaars [35] recently defined and studied the Askey-Wilson function for q on the unit circle using an appropriate analogue of Barnes's integral representation for the ${}_2F_1$-series. It would be desirable to find the corresponding transform properties of the R-function of Ruijsenaars.

Let us note that this paper does not contain rigorous proofs. All limit transitions are considered at a formal level, and since all the transform

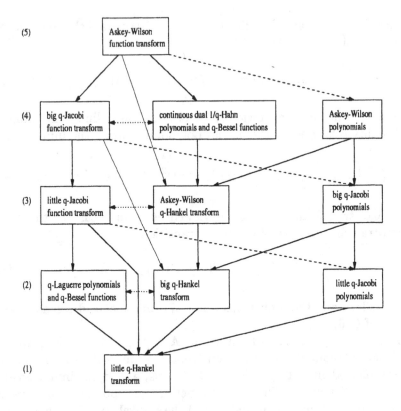

Figure 1.2. Askey-Wilson function scheme.

pairs discussed in this paper have been proved rigorously there is no gain in making the formal limit transitions valid. The transform pairs can be considered as the result of the spectral analysis of a second order q-difference operator that acts symmetrically on a suitable weighted L^2-space, see the references in §§2, 3 and 4. The formal limit transitions are emphasized via the second order q-difference operators involved. We note that all functions involved do not only occur as eigenfunctions of a second order q-difference operator in the geometric parameter, but also as eigenfunctions of another second order q-difference operator in the spectral parameter. So they solve a bispectral problem, and according to Grünbaum and Haine [8] the most general case corresponds to the Askey-Wilson q-difference operator, which is discussed in §2.

So we give three q-analogues of Figure 1.1; one related to the Askey-Wilson polynomials, presented in §2, one related to the big q-Jacobi polynomials, presented in §3, and one related to the little q-Jacobi polynomials, presented in §4. This is depicted in Figure 1.2, where the three boxes on the right hand side are part of Askey's scheme of q-hypergeometric orthogonal polynomials, see [15]. These three q-analogues are related by limit transitions as well, and this is also depicted in Figure 1.2. This is discussed in §5 and §6, where the boxes with the continuous dual q^{-1}-Hahn polynomials and the q-Laguerre polynomials, both corresponding to indeterminate moment problems, are discussed.

In Figure 1.2 the lines denote limit transitions, the dashed lines denote analytic continuation and the dotted lines, which do not appear in Figure 1.1, denote that the results of the two boxes involved are related by duality. In particular, we think of the Askey-Wilson function transform and the little q-Hankel transform as self-dual transforms, i.e. the inverse transform equals the generic Fourier transform itself, possibly for dual parameters. Note that there is a striking difference between Figure 1.1 and Figure 1.2. The dashed line corresponding to analytic continuation in Figure 1.1 is horizontal, i.e. the Jacobi polynomials and the Jacobi function transform both have 2 degrees of freedom. The dashed lines in Figure 1.2 go down, i.e. the q-analogue of the Jacobi function transform has one extra degree of freedom compared to the q-analogue of the Jacobi polynomial. This extra parameter is not contained in the definition of the q-analogues of the Jacobi function, but it appears in the measure for the q-analogue of the Jacobi function transform.

In §7 we discuss the (quantum) group theoretic interpretation of Figure 1.1 and Figure 1.2, which is also the motivation for the names for these transforms. We end with some concluding remarks and some open problems.

Notation. We use the notation for basic hypergeometric series as in the book [7] by Gasper and Rahman. We assume throughout $0 < q < 1$.

2. Askey-Wilson Analogue of the Jacobi Function Scheme

In this section we consider the analogue of Figure 1.1 on the level for the Askey-Wilson case. So the second order difference operator is

$$Lf(x) = A(x)(f(qx) - f(x)) + B(x)(f(q^{-1}x) - f(x)), \qquad (2.1)$$

where

$$A(x) = \frac{(1 - ax)(1 - bx)(1 - cx)(1 - dx)}{(1 - x^2)(1 - qx^2)}, \qquad B(x) = A(x^{-1}). \qquad (2.2)$$

The difference operator (2.1), (2.2) has been introduced by Askey and Wilson [2]. The general set of eigenfunctions for (2.1) with A and B as in (2.2) is given by Ismail and Rahman [12].

2.1. THE ASKEY-WILSON FUNCTIONS

The Askey-Wilson functions are defined by

$$\phi_\gamma(x;a;b,c;d|q) = \frac{(qax\gamma/\tilde{d}, qa\gamma/\tilde{d}x; q)_\infty}{(\tilde{a}\tilde{b}\tilde{c}\gamma, q\gamma/\tilde{d}, q\tilde{a}/\tilde{d}, qx/d, q/dx; q)_\infty}$$
$$\times {}_8W_7\left(\tilde{a}\tilde{b}\tilde{c}\gamma/q; ax, a/x, \tilde{a}\gamma, \tilde{b}\gamma, \tilde{c}\gamma; q, q/\tilde{d}\gamma\right) \quad (2.3)$$

for $|\gamma| > |q/\tilde{d}|$, where $\tilde{a} = \sqrt{q^{-1}abcd}$, $\tilde{b} = ab/\tilde{a}$, $\tilde{c} = ac/\tilde{a}$, and $\tilde{d} = ad/\tilde{a}$. Observe that the Askey-Wilson function in (2.3) is symmetric in b and c, and almost symmetric in a and b;

$$\phi_\gamma(x;a;b,c;d|q) = \frac{(\tilde{c}\gamma, qb/d, \tilde{c}/\gamma; q)_\infty}{(q\gamma/\tilde{d}, qa/d, q/\tilde{d}\gamma; q)_\infty}\phi_\gamma(x;b;a,c;d|q), \quad (2.4)$$

by an application of [7, III.36], or see Suslov [38]. Then $L\phi_\gamma = (-1 - \tilde{a}^2 + \tilde{a}(\gamma + \gamma^{-1}))\phi_\gamma$. Note that the invariance $x \leftrightarrow x^{-1}$ is obvious in (2.3). There exists a meromorphic continuation of the Askey-Wilson function in γ which is invariant under $\gamma \leftrightarrow \gamma^{-1}$ by [7, (III.23)]. By [7, (III.23)] again we see that the Askey-Wilson function is self-dual in the sense that

$$\phi_\gamma(x;a;b,c;d|q) = \phi_x(\gamma; \tilde{a}; \tilde{b}, \tilde{c}; \tilde{d}|q). \quad (2.5)$$

Assume that (i) $0 < b, c \le a < d/q$, (ii) $bd, cd \ge q$ and (iii) $ab, ac < 1$, and let $t < 0$ be an extra parameter. By $\mathcal{H}(a; b, c; d; t)$ we denote the weighted L^2-space of symmetric functions $f(x) = f(x^{-1})$ with respect to the measure $d\nu(x; a, b, c; d|q, t)$, which is defined as follows. We introduce

$$\Delta(x) = \frac{(x^{\pm 2}, qx^{\pm 1}/d; q)_\infty}{\theta(tdx^{\pm 1})(ax^{\pm 1}, bx^{\pm 1}, cx^{\pm 1}; q)_\infty}, \quad (2.6)$$

where $\theta(x) = (x, q/x; q)_\infty$ is the (renormalized) Jacobi theta-function and where we use $(cx^{\pm 1}; q)_\infty = (cx, c/x; q)_\infty$ and similarly for the other $\pm$-signs. Note that this is the standard Askey-Wilson weight function, see [2], [7, Ch. 6], multiplied by q-constant function that can be written as a quotient of theta-functions. The positive measure is now given by

$$\int_{\mathbf{C}^*} f(x)d\nu(x; a; b, c; d|q, t) = \frac{K}{4\pi i}\int_{\mathbf{T}} f(x)\Delta(x)\frac{dx}{x} + K\sum_{s\in S} f(s)\mathrm{Res}_{x=s}\frac{\Delta(x)}{x},$$
$$(2.7)$$

where $S = S_+ \cup S_-$, $S_+ = \{aq^k \mid k \in \mathbf{Z}_{\ge 0}, aq^k > 1\}$ and $S_- = \{tdq^k \mid k \in \mathbf{Z}, tdq^k < 1\}$, under the generic assumption on the parameters that $S \cup S^{-1}$ consists of simple poles of Δ. Note that this condition can be removed by extending the definition of the masses at the discrete set by continuity, see [24] for details. Here $K = K(a; b, c; d; t)$ is a positive constant defined by

$$K = (qabcdt^2)^{-\frac{1}{2}}(ab, ac, bc, qa/d, q; q)_\infty (\theta(qt)\theta(adt)\theta(bdt)\theta(cdt))^{\frac{1}{2}}. \quad (2.8)$$

The Askey-Wilson function transform pair for a sufficiently nice function $u \in \mathcal{H}(a; b, c; d; t)$ is given by

$$\hat{u}(\gamma) = \int_{\mathbb{C}^*} u(x)\phi_\gamma(x; a; b, c; d|q) \, d\nu(x; a; b, c; d|q, t),$$

$$u(x) = \int_{\mathbb{C}^*} \hat{u}(\gamma)\phi_\gamma(x; a; b, c; d|q) \, d\nu(\gamma; \tilde{a}; \tilde{b}, \tilde{c}; \tilde{d}|q, \tilde{t}), \quad (2.9)$$

where $\tilde{t} = 1/qadt$. Furthermore, the dual parameters $(\tilde{a}, \tilde{b}, \tilde{c}, \tilde{d}, \tilde{t})$ satisfy the same conditions as (a, b, c, d, t). So, in view of (2.5), the inverse of the Askey-Wilson function transform is the Askey-Wilson transform for the dual set of parameters. Moreover, the Askey-Wilson function transform extends to an isometric isomorphism from $\mathcal{H}(a, b, c; d; t)$ to $\mathcal{H}(\tilde{a}, \tilde{b}, \tilde{c}; \tilde{d}; \tilde{t})$. Proofs of these results can be found in [24]. See Suslov [37], [38] for Fourier-Bessel type orthogonality relations for the Askey-Wilson functions.

2.2. THE ASKEY-WILSON POLYNOMIALS

The Askey-Wilson polynomials are the eigenfunctions of L as in (2.1), (2.2), which are polynomials in $\frac{1}{2}(x + x^{-1})$. These orthogonal polynomials are very well-known, see [2], [7, §7.5], and they are on top of the Askey scheme of basic hypergeometric orthogonal polynomials, see [15]. See also Brown, Evans and Ismail [4] for an operator approach to the Askey-Wilson polynomials and their orthogonality relations. The Askey-Wilson functions reduce to the Askey-Wilson polynomials for $\gamma^{-1} = \tilde{a}q^n$, $n \in \mathbf{Z}_{\geq 0}$, since the $_8W_7$-series in (2.3) reduces to a terminating series

$$_8W_7(aq^{-n}/d; ax, a/x, q^{-n}, q^{1-n}/cd, q^{1-n}/bd; q, q^n bc) \quad (2.10)$$

$$= \frac{(aq^{1-n}/d, q^{1-n}/ad; q)_n}{(q^{1-n}/dx, q^{1-n}x/d; q)_n} \, _4\varphi_3 \left(\begin{array}{c} q^{-n}, abcdq^{n-1}, ax, a/x \\ ab, ac, ad \end{array} ; q, q \right)$$

by [7, (III.18)]. We can also use [7, (III.36)] to write the $_8W_7$-series as a sum of two balanced $_4\varphi_3$-series, which reduces to a single terminating balanced $_4\varphi_3$-series for $\gamma^{-1} = \tilde{a}q^n$, see e.g. Suslov [37], [38]. This shows that the Askey-Wilson function $\phi_\gamma(x; a; b, c; d|q)$ is the analytic continuation of the Askey-Wilson polynomial.

2.3. THE ASKEY-WILSON q-BESSEL FUNCTIONS

In (2.3) we replace c by $c\epsilon$, d by d/ϵ and γ by $\gamma\epsilon$; then the formal limit transition gives

$$\lim_{\epsilon\downarrow 0} \phi_{\gamma\epsilon}(x; a; b, c\epsilon; \tfrac{d}{\epsilon}|q) = {}_2\varphi_1\left(\begin{matrix} ax, a/x \\ ab \end{matrix}; q, \frac{q}{d\gamma}\right). \qquad (2.11)$$

Taking the limit in the second order q-difference equation shows that the Askey-Wilson q-Bessel function

$$J_\gamma(x; a; b|q) = {}_2\varphi_1\left(\begin{matrix} ax, a/x \\ ab \end{matrix}; q, -\frac{q\gamma}{a}\right) \qquad (2.12)$$

is a solution to $LJ_\gamma(\cdot; a; b|q) = \gamma J_\gamma(\cdot; a; b|q)$, with L as in (2.1) with

$$A(x) = \frac{(1-ax)(1-bx)x}{(1-x^2)(1-qx^2)}, \qquad B(x) = A(x^{-1}). \qquad (2.13)$$

Taking the limit transition (2.11) through the sequence $\epsilon = q^m$, $m \to \infty$, in the Askey-Wilson function transform pair with t replaced by ϵt formally leads to the following orthogonality relations

$$\int_{\mathbf{C}^*} J_{\gamma q^k}(x; a; b|q) J_{\gamma q^l}(x; a; b|q) \, d\nu(x; a; b, q\gamma; \gamma^{-1}|q, -1) \qquad (2.14)$$

$$= \delta_{k,l} a^{-2k} \frac{(-aq^{-k}/\gamma; q)_\infty}{(-bq^{-k}/\gamma; q)_\infty} \frac{K(a; b, q\gamma; \gamma^{-1}; -1)}{((ab; q)_\infty \theta(-a/\gamma))^2}, \qquad k, l \in \mathbf{Z},$$

where $\gamma = -a\tilde{t}$ for $a, b, \gamma > 0$, $ab < 1$, $a > b$ and K as in (2.8). Observe that cancellation in the weight Δ (2.6) occurs in (2.14), since $cd = q$. The orthogonality relations can be obtained from Kakehi [13], see also [23, App. A] where also other ranges of the parameters are considered. Moreover, the functions $J_{\gamma q^k}(\cdot; a; b|q)$ form a complete set in the weighted L^2-space for the measure in (2.14). Note that under the limit transition (2.11) only the infinite set of discrete mass points that tend to $-\infty$ in (2.9) survive.

Using the same limit transition (2.11) in (2.10) we find the Askey-Wilson q-Bessel functions (2.12) with the corresponding orthogonality relations (2.14) from the orthogonality relations of the Askey-Wilson polynomials, see also [16, Ch. 3]. The Askey-Wilson q-Bessel function is also studied by Bustoz and Suslov [5], who derive Fourier series expansions, and by Ismail, Masson and Suslov [11], who derive Fourier-Bessel type orthogonality relations for $J_\gamma(x; a; b|q)$.

3. Big q-Analogue of the Jacobi Function Scheme

In this section we consider the analogue of Figure 1.1 on the level of the big q-Jacobi case. So we consider the second order difference operator L as in (2.1) with

$$A(x) = a^2\left(1 + \frac{1}{abx}\right)\left(1 + \frac{1}{acx}\right), \qquad B(x) = \left(1 + \frac{q}{bcx}\right)\left(1 + \frac{1}{x}\right). \quad (3.1)$$

The general set of eigenfunctions for L in (2.1) with A and B as in (3.1) is given by Gupta, Ismail and Masson [9].

3.1. THE BIG q-JACOBI FUNCTIONS

The big q-Jacobi functions are defined by

$$\phi_\gamma(x; a; b, c; q) = {}_3\varphi_2\left(\begin{matrix} a\gamma, a/\gamma, -1/x \\ ab, ac \end{matrix}; q, -bcx\right), \qquad |bcx| < 1, \quad (3.2)$$

and they satisfy $L\phi_\gamma(\cdot; a; b, c; q) = (-1 - a^2 + a(\gamma + \gamma^{-1}))\phi_\gamma(\cdot; a; b, c; q)$. We can extend the definition of the big q-Jacobi function (3.2) to generic values of x by requiring that this second order q-difference equation remains valid, see [22]. Assume that the parameters a, b and c are positive, a greater than b and c, and that all pairwise products are less than one, and let $z > 0$. The big q-Jacobi function transform is given by the following transform pair

$$\hat{u}(\gamma) = \int_{-1}^{\infty(z)} u(x)\phi_\gamma(x; a; b, c; q)\frac{(-qx, -bcx; q)_\infty}{(-abx, -acx; q)_\infty}\, d_q x,$$

$$u(x) = C\int_{\mathbf{C}^*} \hat{u}(\gamma)\,\phi_\gamma(x; a; b, c; q)\,(\gamma^{\pm 1}abc; q)_\infty^{-1}\, d\nu(\gamma; a; b, c; q/abc|q, -z^{-1}),$$

$$C = \frac{\theta(-abz)\theta(-acz)\theta(-bcz)\,(ab, ac; q)_\infty^2}{(1-q)z\,\theta(-qz)K(a; b, c; q/abc; -1/z)} \quad (3.3)$$

with the notation of (2.7) and where the q-integral is defined by

$$\int_{-1}^{\infty(z)} f(x)\, d_q x = (1-q)\sum_{k=0}^{\infty} f(-q^k)q^k + (1-q)z\sum_{k=-\infty}^{\infty} f(zq^k)q^k.$$

After a suitable scaling the big q-Jacobi function transform extends to an isometric isomorphism of the weighted L^2-space with respect to the q-integral with weight as in the first equality of (3.3) onto the weighted L^2-space of symmetric functions with respect to the weight $C(\gamma^{\pm 1}abc; q)_\infty^{-1}\, d\nu(\gamma; a, b, c; q/abc|q, -1/z)$. See [22] for a proof of these statements.

3.2. THE BIG q-JACOBI POLYNOMIALS

The polynomial eigenfunctions to (2.1), (3.1) are the big q-Jacobi polynomials and they are given by the big q-Jacobi functions (3.2) with $\gamma = aq^n$, $n \in \mathbf{Z}_{\geq 0}$,

$$
{}_3\varphi_2 \left(\begin{matrix} a^2 q^n, q^{-n}, -1/x \\ ab, ac \end{matrix} ; q, -bcx \right) = \frac{(cq^{-n}/a; q)_n}{(ac; q)_n} {}_3\varphi_2 \left(\begin{matrix} q^{-n}, a^2 q^n, -abx \\ ab, qa/c \end{matrix} ; q, q \right),
$$
(3.4)

see [22, Prop. 5.3]. The big q-Jacobi polynomials are orthogonal with respect to a positive discrete measure supported on $-q^{\mathbf{Z}_{\geq 0}} \cup -q^{\mathbf{N}}/bc$ for $ab < 1$, $ac < 1$, $qa/b < 1$, $qa/c < 1$ and $bc < 0$, see [22, §10]. See Andrews and Askey [1], or [7, §7.3], [15] for the standard definition of the big q-Jacobi polynomials.

3.3. THE BIG q-BESSEL FUNCTIONS

If we replace c by $c\epsilon$ and γ by $\gamma\epsilon$ in (3.2), we obtain the following limit

$$
\lim_{\epsilon \downarrow 0} \phi_{\gamma\epsilon}(x; a; b, c\epsilon; q) = {}_1\varphi_1 \left(\begin{matrix} -1/x \\ ab \end{matrix} ; q, -\frac{abcx}{\gamma} \right).
$$
(3.5)

We define the big q-Bessel function by

$$
J_\gamma(x; a; q) = {}_1\varphi_1 \left(\begin{matrix} -1/x \\ a \end{matrix} ; q, a\gamma x \right).
$$
(3.6)

The big q-Bessel function is a solution to $LJ_\gamma(\cdot; a; q) = -\gamma J_\gamma(\cdot; a; q)$, with L as in (2.1) with $A(x) = (1 + 1/ax)/x$ and $B(x) = q(1 + 1/x)/ax$. Taking the limit in the big q-Jacobi function transform pair (3.3) with z fixed, shows that for $\gamma > 0$, $0 < a < 1$,

$$
\int_{-1}^{\infty(q/a\gamma)} (J_{\gamma q^k} J_{\gamma q^l})(x; a; q) \frac{(-qx; q)_\infty}{(-ax; q)_\infty} d_q x
$$
$$
= \delta_{k,l}(1 - q) \frac{(q; q)_\infty^2 \theta(-a\gamma)}{(a; q)_\infty^2 \theta(-\gamma)} a^{-k}(-q^k \gamma; q)_\infty
$$
(3.7)

for $k, l \in \mathbf{Z}$. Here the extra parameter z in the measure of the big q-Jacobi function transform (3.3) is in the limit inverse proportional to γ. Under the limit transition (3.5) the only part of the spectrum of (3.3) that survives is the infinite set of discrete mass points tending to $-\infty$. Moreover, the big q-Bessel functions $J_{\gamma q^l}(\cdot; a; q)$ form a complete orthogonal set in the weighted L^2-space for the measure in (3.7). See [6] for the proof of (3.7).

The big q-Bessel function in (3.5) can also be obtained by taking the limit from the big q-Jacobi polynomials and then the orthogonality relations

(3.7) can be obtained from the orthogonality relations for the big q-Jacobi polynomials in a rigorous way, see [6, §6].

4. Little q-Analogue of the Jacobi Function Scheme

In this section we consider the analogue of Figure 1.1 on the level of the little q-Jacobi case. So the second order difference operator L is now as in (2.1) with

$$A(x) = a^2\left(1 + \frac{1}{ax}\right), \qquad B(x) = \left(1 + \frac{q}{bx}\right). \qquad (4.1)$$

The general set of eigenfunctions for L with A and B as in (4.1) is given by the solutions to second order q-difference equation that is the q-analogue of the hypergeometric differential equation, see [7, Ch. 1].

4.1. THE LITTLE q-JACOBI FUNCTIONS

The little q-Jacobi functions are defined by

$$\phi_\gamma(x; a; b; q) = {}_2\varphi_1\left(\begin{matrix} a\gamma, a/\gamma \\ ab \end{matrix}; q, -bx\right), \qquad (4.2)$$

and they satisfy $L\phi_\gamma(\cdot; a; b; q) = (-1 - a^2 + a(\gamma + \gamma^{-1}))\phi_\gamma(\cdot; a; b; q)$. For $y > 0$, $a > b > 0$, and $ab < 1$ we have the transform pair

$$\hat{u}(\gamma) = \sum_{k=-\infty}^{\infty} u(k)\phi_\gamma(yq^k; a; b; q)\, a^{2k}\frac{(-q^{1-k}/ay; q)_\infty}{(-q^{1-k}/by; q)_\infty}, \qquad (4.3)$$

$$u(k) = C\int_{\mathbf{C}^*} \hat{u}(\gamma)\phi_\gamma(yq^k; a; b; q)\, d\nu(\gamma; a; b, aby; q/aby|q, -1),$$

for $C = (ab; q)_\infty^2 \theta(-by)^2/K(a; b, aby; q/aby; -1)$ and using the notation (2.7). Observe that cancellation in the weight Δ in (2.6) occurs in (4.3), since $cd = q$. For a proof of (4.3) see Kakehi [13] or [23, App. A], where the result is given for more general parameter values.

4.2. THE LITTLE q-JACOBI POLYNOMIALS

The little q-Jacobi polynomials are the polynomial eigenfunctions of L as in (2.1), (4.1), and they occur for $\gamma = aq^n$, $n \in \mathbf{Z}_{\geq 0}$;

$$\phi_{aq^n}(x; a; b; q) = {}_2\varphi_1\left(\begin{matrix} q^{-n}, a^2q^n \\ ab \end{matrix}; q, -bx\right). \qquad (4.4)$$

This is not the standard expression for the little q-Jacobi polynomials as introduced by Andrews and Askey [1], or see [7, §7.3], [15], where the orthogonality relations can be found.

4.3. THE LITTLE q-BESSEL FUNCTIONS

If we replace γ by $\gamma\epsilon$ and x by $x\epsilon$, we obtain the little q-Bessel functions from the little q-Jacobi functions;

$$\lim_{\epsilon\downarrow 0} \phi_{\gamma\epsilon}(x\epsilon; a; b; q) = {}_1\varphi_1\left(\begin{matrix} 0 \\ ab \end{matrix}; q, -\frac{abx}{\gamma}\right). \tag{4.5}$$

The little q-Bessel function is defined by

$$j_\gamma(x; a; q) = {}_1\varphi_1(0; a; q, q\gamma x). \tag{4.6}$$

Note that the little q-Bessel function is self-dual; $j_\gamma(x; a; q) = j_x(\gamma; a; q)$. These q-analogues of the Bessel function are also known under the name ${}_1\varphi_1$ q-Bessel function or Hahn-Exton q-Bessel function, see [29] and [21] for historic references to Hahn, Exton and Jackson. The little q-Bessel functions are eigenfunctions of $Lj_\gamma(\cdot; a; q) = -q\gamma j_\gamma(\cdot; a; q)$ with L as in (2.1) with $A(x) = a/x$ and $B(x) = q/x$. This is the second order q-difference equation for the ${}_1\varphi_1$-series, and we obtain this from the second order q-difference equation for the little q-Bessel function. The little q-Bessel functions satisfy the orthogonality relations for $0 < a < 1$, see Koornwinder and Swarttouw [29, Prop. 2.6],

$$\sum_{k=-\infty}^{\infty} a^k j_{q^n}(q^k; a; q)\, j_{q^m}(q^k; a; q) = \delta_{n,m} a^{-n} \frac{(q; q)_\infty^2}{(a; q)_\infty^2}, \qquad n, m \in \mathbf{Z}. \tag{4.7}$$

In the limit transition (4.5) of the little q-Jacobi function transform (4.3) the only part of the spectrum that survives the contraction is the infinite set of discrete mass points tending to $-\infty$, which leads to a formal derivation of (4.7). We note that the extra degree of freedom in the measure of the little q-Jacobi function transform (4.3) drops out in the limit. By self-duality we see that the little q-Bessel functions form a complete orthogonal set with respect to the discrete measure in (4.7).

The same limit transition of the little q-Jacobi polynomials to the little q-Bessel functions as in (4.5) is valid, and in the limit the orthogonality relations for the little q-Jacobi polynomials tend to the orthogonality relations (4.7), see [29] for a rigorous proof.

5. Limit Transitions .

In the previous sections limits within each level of the q-analogue of the Jacobi function schemes have been discussed. Now there are also limits from the Askey-Wilson polynomials to the big q-Jacobi polynomials and from the big q-Jacobi polynomials to the little q-Jacobi polynomials, see

[27], [36] or [15]. In this section we show that these limit transitions also hold for the appropriate analogues of the Jacobi and Bessel function.

5.1. LIMIT FROM THE ASKEY-WILSON CASE TO THE BIG q-CASE

Replace (a, b, c, d, x) by $(a/\epsilon, b\epsilon, c\epsilon, d/\epsilon, -x/\epsilon)$ in the Askey-Wilson function (2.3). Then, using (2.4), the Askey-Wilson function tends to the big q-Jacobi function (3.2) as $\epsilon \downarrow 0$;

$$
\begin{aligned}
\lim_{\epsilon \downarrow 0} \phi_\gamma(-\frac{x}{\epsilon}; \frac{a}{\epsilon}; b\epsilon, c\epsilon, \frac{d}{\epsilon}|q) &= \frac{(-q\tilde{a}x\gamma/d, \tilde{c}/\gamma; q)_\infty}{(\gamma\tilde{c}, ac, qa/d, -qx/d; q)_\infty} \\
&\times {}_3\varphi_2 \left(\begin{matrix} -bx, \tilde{a}\gamma, \tilde{b}\gamma \\ ab, -q\tilde{a}\gamma x/d \end{matrix} ; q, \frac{\tilde{c}}{\gamma} \right) \\
&= \frac{1}{(qa/d; q)_\infty} \phi_\gamma(\frac{x}{a}; \tilde{a}; \tilde{b}, \tilde{c}; q),
\end{aligned} \tag{5.1}
$$

where we have used [7, (III.9)] in the second equality. Keeping t fixed we can formally take the limit transition within the Askey-Wilson function transform pair (2.9) to recover the big q-Jacobi function transform (3.3) with parameters (a, b, c, z) replaced by $(\tilde{a}, \tilde{b}, \tilde{c}, -td/a)$. Taking $\gamma = \tilde{a}q^n$, $n \in \mathbf{Z}_{\geq 0}$, gives back the limit transition from the Askey-Wilson polynomials to the big q-Jacobi polynomials.

In the Askey-Wilson q-Bessel function replace (a, b, x, γ) by $(a/\epsilon, b\epsilon, -x/\epsilon, \gamma\epsilon)$ in (2.12) and take the limit $\epsilon \downarrow 0$, which gives the big q-Bessel function;

$$
\lim_{\epsilon \downarrow 0} J_{\gamma\epsilon}\left(-\frac{x}{\epsilon}; \frac{a}{\epsilon}; b\epsilon|q\right) = J_{q\gamma/b}\left(\frac{x}{a}; ab; q\right). \tag{5.2}
$$

In this limit transition the second order q-difference equation for the Askey-Wilson q-Bessel function goes over into the second order q-difference equation for the big q-Bessel function. And the orthogonality relations (2.14) go over into the orthogonality relations (3.7), since only the discrete mass points of the measure survive in the limit.

5.2. LIMIT FROM THE BIG q-CASE TO THE LITTLE q-CASE

The big q-Jacobi function of (3.2) tends to the little q-Jacobi function in (4.2) by

$$
\lim_{c \downarrow 0} \phi_\gamma\left(\frac{x}{c}; a; b, c; q\right) = \phi_\gamma(x; a; b; q). \tag{5.3}
$$

In this limit transition the big q-Jacobi function transform (3.3) tends to the little q-Jacobi function after taking $z = y/c$ for the extra parameter in the big q-Jacobi function transform (3.3). The second order q-difference equation for the big q-Jacobi functions tends to the second order q-difference equation for the little q-Jacobi functions under (5.3).

In the big q-Bessel function (3.6) we replace x by x/c and γ by $cq\gamma/a$ and take the limit

$$\lim_{c \downarrow 0} J_{cq\gamma/a}\left(\frac{x}{c}; a; q\right) = j_\gamma(x; a; q) \tag{5.4}$$

to find the little q-Bessel function (4.6). In this limit transition the big q-Hankel orthogonality relations (3.7) tend to the little q-Hankel orthogonality relations (4.7), for which we observe that the γ-dependence drops out in the limit. The second order q-difference equation for the big q-Bessel functions tends to the second order q-difference equation for the little q-Bessel functions under (5.4).

6. Duality and Factoring of Limits

As already observed, the Askey-Wilson function transform and the little q-Hankel transform are self-dual. However, the transforms in between are not self-dual, and the dual transforms are described in this section. It turns out that these are related to indeterminate moment problems, see [6], [22]. Moreover, we find that the limit transition of the little q-Jacobi function to the little q-Hankel transform and of the Askey-Wilson function to the Askey-Wilson q-Hankel transform factors through the transforms dual to the big q-Hankel transform and dual to the big q-Jacobi function transform.

6.1. DUALITY BETWEEN THE LITTLE q-JACOBI FUNCTION TRANSFORM AND THE ASKEY-WILSON q-HANKEL TRANSFORM

The little q-Jacobi function of (4.2) and the Askey-Wilson q-Bessel function of (2.12) are related by interchanging the geometric parameter x and the spectral parameter γ;

$$J_\gamma(x; a; b|q) = \phi_x\left(\frac{q\gamma}{ab}; a, b; q\right). \tag{6.1}$$

The orthogonality relations (2.14) and the transform (4.3) can be obtained from each other by using (6.1). So the little q-Jacobi function and the Askey-Wilson q-Bessel function are also eigenfunctions of a second order q-difference equation acting on the spectral parameter.

6.2. THE DUAL TO THE BIG q-HANKEL TRANSFORM

The big q-Bessel functions form an orthogonal basis for the weighted L^2-space on $[-1, \infty(q/a\gamma))_q$ described in (3.7), see [6]. The corresponding dual orthogonality relations are then labeled by the support of this measure, i.e. by $-q^{\mathbf{Z}_{\geq 0}}$ and by $q^{\mathbf{Z}}/a\gamma$. In the first case, the big q-Bessel functions at $x = -q^n$ are related to the q-Laguerre polynomials in γ;

$$J_\gamma(-q^n; a; q) = {}_1\varphi_1\left(\begin{matrix} q^{-n} \\ a \end{matrix}; q, -a\gamma q^n\right), \tag{6.2}$$

and the dual orthogonality relations for $J_\gamma(-q^n; a; q)$ and $J_\gamma(-q^m; a; q)$, $n, m \in \mathbf{Z}_{\geq 0}$, reduce to the orthogonality relations for the q-Laguerre polynomials related to Ramanujan's ${}_1\psi_1$-sum, see Moak [31, Thm. 2], or [7, Exer. 7.43(ii)]. The big q-Bessel functions are eigenfunctions of a second order q-difference operator in the spectral parameter γ;

$$\left(\left(1 + \frac{1}{\gamma}\right)(T_q^\gamma - 1) + \frac{q}{a\gamma}(T_{q^{-1}}^\gamma - 1)\right) J_\gamma(x; a; q) = -(1 + x)J_\gamma(x; a; q), \tag{6.3}$$

where $(T_{q^{\pm 1}}^\gamma f)(\gamma) = f(q^{\pm 1}\gamma)$. Note that (6.3) is nothing but the second order q-difference equation for the ${}_1\varphi_1$-series.

It is known that the q-Laguerre polynomials correspond to an indeterminate moment problem and that this solution to the moment problem is not extremal in the sense of Nevannlina, meaning that the q-Laguerre polynomials are not dense in the corresponding weighted L^2-space. The functions of γ defined by, $p \in \mathbf{Z}$,

$$J_\gamma\left(\frac{q^{p+1}}{a\gamma}; a; q\right) = {}_1\varphi_1\left(\begin{matrix} -a\gamma q^{-1-p} \\ a \end{matrix}; q, q^{p+1}\right) = \frac{(q^{p+1}; q)_\infty}{(a; q)_\infty} {}_1\varphi_1\left(\begin{matrix} -\gamma \\ q^{p+1} \end{matrix}; q, a\right) \tag{6.4}$$

display q-Bessel coefficient behavior. Here we used a limit case of the transformation formula [7, (III.2)]. These q-Bessel coefficients complement the orthogonal set of q-Laguerre polynomials into an orthogonal basis of the corresponding weighted L^2-space. This is a direct consequence of of the orthogonality relations (3.7) and the completeness. See [6] for details, where also the spectral analysis of (6.3) is given.

6.3. THE DUAL TO THE BIG q-JACOBI FUNCTION TRANSFORM

Evaluating the big q-Jacobi function at the point $-q^k$, $k \in \mathbf{Z}_{\geq 0}$, gives a terminating series in which the base is inverted to q^{-1};

$$\phi_\gamma(-q^k; a; b, c; q) = {}_3\varphi_2\left(\begin{matrix} q^k, \gamma/a, 1/a\gamma \\ 1/ab, 1/ac \end{matrix}; q^{-1}, q^{-1}\right). \tag{6.5}$$

The right hand side is a polynomial of degree k in $\frac{1}{2}(\gamma + \gamma^{-1})$, which is a continuous dual q^{-1}-Hahn polynomial with parameters a^{-1}, b^{-1}, c^{-1}, i.e. an Askey-Wilson polynomial with one parameter set to zero, see [7, §7.5], [15]. Hence, the transform (3.3) gives us an orthogonality measure for the continuous dual q^{-1}-Hahn polynomials, together with a complementing set of orthogonal functions in $\frac{1}{2}(\gamma + \gamma^{-1})$, namely $\phi_\gamma(zq^l; a; b, c; q)$, $l \in \mathbf{Z}$. See [22, §9] for more details. In particular we find that the big q-Jacobi functions are eigenfunctions to a second order q-difference operator in the spectral parameter γ, see [12];

$$
\begin{aligned}
(A(\gamma)(T_q^\gamma - 1) \ &+ \ A(\gamma^{-1})(T_{q^{-1}}^\gamma - 1))\phi_\gamma(x; a; b, c; q) \\
&= \ -(1 + x)\phi_\gamma(x; a; b, c; q), \\
A(\gamma) \ &= \ \frac{(1 - 1/\gamma a)(1 - 1/\gamma b)(1 - 1/\gamma c)}{(1 - \gamma^{-2})(1 - 1/\gamma^2 q)},
\end{aligned}
\tag{6.6}
$$

cf. the second order q-difference equation for the continuous dual q^{-1}-Hahn polynomials [2], [15]. See also Rosengren [34, §4.6] for another orthogonality measure for the continuous dual q^{-1}-Hahn polynomials.

6.4. REMAINING LIMIT TRANSITIONS

Using duality and the limit transitions in §5 we obtain the remaining limit transitions in Figure 1.2, namely the limits from the Askey-Wilson functions to the continuous dual q^{-1}-Hahn polynomials and the associated q-Bessel functions, and from this family to the Askey-Wilson q-Bessel functions, and from the little q-Jacobi functions to the q-Laguerre polynomials and the associated q-Bessel functions, and from this family to the little q-Bessel functions. These limits are formal and can be taken in the second order q-difference equation and in the transform pairs.

As an example, we illustrate the limit from the Askey-Wilson functions to the continuous dual q^{-1}-Hahn polynomials and the associated q-Bessel functions. Using the duality (2.5) we have for $x_k = -aq^k$, $k \in \mathbf{Z}_{\geq 0}$, and $\epsilon > 0$ sufficiently small,

$$
\phi_{-x_k/\epsilon}\left(\gamma; \tilde{a}; \tilde{b}, \tilde{c}; \frac{\tilde{d}}{\epsilon^2}|q\right) = \phi_\gamma\left(-\frac{x_k}{\epsilon}; \frac{a}{\epsilon}; b\epsilon, c\epsilon, \frac{d}{\epsilon}|q\right).
\tag{6.7}
$$

By (5.1) the limit $\epsilon \downarrow 0$ gives

$$
\lim_{\epsilon \downarrow 0} \phi_{-x_k/\epsilon}\left(\gamma; \tilde{a}; \tilde{b}, \tilde{c}; \frac{\tilde{d}}{\epsilon^2}|q\right) = \frac{1}{(qa/d; q)_\infty} \phi_\gamma(-q^k; \tilde{a}; \tilde{b}, \tilde{c}; q)
\tag{6.8}
$$

and the right hand side is a continuous dual q^{-1}-Hahn polynomial of degree k up to a constant, see (6.5). Replacing x_k by the discrete weights

$y_l = -tdq^l$, $l \in \mathbf{Z}$, in (6.7), (6.8) we see that $\phi_{-y_l/\epsilon}(\gamma; \tilde{a}; \tilde{b}, \tilde{c}; \frac{\tilde{d}}{\epsilon^2}|q)$ tend to q-Bessel coefficient type functions which complement the continuous dual q^{-1}-Hahn polynomials to an orthogonal basis in the corresponding weighted L^2-space. The non-extremal measure, which is parametrised by the z-parameter in the big q-Jacobi function transform (3.3), corresponds to $z = -td/a$. In this limit the Askey-Wilson function transform formally tends to the orthogonality relations for the continuous dual q^{-1}-Hahn polynomials and the corresponding q-Bessel functions. For this we only need to remark that, by duality, this reduces to limit transition (5.1) of the Askey-Wilson function transform to the big q-Jacobi function transform.

All the other cases can be considered in a similar manner and for completeness we give the underlying limit transitions;

$$\lim_{c \downarrow 0} \phi_\gamma\left(\frac{x}{c}; a; b, c; q\right) = J_{abx/q}(\gamma; a; b; q), \tag{6.9}$$

$$\lim_{\epsilon \downarrow 0} \phi_{\gamma/\epsilon}\left(\epsilon x; \frac{a}{\epsilon}; \epsilon b; q\right) = J_{ax}\left(-\frac{\gamma}{a}; ab; q\right), \tag{6.10}$$

$$\lim_{\epsilon \downarrow 0} J_{\epsilon\gamma}\left(\frac{x}{\epsilon}; a; q\right) = j_x\left(\frac{\gamma a}{q}; a; q\right). \tag{6.11}$$

Let us finally note that this makes the scheme of limit transitions in Figure 1.2 into a commutative diagram.

7. Concluding Remarks

7.1. QUANTUM GROUP THEORETIC INTERPRETATION

There is a group theoretic interpretation for the scheme of Figure 1.1, see Koornwinder [26]. Here the Jacobi polynomials have an interpretation on the compact real Lie group $SU(2)$ as matrix elements of finite-dimensional irreducible unitary representations, whereas the Jacobi functions have an interpretation as matrix elements of infinite-dimensional irreducible unitary representations of the non-compact real Lie group $SU(1,1)$. The real Lie groups $SU(2)$ and $SU(1,1)$ are both real forms of the complex Lie group $SL(2, \mathbf{C})$. Bessel functions occur as matrix elements of infinite-dimensional irreducible unitary representations of the group $E(2)$ of motions of the Euclidean plane, and then the limit transition can be interpreted on the level of Lie groups as a contraction.

There is also a quantum group theoretic interpretation for the Askey-Wilson function scheme in Figure 1.2. Here the Askey-Wilson, big and little q-Jacobi polynomials are interpreted as matrix elements of irreducible unitary representations of the quantum $SU(2)$ group, see e.g. [20], [27], [32], [33] and further references given there. These interpretations also naturally lead to the limit transitions from Askey-Wilson polynomials to big q-Jacobi polynomials, and from big q-Jacobi polynomials to little q-Jacobi polynomials, see Koornwinder [27]. The various q-Hankel transforms and q-Bessel functions have an interpretation on the quantum $E(2)$ group (or better, its twofold covering); see [17], [39] for the little q-Bessel function; see [3], [19] for the big q-Bessel function, and see [16], [18] for the Askey-Wilson q-Bessel function. The limit transitions from the q-analogues of the Jacobi polynomials to the q-analogues of the Bessel functions as described in §§2, 3 and 4 are motivated from a similar contraction from the quantum $SU(2)$ group to the quantum $E(2)$ group. The quantum $SU(2)$ group is a real form of the quantum $SL(2, \mathbf{C})$ group, and one of the other real forms is the quantum $SU(1, 1)$ group. The interpretation of the little q-Jacobi functions on the quantum $SU(1, 1)$ group as matrix elements of irreducible unitary representations is due to Masuda et al. [30], Kakehi [13], Kakehi, Masuda and Ueno [14] and Vaksman and Korogodskiĭ [40]. For the big q-Jacobi and Askey-Wilson function such an interpretation is given in [23]. This interpretation leads in a natural way to the limit transitions of the Askey-Wilson functions to big q-Jacobi functions, and from big q-Jacobi functions to little q-Jacobi functions as described in §5. Moreover, from a contraction procedure from the quantum $SU(1, 1)$ group to the quantum $E(2)$ group we obtain the limit transition of the q-analogues of the Jacobi function to the corresponding q-analogues of the Bessel function, see §§2, 3 and 4. Due to these representation theoretic interpretations of the three types of q-special functions on the quantum $SU(1, 1)$, $E(2)$ and $SU(2)$ groups, we may view them as q-analogues of the Jacobi functions, Bessel functions and Jacobi polynomials, respectively. This explains the naming of the q-special functions in Figure 1.2.

The second order difference equation for the q-special functions follows from the action of the Casimir element for these quantum groups. In §6.1 we have seen that the Askey-Wilson q-Bessel functions and the little q-Jacobi functions both satisfy a second order q-difference equation in the spectral parameter. The spectral parameter corresponds to the representation label of the irreducible representations of the quantum $E(2)$ group and the quantum $SU(1, 1)$ group respectively. The second order q-difference equation in the spectral parameter can then be obtained from the tensor product decomposition of a three-dimensional (non-unitary) representation with an irreducible infinite dimensional representation into three irreducible infinite dimensional representations of the corresponding quantum groups, see e.g. [20, Remark 7.2] for the corresponding statement for the quantum $SU(2)$ group.

7.2. FURTHER EXTENSIONS

We briefly sketch some possible directions related to the Askey-Wilson function scheme.

1. From the previous subsection we see that the scheme depicted in Figure 1.2 is motivated by the simplest quantum groups; namely for $SU(2)$, $SU(1,1)$, and $E(2)$. We would expect that a greater extension of the scheme in Figure 1.2 is possible using more complicated quantum groups, especially higher rank quantum groups, or other interpretations of special functions on quantum groups, such as Clebsch-Gordan, Racah or other type of overlap coefficients. See e.g. Koornwinder [26] for a further extension of the Jacobi function scheme in Figure 1.1. In such an extension of Figure 1.2 the other q-analogues of the Bessel function, as studied by Ismail [10], should also find a place.

2. For the Hankel transform and Jacobi function transform there are systems of orthogonal polynomials that are mapped onto each other, such as the Laguerre polynomials for the Hankel transform or the Jacobi polynomials that are mapped onto the Wilson polynomials by the Jacobi function transform, see [26]. It would be interesting to know what the corresponding results for the q-analogues of these transforms are.

Acknowledgement. The second author is supported by a fellowship from the Royal Netherlands Academy of Arts and Sciences (KNAW). Part of the research was done during the second author's stay at Université Pierre et Marie Curie (Paris VI) and Institut de Recherche Mathématique Avancée (Strasbourg) in France, supported by an NWO-TALENT stipendium of the Netherlands Organization for Scientific Research (NWO) and by the EC TMR network "Algebraic Lie Representations", grant no. ERB FMRX-CT97-0100.

References

1. G.E. Andrews and R. Askey, *Classical orthogonal polynomials*, in: "Polynômes Orthogonaux et Applications", Lecture Notes in Mathematics, Vol. 1171, Springer-Verlag, 1985, pp. 36–62.
2. R. Askey and J. Wilson, *Some basic hypergeometric orthogonal polynomials that generalize Jacobi polynomials*, Mem. Amer. Math. Soc. **54**, no. 319 (1985).
3. F. Bonechi, N. Ciccoli, R. Giachetti, E. Sorace, and M. Tarlini, *Free q-Schrödinger equation from homogeneous spaces of the 2-dim Euclidean quantum group*, Comm. Math. Phys. **175** (1996), 161–176.
4. B. M. Brown, W. D. Evans and M. E. H. Ismail, *The Askey-Wilson polynomials and q-Sturm-Liouville problems*, Math. Proc. Cambridge Philos. Soc. **119** (1996), 1–16.
5. J. Bustoz and S. K. Suslov, *Basic analog of Fourier series on a q-quadratic grid*, Methods Appl. Anal. **5** (1998), 1–38.
6. N. Ciccoli, E. Koelink and T. H. Koornwinder, *q-Laguerre polynomials and big q-Bessel functions and their orthogonality relations*, Methods Appl. Anal. **6** (1999), 109–127.

7. G. Gasper and M. Rahman, *Basic Hypergeometric Series*, Cambridge Univ. Press, 1990.
8. F. A. Grünbaum and L. Haine, *Some functions that generalize the Askey-Wilson polynomials*, Comm. Math. Phys. **184** (1997), 173–202.
9. D. P. Gupta, M. E. H. Ismail and D. R. Masson, *Contiguous relations, basic hypergeometric functions, and orthogonal polynomials III. Associated continuous dual q-Hahn polynomials*, J. Comput. Appl. Math. **68** (1996), 115–149.
10. M. E. H. Ismail, *The zeros of basic Bessel functions, the function $J_{\nu+ax}(x)$, and associated orthogonal polynomials*, J. Math. Anal. Appl. **86** (1982), 1–19.
11. M. E. H. Ismail, D.R. Masson and S.K. Suslov, *The q-Bessel function on a q-quadratic grid*, in: "Algebraic Methods and q-Special Functions", J. F. van Diejen and L. Vinet, eds., CRM Proc. Lect. Notes, Vol. 22, Amer. Math. Soc., 1999, pp. 183–200.
12. M. E. H. Ismail and M. Rahman, *The associated Askey-Wilson polynomials*, Trans. Amer. Math. Soc. **328** (1991), 201–237.
13. T. Kakehi, *Eigenfunction expansion associated with the Casimir operator on the quantum group $SU_q(1,1)$*, Duke Math. J. **80** (1995), 535–573.
14. T. Kakehi, T. Masuda and K. Ueno, *Spectral analysis of a q-difference operator which arises from the quantum $SU(1,1)$ group*, J. Operator Theory **33** (1995), 159–196.
15. R. Koekoek and R. F. Swarttouw, *The Askey-scheme of hypergeometric orthogonal polynomials and its q-analogue*, Delft University of Technology Report no. 98-17, 1998.
16. H. T. Koelink, *On quantum groups and q-special functions*, dissertation, Rijksuniversiteit Leiden, 1991.
17. H. T. Koelink, *The quantum group of plane motions and the Hahn-Exton q-Bessel function*, Duke Math. J. **76** (1994), 483–508.
18. H. T. Koelink, *The quantum group of plane motions and basic Bessel functions*, Indag. Math. (N.S.) **6** (1995), 197–211.
19. H.T. Koelink, *Yet another basic analogue of Graf's addition formula*, J. Comput. Appl. Math. **68** (1996), 209–220.
20. H. T. Koelink, *Askey-Wilson polynomials and the quantum $SU(2)$ group: survey and applications*, Acta Appl. Math. **44** (1996), 295–352.
21. H. T. Koelink, *Some basic Lommel polynomials*, J. Approx. Theory **96** (1999), 345–365.
22. E. Koelink and J. V. Stokman, *The big q-Jacobi function transform*, preprint 40 p., math.CA/9904111, 1999.
23. E. Koelink and J. V. Stokman, with an appendix by M. Rahman, *Fourier transforms on the quantum $SU(1,1)$ group*, preprint 77 p., math.QA/9911163, 1999.
24. E. Koelink and J. V. Stokman, *The Askey-Wilson function transform*, preprint 19 p., math.CA/0004053, 2000.
25. T. H. Koornwinder, *Jacobi functions and analysis on noncompact semisimple Lie groups*, in: "Special Functions: Group Theoretical Aspects and Applications", R. A. Askey, T. H. Koornwinder and W. Schempp, eds., Reidel, 1984, pp. 1–85.
26. T. H. Koornwinder, *Group theoretic interpretations of Askey's scheme of hypergeometric orthogonal polynomials*, in: "Orthogonal Polynomials and their Applications", Lecture Notes in Mathematics, Vol. 1329, Springer, 1988, pp. 46–72.
27. T. H. Koornwinder, *Askey-Wilson polynomials as zonal spherical functions on the $SU(2)$ quantum group*, SIAM J. Math. Anal. **24** (1993), 795–813.
28. T. H. Koornwinder, *Compact quantum groups and q-special functions*, in: "Representations of Lie Groups and Quantum Groups", Pitman Res. Notes Math. Ser., Vol. 311, Longman Sci. Tech., 1994, pp. 46–128.
29. T. H. Koornwinder and R. F. Swarttouw, *On q-analogues of the Fourier and Hankel transforms*, Trans. Amer. Math. Soc. **333** (1992), 445–461.
30. T. Masuda, K. Mimachi, Y. Nakagami, M. Noumi, Y. Saburi and K. Ueno, *Uni-*

tary representations of the quantum group $SU_q(1,1)$: Structure of the dual space of $U_q(sl(2))$, Lett. Math. Phys. **19** (1990), 197–194; II: *Matrix elements of unitary representations and the basic hypergeometric functions*, 195–204.

31. D. S. Moak, *The q-analogue of the Laguerre polynomials*, J. Math. Anal. Appl. **81** (1981), 20–47.

32. M. Noumi, *Quantum groups and q-orthogonal polynomials. Towards a realization of Askey-Wilson polynomials on $SU_q(2)$*, in: "Special Functions", M. Kashiwara and T. Miwa, eds., ICM-90 Satellite Conference Proceedings, Springer-Verlag, 1991, pp. 260–288.

33. M. Noumi and K. Mimachi, *Askey-Wilson polynomials as spherical functions on $SU_q(2)$*, "Quantum Groups," LNM, vol. 1510, Springer Verlag, 1992, 98–103.

34. H. Rosengren, *A new quantum algebraic interpretation of the Askey-Wilson polynomials*, Contemp. Math. **254** (2000), 371–394.

35. S. N. M. Ruijsenaars, *A generalized hypergeometric function satisfying four analytic difference equations of Askey-Wilson type*, Comm. Math. Phys. **206** (1999), 639–690.

36. J. V. Stokman and T. H. Koornwinder, *On some limit cases of Askey-Wilson polynomials*, J. Approx. Theory **95** (1998), 310–330.

37. S. K. Suslov, *Some orthogonal very well poised $_8\phi_7$-functions*, J. Phys. A **30** (1997), 5877–5885.

38. S. K. Suslov, *Some orthogonal very-well-poised $_8\varphi_7$-functions that generalize Askey-Wilson polynomials*, MSRI preprint, 1997; to appear in The Ramanujan Journal.

39. L. L. Vaksman and L. I. Korogodskiĭ, *The algebra of bounded functions on the quantum group of motions of the plane and q-analogues of Bessel functions*, Soviet Math. Dokl. **39** (1989), 173–177.

40. L. L. Vaksman and L. I. Korogodskiĭ, *Spherical functions on the quantum group $SU(1,1)$ and a q-analogue of the Mehler-Fock formula*, Funct. Anal. Appl. **25** (1991), 48–49.

41. G. N. Watson, *A Treatise on the Theory of Bessel Functions*, Cambridge Univ. Press, 1944.

ARITHMETIC OF THE PARTITION FUNCTION

KEN ONO
Department of Mathematics
University of Wisconsin at Madison
Madison, Wisconsin 53706 USA

1. Introduction

Here we describe some recent advances that have been made regarding the arithmetic of the unrestricted partition function $p(n)$. A *partition* of a nonnegative integer n is any nonincreasing sequence of positive integers whose sum is n. As usual, we let $p(n)$ denote the number of partitions of n. For example, it is easy to see that $p(4) = 5$ since the partitions of 4 are:

$$4, \quad 3+1, \quad 2+2, \quad 2+1+1, \quad 1+1+1+1.$$

Partitions have played an important role in many aspects of combinatorics, Lie theory, physics, and representation theory. Here we describe some of the recent discoveries regarding the arithmetic of the partition function, including a relationship between partitions and Tate-Shafarevich groups of modular motives in arithmetic algebraic geometry.

Euler [4] showed that the generating function for $p(n)$ is given by the convenient infinite product

$$\sum_{n=0}^{\infty} p(n)q^n = \prod_{n=1}^{\infty} \frac{1}{1-q^n} = 1 + q + 2q^2 + 3q^3 + 5q^4 + \cdots, \qquad (1)$$

and his Pentagonal Number Theorem asserts that

$$\prod_{n=1}^{\infty}(1-q^n) = \sum_{n=-\infty}^{\infty} (-1)^n q^{(3n^2+n)/2}. \qquad (2)$$

It is easy to see that (1) and (2) together imply, for every positive n, that

$$p(n) = \sum_{k=1}^{\infty} \left((-1)^{k+1} p(n - 3(k^2 + k)/2) + (-1)^{k+1} p(n - (3k^2 - k)/2) \right).$$

$$(3)$$

J. Bustoz et al. (eds.), Special Functions 2000, 243–253.

Although (3) is an efficient recursive device for computing $p(n)$, it is not a formula. Fortunately, an improvement, by Rademacher, of the Hardy-Ramanujan asymptotic formula

$$p(n) \sim \frac{1}{4n\sqrt{3}} \cdot e^{\pi \sqrt{\frac{2n}{3}}}$$

leads to an 'exact formula' for $p(n)$.

Although one would hope that these formulas would be powerful tools for proving theorems about $p(n)$, it has turned out that many of the most basic questions remain open. In this section we review some of the remaining classical problems, and in the next section we describe the recent progress that has been made on these questions. In the last section we describe the relationship between $p(n)$ and certain families of modular L-functions in the context of the Bloch-Kato Conjecture.

1.1. PARITY

One of the simplest questions concerns the parity of $p(n)$. Using (1) and (2), Kolberg [21] proved that there are infinitely many even (resp. odd) values of $p(n)$. However, much more is conjectured to be true. The following widely believed conjecture is one of the most notorious problems in the subject.

Conjecture 1 (Parkin and Shanks [30]). *As $n \to +\infty$ we have*

$$\lim_{X \to +\infty} \frac{\#\{n \le X \; : \; p(n) \equiv 0 \pmod 2\}}{X} = \frac{1}{2}.$$

In 1983 Mirsky [23] proved that

$$\#\{n \le X \; : \; p(n) \text{ is even (resp. odd)}\} \gg \log \log X,$$

and in 1995 Nicolas and Sárközy [25] improved this estimate and proved that there is a constant $c > 0$ for which

$$\#\{n \le X \; : \; p(n) \text{ is even (resp. odd)}\} \gg \log^c X. \tag{4}$$

Subbarao made the following more accessible conjecture [33].

Conjecture 2 (Subbarao). *In an arithmetic progression $r \pmod t$, there are infinitely many integers $M \equiv r \pmod t$ for which $p(M)$ is odd, and there are infinitely many integers $N \equiv r \pmod t$ for which $p(N)$ is even.*

Garvan, Kolberg, Hirschhorn, Stanton and Subbarao ([15], [18], [19], [20], [21]) have verified this conjecture for every arithmetic progression with modulus

$$t \in \{1, 2, 3, 4, 5, 6, 8, 10, 12, 16, 20, 40\}. \tag{5}$$

1.2. ARBITRARY MODULI

One is also naturally interested in the reduction of $p(n)$ modulo arbitrary integers M. In this direction Newman [24] made the following similar conjecture regarding the behavior of $p(n) \pmod{M}$, as one varies n, for arbitrary M.

Conjecture 3 (Newman) *If M is a positive integer, then in every residue class $r \pmod{M}$ there are infinitely many integers N for which*

$$p(N) \equiv r \pmod{M}.$$

Works by Atkin, Kolberg and Newman ([7], [21], [24]) have verified Newman's conjecture for every

$$M \in \{2, 5, 7, 13\}. \tag{6}$$

To clarify the nature of these problems, consider the following conjecture due to Erdös [16]:

Conjecture 4 (Erdös) *If M is prime, then there is at least one nonnegative integer N_M for which*

$$p(N_M) \equiv 0 \pmod{M}.$$

Obviously, Erdös' conjecture is implied by Newman's Conjecture.

Erdös' Conjecture has also been very difficult to handle. Using the asympotic formula for $p(n)$, Schinzel (see [12]) proved that there are infinitely many prime divisors among the values of $p(n)$, and later Schinzel and Wirsing [32] proved that the number of primes $M \leq X$ which divide at least one value of $p(n)$ is $\gg \log \log X$. In view of the Prime Number Theorem (i.e. that the number of primes $p \leq X$ is asymptotically $X/\log X$), these results fall far short of Conjecture 4.

1.3. RAMANUJAN-TYPE CONGRUENCES

The most striking results concerning the congruence properties of $p(n)$ are due to Ramanujan, the legendary Indian mathematician. Ramanujan's findings are particularly shocking in view of the conjectures and problems above. For instance, Ramanujan [31] proved that

$$p(5n + 4) \equiv 0 \pmod{5}, \tag{7}$$

$$p(7n + 5) \equiv 0 \pmod{7}, \tag{8}$$

$$p(11n + 6) \equiv 0 \pmod{11} \tag{9}$$

for every non-negative integer n.

These three congruences are the simplest cases of three infinite families of congruences which were conjectured by Ramanujan. By the works of Atkin, Ramanujan, and Watson (see [6], [10], [34]) it is now known that for every integer k we have

$$p(5^k n + \delta_{5,k}) \equiv 0 \pmod{5^k}, \tag{10}$$

$$p(7^k n + \delta_{7,k}) \equiv 0 \pmod{7^{[k/2]}}, \tag{11}$$

$$p(11^k n + \delta_{11,k}) \equiv 0 \pmod{11^k} \tag{12}$$

for every non-negative integer n where $24\delta_{\ell,k} \equiv 1 \pmod{\ell^k}$.

Such congruences are particularly suprising since a cursory examination of values of $p(n)$ fails to reveal any further congruences. In fact, a study of the values of $p(n)$ suggests that $p(n) \pmod{M}$ is random apart from progressions where there are Ramanujan-type congruences. Therefore, it is natural to ask the following two questions:

Question 1 *How rare are congruences of the form*

$$p(an + b) \equiv 0 \pmod{M}?$$

Question 2 *If $M > 1$ is an integer, is there a progression $b \pmod{a}$ with the property that*

$$p(an + b) \equiv 0 \pmod{M}$$

for every non-negative integer n?

There is some evidence that supports the view that there might be many congruences. In the 1960s, some further congruences for $p(n)$, such as

$$p(11^3 \cdot 13n + 237) \equiv 0 \pmod{13}$$

were discovered by Atkin, O'Brien and Swinnerton-Dyer (see [7], [8], [9]).

Since partitions are combinatorial objects, it is natural to ask whether there are combinatorial explanations for Ramanujan's congruences. In 1944, Dyson [11] conjectured that the 'rank' provides such an explanation for (7) and (8). The rank of a partition is the difference between the number of its parts and its largest part. If $\ell = 5$ or 7 and $0 \leq i \leq \ell - 1$, then Dyson conjectured, for every non-negative integer n, that $p(\ell n + \delta_{\ell,1})/\ell$ equals the number of partitions of $\ell n + \delta_{\ell,1}$ with rank congruent to i (mod ℓ). In 1954, Atkin and Swinnerton-Dyer [7] proved Dyson's conjecture. More recent works by Andrews, Garvan, Kim, and Stanton ([5], [13], [14]) have produced a number of further statistics (a.k.a. 'cranks') which explain some of the other Ramanujan congruences.

Question 3 *Are there systematic statistics which uniformly describe* (10), (11) *and* (12)?

We conclude this sections with the following question.

Question 4 *Do the numbers $p(n)$* (mod M) *play a fundamental role in other areas of mathematics?*

2. Recent Results

In this section we highlight the recent advances that have been made on the problems described in the previous section.

2.1. PARITY

Conjecture 1 remains wide open. However, recent work by Nicolas, Ruzsa, and Sárközy [26] have made a substantial improvement on (4) using a careful analysis of (3).

Theorem 2.1 (Nicolas-Ruzsa-Sárközy)
a) For large X, we have $\#\{n \leq X \ : \ p(n) \text{ is even}\} \gg \sqrt{X}$.
b) If $\epsilon > 0$, then

$$\#\{n \leq X \ : \ p(n) \text{ is odd}\} \gg \sqrt{X} \cdot \exp\left((-\log 2 + \epsilon) \cdot \frac{\log X}{\log \log X}\right).$$

Regarding Subbarao's Conjecture, much more is now known. Using the theory of modular forms and Galois representations, as developed by Deligne and Serre, the author was able to prove the following theorem [27].

Theorem 2.2 (Ono)
a) In any arithmetic progression r (mod t), *there are infinitely many integers n such that $p(n)$ is even.*
b) In any arithmetic progression r (mod t), *there are infinitely many integers n such that $p(n)$ is odd, provided that there is at least one such n. Furthermore, if such an n exists, then the smallest such n is $< 10^{10}t^7$.*

Therefore, the 'even' case of Subbarao's Conjecture is always true, and there is now a simple algorithm which determines the truth of the 'odd part' of the conjecture for any given arithmetic progression. Using this algorithm, the odd case has now been verified for every arithmetic progression with modulus $t \leq 10^5$.

It is natural to ask for quantitative forms of Theorem 2.2. Ahlgren and Serre (see [1], [26]) have obtained such results.

Theorem 2.3 (Ahlgren and Serre)
In any arithmetic progression r *(mod t) we have*

$$\#\{n \leq X \ : \ n \equiv r \pmod{t} \text{ and } p(n) \text{ even}\} \gg_{r,t} \sqrt{X}.$$

Theorem 2.4 (Ahlgren) *If there is an integer* $n \equiv r$ *(mod t) for which* $p(n)$ *is odd, then*

$$\#\{n \leq X \ : \ n \equiv r \pmod{t} \text{ and } p(n) \text{ odd}\} \gg_{r,t} \sqrt{X}/\log X.$$

(Note. Ahlgren [2] has obtained a generalization of Theorem 2.4 which holds for arbitrary prime modulus M, not just $M = 2$.)

2.2. ARBITRARY MODULI AND RAMANUJAN CONGRUENCES

We begin by considering the rarity of Ramanujan-type congruences. The first author was able to quantify [27] their rarity using the fact that the Hecke operators commute with the action of twisting a modular form.

Here we describe a typical result in this direction. Let ℓ be prime, and let S_ℓ denote the set of primes t with the property that in every arithmetic progression

$$24^{-1} \not\equiv r \pmod{t}$$

there are infinitely many integers $n \equiv r$ (mod t) for which

$$p(n) \not\equiv 0 \pmod{\ell}.$$

Theorem 2.5 (Ono) *If ℓ is prime, then the set of primes S_ℓ has density exceeding $1 - 10^{-100}$ within the set of prime numbers.*

Such results clarify the rarity of Ramanujan-type congruences and provides a reasonable resolution to Question 1.

Recently, we have learned a lot about Question 2. Using the theory of modular Galois representations and the commutativity of Hecke algebras across Shimura's correspondence, the author was able to prove the following theorem [29].

Theorem 2.6 (Ono) *Let $M \geq 5$ be prime and let k be any positive integer. A positive proportion of the primes ℓ have the property that*

$$p\left(\frac{m^k \ell^3 n + 1}{24}\right) \equiv 0 \pmod{M}$$

for every non-negative integer n coprime to ℓ.

It is easy to see that this theorem immediately implies that if $M \geq 5$ is prime, then there are infinitely many distinct arithmetic progressions b (mod a) for which

$$p(an + b) \equiv 0 \pmod{M}$$

for every non-negative integer n. Weaver [35] has computed over 70,000 explicit examples of such congruences. For instance, she has found that

$$p(48037937 \cdot N + 1122838) \equiv 0 \pmod{17}, \tag{13}$$
$$p(1977147619 \cdot N + 815655) \equiv 0 \pmod{19}, \tag{14}$$
$$p(14375 \cdot N + 3474) \equiv 0 \pmod{23}, \tag{15}$$
$$p(348104768909 \cdot N + 43819835) \equiv 0 \pmod{29}, \tag{16}$$
$$p(4063467631 \cdot N + 30064597) \equiv 0 \pmod{31}. \tag{17}$$

Furthermore, Theorem 2.6 implies the truth of Erdös' Conjecture (Conjecture 4). Unfortunately, the following problem remains open.

Question 5 *Show that there are infinitely many integers n for which*

$$p(n) \equiv 0 \pmod{3}.$$

The methods which proved Theorem 2.6 are also useful in attacking Newman's Conjecture 3. In particular, the author [29] proved Conjecture 3 for every prime modulus $M < 1000$ with the possible exception of $M = 3$.

Recently, Ahlgren [3] has extended Theorem 2.6 to include composite moduli using an elegant p-adic completion of the forms employed by the author in [29]. The most elegant consequence of Ahlgren's theorem is the following result.

Theorem 2.7 (Ahlgren) *If M is a positive integer coprime to 6, then there are infinitely many distinct arithmetic progressions b (mod a) for which*

$$p(an + b) \equiv 0 \pmod{M}$$

for every non-negative integer n.

Therefore, by Theorems 2.6 and 2.7, it is now apparent that Ramanujan-type congruences are plentiful. However, it is typical that such congruences are monstrous like those appearing in (13-17). We conclude this section by noting that there are some new congruences which are elegant and systematic. In a recent preprint [22], the author and Lovejoy have extended (10) in infinitely many ways. The following result is one special case of these general results.

Theorem 2.8 (Lovejoy and Ono) *If j is a positive integer and*

$$\beta(j) = (3887 \cdot 5^{2j} + 1)/24,$$

then for every non-negative integer N we have

$$p(25^j \cdot 13^3 N + 25^j \cdot 13^2 + \beta(j)) \equiv 0 \pmod{5^{2j+1}}.$$

Although many questions such as Conjecture 1 remain open, it is refreshing to see that progress is being made on some of the questions described in Section 1.

3. Modular L-functions and $p(n)$

In this last section we address Question 4 regarding the role that $p(n)$ plays in other areas of mathematics. To the author's surprise, it turns out that the residues of $p(n) \pmod{M}$ play a fundamental role in the arithmetic of certain modular motives [17]. Here we briefly describe the results proved in [17]. If $13 \leq \ell \leq 31$ is prime, then let $G_\ell(z) = \sum_{n=1}^{\infty} a_\ell(n) q^n$ ($q := e^{2\pi i z}$) be the unique newform in $S_{\ell-3}(\Gamma_0(6))$ whose Fourier expansion begins with the terms

$$G_\ell(z) = q + \left(\frac{2}{\ell}\right) \cdot 2^{(\ell-5)/2} q^2 + \left(\frac{3}{\ell}\right) \cdot 3^{(\ell-5)/2} q^3 + \cdots. \tag{18}$$

Here $\left(\frac{x}{\ell}\right)$ denotes the Legendre symbol modulo ℓ. If D is a fundamental discriminant of a quadratic field, then let χ_D denote the usual Kronecker character for the quadratic field $Q(\sqrt{D})$ and let $L(G_\ell \otimes \chi_D, s)$ denote the L-function of the twisted form given by

$$L(G_\ell \otimes \chi_D, s) = \sum_{n=1}^{\infty} \frac{\chi_D(n) a_\ell(n)}{n^s}. \tag{19}$$

Define integers $1 \leq \delta_\ell \leq \ell - 1$ and $1 \leq r_\ell \leq 23$ by

$$24\delta_\ell \equiv 1 \pmod{\ell}, \tag{20}$$

$$r_\ell \equiv -\ell \pmod{24}. \tag{21}$$

For every non-negative integer n let $D(\ell, n)$ be the integer given by

$$D(\ell, n) := (-1)^{(\ell-3)/2} \cdot (24n + r_\ell). \qquad (22)$$

Using Shimura's correspondence and a deep theorem of Waldspurger, the author and Guo proved the following theorem which relates the values of the partition function to these L-functions.

Theorem 3.1 *If $13 \leq \ell \leq 31$ is prime and $n \geq 0$ is an integer for which $D(\ell, n)$ is square-free, then*

$$\frac{L\left(G_\ell \otimes \chi_{D(\ell,n)}, \frac{\ell-3}{2}\right)(24n + r_\ell)^{(\ell-4)/2}}{L\left(G_\ell \otimes \chi_{D(\ell,0)}, \frac{\ell-3}{2}\right) r_\ell^{(\ell-4)/2}} \equiv \frac{p(\ell n + \delta_\ell)^2}{p(\delta_\ell)^2} \pmod{\ell}.$$

This theorem has deep implications regarding the arithmetic of certain motives. If $13 \leq \ell \leq 31$ is prime, then let $M^{(\ell)}$ be the $(\ell-3)/2$-th Tate twist of the motive associated to $G_\ell(z)$ by the work of Scholl. Similarly, let $M^{(\ell,n)}$ denote the twisted motive obtained by twisting $M^{(\ell)}$ by $\chi_{D(\ell,n)}$. For each $D(\ell, n)$, let $\text{III}(M^{(\ell,n)})$ denote the Tate-Shafarevich group of $M^{(\ell,n)}$. The celebrated conjectures of Bloch and Kato asserts that if $L\left(G_\ell \otimes \chi_{D(\ell,n)}, \frac{\ell-3}{2}\right) \neq 0$, then

$$L\left(G_\ell \otimes \chi_{D(\ell,n)}, \frac{\ell-3}{2}\right) = \Gamma_{\ell,n} \times \#\text{III}(M^{(\ell,n)}), \qquad (23)$$

where $\Gamma_{\ell,n}$ is an explicit non-zero number depending on n and ℓ.

By a careful analysis of the factor $\Gamma_{\ell,n}$ in (23), the author and Guo have been able to prove [17] the following theorem which gives further importance to the partition function.

Theorem 3.2 (Guo and Ono) *Suppose that $13 \leq \ell \leq 31$ is prime and $n \geq 0$ is an integer for which*

(i) $n \not\equiv -[(\ell+1)/12] \pmod{\ell}$,

(ii) $D(\ell, n)$ is square-free,

(iii) $L\left(G_\ell \otimes \chi_{D(\ell,n)}, \frac{\ell-3}{2}\right) \neq 0$.

Assuming the truth of the Bloch-Kato Conjecture, we have that

$$\text{ord}_\ell\left(\frac{\#\text{III}(M^{(\ell,n)})}{\#\text{III}(M^{(\ell,0)})}\right) \iff p(\ell n + \delta_\ell) \equiv 0 \pmod{\ell}.$$

Theorem 3.2 is a new role for the partition function in mathematics. The divisibility of $p(n)$ now dictates, subject to the truth of the Bloch-Kato Conjecture, the presence of elements of order ℓ in certain Galois cohomology groups which are central in arithmetic geometry.

4. Acknowledgements

The author thanks the Alfred P. Sloan Foundation, the David and Lucile Packard Foundation, and the National Science Foundation for their generous support.

References

1. S. Ahlgren, *Distribution of parity of the partition function in arithmetic progressions*, Indagationes Math. **10** (1999), 173–181.
2. S. Ahlgren, *The partition function modulo odd primes ℓ*, Mathematika **46** (1999), 185–192.
3. S. Ahlgren, *Distribution of the partition function modulo composite integers M*, Mathematische Annalen, **318** (2000), 795–803.
4. G. E. Andrews, *The Theory of Partitions*, Cambridge University Press, Cambridge, 1998.
5. G. E. Andrews and F. Garvan, *Dyson's crank of a partition*, Bull. Amer. Math. Soc. **18** (1988), 167–171.
6. A. O. L. Atkin, *Proof of a conjecture of Ramanujan*, Glasgow J. Math. **7** (1967), 14–32.
7. A. O. L. Atkin, *Multiplicative congruence properties and density problems for $p(n)$*, Proc. London Math. Soc. **18** (1968), 563–576.
8. A. O. L. Atkin and J.N. O'Brien, *Some properties of $p(n)$ and $c(n)$ modulo powers of 13*, Trans. Amer. Math. Soc. **126** (1967), 442–459.
9. A. O. L. Atkin and H. P. F. Swinnerton Dyer, *Modular forms on noncongruence subgroups*, Combinatorics **19** (1971), 1–25.
10. B. C. Berndt and K. Ono, *Ramanujan's unpublished manuscript on the partition and tau functions with commentary*, Seminaire Lotharingien de Combinatoire **42** (1999).
11. F. Dyson, *Some guesses in the theory of partitions*, Eureka **8** (1944), 10–15.
12. P. Erdös and A. Ivić, *The distribution of certain arithmetical functions at consecutive integers*, Proc. Budapest Conf. Number Theory, Coll. Math. Soc. J. Bolyai, North-Holland, Amsterdam, **51** (1989), 45–91.
13. F. Garvan, *New combinatorial interpretations of Ramanujan's partition congruences mod 5, 7, and 11*, Trans. Amer. Math. Soc. **305** (1988), 47–77.
14. F. Garvan, D. Kim and D. Stanton, *Cranks and t-cores*, Inventiones Mathematicae **101** (1990), 1–17.
15. F. Garvan and D. Stanton, *Sieved partition functions and q-binomial coefficients*, Mathematics of Computation **55** (1990), 299–311.
16. B. Gordon, Private communication (1999).
17. L. Guo and K. Ono, *The partition function and the arithmetic of certain modular L-functions*, International Mathematics Research Notices **21** (1999), 1179–1197.
18. M. Hirschhorn, *On the residue mod 2 and mod 4 of $p(n)$*, Acta Arith. **38** (1980), 105–109.
19. M. Hirschhorn, *On the parity of $p(n)$*, II, J. Comb. Th. (A) **62** (1993), 128–138.
20. M. Hirschhorn and M. Subbarao, *On the parity of $p(n)$*, Acta Arith. **50** (1980), 355–356.

21. O. Kolberg, *Note on the parity of the partition function*, Math. Scand. **7** (1959), 377–378.
22. J. Lovejoy and K. Ono, *Ramanujan's congruences for the partition function modulo powers of 5*, J. reine. angew. math., accepted for publication.
23. L. Mirsky, *The distribution of values of the partition function in residue classes*, J. Math. Anal. Appl. **93** (1983), 593–598.
24. M. Newman, *Periodicity modulo m and divisibility properties of the partition function*, Trans. Amer. Math. Soc. **97** (1960), 225–236.
25. J.-L. Nicolas and A. Sárközy, *On the parity of partition functions*, Illinois J. Math. **39** (1995), 586–597.
26. J.-L. Nicolas, I.Z. Ruzsa and A. Sárközy (with an appendix by J.-P. Serre), *On the parity of additive representation functions*, J. Number Theory **73** (1998), 292–317.
27. K. Ono, *Parity of the partition function in arithmetic progressions*, J. reine angew. Math. **472** (1996), 1–15.
28. K. Ono, *The partition function in arithmetic progressions*, Mathematische Annalen **312** (1998), 251–260.
29. K. Ono, *Distribution of the partition function modulo m*, Annals of Mathematics **151** (2000), 293–307.
30. T. R. Parkin and D. Shanks, *On the distribution of parity in the partition function*, Mathematics of Computation **21** (1967), 466–480.
31. S. Ramanujan, *Congruences properties of partitions*, Proceedings of the London Mathematical Society **19** (1919), 207–210.
32. A. Schinzel and E. Wirsing, *Multiplicative properties of the partition function*, Proc. Indian Acad. Sci. Math. Sci. **97** (1987), 297–303.
33. M. Subbarao, *Some remarks on the partition function*, Amer. Math. Monthly **73** (1966), 851–854.
34. G. N. Watson, *Ramanujan's vermutung über zerfällungsanzahlen*, J. reine angew. Math. **179** (1938), 97–128.
35. R. L. Weaver, *New congruences for the partition function*, The Ramanujan Journal **5** (2001), 53–64.

THE ASSOCIATED CLASSICAL ORTHOGONAL POLYNOMIALS

MIZAN RAHMAN
School of Mathematics & Statistics
Carleton University, Ottawa K1S 5B6
Canada

Abstract. The associated orthogonal polynomials $\{p_n(x;c)\}$ are defined by the 3-term recurrence relation with coefficients A_n, B_n, C_n for $\{p_n(x)\}$ with $c = 0$, replaced by A_{n+c}, B_{n+c} and C_{n+c}, c being the association parameter. Starting with examples where such polynomials occur in a natural way some of the well-known theories of how to determine their measures of orthogonality are discussed. The highest level of the family of classical orthogonal polynomials, namely, the associated Askey-Wilson polynomials which were studied at length by Ismail and Rahman in 1991 is reviewed with special reference to various connected results that exist in the literature.

Keywords. Classical orthogonal polynomials; Associated orthogonal polynomials; Associated Legendre, Laguerre, Hermite, Jacobi, q-ultraspherical, q-Jacobi and Askey-Wilson polynomials; Continued fractions, Stieltjes transform, Perron-Stieltjes inversion formula; Hypergeometric and basic hypergeometric series.

1. Introduction

For problems in heat conduction or potential theory that have axial symmetry the Laplace equation in cylindrical coordinates is

$$\frac{1}{r}\frac{\partial}{\partial r}\left(r\frac{\partial V}{\partial r}\right) + \frac{\partial^2 V}{\partial z^2} = 0. \tag{1.1}$$

Discrete analogues of Laplace equation were considered by Courant, Friedricks and Lewy [11]. Boyer [7] studied the solutions of the following dis-

J. Bustoz et al. (eds.), Special Functions 2000, 255–279.

cretization of (1.1):

$$\left(m + \frac{1}{2}\right)k \left(\frac{V(mh+h, nk) - V(mh, nk)}{h}\right) \tag{1.2}$$
$$- \left(m - \frac{1}{2}\right)k \left(\frac{V(mh, nk) - V(mh-h, nk)}{h}\right)$$
$$+ mh \left(\frac{V(mh, nk+k) - 2V(mh, nk) + V(mh, nk-k)}{k}\right) = 0,$$

$m, n = 1, 2, \ldots$, (h, k) represents the mesh-size, and the "boundary" condition is

$$\frac{V(h, nk) - V(0, nk)}{h} + \frac{h}{4}\left(\frac{V(0, nk+k) - 2V(0, nk) + V(0, nk-k)}{k^2}\right) = 0. \tag{1.3}$$

It is well-known that by separating the r and z variables in (1.1) one gets the Bessel function for the r-equation. Analogously, Boyer [7] sought solutions of (1.2) in the form

$$V(mh, nk) = \phi_m e^{n\xi}, \tag{1.4}$$

where ξ is related to the separation constant λ by

$$\lambda = \left(\frac{h}{k} \sinh \frac{\xi}{2}\right)^2. \tag{1.5}$$

Assuming that $\phi_0 = 1$ this leads to the 3-term recurrence relation

$$(2m+1)\phi_{m+1} + (2m-1)\phi_{m-1} - 4m(1-2\lambda)\phi_m = 0 \tag{1.6}$$

with

$$\phi_0 = 1, \quad \phi_1 = 1 - \lambda. \tag{1.7}$$

Changing the variable by $\lambda = \frac{1-x}{2}$, (1.6) and (1.7) become

$$\left(n + \frac{1}{2}\right)p_{n+1}(x) = 2nx p_n(x) - \left(n - \frac{1}{2}\right)p_{n-1}(x), \quad n \geq 1, \tag{1.8}$$

$$p_0(x) = 1, \quad p_1(x) = \frac{1+x}{2}, \quad \phi_n(\lambda) = p_n\left(\frac{1-x}{2}\right). \tag{1.9}$$

Boyer calls any solution of (1.8) a discrete Bessel function. However, it is clear that (1.8) is the $\nu = -\frac{1}{2}$ special case of the 3-term recurrence relation of the so-called associated Legendre polynomials:

$$(n+\nu+1)P_{n+1}(\nu, x) = (2n+2\nu+1)xP_n(\nu, x) - (n+\nu)P_{n-1}(\nu, x), \tag{1.10}$$

see also Chihara [9].

In general, the 3-term recurrence relation satisfied by an orthogonal polynomial system $\{p_n(x)\}$ (OPS) is of the form

$$p_{n+1}(x) = (A_n x + B_n)p_n(x) - C_n p_{n-1}(x), \quad n = 0, 1, \ldots, \qquad (1.11)$$

with $p_{-1}(x) = 0$, $p_0(x) = 1$. If $A_n A_{n+1} C_{n+1} > 0$, $n = 0, 1, \ldots$, then there exists a positive measure $d\mu(x)$ with finite moments such that the following orthogonality relation

$$\int_{-\infty}^{\infty} p_n(x)p_m(x)d\mu(x) = \lambda_n \delta_{m,n}, \quad m, n = 0, 1, \ldots, \quad \lambda = A_0^{-1}, \qquad (1.12)$$

holds.

Favard's theorem [9] assures us that the converse is also true. Finding the measure and the explicit solutions of (1.11) are a formidable task except in some special cases including the very important class of what is called the classical orthogonal polynomials. For instance, the Jacobi polynomials, $P_n^{(\alpha,\beta)}(x)$, the ultraspherical polynomial $C_n^\lambda(x)$, the Laguerre polynomial, $L_n^\alpha(x)$, and the Hermite polynomials, $H_n(x)$, satisfy, respectively, the recurrence relations

$$P_{n+1}^{(\alpha,\beta)}(x) = \left\{ \frac{(2n+\alpha+\beta+1)(2n+\alpha+\beta+2)}{2(n+1)(n+\alpha+\beta+1)} x \right. \qquad (1.13)$$

$$\left. + \frac{(\alpha^2 - \beta^2)(2n+\alpha+\beta+1)}{2(n+1)(n+\alpha+\beta+1)(2n+\alpha+\beta)} \right\} P_n^{(\alpha,\beta)}(x)$$

$$- \frac{(n+\alpha)(n+\beta)(2n+\alpha+\beta+2)}{(n+1)(n+\alpha+\beta+1)(2n+\alpha+\beta)} P_{n-1}^{(\alpha,\beta)}(x),$$

$$C_{n+1}^\lambda(x) = \frac{2(n+\lambda)}{n+1} x C_n^\lambda(x) - \frac{n+2\lambda-1}{n+1} C_{n-1}^\lambda(x), \qquad (1.14)$$

$$L_{n+1}^\alpha(x) = \frac{2n+\alpha+1-x}{n+1} L_n^\alpha(x) - \frac{n+\alpha}{n+1} L_{n-1}^\alpha(x), \qquad (1.15)$$

$$H_{n+1}(x) = 2x\, H_n(x) - 2n\, H_{n-1}(x), \qquad (1.16)$$

with

$$P_n^{(\alpha,\beta)}(x) = \frac{(\alpha+1)_n}{n!} \,{}_2F_1(-n, n+\alpha+\beta+1; \alpha+1; (1-x)/2), \qquad (1.17)$$

$$C_n^\lambda(x) = \sum_{k=0}^{n} \frac{(\lambda)_k (\lambda)_{n-k}}{k!\,(n-k)!} \cos(n-2k)\theta \qquad (1.18)$$

$$= \frac{(2\lambda)_n}{n!} \,{}_2F_1(-n, n+2\lambda; \lambda+\tfrac{1}{2}; (1-x)/2), \quad \cos\theta = x;$$

$$L_n^\alpha(x) = \frac{(\alpha+1)_n}{n!} \, {}_1F_1(-n; \alpha+1; x), \qquad (1.19)$$

$$H_n(x) = n! \sum_{k=0}^{\left[\frac{n}{2}\right]} \frac{(-1)^k (2x)^{n-2k}}{k! \, (n-2k)!}. \qquad (1.20)$$

Discrete classical OPS that include the Krawtchouk, Meixner and Charlier polynomials, see [13], also have similar simple expressions in terms of hypergeometric functions. Usually the measures of orthogonality of the classical OPS can be determined in a fairly straightforward manner, often by quite elementary methods. However, things can change quite dramatically even on an apparently harmless and innocent-looking alteration of the 3-term recurrence relation:

$$p_{n+1}(x) = (A_{n+c}x + B_{n+c})p_n(x) - C_{n+c}p_{n-1}(x), \quad n = 0, 1, \ldots, \quad (1.21)$$

with $p_{-1}(x) = 0$, $p_0(x) = 1$, $A_{n+c}A_{n+c+1}C_{n+c+1} > 0$, $n = 0, 1, \ldots$, where c is an additional parameter such as $c = -1/2$ that was used in (1.8). Of course, these changed coefficients make sense if there is a pattern, e.g., if A_n, B_n, C_n are rational functions or quotients of exponential functions. Polynomial solutions of (1.21) satisfying the given initial conditions are called the associated orthogonal polynomials (i.e. associated with the ones with $c = 0$).

Apart from their possible use in physical models, exemplified by the discretized Laplace eqn. (1.2), they occur in a natural way, for positive integer values of c, in the theory of the solutions of the original system satisfying (1.11).

It is well-known that closely related to (1.11) is the continued J–fraction

$$F(x) = \frac{1}{\left| A_0 x + B_0 \right.} - \frac{C_1}{\left| A_1 x + B_1 \right.} - \frac{C_2}{\left| A_2 x + B_2 \right.} - \cdots, \qquad (1.22)$$

see, for example, [9] and [2]. The nth convergent of this continued fraction, say, $P_n^{(1)}(x)/P_n(x)$, is a rational function where the denominator polynomial $P_n(x)$ satisfies (1.11) with the same initial conditions, but the numerator polynomial $P_n^{(1)}(x)$ satisfies the same recurrence relation, but a different initial condition, namely,

$$P_0^{(1)}(x) = 0, \quad P_1^{(1)}(x) = 1. \qquad (1.23)$$

Hence $P_n^{(1)}(x)$ is a polynomial of precise degree $n-1$, and so it satisfies the associated equation

$$r_{n+1}(x) = (A_{n+1}x + B_{n+1})r_n(x) - C_{n+1}r_{n-1}(x), \qquad (1.24)$$

with $r_{-1}(x) = 0$, $r_0(x) = 1$, $r_{n-1}(x) = P_n^{(1)}(x)$. In this manner one can generate a whole sequence of polynomial systems $P_n^{(k)}(x)$, the so-called numerator polynomials, [9], satisfying the recurrence relation

$$r_{n+1}(x) = (A_{n+k}x + B_{n+k})\, r_n(x) - C_{n+k}r_{n-1}(x), \qquad (1.25)$$

with $r_{-1}(x) = 0$, $r_0(x) = 1$, $k = 1, 2, \ldots$.

Whether or not the continued fraction in (1.22) converges it is clear that $P_n^{(1)}(x)/P_n(x)$ provides an $[n-1/n]$ Padé approximant to $F(x)$. However, when the continued fraction does converge, we write

$$F(x) = \lim_{n \to \infty} \frac{P_n^{(1)}(x)}{P_n(x)}. \qquad (1.26)$$

Wimp [5] considered the $[n/n]$ Padé approximant, $A_n(z)/B_n(z)$, for the logarithmic derivative of the confluent hypergeometric function ${}_1F_1(a; b; z)$ and used the polynomials $A_n(z)$ and $B_n(z)$ to derive a discrete orthogonality relation for the Lommel and the corresponding associated polynomials.

The associated orthogonal polynomials also appear in a natural way in the birth-and-death processes, studied by Karlin and McGregor [29], [30], see also Ismail, Letessier and Valent [23]. Let

$$p_{mn}(t) = \text{Prob}\{X(t) = n \,|\, X(0) = m\}, \qquad (1.27)$$

the transition probabilities of a birth-and-death process, be given by

$$p_{mn}(t) = \begin{cases} \lambda_m t + o(t), & n = m+1 \\ \mu_m t + o(t), & n = m-1 \\ 1 - (\lambda_m + \mu_m)t + o(t), & n = m, \end{cases} \qquad (1.28)$$

where λ_n and μ_n are the birth and death rates at the state n, respectively. With $\lambda_n > 0$, $\mu_{n+1} > 0$, $n \geq 0$, $\mu_0 \geq 0$, Karlin and McGregor [29] proved that

$$p_{mn}(t) = \pi_n \int_0^\infty e^{-xt} Q_m(x) Q_n(x) d\mu(x), \qquad (1.29)$$

where $\pi_0 = 1$, $\pi_n = (\lambda_0 \lambda_1 \ldots \lambda_{n-1})/(\mu_1 \mu_2 \ldots \mu_n)$, $n > 0$, and $\{Q_n(x)\}$ are polynomials orthogonal with respect to $d\mu$. If we set $F_n(x) = \pi_n Q_n(x)$, then the orthogonality relation for $\{F_n(x)\}$ is

$$\int_0^\infty F_n(x)F_m(x) d\mu(x) = \pi_n \delta_{mn}, \qquad (1.30)$$

where the polynomials $F_n(x)$ satisfy the 3-term recurrence relation

$$F_{n+1}(x) = \frac{\lambda_n + \mu_n - x}{\mu_{n+1}} F_n(x) - \frac{\lambda_{n-1}}{\mu_{n+1}} F_{n-1}(x), \qquad (1.31)$$

with

$$F_0(x) = 1, \quad F_1(x) = (\lambda_0 + \mu_0 - x)/\mu_1. \qquad (1.32)$$

For linear birth-and-death models Ismail, Letessier and Valent [23] use the two cases: (i) $\lambda_n = n + \alpha + c + 1$, $\mu_n = n + c$, $n \geq 0$, $c > 0$; (ii) $\lambda_n = n + \alpha + c + 1$, $n \geq 0$; $\mu_0 = 0$, $\mu_n = n + c$, $n \geq 1$. In case (i) the recurrence relation (1.31) can be rewritten in the form

$$L_{n+1}^\alpha(x; c) = \frac{2n + 2c + \alpha + 1 - x}{n + c + 1} L_n^\alpha(x; c) - \frac{n + \alpha + c}{n + c + 1} L_{n-1}^\alpha(x; c), \quad (1.33)$$

which is the relation associated to (1.15) with n replaced by $n + c$ in the coefficients. Askey and Wimp [5] found the measure of orthogonality for these associated Laguerre polynomials $L_n^\alpha(x; c)$, as well as their explicit polynomial form:

$$L_n^\alpha(x; c) = \frac{(\alpha + 1)_n}{n!} \sum_{k=0}^{n} \frac{(-n)_k x^k}{(c + 1)_k (\alpha + 1)_k} \, {}_3F_2 \left[\begin{matrix} k - n, \, k + 1 - \alpha, \, c \\ k + c + 1, \, -\alpha - n \end{matrix} ; 1 \right].$$
$$(1.34)$$

By a different method the measures of orthogonality were found in [23], in addition to the explicit form of the second system of associated Laguerre polynomials that corresponds to case (ii):

$$\mathcal{L}_n^\alpha(x; c) = \frac{(\alpha + 1)_n}{n!} \sum_{k=0}^{n} \frac{(-n)_k x^k}{(c + 1)_k (\alpha + 1)_k} \, {}_3F_2 \left[\begin{matrix} k - n, \, k - \alpha, \, c \\ k + c + 1, \, -\alpha - n \end{matrix} ; 1 \right].$$
$$(1.35)$$

In a later paper the same authors [24] considered the polynomials associated with symmetric birth and death processes with quadratic rates: $\lambda_n = (n + a)(n + b)$, $n \geq 0$, $\mu_n = (n + \alpha)(n + \beta)$, $n > 0$, $\mu_0 = 0$ or $\mu_0 = \alpha\beta$. They obtained the absolutely continuous measure for the resulting associated continuous dual Hahn orthogonal polynomials as well as the following explicit formula:

$$P_n(x; \, a, b, \alpha, \beta, \eta) \qquad (1.36)$$
$$= \frac{(a)_n}{n!} \sum_{k=0}^{n} \frac{(-n)_k (a + \gamma + i\sqrt{x - \gamma^2})_k (a + \gamma - i\sqrt{x - \gamma^2})_k}{(a)_k (\alpha + 1)_k (\beta + 1)_k}$$
$$\times \sum_{j=0}^{k} \frac{(\alpha)_j (\beta)_j (a + \eta - 1)_j}{j! (a + \gamma + i\sqrt{x - \gamma^2})_j (a + \gamma - i\sqrt{x - \gamma^2})_j},$$

where $\gamma = (1 + \alpha + \beta - a - b)/2$ and $\eta = 0$ when $\mu_0 = 0$ and $\eta = 1$ when $\mu_0 = \alpha\beta \neq 0$. Note that $P_n(x; a, b, \alpha, \beta, \eta)$ reduces to the continuous dual Hahn polynomials, see Andrews and Askey [1] and Askey and Wilson [3] when $\alpha = 0$ or $\beta = 0$.

It may be hazardous to speculate who was the first one to study the associated orthogonal polynomials or who used the adjective "associated" to describe them, but Humbert's 1918 paper [21] is the earliest work that I could find in the literature. Hahn [20] seems to be the first to study the associated Laguerre polynomials. Although he did not find their orthogonality relation he found the fourth order differential equation that they satisfy by first expressing them as sums of products of confluent hypergeometric functions. Of all the earlier works on associated orthogonal polynomials, however, Pollaczek's work [39]–[42] was probably the most important because his results constituted a significant departure from the classical orthogonal polynomials (belonging to the Szegő class) on one hand, and a generalization to the associated ones on the other. In its full generality Pollaczek [42] gave the following 3-term recurrence relation

$$(n+c+1)P_{n+1}^\lambda(x) = 2[(n+c+\lambda+a)x + b]P_n^\lambda(x) - (n+c+2\lambda-1)P_{n-1}^\lambda(x),$$
(1.37)

with $P_{-1}^\lambda(x) = 0$, $P_0^\lambda(x) = 1$, $|x| \leq 1$, and either $a > |b|$, $2\lambda + c > 0$, $c \geq 0$ or $a > |b|$, $2\lambda + c \geq 1$, $c > -1$. Pollaczek proved the orthogonality of $\{P_n^\lambda(x)\}$ on $[-1, 1]$ with respect to the weight function

$$w^{(\lambda)}(x; a, b, c) = \frac{(2\sin\theta)^{2\lambda-1}e^{(2\theta-\pi)t}}{2\pi\Gamma(2\lambda+c)\Gamma(c+1)}|\Gamma(\lambda+c+it)|^2|$$ (1.38)
$$\times\, {}_2F_1(1 - \lambda + it, c; c + \lambda + it; e^{2i\theta})|^{-2},$$

where $t = (a\cos\theta + b)/\sin\theta$, $x = \cos\theta$, and derived the following expression:

$$P_n^\lambda(x) = P_n^\lambda(x; a, b, c) \tag{1.39}$$
$$= \frac{A_{-1}B_n - A_n B_{-1}}{A_{-1}B_0 - B_{-1}A_0},$$

with

$$A_n = \frac{\Gamma(2\lambda+c+n)e^{i(c+n)\theta}}{\Gamma(c+n+1)\Gamma(2\lambda)}\, {}_2F_1(-c-n, \lambda+it; 2\lambda; 1-e^{-2i\theta}),$$

$$B_n = \frac{\Gamma(1-\lambda+it)\Gamma(1-\lambda-it)}{\Gamma(2-2\lambda)}(2\sin\theta)^{1-2\lambda}e^{i(2\lambda+c+n-1)\theta} \tag{1.40}$$
$$\times\, {}_2F_1(1-2\lambda-c-n, 1-\lambda+it; 2-2\lambda; 1-e^{-2i\theta}),$$

$n = -1, 0, 1, \ldots$, provided 2λ is not an integer, see also [13].

Taking the limit $a \to 0$, replacing θ and t by ϕ and x, respectively, one obtains the limiting relation:

$$(n + c + 1)P_{n+1}^\lambda(x) \tag{1.41}$$
$$= 2[(n + c + \lambda)\cos\phi + x\sin\phi]P_n^\lambda(x) - (n + c + 2\lambda - 1)P_{n-1}^\lambda(x),$$

with $P_{-1}^\lambda(x) = 0$, $P_0^\lambda(x) = 1$, $0 < \phi < \pi$, $2\lambda + c > 0$, $c \geq 0$, or $0 < \phi < \pi$, $2\lambda + c \geq 1$, $c > -1$. The resulting polynomials, called by Askey and Wimp [5] the associated Meixner-Pollaczek polynomials, (Meixner [35] had also found them and their orthogonality relation when $c = 0$) are orthogonal on the infinite interval $-\infty < x < \infty$ with respect to the same weight function (1.38) with appropriate replacement of the variables. If we replace λ and x in (1.41) by $\frac{1}{2}(\alpha + 1)$ and $-x/2\sin\phi$, respectively, and take the limit $\phi \to 0$, then (1.41) becomes the 3-term recurrence relation for the associated Laguerre polynomials, $L_n^\alpha(x; c)$, as was observed by Pollaczek in [41]. Askey and Wimp [5] took advantage of this property to obtain the weight function for the orthogonality of $\{L_n^\alpha(x; c)\}$ on $0 < x < \infty$, namely,

$$W^\alpha(x; c) = \frac{x^\alpha e^{-x}}{|\Psi(c, 1 - \alpha; xe^{-\pi i})|^2}, \tag{1.42}$$

where

$$\Psi(a, b; x) = \frac{1}{\Gamma(a)} \int_0^\infty e^{-xt}t^{a-1}(1 + t)^{b-a-1}dt, \quad \text{Re } a > 0 \tag{1.43}$$

is the second solution of the confluent hypergeometric equation. Askey and Wimp [5] also found the weight function for the associated Hermite polynomials $H_n(x; c)$:

$$w(x; c) = |D_{-c}(ix\sqrt{2})|^{-2}, \tag{1.44}$$

where $\lim_{b\to\infty} b^{a/2}\Psi(a, b + 1; x\sqrt{b} + b) = e^{x^2/4}D_{-a}(x)$, and the expression

$$H_n(x; c) = \sum_{k=1}^{\lfloor \frac{n}{2} \rfloor} \frac{(-2)^k(c)_k(n - k)!}{k!(n - 2k)!}H_{n-2k}(x). \tag{1.45}$$

These results were derived by a different method in [23].

In section 2 we shall give a brief summary of the methods of determining the spectral measures $d\mu(x)$ and in section 3 we discuss in some details the case of the two families of Askey-Wilson polynomials. Finally we consider some special and limiting cases in section 4.

2. Determination of the Spectral Measure $d\mu(x)$

As we saw in the previous section the existence of a positive measure $d\mu(x)$ for an OPS $\{p_n(x)\}$ defined by the 3-term recurrence relation (1.11) or

(1.21) along with the stated conditions is guaranteed by Favard's theorem, but nothing else can be said about this measure without a good deal of additional work. First, one has to determine the bounds of the support of $d\mu$ for which one can use a characterization theorem of Wall and Wetzel [47] in terms of the so-called chain sequences to find the true interval of orthogonality, or an alternate approach suggested by Chihara [9], [10]. See [22] for a brief summary of Chihara's ideas. If the support turns out to be $(0, \infty)$ or $(-\infty, \infty)$ then the unboundedness of it poses the additional problem of whether or not one has a determined or an undetermined moment problem. For a comprehensive analysis of this problem see [45]. However, for a finite interval $[a, b]$ the measure is necessarily unique, see [46], and can be obtained, in principle, in four different ways.

2.1. METHOD OF MOMENTS

The moments of the measure $d\mu(x)$ are defined by $\mu_n = \int_a^b x^n d\mu(x)$. Assuming that the measure is normalized to unity so that $\mu_0 = 1$, the first few moments can be computed by using the recurrence relation (1.11) and the orthogonality property (1.12), whether or not they are for the associated polynomials. A pattern may or may not emerge from these calculations. If it does then a guess can be made about the general formula and, hopefully, proved by induction. Then one can compute the function

$$F(z) = \sum_{n=0}^{\infty} \mu_n z^{-n-1}, \quad |z| > r, \tag{2.1}$$

which is really an asymptotic expansion of $F(z)$ as $z \to \infty$, that converges in the exterior of any circle $|z| = r$ containing the interval of orthogonality in its interior. For $z \notin [a, b]$, it is well-known that

$$F(z) = \int_a^b \frac{d\mu(t)}{z - t} \tag{2.2}$$

is the Stieltjes transform of $d\mu(t)$. Then the familiar Perron-Stieltjes inversion formula tells us that

$$\mu(t_2) - \mu(t_1) = \frac{1}{2\pi i} \lim_{\epsilon \to 0^+} \int_{t_1}^{t_2} \{F(t - i\epsilon) - F(t + i\epsilon)\} dt \tag{2.3}$$

if and only if (2.2) holds, see [22], [45].

2.2. THE GENERATING FUNCTION METHOD

When the interval of orthogonality is bounded (or, more generally, when the moment problem is determined) one can show that

$$F(z) = \lim_{n \to \infty} \frac{P_n^{(1)}(z)}{P_n(z)} = \int_{-\infty}^{\infty} \frac{d\mu(t)}{z - t}, \tag{2.4}$$

which may be compared with (1.26), where z is in the complex plane cut along the support of $d\mu$, and the convergence is uniform on compact subsets of this cut plane, see [45].

Note that it is not necessary to compute the denominator and numerator polynomials explicitly, only their asymptotic behaviour as $n \to \infty$. It was pointed out in [2] that the asymptotic method of Darboux [46] can be applied to the generating functions of both $\{P_n(x)\}$ and $\{P_n^{(1)}(x)\}$ to determine their asymptotic behaviour. Darboux's theorem, as stated in [37], is as follows.

Let

$$f(z) = \sum_{n=-\infty}^{\infty} a_n z^n \tag{2.5}$$

be the Laurent expansion of an analytic function $f(z)$ in an annulus $0 < |z| < r < \infty$. By Cauchy's formula

$$a_n = \frac{1}{2\pi i} \int_C \frac{f(z)}{z^{n+1}} dz, \tag{2.6}$$

where C is a simple closed contour in the annulus around $r = 0$. Let r be the distance from the origin of the nearest singularity of $f(z)$, and suppose we can find a "comparison" function $g(z)$ such that

(i) $g(z)$ is holomorphic in $0 < |z| < r$;

(ii) $f(z) - g(z)$ is continuous in $0 < |z| < r$ (this can be weakened, see [37]);

(iii) the coefficients b_n in the Laurent expansion

$$g(z) = \sum_{n=-\infty}^{\infty} b_n z^n \tag{2.7}$$

have known asymptotic behaviour. Then

$$a_n = b_n + o(r^{-n}), \quad n \to \infty. \tag{2.8}$$

Since $G^{(k)}(z,t) := \sum_{n=0}^{\infty} P_n^{(k)}(z) t^n$, the generating function of the orthogonal polynomials $P_n^{(k)}(z)$, can be computed from their 3-term recurrence relations (at least in principle) without the explicit knowledge of the polynomials themselves, one can usually locate its singularities in the complex plane which enables one to construct simpler "comparison" functions with known asymptotic behaviour. Therefore, this method is inherently simpler to use than the other methods, and have been used in a number of works on associated OPS. See, for example, [2], [23], [24], [22], [8].

2.3. SPECIAL FUNCTION METHODS

These methods are usually quite computational and rely more on inspired guesses than on the elaborate machinery of the theory of general OPS. The first step is to find an explicit form of the polynomial solution of the recurrence relation. In the case of classical orthogonal polynomials this solution also satisfies a second-order differential or difference equation which can be transformed, if necessary, into a self-adjoint form. The weight function usually reveals itself from that form. Consider, for example, the Askey-Wilson polynomials:

$$
\begin{aligned}
& p_n(x;\, a, b, c, d | q) \qquad\qquad\qquad\qquad\qquad\qquad\qquad (2.9) \\
& = {}_4\phi_3\left[\begin{matrix} q^{-n},\ abcdq^{n-1},\ ae^{i\theta},\ ae^{-i\theta} \\ ab,\ ac,\ ad \end{matrix}\; ; q, q\right], \quad n = 0, 1, \ldots,
\end{aligned}
$$

where the ${}_4\phi_3$ series is a balanced and terminating basic hypergeometric series defined in [14], see also [3]. Askey and Wilson [4] had shown in an earlier paper that if one of a, b, c, d is of the form q^{-n}, $n = 0, 1, \ldots$, then

$$
\phi(a, b) := {}_4\phi_3\left[\begin{matrix} a,\ b,\ c,\ d \\ e,\ f,\ g \end{matrix}\; ; q, q\right], \quad efg = abcdq, \qquad (2.10)
$$

satisfies the contiguous relation

$$
A\phi(aq^{-1}, bq) + B\phi(a, b) + C\phi(aq, bq^{-1}) = 0, \qquad (2.11)
$$

where

$$
\begin{aligned}
A &= b(1 - b)(aq - b)(a - e)(a - f)(a - g), \\
C &= -a(1 - a)(bq - a)(b - e)(b - f)(b - g), \qquad (2.12) \\
B &= C - A + ab(a - bq)(a - b)(aq - b)(1 - c)(1 - d).
\end{aligned}
$$

So, if we replace a, b by q^{-n} and $abcdq^{n-1}$, respectively, and then c, d, e, f, g by $ae^{i\theta}$, $ae^{-i\theta}$, ab, ac, ad, in that order, then we obtain the 3-term recurrence relation for $p_n(x;\, a, b, c, d \,|\, q)$:

$$
2x\, p_n(x) = A_n\, p_{n+1}(x) + B_n\, p_n(x) + C_n\, p_{n-1}(x), \qquad (2.13)
$$

with

$$A_n = \frac{a^{-1}(1 - abq^n)(1 - acq^n)(1 - adq^n)(1 - abcdq^{n-1})}{(1 - abcdq^{2n-1})(1 - abcdq^{2n})},$$

$$C_n = \frac{a(1 - bcq^{n-1})(1 - bdq^{n-1})(1 - cdq^{n-1})(1 - q^n)}{(1 - abcdq^{2n-2})(1 - abcdq^{2n-1})}, \qquad (2.14)$$

$$B_n = a + a^{-1} - A_n - C_n.$$

However, if we had replaced a, b, c, d, e, f, g by $ae^{i\theta}$, $ae^{-i\theta}$, q^{-n}, $abcdq^{n-1}$, ab, ac, ad, respectively, with $x = \cos\theta$, then we would obtain the following relation from (2.11) and (2.12):

$$q^{-n}(1 - q^n)(1 - abcdq^{n-1})r_n(e^{i\theta}) \qquad (2.15)$$
$$= \lambda(-\theta)\{r_n(q^{-1}e^{i\theta}) - r_n(e^{i\theta})\} + \lambda(\theta)\{r_n(qe^{i\theta}) - r_n(e^{i\theta})\},$$

where $r_n(e^{i\theta}) := p_n(\cos\theta; a, b, c, d|q)$, and

$$\lambda(\theta) = \frac{(1 - ae^{i\theta})(1 - be^{i\theta})(1 - ce^{i\theta})(1 - de^{i\theta})}{(1 - e^{2i\theta})(1 - qe^{2i\theta})}. \qquad (2.16)$$

Askey and Wilson [3] manipulated (2.15) to obtain the second-order divided-difference equation in a "self-adjoint" form:

$$D_q\Big[w(x; aq^{1/2}, bq^{1/2}, cq^{1/2}, dq^{1/2})D_q p_n(x)\Big] \qquad (2.17)$$
$$+\lambda_n w(x; a, b, c, d)p_n(x) = 0$$

where

$$w(x; a, b, c, d)dx = \frac{(e^{2i\theta}, e^{-2i\theta}; q)_\infty}{|(ae^{i\theta}, be^{i\theta}, ce^{i\theta}, de^{i\theta}; q)_\infty|^2} \frac{dx}{\sqrt{1 - x^2}} \qquad (2.18)$$

is the absolutely continuous measure on $[-1, 1]$.

Unfortunately, however, this simple device does not seem to work for nonclassical OPS, and certainly not for the associated ones, classical or otherwise. Magnus [31] showed that the OPS belonging to what he calls the Laguerre-Hahn class, defined in terms of a Riccati-type difference equation, that includes the classical OPS along with the associated ones, generally satisfy linear difference or differential equations of fourth order. Magnus did not attempt to obtain the measure of orthogonality for the associated Askey-Wilson polynomials, $p_n^\alpha(x; a, b, c, d|q)$, which satisfy the 3-term recurrence relation (2.13) with A_n, B_n, C_n replaced by $A_{n+\alpha}$, $B_{n+\alpha}$, $C_{n+\alpha}$, but gave a scheme of how to derive the corresponding fourth order difference equation. Wimp [50], however, was able to find an explicit fourth order differential equation for the associated Jacobi polynomials which are the $q \to 1$ limit cases of $p_n^{\alpha}(x; q^{1/2}, q^{\alpha+1/2}, -q^{\beta+1/2}, -q^{1/2}|q)$. In an as yet unpublished work Ismail and Wimp were able to give a general approach of finding a fourth order differential equation for the associated polynomials, $p_n^{(c)}(x)$ ($q = 1$ case).

Considering the complexity of the weight function $w^{\lambda}(x; a, b, c)$ in (1.37) for the Pollaczek polynomials it would appear that guessing the measure of orthogonality for the associated OPS or to derive it from special function formulas would be a daunting task indeed. Barrucand and Dickinson [6] discovered the weight function for the associated Legendre polynomials satisfying eqn. (1.10) by a clever manipulation of the Legendre functions of both kinds, without realizing that their results follow as special cases of Pollaczek's formulas given in (1.36)–(1.39), see also [12], [39].

One may describe the special function methods and even the moment methods pretty adhoc and simple-minded, but they can be quite effective in determining the measure of orthogonality, sometimes even when it is not unique, for example, the $V_n(x; a)$ polynomials of Al-Salam and Carlitz, see [9].

2.4. METHOD OF MINIMAL SOLUTIONS

This method, extensively used by Masson [32]–[34] and his collaborators [18]–[19], [22], [27], relies on the following ideas. A solution $X_n^{(s)}$ of the 3-term recurrence relation

$$X_{n+1} = A_n X_n + B_n X_{n-1}, \quad n \geq 0, \tag{2.19}$$

is a minimal (or subdominant) solution if for any other linearly independent solution $X_n^{(d)}$ (dominant) one has the property

$$\lim_{n \to \infty} \frac{X_n^{(s)}}{X_n^{(d)}} = 0. \tag{2.20}$$

Pincherle's theorem (see [32] and the references therein) states that (2.20) is a necessary and sufficient condition for the convergence of the continued fraction that corresponds to (2.19), namely,

$$\frac{B_0}{A_0+} \frac{B_1}{A_1+} \frac{B_2}{A_2+} \cdots . \tag{2.21}$$

In case of convergence the limit is simply given by $-\dfrac{X_0^{(s)}}{X_{-1}^{(s)}}$, assuming, of course, that $B_n \neq 0$, $n \geq 0$. When a minimal solution exists it is unique up to a multiple independent of n. However, it was noted in [34] by means of an example that the minimal solution may change in different parts of the complex plane or from $n \geq 0$ to $n \leq 0$. An example of how to construct a minimal solution, when it exists, by first deriving a set of solutions of (2.19) (any two of them being linearly independent) is given in [18].

There is yet another approach to the associated polynomials, due to Grünbaum and Haine [15], [16], who consider a discrete-continuous version of the bispectral problem, i.e., they ask for a description of all families of functions $f_n(z)$, $n \in \mathbf{Z}$, $z \in \mathbf{C}$, such that

$$(Lf)_n(z) = z f_n(z), \quad B f_n(z) = \lambda_n f_n(z), \tag{2.22}$$

with L some tridiagonal bi-infinite matrix and B some second order differential operator acting on z. This can be seen as a version of the classical problem of Bochner when the nonnegative integers have been replaced by all integers, and the usual boundary condition $f_{-1}(z) = 0$ has been removed. The solutions of this problem are then exactly given by the coefficients of the recurrence relations satisfied by the associated classical orthogonal polynomials, extended now over all the integers.

A similar approach with B a second-order q–difference operator leads to the associated Askey-Wilson polynomials, [17]. In [17] one also finds an explicit expression for the matrix valued spectral measure corresponding to the associated doubly infinite Jacobi matrix. This is first done in terms of a standard basis. It is well known that the polynomials in this basis do not satisfy nice second order differential equations. It is then observed that by using a "bispectral basis" i.e. one made up of functions (not polynomials) that satisfy not only the three term recursion relation but also a simple second order differential equation, the spectral measure takes a much simpler form.

3. Associated Askey-Wilson Polynomials

The associated polynomials that generalize the Askey-Wilson polynomials given in (2.9) are solutions of the 3-term recurrence relation

$$(z + z^{-1} - a - a^{-1} + A_{n+\alpha} + C_{n+\alpha})p_n^\alpha(x) \tag{3.1}$$
$$= A_{n+\alpha}p_{n+1}^\alpha(x) + C_{n+\alpha}p_{n-1}^\alpha(x), \qquad n = 0, 1, 2, \ldots,$$

with $p_{-1}^\alpha(x) = 0$, $p_0^\alpha(x) = 1$; $A_{n+\alpha}$, $C_{n+\alpha}$ being the same as in (2.14) with n replaced by $n + \alpha$. It was shown in [28] that the two linearly independent solutions of (3.1) are

$$r_{n+\alpha} = \frac{(abq^{n+\alpha}, acq^{n+\alpha}, adq^{n+\alpha}, bcdq^{n+\alpha}/z; q)_\infty}{(bcq^{n+\alpha}, bdq^{n+\alpha}, cdq^{n+\alpha}, azq^{n+\alpha}; q)_\infty} \left(\frac{a}{z}\right)^{n+\alpha}$$

$$\times \; {}_8W_7(bcd/qz; \; b/z, \; c/z, \; d/z, \; abcdq^{n+\alpha-1}, \qquad (3.2)$$
$$q^{-\alpha-n}; \; q, qz/a)$$

and

$$s_{\alpha+n} \qquad (3.3)$$
$$= \frac{(abcdq^{2n+2\alpha}, bzq^{\alpha+n+1}, czq^{n+\alpha+1}, dzq^{n+\alpha+1}, bcdzq^{n+\alpha}; \; q)_\infty}{(bcq^{n+\alpha}, bdq^{n+\alpha}, cdq^{n+\alpha}, q^{n+\alpha+1}, bcdzq^{2n+2\alpha+1}; \; q)_\infty} \cdot$$
$$\times (az)^{n+\alpha} \; {}_8W_7(bcdzq^{2n+2\alpha}; \; bcq^{n+\alpha}, bdq^{n+\alpha}, cdq^{n+\alpha}, q^{n+\alpha+1}, zq/a;$$
$$q, az),$$

where

$$\begin{aligned} & {}_{2r+1}W_{2r}(a; \; a_1, \; a_2, \; \ldots, \; a_{2r-2}; \; q, z) \\ &:= {}_{2r+1}\phi_{2r}\left[\begin{matrix} a, \; qa^{1/2}, \; -qa^{1/2}, \; a_1, \; \ldots, \; a_{2r-2} \\ a^{1/2}, \; -a^{1/2}, \; qa/a_1, \ldots, qa/a_{2r-2} \end{matrix} \; ; \; q, z \right], \qquad (3.4) \end{aligned}$$

is a very-well-poised basic hypergeometric series, see [14] for definitions, properties and relevant formulas. Taking a linear combination of these solutions along with the given initial conditions ($p_{-1}^\alpha = 0$, $p_0^\alpha = 1$) leads to the following expression:

$$p_n^\alpha(x) \qquad (3.5)$$
$$= \frac{(bc, bd, cd, bcq^{\alpha-1}, bdq^{\alpha-1}, cdq^{\alpha-1}, q^\alpha, az; \; q)_\infty \; za^{1-2\alpha}}{(1 - abcdq^{2\alpha-2})(abq^\alpha, acq^\alpha, adq^\alpha, abcdq^{\alpha-1}, bz, cz, dz, bcd/z; \; q)_\infty}$$
$$\times \; \{s_{\alpha-1}(z)r_{n+\alpha}(z) - r_{\alpha-1}(z)s_{n+\alpha}(z)\},$$

where $2x = z + z^{-1}$, from which one gets the asymptotic formula:

$$\lim_{n\to\infty} (z/a)^n p_n^\alpha(x) \qquad (3.6)$$
$$= \frac{(az)^{1-\alpha}(az, bcq^{\alpha-1}, bdq^{\alpha-1}, cdq^{\alpha-1}, q^\alpha; \; q)_\infty s_{\alpha-1}(z)}{(1 - abcdq^{2\alpha-2})(abq^\alpha, acq^\alpha, adq^\alpha, abcdq^{\alpha-1}, z^2; \; q)_\infty}$$

for $|z| < 1$, i.e. $x \in \mathbf{C}/[-1,1]$, and $|a| < 1$.

A second OPS satisfying (3.1) but a different initial condition

$$q_0^\alpha(x) = 1, \quad q_1^\alpha(x) = 1 + A_\alpha^{-1}(z + z^{-1} - a - a^{-1}), \qquad (3.7)$$

which corresponds to the case of zero initial death rate in the birth-and-death model discussed in section 1, was found to be, see [28]:

$$q_n^\alpha(x) \qquad (3.8)$$
$$= \frac{(bc, bd, cd, bcq^{\alpha-1}, bdq^{\alpha-1}, cdq^{\alpha-1}, q^\alpha, az; \; q)_\infty za^{1-2\alpha}}{(1 - abcdq^{2\alpha-2})(abq^\alpha, acq^\alpha, adq^\alpha, abcdq^{\alpha-1}, bz, cz, dz, bcd/z; \; q)_\infty}$$
$$\times \; \{(s_{\alpha-1}(z) - s_\alpha(z))r_{n+\alpha}(z) - (r_{\alpha-1}(z) - r_\alpha(z))s_{n+\alpha}(z)\},$$

and hence, for $|z| < 1$,

$$\lim_{n\to\infty} (z/a)^n q_n^\alpha(x) \tag{3.9}$$

$$= \frac{(az)^{1-\alpha}(bcq^{\alpha-1}, bdq^{\alpha-1}, cdq^{\alpha-1}, q^\alpha, az; q)_\infty (az)^{1-\alpha}}{(1 - abcdq^{2\alpha-2})(abq^\alpha, acq^\alpha, adq^\alpha, abcdq^{\alpha-1}, z^2; q)_\infty}$$
$$\times (s_{\alpha-1}(z) - s_\alpha(z)).$$

Let us denote the measures of orthogonality for $\{p_n^\alpha(x)\}$ and $\{q_n^\alpha(x)\}$ by $d\mu^{(1)}(x; \alpha)$ and $d\mu^{(2)}(x; \alpha)$, respectively, with

$$\int_{-\infty}^{\infty} d\mu^{(i)}(x; \alpha) = 1, \tag{3.10}$$

$$\int_{-\infty}^{\infty} p_n(x)p_m(x)d\mu^{(1)}(x; \alpha) = \pi_n^{(1)}\delta_{mn},$$

$$\int_{-\infty}^{\infty} q_n(x)q_m(x)d\mu^{(2)}(x; \alpha) = \pi_n^{(2)}\delta_{m,n},$$

$i = 1, 2$.

Since $A_{n+\alpha}$, $B_{n+\alpha}$ and $C_{n+\alpha}$ of (3.1) and (2.14) are bounded if $|a|, |b|, |c|, |d| < 1$ and $\alpha \geq 0$ (we shall assume these restrictions to hold) the supports of both $d\mu^{(i)}(t; \alpha)$, $i = 1, 2$, are bounded, by [9, Th. IV 2.2] and consequently the moment problem is determined. Since the numerator polynomial corresponding to both $p_n^\alpha(x)$ and $q_n^\alpha(x)$ is a multiple of $p_{n-1}^{\alpha+1}(x)$ (which can be easily proved) formula (2.4) now takes the form

$$\int_{-\infty}^{\infty} \frac{d\mu^{(1)}(t; \alpha)}{x - t}$$
$$= \frac{2z(1 - bcdzq^{2\alpha-1})(1 - bcdzq^{2\alpha})}{(1 - bzq^\alpha)(1 - czq^\alpha)(1 - dzq^\alpha)(1 - bcdzq^{\alpha-1})} \tag{3.11}$$
$$\times \frac{{}_8W_7(bcdzq^{2\alpha}; bcq^\alpha, bdq^\alpha, cdq^\alpha, q^{\alpha+1}, zq/a; q, az)}{{}_8W_7(bcdzq^{2\alpha-2}; bcq^{\alpha-1}, bdq^{\alpha-1}, cdq^{\alpha-1}, q^\alpha, zq/a; q, az)},$$

and

$$\int_{-\infty}^{\infty} \frac{d\mu^{(2)}(t; \alpha)}{x - t}$$
$$= \frac{2z(1 - bcdzq^{2\alpha-1})(1 - bcdzq^{2\alpha})}{(1 - bzq^\alpha)(1 - czq^\alpha)(1 - dzq^\alpha)(1 - bcdzq^{\alpha-1})} \tag{3.12}$$
$$\times \frac{{}_8W_7(bcdzq^{2\alpha}; bcq^\alpha, bdq^\alpha, cdq^\alpha, q^{\alpha+1}, zq/a; q, az)}{{}_8W_7(bcdzq^{2\alpha-2}; bcq^{\alpha-1}, bdq^{\alpha-1}, cdq^{\alpha-1}, q^\alpha, z/a; q, aqz)},$$

for $x \in \mathbf{C}/[-1, 1]$. Using a very useful theorem of Nevai [36, Corollary 36, p. 141, Theorem 40, p. 143], one can show that the support of the absolutely continuous part of the measure is $(-1, 1)$ and that the jump function $j(x)$ for the discrete part $dj(x)$ is constant in $(-1, 1)$. Furthermore it was shown in [28] that there are no discrete masses outside $(-1, 1)$ or at the points $x = \pm 1$. So from (2.3) one deduces that

$$
\frac{d\mu^{(1)}(\cos\theta; \alpha)}{d\theta} \tag{3.13}
$$

$$
= \frac{(abq^\alpha, acq^\alpha, adq^\alpha, bcq^\alpha, bdq^\alpha, cdq^\alpha, q^{\alpha+1}; q)_\infty}{2\pi(1 - abcdq^{\alpha-2})(abcdq^{\alpha-2}, abcdq^{2\alpha}; q)_\infty}(1 - abcdq^{2\alpha-2})
$$

$$
\times (abcdq^{2\alpha-2}; q)_\infty \left| \frac{(e^{2i\theta}, q^{\alpha+1}e^{2i\theta}; q)_\infty}{(aq^\alpha e^{i\theta}, bq^\alpha e^{i\theta}, cq^\alpha e^{i\theta}, dq^\alpha e^{i\theta}, qe^{2i\theta}; q)_\infty} \right|^2
$$

$$
\times \left| {}_8W_7(q^\alpha e^{2i\theta}; qe^{i\theta}/a, qe^{i\theta}/b, qe^{i\theta}/c, qe^{i\theta}/d, q^\alpha; q, abcdq^{\alpha-2}) \right|^{-2}.
$$

It was also shown in [28] that $d\mu^{(2)}(x; \alpha)$ has at most one discrete mass for $0 < q < 1$, $\alpha \geq 0$ and $-1 < a, b, c, d < 1$, and that

$$
\frac{d\mu^{(2)}(\cos\theta; \alpha)}{d\theta}
$$

$$
= \frac{(abq^\alpha, acq^\alpha, adq^\alpha, bcq^\alpha, bdq^\alpha, cdq^\alpha, q^{\alpha+1}; q)_\infty}{2\pi(abcdq^{2\alpha}; q)_\infty}
$$

$$
\times \frac{(abcdq^{2\alpha-1}; q)_\infty}{(abcdq^{\alpha-1}; q)_\infty} \frac{1 - 2axq^\alpha + a^2q^{2\alpha}}{1 - 2ax + a^2} \tag{3.14}
$$

$$
\times \frac{|(e^{2i\theta}, q^{\alpha+1}e^{2i\theta}; q)_\infty|^2}{|(aq^\alpha e^{i\theta}, bq^\alpha e^{i\theta}, cq^\alpha e^{i\theta}, dq^\alpha e^{i\theta}, qe^{2i\theta}; q)_\infty|^2}
$$

$$
\times \left| {}_8W_7(q^\alpha e^{2i\theta}; e^{i\theta}/a, qe^{i\theta}/b, qe^{i\theta}/c, qe^{i\theta}/d, q^\alpha; q, abcdq^{\alpha-1}) \right|^{-2}.
$$

Explicit forms of $p_n^\alpha(x)$ and $q_n^\alpha(x)$ found in [28] are:

$$
p_n^\alpha(x) = \sum_{k=0}^n \frac{(q^{-n}, abcdq^{2\alpha+n-1}, abcdq^{2\alpha-1}, ae^{i\theta}, ae^{-i\theta}; q)_k}{(q, abq^\alpha, acq^\alpha, adq^\alpha, abcdq^{\alpha-1}; q)_k} q^k \tag{3.15}
$$

$$
\times {}_{10}W_9\left(abcdq^{2\alpha+k-2}; q^\alpha, bcq^{\alpha-1}, bdq^{\alpha-1}, cdq^{\alpha-1}, q^{k+1}, \right.
$$

$$
\left. abcdq^{2\alpha+n+k-1}, q^{k-n}; q, a^2\right),
$$

and

$$
q_n^\alpha(x) = \sum_{k=0}^n \frac{(q^{-n}, abcdq^{2\alpha+n-1}, abcdq^{2\alpha-1}, ae^{i\theta}, ae^{-i\theta}; q)_k}{(q, abq^\alpha, acq^\alpha, adq^\alpha, abcdq^{\alpha-1}; q)_k} q^k
$$

$$
\times {}_{10}W_9\left(abcdq^{2\alpha+k-2}; q^\alpha, bcq^{\alpha-1}, bdq^{\alpha-1}, cdq^{\alpha-1}, q^k, \right.
$$

$$
\left. abcdq^{2\alpha+n+k-1}, q^{k-n}; q, qa^2\right). \tag{3.16}
$$

Masson's exceptional Askey-Wilson polynomials, see [32], [26] that correspond to the indeterminate cases $abcd = q$ or q^2 (see eqn. (2.14)) turn out to be the limiting case $\alpha \to 0^+$ of $p_n^\alpha(x)$, while $q_n^\alpha(x)$ approaches the Askey-Wilson polynomials.

By using the transformation theory of basic hypergeometric series a simpler form of (3.14) was found in [43]

$$
\begin{aligned}
p_n^\alpha(x) &\equiv p_n^\alpha(x;\,a,b,c,d|q) \qquad\qquad\qquad\qquad\qquad (3.17)\\
&= \frac{(abcdq^{2\alpha-1}, q^{\alpha+1}; q)_n}{(q, abcdq^{\alpha-1}; q)_n}\\
&\times\; q^{-\alpha n} \sum_{k=0}^{n} \frac{(q^{-n}, abcdq^{2\alpha+n-1}; q)_k}{(q^{\alpha+1}, abq^\alpha; q)_k}\\
&\times \frac{(aq^\alpha e^{i\theta}, aq^\alpha e^{-i\theta}; q)_k}{(acq^\alpha, acq^\alpha; q)_k}\\
&\times\; q^k \sum_{j=0}^{k} \frac{(q^\alpha, abq^{\alpha-1}, acq^{\alpha-1}, adq^{\alpha-1}; q)_j}{(q, abcdq^{2\alpha-2}, aq^\alpha e^{i\theta}, aq^\alpha e^{-i\theta}; q)_j}\, q^j.
\end{aligned}
$$

Assuming that $\max(|a|, |b|, |c|, |d|) < q^{(1-\alpha)/2}$, $\alpha > 0$, one can derive the following integral representation [43]:

$$
\begin{aligned}
p_n^\alpha(x;&\, a, b, c, d|q) \qquad\qquad\qquad\qquad\qquad\qquad (3.18)\\
&= \int_{-1}^{1} K(x,y) \frac{(abcdq^{2\alpha-1}, q^{\alpha+1}; q)_n}{(q, abcdq^{\alpha-1}; q)_n}\\
&\times q^{-\alpha n}\, p_n(y;\, aq^{\alpha/2}, bq^{\alpha/2}, cq^{\alpha/2}, dq^{\alpha/2}|q)\,dy,
\end{aligned}
$$

where

$$
\begin{aligned}
K(x,y) &= \frac{(q, q, q, abq^{\alpha-1}, acq^{\alpha-1}, adq^{\alpha-1}; q)_\infty}{4\pi^2 (abcdq^{2\alpha-2}, q^{\alpha+1}; q)_\infty \sqrt{1-y^2}}\\
&\times\; (bcq^{\alpha-1}, bdq^{\alpha-1}, cdq^{\alpha-1}, q^\alpha; q)_\infty\\
&\times\; \frac{(e^{2i\phi}, e^{-2i\phi}; q)_\infty}{h(y; q^{\alpha/2}e^{i\theta}, q^{\alpha/2}e^{-i\theta})} \qquad\qquad\qquad (3.19)\\
&\times\; \int_0^\pi \frac{(e^{2i\psi}, e^{-2i\psi}; q)_\infty}{h(\cos\psi; aq^{\frac{\alpha-1}{2}}, bq^{\frac{\alpha-1}{2}}, cq^{\frac{\alpha-1}{2}}, dq^{\frac{\alpha-1}{2}})}\\
&\qquad\times \frac{h(\cos\psi; q^{\frac{\alpha+1}{2}}e^{i\theta}, q^{\frac{\alpha+1}{2}}e^{-i\theta})}{h(\cos\psi; q^{1/2}e^{i\phi}, q^{1/2}e^{-i\phi})}\,d\psi,
\end{aligned}
$$

where $y = \cos\phi$, $0 \le \phi \le \pi$, and

$$
h(\cos\theta;\, a_1, a_2, \ldots, a_k) := \prod_{j=1}^{k} h(\cos\theta; a_j) \qquad\qquad (3.20)
$$

$$h(\cos\theta;\, a) = (ae^{i\theta}, ae^{-i\theta}; q)_\infty \; = \; \prod_{k=0}^{\infty}(1 - 2aq^k\cos\theta + a^2q^{2k}).$$

This representation enables us to compute many formulas for the associated polynomials $p_n^\alpha(x; a, b, c, d|q)$ from the corresponding formulas for the ordinary Askey-Wilson polynomials $p_n(x; aq^{\alpha/2}, bq^{\alpha/2}, cq^{\alpha/2}, dq^{\alpha/2}|q)$, see, for example, [44].

By replacing a, b, c, d and α in (3.17) by $q^{1/2}$, $q^{\alpha+1/2}$, $-q^{\beta+1/2}$, $-q^{1/2}$ and c, respectively, then taking the limit $q \to 1$ and defining

$$P_n^{(\alpha,\beta)}(x; c) \; = \; \frac{(\alpha+c+1)_n(\alpha+\beta+c+1)_n}{(c+1)_n(\alpha+\beta+2c+1)_n} \tag{3.21}$$
$$\times \; \lim_{q\to 1} p_n^c\Big(x;\, q^{1/2}, q^{\alpha+1/2}, -q^{\beta+1/2}, -q^{1/2}\,\big|\,q\Big),$$

one can show that

$$P_n^{(\alpha,\beta)}(x; c) \; = \; \frac{(\alpha+c+1)_n}{n!}$$
$$\times \; \sum_{k=0}^{n} \frac{(-n, \alpha+\beta+2c+n+1)_k}{(c+1, \alpha+c+1)_k}\Big(\frac{1-x}{2}\Big)^k$$
$$\times \; \sum_{j=0}^{k} \frac{(-k, c, \alpha+c)_j}{j!\,(-k, \alpha+\beta+2c)_j}\Big(\frac{2}{1-x}\Big)^j \tag{3.22}$$
$$= \; \frac{c\Gamma(\alpha+\beta+2c)}{\Gamma(\alpha+c)\Gamma(\beta+c)}2^{1-\alpha-\beta-3c}$$
$$\times \; \int_{-1}^{1}\int_{-1}^{1} dy\,dz(1-y)^{\alpha+c-1}(1+y)^{\beta+c-1}(1-z)^{c-1}$$
$$\times \; P_n^{(\alpha+c,\beta+c)}\Big(\frac{x(1+z)+y(1-z)}{2}\Big),$$

$\mathrm{Re}(\alpha+c) > 0$, $\mathrm{Re}(\beta+c) > 0$, $\mathrm{Re}\,c > 0$.

4. Special and Limiting Cases

4.1. CONTINUOUS q-JACOBI POLYNOMIALS

Let us take

$$a = q^{1/2}, \quad b = q^{\beta+1/2}, \quad c = -q^{\alpha+1/2}, \quad d = -q^{1/2} \tag{4.1}$$

and replace α by c in (3.14). The $_{10}\phi_9$ then is a balanced, terminating and very-well-poised series which can be transformed by [14, III. 28]. Thus

$$_{10}W_9\Big(q^{\alpha+\beta+2c+k};\, q^c, -q^{\alpha+\beta+c}, -q^{\beta+c}, q^{\alpha+c}, q^{k+1}, \tag{4.2}$$

$$q^{\alpha+\beta+2c+n+k+1}, q^{k-n}; q, q\Big)$$

$$= \frac{(q^{c+1}, q^{\alpha+\beta+2c+1}, -q, -q^{\alpha+\beta+c+1}; q)_n}{(-q^{c+1}, -q^{\alpha+\beta+2c+1}, q, q^{\alpha+\beta+c+1}; q)_n}$$

$$\times \frac{(q^{\alpha+\beta+c+1}, -q^{c+1}, q, -q^{\alpha+\beta+2c+1}; q)_k}{(-q^{\alpha+\beta+c+1}, q^{c+1}, -q, q^{\alpha+\beta+2c+1}; q)_k}$$

$$\times {}_{10}W_9\Big(-q^{\alpha+\beta+2c+k};\ q^c, -q^{\alpha+\beta+c}, q^{\beta+c}, -q^{\alpha+c},$$

$$-q^{k+1}, q^{\alpha+\beta+2c+n+k+1}, q^{k-n}; q, q\Big).$$

Hence the explicit form of the associated continuous q-Jacobi polynomials is:

$$p_n^c(x;\ q^{1/2}, q^{\beta+1/2}, -q^{\alpha+1/2}, -q^{1/2}|q)$$

$$= \frac{(q^{c+1}, q^{\alpha+\beta+2c+1}, -q, -q^{\alpha+\beta+c+1}; q)_n}{(q, q^{\alpha+\beta+c+1}, -q^{c+1}, -q^{\alpha+\beta+2c+1}; q)_n} \tag{4.3}$$

$$\times \sum_{k=0}^{n} \frac{(q^{-n}, q^{\alpha+\beta+2c+n+1}, -q^{\alpha+\beta+2c+1}; q)_k}{(q^{c+1}, q^{\beta+c+1}, -q^{\alpha+c+1}; q)_k}$$

$$\times \frac{(q^{1/2}e^{i\theta}, q^{1/2}e^{-i\theta}; q)_k}{(-q, -q^{\alpha+\beta+c+1}; q)_k}q^k$$

$$\times {}_{10}W_9\Big(-q^{\alpha+\beta+2c+k};\ q^c, -q^{\alpha+\beta+c}, q^{\beta+c}, -q^{\alpha+c}, -q^{k+1},$$

$$q^{\alpha+\beta+2c+n+k+1}, q^{k-n}; q, q\Big)$$

As $q \to 1$ this approaches a multiple of Wimp's [50, eqn. (19)] expression for the associated Jacobi polynomials:

$$\frac{(c+1)_n(\alpha+\beta+2c+1)_n}{n!\,(\alpha+\beta+c+1)_n}$$

$$\times \sum_{k=0}^{n} \frac{(-n)_k(\alpha+\beta+2c+n+1)_k}{(c+1)_k(\beta+c+1)_k}\Big(\frac{1-x}{2}\Big)^k$$

$$\tag{4.4}$$

$$\times {}_4F_3\left[\begin{array}{c} k-n,\ \alpha+\beta+2c+n+k+1,\ c,\ \beta+c \\ c+k+1,\ \beta+c+k+1,\ \alpha+\beta+2c \end{array}; 1\right].$$

4.2. CONTINUOUS q-ULTRASPHERICAL POLYNOMIALS

Instead of (4.1) let us now set $b = aq^{1/2}$, $c = -a$, $d = -aq^{1/2}$ and denote

$$p_n^\alpha(x; a, aq^{1/2}, -a - aq^{1/2}|q) = \frac{(q^{\alpha+1}; q)_n}{(a^4 q^\alpha; q)_n} a^n C_n^\alpha(x; a^2|q) \tag{4.5}$$

where $C_n^\alpha(x; a^2|q)$ is the associated q-ultraspherical polynomials which reduces to the q-ultraspherical polynomials

$$C_n(x; a^2|q) = \sum_{k=0}^n \frac{(a^2; q)_k (a^2; q)_{n-k}}{(q; q)_k (q; q)_{n-k}} e^{i(n-2k)\theta}, \tag{4.6}$$

when $\alpha = 0$.

Use of the transformation formula [14, (3.4.7)] and some simplifications gives

$$
\begin{aligned}
C_n^\alpha(x; a^2|q) &= \frac{(a^4 q^{\alpha-1}; q)_\infty}{(q^{\alpha+1}; q)_\infty} \frac{z^n}{1 - z^{-2}}\ {}_2\phi_1\left[\begin{matrix} q/a^2 z^2,\ q/a^2 \\ q/z^2 \end{matrix};\ q, a^4 q^{\alpha-1}\right] \\
&\quad \times\ {}_2\phi_1\left[\begin{matrix} qz^2/a^2,\ q/a^2 \\ qz^2 \end{matrix};\ q, a^4 q^{\alpha+n}\right] + (z \leftrightarrow z^{-1}),
\end{aligned}
\tag{4.7}
$$

where $z = e^{i\theta}$, $x = \cos\theta$, $0 < \theta < \pi$. From this the generating function follows immediately:

$$
\begin{aligned}
G_t^\alpha(x; \beta|q) &= \sum_{n=0}^\infty C_n^\alpha(x; \beta|q) t^n \tag{4.8} \\
&= \frac{(\beta^2 q^{\alpha-1}; q)_\infty}{(q^{\alpha+1}; q)_\infty (1 - zt)(1 - z^{-2})} \\
&\quad \times\ {}_2\phi_1\left[\begin{matrix} q/\beta,\ q/\beta z^2 \\ q/z^2 \end{matrix};\ q, \beta^2 q^{\alpha-1}\right] \\
&\quad \times\ {}_3\phi_2\left[\begin{matrix} q/\beta,\ qz^2/\beta,\ zt \\ qz^2, qzt \end{matrix};\ q, \beta^2 q^\alpha\right] + (z \leftrightarrow z^{-1}),
\end{aligned}
$$

$|t| < 1$. We can now combine the two terms on the right hand side of (4.8) in the following way. First, note that

$$
\begin{aligned}
G_t^\alpha(x; \beta|q) &= \frac{(\beta^2 q^{\alpha-1}; q)_\infty}{(q^{\alpha+1}; q)_\infty (1 - z^{-2})} \sum_{j=0}^\infty \sum_{k=0}^\infty \frac{(q/\beta, q/\beta z^2; q)_j}{(q, q/z^2; q)_j} \tag{4.9} \\
&\quad \times \frac{(q/\beta, qz^2/\beta; q)_k}{(q, qz^2; q)_k} (\beta^2 q^\alpha)^{j+k} \frac{q^{-j}(1 - z^{-2} q^{j-k})}{(1 - ztq^k)(1 - tq^j/z)}.
\end{aligned}
$$

Setting $k = n - j$ we find, after some simplifications, that

$$
\begin{aligned}
G_t^\alpha(x; \beta|q) &= \frac{(\beta^2 q^{\alpha-1}; q)_\infty}{(q^{\alpha+1}; q)_\infty (1 - zt)(1 - t/z)} \tag{4.10} \\
&\quad \times \sum_{n=0}^\infty \frac{(q/\beta, qz^2/\beta, zt; q)_n}{(q, z^2, qzt; q)} (\beta^2 q^{\alpha-1})^n \\
&\quad \times {}_8W_7(q^{-n}/z^2; q/\beta, q/\beta z^2, t/z, q^{-n}/zt, q^{-n}; q, \beta^2).
\end{aligned}
$$

However, using [14, III. 16] and then [14, III.15] we get

$$_8W_7(q^{-n}/z^2;q/\beta,q/\beta z^2,t/z,q^{-n}/zt,q^{-n};q,\beta^2) \tag{4.11}$$
$$= \frac{(z^2,q/\beta^2,qzt;\,q)_n}{(qz^2/\beta,q/\beta,zt;\,q)_n} \,_4\phi_3\left[\begin{array}{c} q^{-n},\ \beta zt,\ \beta t/z,\ q \\ qzt,\ qt/z,q^{-n}/\beta^2 \end{array};\,q,q\right].$$

Hence

$$G_t^\alpha(x;\beta\ q) = \frac{(\beta^2 q^{\alpha-1};q)_\infty}{(q^{\alpha+1};q)_\infty(1-zt)(1-t/z)} \sum_{n=0}^\infty \frac{(q/\beta^2;q)_n}{(q;q)_n}(\beta^2 q^{\alpha-1})^n$$
$$\times\,_4\phi_3\left[\begin{array}{c} q^{-n},\ \beta zt,\ \beta t/z,\ q \\ qzt,\ qt/z,q^{-n}/\beta^2 \end{array};\,q,q\right] \tag{4.12}$$
$$= \frac{1-q^\alpha}{(1-2xt+t^2)}\,_3\phi_2\left[\begin{array}{c} \beta te^{i\theta},\ \beta te^{-i\theta},\ q \\ qte^{i\theta},\ qte^{-i\theta} \end{array};\,q,q^\alpha\right],$$

which is the same as [8, eqn. (2.8)].

4.3. ASSOCIATED WILSON POLYNOMIALS

The Wilson polynomials $P_n(x;a,b,c,d)$ satisfy the 3-term recurrence relation

$$(\lambda_n + \mu_n - x)P_n(x) = \lambda_n P_{n+1}(x) + \mu_n P_{n-1}(x), \tag{4.13}$$
$$P_{-1}(x) = 0, \quad P_0(x) = 1, \quad x = a^2 + t^2,$$

where

$$\lambda_n = \frac{(n+a+b)(n+a+c)(n+a+d)(n+s-1)}{(2n+s-1)(2n+s)}, \tag{4.14}$$
$$\mu_n = \frac{n(n+b+c-1)(n+b+d-1)(n+c+d-1)}{(2n+s-2)(2n+s-1)},$$
$$s = a+b+c+d,$$

see [48], [49] and [3].

Wilson [48] found that they are orthogonal on $(-\infty,\infty)$ with respect to the weight function

$$\left|\frac{\Gamma(a+it)\Gamma(b+it)\Gamma(c+it)\Gamma(d+it)}{\Gamma(2it)}\right|^2$$

and that

$$P_n(x;a,b,c,d) = \,_4F_3\left[\begin{array}{c} -n,\ n+s-1,\ a-it,\ a+it \\ a+b,\ a+c,\ a+d \end{array};\,1\right]. \tag{4.15}$$

The associated Wilson polynomials are, of course, the solutions of (4.13) with n replaced by $n + \alpha$ in λ_n and μ_n. Their weight function and the orthogonality properties were worked out by Ismail et al. [27], see also [36] and [25]. In fact, the authors in [27] found two families of polynomials that correspond to the two cases we have considered in the previous section. Their weight function [27, (3.39)] for the first family can be found as the $q \to 1^-$ limit of our formula (3.12) after having replaced $a, b, c, d, \alpha, abcd, e^{i\theta}$ by $q^a, q^b, q^c, q^d, q^\gamma$ and $q^{i\tau}$, respectively. Similarly, formula [27, (4.25)] follows as the $q \to 1^-$ limit of (3.13) with the same replacements. Furthermore, the functional forms for the two polynomial systems are given by

$$P_n^\gamma(x; a, b, c, d) \tag{4.16}$$

$$= \sum_{k=0}^{n} \frac{(-n, s + 2\gamma + n - 1, s + 2\gamma - 1, a + i\tau, a - i\tau)_k}{k!(a + b + \gamma, a + c + \gamma, a + d + \gamma, s + \gamma - 1)_k}$$

$$\times \ _9F_8\Big(s + 2\gamma + k - 2; \gamma, b + c + \gamma - 1, b + d + \gamma - 1,$$

$$\times \ c + d + \gamma - 1, k + 1, s + 2\gamma + n + k - 1, k - n; 1\Big)$$

as the $q \to 1^-$ limit of (3.14), and

$$Q_n^\gamma(x; \ a, b, c, d) \tag{4.17}$$

$$= \sum_{k=0}^{n} \frac{(-n, s + 2\gamma + n - 1, s + 2\gamma - 1, a + i\tau, a - i\tau)_k}{k!(a + b + \gamma, a + c + \gamma, a + d + \gamma, s + \gamma - 1)_k}$$

$$\times {}_9F_8(s + 2\gamma + k - 2; \gamma, b + c + \gamma - 1, b + d + \gamma - 1,$$

$$\times \ c + d + \gamma - 1, k,$$

$$s + 2\gamma + n + k - 1, k - n; \ 1),$$

as the $q \to 1^-$ limit of (3.15), where

$$_9F_8(a; \ b, c, d, e, f, g, h; 1) \tag{4.18}$$

$$:= \sum_{n=0}^{\infty} \frac{a + 2n}{a} \frac{(a, \ b, \ c, \ d, \ e, \ f, \ g, \ h)_n}{n! \ (1 + a - b, 1 + a - c, 1 + a - d, 1 + a - e,}$$

$$\times \frac{1}{1 + a - f, 1 + a - g, 1 + a - h)_n}.$$

Acknowledgement. I wish to express my thanks and appreciation to Mourad Ismail for his help and suggestions in preparing the final version of this paper.

References

1. G. Andrews and R. Askey, *Classical orthogonal polynomials*, in: "Polynomes Orthogonaux et applications", Lecture Notes in Mathematics, Vol. 1171, Springer-Verlag, 1985, pp. 36–82.

2. R. Askey and M.E.H. Ismail, *Recurrence relations, continued fractions and orthogonal polynomials*, Mem. Amer. Math. Soc., **300** (1984).

3. R. Askey and J. A. Wilson, *Some basic hypergeometric polynomials that generalize Jacobi polynomials*, Mem. Amer. Math. Soc. **319** (1985).

4. R. Askey and J. A. Wilson, *A set of orthogonal polynomials that generalize the Racah coefficients or 6 − j symbols*, SIAM J. Math. Anal. **10** (1979), 1008–1016.

5. R. Askey and J. Wimp, *Associated Laguerre and Hermite polynomials*, Proc. Roy. Soc. Edn. **96A** (1984), 15–37.

6. P. Barrucand and D. Dickinson, *On the associated Legendre polynomials*, in: "Orthogonal Expansions and Their Continual Analogues", D. T. Haimo, ed., Southern Illinois University Press, Edwardsville, 1968, pp. 43–50.

7. R. H. Boyer, *Discrete Bessel functions*, J. Math. Anal. Appl. **2** (1961), 509–524.

8. J. Bustoz and M. E. H. Ismail, *The associated ultraspherical polynomials and their q-analogues*, Can. J. Math. **34** (1982), 718–736.

9. T. S. Chihara, *An Introduction to Orthogonal Polynomials*, Gordon and Breach, New York, 1978.

10. T. S. Chihara, *Chain sequences and orthogonal polynomials*, Trans. Amer. Math. Soc. **104** (1962), 1–16.

11. R. Courant, K. O. Friedrichs, and H. Lewy, *Über die partiellen Differenzengleichungen der Physik*, Math. Ann., **100** (1928), 32–74.

12. D. J. Dickinson, *On certain polynomials associated with orthogonal polynomials*, Boll. Un. Mat. Ital. (3) **13** (1958), 116–124.

13. A. Erdélyi, et al., eds., *Higher Transcendental Functions*, Vol. II, Bateman Manuscript Project, McGraw-Hill, 1953.

14. G. Gasper and M. Rahman, *The Basic Hypergeometric Series*, Cambridge University Press, Cambridge, U.K., 1990.

15. F. A. Grünbaum and L. Haine, *A theorem of Bochner, revisited*, in: "Algebraic Aspects of Integrable Systems", A. S. Fokas and I. M. Gelfand, eds., Progr. Nonlinear Differential Equations Appl., Vol. **26**, Birkhäuser, Boston, 1997, pp. 143–172.

16. F. A. Grünbaum and L. Haine, *Some functions that generalize the Askey-Wilson polynomials*, Commun. Math. Phys. **184** (1997), 173–202.

17. F. A. Grünbaum and L. Haine, *Associated polynomials, spectral matrices and the bispectral problem*, Methods and Appl. Anal. **6** (2) (1999), 209–224.

18. D. P. Gupta, M. E. H. Ismail, and D. R. Masson, *Contiguous relations, basic hypergeometric functions and orthogonal polynomials II: Associated big q-Jacobi polynomials*, J. Math. Anal. Appl. **171** (1992), 477–497.

19. D. P. Gupta, M. E. H. Ismail, and D. R. Masson, *Associated continuous Hahn polynomials*, Canad. J. Math. **43** (1991), 1263–1280.

20. W. Hahn, *Über Orthogonalpolynome mit drei Parametern*, Deutche. Math. **5** (1940-41), 373–378.

21. P. Humbert, *Sur deux polynomes associés aux polynomes de Legendre*, Bull. Soc. Math. de France **XLVI** (1918), 120–151.

22. M. E. H. Ismail, J. Letessier, D.R. Masson, and G. Valent, *Birth and death processes and orthogonal polynomials*, in: "Orthogonal Polynomials: Theory and Practice", P. Nevai, ed., Kluwer Academic Publishers, The Netherlands, 1990, pp. 225-229.

23. M. E. H. Ismail, J. Letessier, and G. Valent, *Linear birth and death models and associated Laguerre and Meixner polynomials*, J. Approx. Theory **56** (1988), 337–348.

24. M. E. H. Ismail, J. Letessier, and G. Valent, *Quadratic birth and death processes and associated continuous dual Hahn polynomials*, SIAM J. Math. Anal. **20** (1989), 727–737.

25. M. E. H. Ismail, J. Letessier, G. Valent, and J. Wimp, *Some results on associated Wilson polynomials*, in: "Orthogonal Polynomials and Their Applications", C. Brezinski, L. Gori and A. Ronveaux, eds., IMACS Annals on Computing and

Applied Mathematics, Vol. 9 (1991), pp. 293–298.

26. M. E. H. Ismail, J. Letessier, G. Valent, and J. Wimp, *Two families of associated Wilson polynomials*, Can. J. Math. **42** (1990), 659–695.

27. M. E. H. Ismail and D. R. Masson, *Two families of orthogonal polynomials related to Jacobi polynomials*, Rocky Mountain J. Math. **21** (1991), 359–375.

28. M. E. H. Ismail and M. Rahman, *Associated Askey-Wilson polynomials*, Trans. Amer. Math. Soc. **328** (1991), 201–237.

29. S. Karlin and J. McGregor, *The differential equations of birth-and-death processes, and the Stieltjes moment problem*, Trans. Amer. Math. Soc. **85** (1957), 489–546.

30. S. Karlin and J. McGregor, *Linear growth, birth-and-death processes*, J. Math. Mech. (now *Indiana Univ. Math. J.*) **7** (1958), 643–662.

31. A. P. Magnus, *Associated Askey-Wilson polynomials as Laguerre-Hahn orthogonal polynomials*, in: "Orthogonal Polynomials and Their Applications", M. Afaro et al., eds., Springer Lecture Notes Math. #1329 (1988), pp. 261–278.

32. D. Masson, *Convergence and analytic continuation for a class of regular C-fractions*, Canad. Math. Bull. **28** (4) (1985), 411–421.

33. D. Masson, *Wilson polynomials and some continued fractions of Ramanujan*, Rocky Mountain J. Math., **21** (1991), 489–499.

34. D. Masson, *Associated Wilson polynomials*, Const. Approx. **7** (1991), 521–534.

35. J. Meixner, *Orthogonale Polynomsysteme mit einem besonderen Gestalt der erzeugenden Funktion*, J. Lond. Math. Soc. **9** (1934), 6–13.

36. P. Nevai, *Orthogonal Polynomials*, Mem. Amer. Math. Soc. #213 (1979), Providence, R.I.

37. F. W. J. Olver, *Asymptotics and Special Functions*, Academic Press, New York, 1974.

38. G. Palama, *Polinomi più generali di altri classici e dei loro associati e relazioni tra essi funzioini di seconda specie*, Riv. Mat. Univ. Parma **4** (1953), 363–383.

39. F. Pollaczek, *Sur une généralisation des polynômes de Legendre*, C.R. Acad. Sci. Paris **228** (1949), 1363–1365.

40. F. Pollaczek, *Systèmes des polynômes biorthogonaux que généralisent les polynômes ultrasphériques*, C. R. Acad. Sci. Paris **228**, (1949), 1998–2000.

41. F. Pollaczek, *Sur une famille de polynômes orthogonaux qui contient les polynômes d'Hermite et de Laguerre comme cas limites*, C.R. Acad. Sci. Paris **230** (1950), 1563–2256.

42. F. Pollaczek, *Sur une famille de polynômes orthogonaux à quatre paramitres*, C.R. Acad. Sci. Paris **230** (1950), 2254–2256.

43. M. Rahman, *Some generating functions for the associated Askey-Wilson polynomials*, J. Comp. Appl. Math. **68** (1996), 287–296.

44. M. Rahman and Q. M. Tariq, *Poisson kernel for the associated continuous q-ultraspherical polynomials*, Meth. Appl. Anal. **4** (1997), 77–90.

45. J. A. Shohat and J. D. Tamarkin, *The Problem of Moments*, revised edition, Math. Surveys, Vol. 1, Amer. Math. Soc., Providence, R.I., 1950.

46. G. Szegő, *Orthogonal Polynomials*, 4th edn., Amer. Math. Soc. Coll. Publ., Vol. 23, Providence, R.I., 1975.

47. H. S. Wall and M. Wetzel, *Quadratic forms and convergence regions for continued fractions*, Duke Math. J. **11** (1944), 89–102.

48. J. Wilson, *Hypergeometric series, recurrence relations and some orthogonal functions*, Doctoral Dissertation, University of Wisconsin, Madison, 1978.

49. J. Wilson, *Some hypergeometric orthogonal polynomials*, SIAM J. Math. Anal. **11** (1980), 690–701.

50. J. Wimp, *Explicit formulas for the associated Jacobi polynomials and some applications*, Can. J. Math. **39** (1987), 983–1000.

51. J. Wimp, *Some explicit Padé approximants for the function Φ'/Φ and a related quadrature formula involving Bessel functions*, SIAM J. Math. Anal. **16** (1985), 887–895.

SPECIAL FUNCTIONS DEFINED BY ANALYTIC DIFFERENCE EQUATIONS

S. N. M. RUIJSENAARS
Centre for Mathematics and Computer Science
P. O. Box 94079, 1090 GB Amsterdam
The Netherlands

Abstract. We survey our work on a number of special functions that can be viewed as solutions to analytic difference equations. In the infinite-dimensional solution spaces of the pertinent equations, these functions are singled out by various distinctive features. In particular, starting from certain first order difference equations, we consider generalized gamma and zeta functions, as well as Barnes' multiple zeta and gamma functions. Likewise, we review the generalized hypergeometric function we introduced in recent years, emphasizing the four second order Askey-Wilson type difference equations it satisfies. Our results on trigonometric, elliptic and hyperbolic generalizations of the Hurwitz zeta function are presented here for the first time.

1. Introduction

In the following we survey various special functions whose common feature is that they can be viewed as analytic solutions to (ordinary linear) analytic difference equations with analytic coefficients. Some of these functions have been known and studied for centuries, whereas a few others are of quite recent vintage.

The difference equations that will be relevant for our account are of three types. The first type is the first order equation

$$\frac{F(z + ia/2)}{F(z - ia/2)} = \Phi(z), \quad a > 0, \tag{1.1}$$

and the second type its logarithmic version,

$$f(z + ia/2) - f(z - ia/2) = \phi(z). \tag{1.2}$$

J. Bustoz et al. (eds.), Special Functions 2000, 281–333.
© 2001 *Kluwer Academic Publishers. Printed in the Netherlands.*

The third type is the second order equation

$$C^{(+)}(z)F(z+ia) + C^{(-)}(z)F(z-ia) + C^{(0)}(z)F(z) = 0, \quad a > 0. \tag{1.3}$$

Introducing the spaces

$$\mathcal{M} \equiv \{F(z) \mid F \text{ meromorphic}\}, \tag{1.4}$$

$$\mathcal{M}^* \equiv \mathcal{M} \setminus \{F(z) = 0, \forall z \in \mathbb{C}\}, \tag{1.5}$$

we require

$$\Phi, C^{(\pm)}, C^{(0)} \in \mathcal{M}^*, \tag{1.6}$$

and we only consider solutions $F \in \mathcal{M}$ to the equations (1.1) and (1.3). The right-hand side functions $\phi(z)$ in (1.2) are allowed to have branch points, giving rise to solutions $f(z)$ that have branch points, as well.

We are primarily interested in analytic difference equations (henceforth AΔEs) admitting solutions with various properties that render them unique. In this connection a crucial point to be emphasized at the outset is that the solutions form an infinite-dimensional space (assuming at least one non-trivial solution exists). More specifically, introducing spaces of α-periodic multipliers,

$$\mathcal{P}_\alpha^{(*)} \equiv \{\mu(z) \in \mathcal{M}^{(*)} \mid \mu(z+\alpha) = \mu(z)\}, \quad \alpha \in \mathbb{C}^*, \tag{1.7}$$

and assuming $F \in \mathcal{M}^*$ solves (1.1) or (1.3), it is obvious that $\mu(z)F(z), \mu \in \mathcal{P}_{ia}^*$, is also a solution. Likewise, adding any $\mu \in \mathcal{P}_{ia}^*$ to a given solution $f(z)$ of (1.2), one obtains another solution.

In view of this infinite-dimensional ambiguity, it is natural to try and single out solutions by requiring additional properties. In a nutshell, this is how the special functions at issue in this contribution will arise. The extra properties alluded to will be made explicit in Section 2, with Subsection 2.1 being devoted to the first order equations (1.1) and (1.2), and Subsection 2.2 to the second order one (1.3).

Section 2 is the only section of a general nature. In the remaining sections we focus on special AΔEs, giving rise to generalized gamma functions (Section 3), generalized zeta functions, (Section 4), the multiple zeta and gamma functions introduced and studied by Barnes (Section 5), and a novel generalization of the hypergeometric function $_2F_1$ (Section 6). This involves special AΔEs of the multiplicative type (1.1) for the various gamma functions, of the additive type (1.2) for the zeta functions, and of the second order type (1.3) for our generalized hypergeometric function.

The special functions surveyed in Sections 3 and 6 play an important role in the context of quantum Calogero-Moser systems of the relativistic variety. Here we will not address such applications. We refer the interested reader to our lecture notes Refs. [1, 2] for an overview of Calogero-Moser type integrable systems and the special functions arising in that setting. The present survey overlaps to some extent with Sections 2 and 3 of Ref. [2], but for lack of space we do not consider generalized Lamé functions. The latter can also be regarded as special solutions to second order A$\triangle$Es. They are surveyed in Section 4 of Ref. [2] and in Ref. [3].

2. Ordinary Linear A$\triangle$Es

In most of the older literature, nth order ordinary linear A$\triangle$Es are written in the form

$$\sum_{k=0}^{n} c_k(w)u(w+k) = 0, \quad w \in \mathbb{C}. \tag{2.1}$$

Here, the coefficients $c_0(w), \dots, c_n(w)$ have some specified analyticity properties, and one is looking for solutions $u(w)$ with corresponding properties. We have adopted a slightly different form for the A$\triangle$Es (1.1), (1.3), since this is advantageous for most of the special cases at issue. (Note that these A$\triangle$Es can be written in the form (2.1) by taking $z \to iw$, shifting and scaling.)

To provide some historical perspective, we mention that the subject of analytic difference equations burgeoned in the late 18th century and was vigorously pursued in the 19th century. In the course of the 20th century, this activity subsided considerably. Nörlund was one of the few early 20th century authors still focussing on (ordinary, linear) A$\triangle$Es. His well-known 1924 monograph Ref. [4] summarized the state of the art, in particular as regards his own extensive work. Later monographs from which further developments can be traced include Milne-Thompson (1933) [5], Meschkowski (1959) [6], and Immink (1984) [7].

Toward the end of the 20th century, it became increasingly clear that a special class of partial analytic difference equations plays an important role in the two related areas of quantum integrable N-particle systems and quantum groups. An early appraisal of this state of affairs (mostly from the perspective of integrable systems) can be found in our survey Ref. [8]. The most general class of A$\triangle$Es involved can be characterized by the coefficient functions that occur: They are combinations of Weierstrass σ-functions (equivalently, Jacobi theta functions).

Even for the case of ordinary A$\triangle$Es (to which we restrict attention in this contribution), this class of coefficients, referred to as 'elliptic' for brevity, had not been studied in previous literature. Our outlook on the general first order case, which we present in Subsection 2.1, arose from the need to handle the elliptic case and its specializations. In the same vein, the aspects of the second order case dealt with in Subsection 2.2 are anticipating later specializations. In particular, we point out some questions of a general nature that, to our knowledge, have not been addressed in the literature, and whose answers would have an important bearing on the special second order AΔEs for which explicit solutions are known.

2.1. FIRST ORDER AΔES

In this subsection we summarize some results concerning AΔEs of the forms (1.1) and (1.2), which we will use for the special cases in Sections 3–5. The pertinent results are taken from our paper Ref. [9], where proofs and further details can be found.

To start with, let us point out that all solutions to (1.1) satisfy

$$F(z + ika) \equiv \prod_{j=1}^{k} \Phi(z + (j - 1/2)ia) \cdot F(z), \qquad (2.2)$$

$$F(z - ika) \equiv \prod_{j=1}^{k} \frac{1}{\Phi(z - (j - 1/2)ia)} \cdot F(z). \qquad (2.3)$$

Whenever $\Phi(x + iy), x, y \in \mathbb{R}$, converges to 1 for $y \to \infty$, uniformly for x varying over arbitrary compact subsets of $\mathbb{R}$ and sufficiently fast, the infinite product

$$F_+(z) \equiv \prod_{j=1}^{\infty} \frac{1}{\Phi(z + (j - 1/2)ia)} \qquad (2.4)$$

defines a solution referred to as the upward iteration solution. Similarly, the downward iteration solution

$$F_-(z) \equiv \prod_{j=1}^{\infty} \Phi(z - (j - 1/2)ia) \qquad (2.5)$$

exists provided $\Phi(x + iy) \to 1$ for $y \to -\infty$ (uniformly on x-compacts and sufficiently fast).

The restrictions on $\Phi(z)$ for iteration solutions to exist are clearly quite strong. Indeed, they are violated in most of the special cases considered below. It is however possible to modify the iteration procedure, so as to handle larger classes of right-hand side functions $\Phi(z)$. This is the approach taken by Nörlund [4], which gives rise to the special solution he refers to as the 'Hauptlösung' (cf. also Ref. [10]).

Nörlund's methods do not apply to the case where $\Phi(z)$ is an elliptic or hyperbolic function, whereas the solution methods we consider next do not apply to all of Nörlund's class of $\Phi(z)$. The latter methods, however, can be used in particular for elliptic and hyperbolic right-hand sides $\Phi(z)$ (specifically, by exploiting the elliptic and hyperbolic gamma functions of Subsections 3.3 and 3.4).

We begin by imposing a rather weak requirement: We restrict attention to functions $\Phi(z)$ without poles and zeros in a strip $|\operatorname{Im} z| < c, c > 0$. (Usually, this can be achieved by a suitable shift of z.) Then we can take logarithms to trade the multiplicative AΔE (1.1) for the additive one (1.2), with z restricted to the strip $|\operatorname{Im} z| < c$. Let us next require that $\phi(z)$ have at worst polynomial increase on the real axis. Then we may and will restrict attention to solutions $f(z)$ that are polynomially bounded in the strip $|\operatorname{Im} z| \leq a/2$ and that are moreover analytic for $|\operatorname{Im} z| < c + a/2$.

We call solutions $f(z)$ of the form just delineated *minimal* solutions: Their analytic behavior and asymptotics for $|\operatorname{Re} z| \to \infty$ are optimal. The polynomial boundedness is critical in proving that minimal solutions are unique up to an additive constant, whenever they exist. (Note that entire ia-periodic functions such as $\operatorname{ch}(2\pi z/a)$ are not polynomially bounded.)

It is readily verified that the resulting function $F(z) = \exp(f(z))$ extends to a meromorphic solution of (1.1) that has no poles and zeros for $|\operatorname{Im} z| < c + a/2$, and whose logarithm is polynomially bounded for $|\operatorname{Im} z| \leq a/2$. Once more, solutions to (1.1) with the latter properties are termed *minimal*, and now they are unique up to a multiplier $\alpha \exp(2\pi k z/a), \alpha \in \mathbb{C}^*, k \in \mathbb{Z}$. (The exponential ambiguity reflects the ambiguity in the branch choice for $\ln \Phi(z)$.)

A large class of right-hand sides arises as follows. Assume that $\phi(z)$ satisfies (in addition to the above)

$$\phi(x) \in L^1(\mathbb{R}), \quad \hat{\phi}(y) \in L^1(\mathbb{R}), \quad \hat{\phi}(y) = O(y), \quad y \to 0. \qquad (2.6)$$

Here, $\hat{\phi}(y)$ denotes the Fourier transform, normalized by

$$\hat{\phi}(y) \equiv \frac{1}{2\pi} \int_{-\infty}^{\infty} dx \phi(x) e^{ixy}. \qquad (2.7)$$

Then the function

$$f(z) = \int_{-\infty}^{\infty} dy \frac{\hat{\phi}(2y)}{\operatorname{sh} ay} e^{-2iyz}, \quad |\operatorname{Im} z| \leq a/2, \qquad (2.8)$$

is clearly well defined, and analytic for $|\text{Im}\,z| < a/2$. Proceeding formally, it is also obvious that $f(z)$ satisfies (1.2) (by virtue of the Fourier inversion formula).

As a matter of fact, it can be shown that the function $f(z)$ defined by (2.8) analytically continues to $|\text{Im}\,z| < c + a/2$ and indeed obeys the AΔE (1.2). Moreover, $f(z)$ is bounded in the strip $|\text{Im}\,z| \le a/2$ and goes to 0 for $|\text{Re}\,z| \to \infty$ in the latter strip. (Therefore, it is a *minimal* solution.) The solution $f(z)$ is uniquely determined by these properties. A useful alternative representation reads

$$f(z) = \frac{1}{2ia} \int_{-\infty}^{\infty} du\phi(u)\text{th}\frac{\pi}{a}(z - u), \quad |\text{Im}\,z| < a/2. \qquad (2.9)$$

(See Theorem II.2 in Ref. [9] for proofs of the facts just mentioned.)

Another extensive class of right-hand sides $\Phi(z)$ admitting minimal solutions arises by assuming that $\Phi(z)$ has a real period, in addition to the standing assumption of absence of zeros and poles in a strip $|\text{Im}\,z| < c$. To anticipate our later needs, we denote this period by $\pi/r, r > 0$. Now $\phi(x) = \ln\Phi(x), x \in \mathbb{R}$, is well defined up to a multiple of $2\pi i$. We assume first that $\Phi(x)$ has zero winding number around 0 in the period interval $[-\pi/2r, \pi/2r]$. Then $\phi(x)$ is a smooth π/r-periodic function, too. Defining its Fourier coefficients by

$$\hat{\phi}_n \equiv \frac{r}{\pi} \int_{-\pi/2r}^{\pi/2r} dx\phi(x)e^{2inrx}, \quad n \in \mathbb{Z}, \qquad (2.10)$$

we also assume at first $\hat{\phi}_0 = 0$.

With these assumptions in effect, it is clear that the function

$$f(z) = \frac{1}{2} \sum_{n \in \mathbb{Z}^*} \frac{\hat{\phi}_n e^{-2inrz}}{\text{sh}nra}, \quad |\text{Im}\,z| \le a/2, \qquad (2.11)$$

is well defined, and analytic for $|\text{Im}\,z| < a/2$. Again, it is formally obvious (by Fourier inversion) that $f(z)$ satisfies the AΔE (1.2). Once more, it can be shown that $f(z)$ (2.11) has an analytic continuation to $|\text{Im}\,z| < c + a/2$ and solves (1.2) (cf. Theorem II.5 in Ref. [9]). Thus one obtains minimal solutions to the AΔE (1.1).

Next, we point out that the assumption $\hat{\phi}_0 = 0$ is not critical. Indeed, when $\hat{\phi}_0 \ne 0$, one need only add the function $\hat{\phi}_0 z/ia$ to $f(z)$ (2.11) to obtain a solution. Of course, this entails that $f(z)$ is no longer π/r-periodic.

The obvious generalization detailed in the previous paragraph is relevant to the case in which $\Phi(z)$ has winding number $l \in \mathbb{Z}^*$ on $[-\pi/2r, \pi/2r]$. Then one needs to take the z-derivative of the AΔE (1.2) to obtain a rhs $\phi'(z)$

that is π/r-periodic, but for which the zeroth Fourier coefficient of $\phi'(x)$ equals $2irl$, cf. (2.10). Thus the solution $f'(z)$ obtained as just sketched has a term linear in z, and so $f(z)$ is the sum of a π/r-periodic function and a quadratic function. Clearly, this still gives rise to a minimal solution $F(z) = \exp f(z)$ to the AΔE (1.1).

The idea underlying this construction is easily generalized to functions $\Phi(z)$ for which a suitable derivative of $\ln \Phi(z)$ has period π/r. Likewise, the Fourier transform method yielding the minimal solution (2.8) can be generalized to $\Phi(z)$ for which $\phi(z) \equiv \partial_z^k \ln \Phi(z)$ has the three properties (2.6) for a suitable $k \in \mathbb{N}^*$. (See Theorems II.3 and II.6 in Ref. [9] for the details.)

We have thus far focussed on the AΔE (1.1) with meromorphic rhs $\Phi(z)$ and its logarithmic version (1.2). But we may also consider (1.2) in its own right, relaxing the requirement that $\exp \phi(z)$ be meromorphic to the requirement that $\phi(z)$ be analytic for $|\mathrm{Im}\, z| < c$. In that case the above, inasmuch as it deals with (1.2), is still valid. Moreover, the analytic continuation behavior of the minimal solution $f(z)$ *follows* from the AΔE (1.2), in the sense that one can extend $f(z)$ beyond the strip $|\mathrm{Im}\, z| < c+a/2$, whenever $\phi(z)$ can be extended beyond the strip $|\mathrm{Im}\, z| < c$. For example, a function $\phi(z)$ that has branch points at the edges of the latter strip gives rise to a minimal solution $f(z)$ that has the same type of branch points at the edges of the former strip.

This more general version of (1.2) is relevant for the various zeta functions encountered below. In those cases the pertinent functions $\phi(z)$ are analytic in a half plane $\mathrm{Im}\, z > -c$ or in a strip $|\mathrm{Im}\, z| < c$, and they extend to multi-valued functions on all of $\mathbb{C}$ due to certain branch points for $|\mathrm{Im}\, z| \geq c$. Using the AΔE, the zeta functions admit a corresponding extension. We will not spell that out, however, but focus on properties pertaining to the analyticity half plane or strip.

2.2. SECOND ORDER AΔES

We begin by recalling some well-known facts concerning the general second order AΔE (1.3), cf. Ref. [4]. Let $F_1, F_2 \in \mathcal{M}^*$ be two solutions. Then their Casorati determinant,

$$\mathcal{D}(F_1, F_2; z) \equiv F_1(z + ia/2)F_2(z - ia/2) - F_1(z - ia/2)F_2(z + ia/2), \tag{2.12}$$

vanishes identically if and only if $F_1/F_2 \in \mathcal{P}_{ia}$. Assuming from now on $F_1/F_2 \notin \mathcal{P}_{ia}$, it is readily verified that the function (2.12) solves the first order AΔE

$$\frac{\mathcal{D}(z + ia/2)}{\mathcal{D}(z - ia/2)} = \frac{C^{(-)}(z)}{C^{(+)}(z)}. \tag{2.13}$$

Next, assume $F_3(z)$ is a third solution to (1.3). Then the functions

$$\mu_j(z) \equiv \mathcal{D}(F_j, F_3; z + ia/2)/\mathcal{D}(F_1, F_2; z + ia/2), \quad j = 1, 2, \tag{2.14}$$

belong to $\mathcal{P}_{ia}$ (1.7). (Indeed, quotients of Casorati determinants are ia-periodic in view of the AΔE (2.13).) It is routine to check that one has

$$F_3(z) = \mu_1(z)F_2(z) - \mu_2(z)F_1(z). \tag{2.15}$$

Conversely, any function of this form with $\mu_1, \mu_2 \in \mathcal{P}_{ia}$ solves (1.3). Therefore, whenever two solutions F_1, F_2 exist with $\mathcal{D}(F_1, F_2; z) \in \mathcal{M}^*$, the solution space may be viewed as a two-dimensional vector space over the field $\mathcal{P}_{ia}$ of ia-periodic meromorphic functions.

In contrast to ordinary second order differential and discrete difference equations, various natural existence questions have apparently not been answered in the literature. As a first example, the existence of a solution basis as just considered seems not to be known in general.

Further questions arise in the following situation. Assume that two AΔEs of the above form,

$$C_\delta^{(+)}(z)F(z + ia_{-\delta}) + C_\delta^{(-)}(z)F(z - ia_{-\delta}) + C_\delta^{(0)}(z)F(z) = 0, \tag{2.16}$$
$$a_\delta > 0, \quad \delta = +, -,$$

are given. Then one may ask for conditions on the two sets of coefficients such that joint solutions $F \in \mathcal{M}^*$ exist. Next, assume there do exist two joint solutions to (2.16) whose Casorati determinants w.r.t. a_+ and a_- belong to $\mathcal{M}^*$. When a_+/a_- is rational, this entails that the joint solution space is infinite-dimensional. (Indeed, letting $a_+ = pa$ and $a_- = qa$ with p, q coprime integers, one can allow arbitrary multipliers in $\mathcal{P}_{ia}$.) Now it is not hard to see that one has

$$\mathcal{P}_{ia_+} \cap \mathcal{P}_{ia_-} = \mathbb{C}, \quad a_+/a_- \notin \mathbb{Q}. \tag{2.17}$$

But it is not clear whether this entails that the joint solution space is two-dimensional for irrational a_+/a_-.

On the other hand, assuming $F_1, F_2 \in \mathcal{M}^*$ are two joint solutions to (2.16) with a_+/a_- irrational, there is a quite useful extra assumption guaranteeing that the solution space is two-dimensional with basis $\{F_1, F_2\}$. Specifically, one need only assume

$$\lim_{\text{Im } z \to \infty} F_1(z)/F_2(z) = 0. \tag{2.18}$$

for all Re z in some open interval. Indeed, the sufficiency of this condition can be easily gleaned from the proof of Theorem B.1 in Ref. [11], where certain special cases are considered.

Questions about joint solutions arise from commuting operator pairs, not only in the elliptic context of Ref. [11], but also in the hyperbolic one of Ref. [12] and Section 6. In order to detail the latter setting in general terms, consider the eigenvalue problems for the analytic difference operators

$$A_\delta \equiv C_\delta(z)T_{ia_{-\delta}} + C_\delta(-z)T_{-ia_{-\delta}} + V_\delta(z), \quad \delta = +, -,$$

$$(2.19)$$

on $\mathcal{M}$, with $C_\delta \in \mathcal{P}_{ia_\delta}^*, V_\delta \in \mathcal{P}_{ia_\delta}$, and $T_{\pm ia_\delta}$ given by

$$(T_\alpha F)(z) \equiv F(z - \alpha), \quad \alpha \in \mathbb{C}^*.$$

$$(2.20)$$

Thus we are interested in solutions to the AΔEs

$$A_\delta F = E_\delta F, \quad E_\delta \in \mathbb{C}, \quad \delta = +, -.$$

$$(2.21)$$

Fixing E_+ and E_-, these are of the form considered before. Moreover, since A_+ and A_- clearly commute, it is reasonable to ask for *joint* solutions.

In Section 6 we encounter an operator pair with this structure. Furthermore, we have one joint solution $R(z)$ available for a family of joint eigenvalues $(E_+(p), E_-(p)), p \in \mathbb{C}$. Fixing p, the three solutions $R(z + ia_+), R(z - ia_+)$ and $R(z)$ to the AΔE $A_+F = E_+(p)F$ are related via the AΔE $(A_-R)(z) = E_-(p)R(z)$ with coefficients in $\mathcal{P}_{ia_-}$, in agreement with the general theory. Put differently, when one fixes attention on one of the AΔEs, the other AΔE can be viewed as an extra requirement of monodromy type.

The coefficients occurring in Section 6 are analytic in the shift parameters a_+, a_- for $a_+, a_- \in \mathbb{C}^*$, and entire in four extra parameters. The joint solution $R(z)$ is real-analytic in a_+, a_- for $a_+, a_- \in (0, \infty)$ and meromorphic in the extra parameters. But we do not know whether another joint solution exists for general parameters. On the other hand, upon *specializing* the extra parameters, we do obtain two joint solutions for all $a_+, a_- \in (0, \infty)$ and $p \in \mathbb{C}$, which moreover satisfy the extra condition (2.18) for all Re $z > 0$ and p in the right half plane, cf. Ref. [12].

In the latter article we also arrive at a commuting analytic difference operator pair $B_\delta(b), b \in \mathbb{C}$, as considered above, which depends analytically on the extra parameter b, and which admits joint eigenfunctions for a set of (a_+, a_-, b) that is *dense* in $(0, \infty)^2 \times \mathbb{R}$. But from the properties of these eigenfunctions one can deduce that the pair does not admit joint eigenfunctions depending continuously on a_+, a_-, b. (See the paragraphs between Eqs. (3.11) and (3.12) in Ref. [12].) From such concrete examples one sees that expectations based on experience with analytic differential equations need not be borne out for analytic difference equations.

3. Generalized Gamma Functions

When $F(z)$ solves the first order AΔE (1.1), it is evident that $1/F(z)$ solves the AΔE with rhs $1/\Phi(z)$. Likewise, when $F_1(z), F_2(z)$ are solutions to the AΔEs with right-hand sides $\Phi_1(z), \Phi_2(z)$, then $F_1(z)F_2(z)$ clearly solves the AΔE with rhs $\Phi_1(z)\Phi_2(z)$. Since elliptic right-hand sides $\Phi(z)$ and their various degenerations admit a factorization in terms of the Weierstrass σ-function and its degenerations, it is natural to construct 'building block' solutions corresponding to the σ-function and its trigonometric, hyperbolic and rational specializations.

We refer to the building blocks reviewed below as generalized gamma functions. Indeed, in the rational case considered in Subsection 3.1, the pertinent gamma function amounts to Euler's gamma function. Trigonometric, elliptic and hyperbolic gamma functions are surveyed in Subsections 3.2–3.4. As we will explain, these functions can all be viewed as minimal solutions, rendered unique by normalization requirements. They were introduced and studied from this viewpoint in Ref. [9]. The hyperbolic gamma function will be a key ingredient in Section 6, where we need a great many of its properties sketched in Subsection 3.4.

3.1. THE RATIONAL GAMMA FUNCTION

We define the rational gamma function by

$$G_{\mathrm{rat}}(a; z) \equiv (2\pi)^{-1/2} \exp\left(-\frac{iz}{a}\ln a\right)\Gamma\left(-\frac{iz}{a}+\frac{1}{2}\right). \tag{3.1}$$

(This definition deviates from our previous one in Ref. [2] by the first two factors on the rhs. We feel that the present definition is more natural from the perspective of the zeta functions in Sections 4 and 5.) The Γ-function AΔE $\Gamma(x+1) = x\Gamma(x)$ entails that G_{rat} solves the AΔE

$$\frac{G(z+ia/2)}{G(z-ia/2)} = -iz. \tag{3.2}$$

More generally, when $\Phi(z)$ is an arbitrary rational function, we can solve the AΔE (1.1) by a function $F(z)$ of the form

$$F(z) = \exp(c_0 + c_1 z)\frac{\prod_{j=1}^{M} G_{\mathrm{rat}}(a; z - \alpha_j)}{\prod_{k=1}^{N} G_{\mathrm{rat}}(a; z - \beta_k)}. \tag{3.3}$$

Indeed, when we let the integers N, M vary over $\mathbb{N}$ and c_1, α_j, β_k over $\mathbb{C}$, then we obtain *all* rational functions on the rhs of (1.1) (save for the zero function, of course).

We now explain the relation of G_{rat} to the notion of 'minimal solution'. To this end we first of all need a shift of z on the rhs of (3.2) in order to get a function $\Phi(z)$ that is analytic and free of zeros in a strip around the real axis. Taking

$$\Phi(z) \equiv -i(z + ia/2), \tag{3.4}$$

a simple contour integration yields

$$\frac{1}{2\pi} \int_{-\infty}^{\infty} dx e^{ixy} \partial_x^2 \ln \Phi(x) = \begin{cases} -ay \exp(ay/2), & y < 0, \\ 0, & y \geq 0. \end{cases} \tag{3.5}$$

Therefore $\phi(z) \equiv \partial_z^2 \ln \Phi(z)$ has the three properties (2.6), and the corresponding minimal solution (2.8) is then given by

$$f(z) = -2 \int_{-\infty}^{0} dy \frac{y e^{ay}}{\operatorname{sh} ay} e^{-2iyz}. \tag{3.6}$$

The point is now that the rhs of (3.6) equals $\partial_z^2 \ln G_{\text{rat}}(a; z + ia/2)$, so that $G_{\text{rat}}(a; z + ia/2)$ may be characterized as a *minimal* solution to the AΔE with rhs (3.4). It is not hard to check the asserted equality via Gauss' formula for the psi function (the logarithmic derivative of $\Gamma(x)$). But one may in fact reobtain various results on the gamma function (such as Gauss' formula) by taking the function (3.6) as a starting point. Put differently, if one would not have been familiar with the gamma function beforehand, one would have been led to it (and to a substantial part of its theory) via the 'minimal solution' approach sketched above.

We have worked out the details of this useful perspective on Euler's gamma function in Appendix A of our paper Ref. [9]. We cite in particular one Γ-function representation derived there, which we have occasion to invoke in Subsection 4.1. It reads

$$\Gamma(w) = (2\pi)^{1/2} \exp\left(\int_0^\infty \frac{dt}{t} \left((w - \frac{1}{2}) e^{-t} - \frac{1}{t} + \frac{e^{-wt}}{1 - e^{-t}} \right) \right), \quad \operatorname{Re} w > 0, \tag{3.7}$$

cf. Eq. (A37) in Ref. [9]. Here we only add a few more remarks on several features that are important with an eye on the generalized gamma functions introduced and studied in later subsections.

First, it should be observed that the shift of z by $ia/2$ is arbitrary; any shift $z \to z + ic, c > 0$, would yield substantially the same conclusions.

Second, a shift $z \to z-ic, c > 0$, on the rhs of (3.2) yields in the same way as sketched above solutions to (3.2) of the form $\alpha \exp(-iza^{-1}\ln a) \exp((2k+1)\pi z/a)/\Gamma(iz/a + 1/2)$, with $\alpha \in \mathbb{C}^*, k \in \mathbb{Z}$. Hence the quotient of such a solution and $G_{\text{rat}}(a; z)$ is ia-periodic. This is in agreement with the well-known reflection equation, which becomes here

$$\Gamma\left(\frac{iz}{a} + \frac{1}{2}\right)\Gamma\left(-\frac{iz}{a} + \frac{1}{2}\right) = \pi \text{ch}\left(\frac{\pi z}{a}\right)^{-1}. \tag{3.8}$$

(This identity can also be derived from the AΔE-viewpoint, cf. Appendix A in Ref. [9].)

Third, we recall that the minimal solutions to the AΔE (1.1) with rhs (3.4) (obtained via twofold integration of $f(z)$ (3.6)) are analytic and zero-free for $|\text{Im } z| < a$. From the AΔE one then sees that they are analytic and zero-free for all z not in $-ia\mathbb{N}^*$, whereas they have simple poles for $z = -iak, k \in \mathbb{N}^*$. (Similarly, the shift $z \to z - ic, c > 0$, in (3.2) yields minimal solutions without poles and with a zero sequence in the upper half plane.) Of course, this is once again well known for the special function $G_{\text{rat}}(a; z)$ (3.1). It is illuminating to compare its pole sequence

$$z = -i(k + 1/2)a, \quad k \in \mathbb{N}, \quad \text{(poles)}, \tag{3.9}$$

to those of the generalized gamma functions introduced below.

3.2. THE TRIGONOMETRIC GAMMA FUNCTION

In the same way as $G_{\text{rat}}(a; z)$ (3.1) serves as a building block to solve AΔEs with rational right-hand sides, the trigonometric gamma function can be used to handle trigonometric functions $\Phi(z)$ with period π/r. (More precisely, functions in the field $\mathbb{C}(\exp(2irz))$.) Specifically, letting

$$F(z) = \exp(c_0 + c_1 z + c_2 z^2)\frac{\prod_{j=1}^{M} G_{\text{trig}}(r, a; z - \alpha_j)}{\prod_{k=1}^{N} G_{\text{trig}}(r, a; z - \beta_k)}, \tag{3.10}$$

with $G_{\text{trig}}(r, a; z)$ solving

$$\frac{G(z + ia/2)}{G(z - ia/2)} = 1 - \exp(2irz), \tag{3.11}$$

one obtains all trigonometric functions by letting N, M vary over $\mathbb{N}$ and $c_1, c_2, \alpha_j, \beta_k$ over $\mathbb{C}$ in the quotient $F(z - ia/2)/F(z + ia/2)$.

The obvious solution to (3.11) is the upward iteration solution

$$G_{\text{trig}}(r, a; z) \equiv \prod_{k=1}^{\infty}(1 - q^{2k-1}e^{2irz})^{-1}, \quad q \equiv e^{-ar}. \tag{3.12}$$

Indeed, the infinite product clearly converges, yielding a meromorphic solution without zeros and with poles for

$$z = j\pi/r - i(k+1/2)a, \quad j \in \mathbb{Z}, \quad k \in \mathbb{N}, \quad \text{(poles)}.$$

$$(3.13)$$

Another representation for G_{trig} arises by writing

$$\prod_{k=1}^{\infty}(1 - q^{2k-1}e^{2irz})^{-1} = \exp\left(-\sum_{k=1}^{\infty}\ln(1 - q^{2k-1}e^{2irz})\right),$$

$$(3.14)$$

and using the elementary Fourier series

$$\ln(1 - e^{2ir(z+ic)}) = -\sum_{n=1}^{\infty}\frac{e^{2inr(z+ic)}}{n}, \quad c > 0.$$

$$(3.15)$$

Indeed, this gives rise to the formula

$$G_{\text{trig}}(r, a; z) = \exp\left(\sum_{n=1}^{\infty}\frac{e^{2inrz}}{2n\operatorname{sh}nra}\right), \quad \operatorname{Im} z > -a/2,$$

$$(3.16)$$

as is easily checked.

Taking $z \to z + ic, c > 0$, in the AΔE (3.11), it is of the form discussed in Subsection 2.1. Taking next logarithms and using (3.15), one deduces that (3.16) amounts to the special solution (2.11). That is, $G_{\text{trig}}(r, a; z)$ is once more a *minimal* solution to the (shifted) AΔE (3.11). As such, it is uniquely determined by the asymptotics

$$G_{\text{trig}}(r, a; z) \sim 1, \quad \operatorname{Im} z \to \infty,$$

$$(3.17)$$

which can be read off from (3.16).

Our trigonometric gamma function is closely related to Thomae's q-gamma function $\Gamma_q(x)$ [13]: One has

$$G_{\text{trig}}(r, a; z) = \exp(c_0 + c_1 z)\Gamma_{\exp(-2ar)}\left(-\frac{iz}{a} + \frac{1}{2}\right),$$

$$(3.18)$$

for suitable constants c_0, c_1. The AΔE-perspective from which G_{trig} arises leads to various other features that are detailed in Ref. [9]. A useful limit that is not mentioned explicitly there reads

$$\lim_{r\downarrow 0}\exp\left(-\frac{\pi^2}{12ar}\right)G_{\text{trig}}(r, a; 0) = 2^{-1/2}.$$

$$(3.19)$$

(This follows by combining Eqs. (3.128), (3.129) and (3.154) in *loc. cit.*, taking $z = 0$.) Using Proposition III.20 in *loc. cit.*, we then deduce

$$\lim_{r \downarrow 0} \exp\left(-\frac{\pi^2}{12ar} + \frac{iz}{a} \ln(2r) \right) G_{\text{trig}}(r, a; z) = G_{\text{rat}}(a; z).$$

(3.20)

Finally, we point out that it is evident from the infinite product representation (3.12) that one can allow r to vary over the (open) right half plane, whereas one cannot take $r \in i\mathbb{R}$. This is why one needs another building block function to handle hyperbolic right-hand side functions $\Phi(z)$ in the AΔE (1.1), cf. Subsection 3.4.

3.3. THE ELLIPTIC GAMMA FUNCTION

The elliptic gamma function is the minimal solution to the AΔE

$$\frac{G(z + ia/2)}{G(z - ia/2)} = \exp\left(-\sum_{n=1}^{\infty} \frac{\cos(2nrz)}{n \operatorname{sh}(nrb)} \right), \quad |\operatorname{Im} z| < b,$$

(3.21)

obtained via the formula (2.11):

$$G_{\text{ell}}(r, a, b; z) \equiv \exp\left(i \sum_{n=1}^{\infty} \frac{\sin(2nrz)}{2n \operatorname{sh}(nra) \operatorname{sh}(nrb)} \right), \quad |\operatorname{Im} z| < (a + b)/2.$$

(3.22)

Indeed, our definition entails that the functions

$$F(z) = \exp(c_0 + c_1 z + c_2 z^2 + c_3 z^3) \frac{\prod_{j=1}^{M} G_{\text{ell}}(r, a, b; z - \alpha_j)}{\prod_{k=1}^{N} G_{\text{ell}}(r, a, b; z - \beta_k)},$$

(3.23)

give rise to all elliptic right-hand side functions $\Phi(z)$ with periods π/r and ib.

To explain why this is true, we recall first that any elliptic function with periods π/r and ib admits a representation as

$$\alpha \prod_{j=1}^{N} \frac{\sigma(z - \gamma_j; \frac{\pi}{2r}, \frac{ib}{2})}{\sigma(z - \delta_j; \frac{\pi}{2r}, \frac{ib}{2})},$$

(3.24)

where $\alpha, \gamma_j, \delta_j \in \mathbb{C}$. The crux is now that the function on the rhs of (3.21) is of the form

$$\exp(d_0 + d_1 z + d_2 z^2) \sigma\left(z + ib/2; \frac{\pi}{2r}, \frac{ib}{2} \right).$$

(3.25)

Hence the functions $F(z + ia/2)/F(z - ia/2)$, with $F(z)$ given by (3.23), yield all functions (3.24) (with $\alpha \neq 0$) by choosing

$$N = M, \quad \alpha_j = \gamma_j + ib/2, \quad \beta_j = \delta_j + ib/2, \quad j = 1, \ldots, N,$$

(3.26)

and appropriate constants c_1, c_2, c_3 determined by the constants α, d_0, d_1 and d_2.

Next, we mention that the elliptic gamma function can also be written as an infinite product

$$G_{\text{ell}}(r, a, b; z) = \prod_{m,n=1}^{\infty} \frac{1 - q_a^{2m-1} q_b^{2n-1} e^{-2irz}}{1 - q_a^{2m-1} q_b^{2n-1} e^{2irz}}, \quad q_a \equiv e^{-ar}, \quad q_b \equiv e^{-br}.$$

(3.27)

To check this, one need only proceed as in the trigonometric case: The infinite product can be written as the exponential of a series; using the Fourier series (3.15) and summing the resulting geometric series yields (3.22).

From (3.27) one can read off meromorphy in z and the locations of poles and zeros. Specifically, one obtains the doubly-infinite sequences

$$z = j\pi/r - i(k + 1/2)a - i(l + 1/2)b, \quad j \in \mathbb{Z}, \; k, l \in \mathbb{N}, \quad \text{(poles)},$$

(3.28)

$$z = j\pi/r + i(k + 1/2)a + i(l + 1/2)b, \quad j \in \mathbb{Z}, \; k, l \in \mathbb{N}, \quad \text{(zeros)}.$$

(3.29)

It is immediate from the product representations (3.12) and (3.27) that one has

$$G_{\text{trig}}(r, a; z) = \lim_{b \uparrow \infty} G_{\text{ell}}(r, a, b; z - ib/2).$$

(3.30)

It is also not difficult to see that for the renormalized function

$$\tilde{G}_{\text{ell}}(r, a, b; z) \equiv \exp\left(\frac{\pi^2 z}{6irab}\right) G_{\text{ell}}(r, a, b; z),$$

(3.31)

the $r \downarrow 0$ limit exists. This yields the function

$$G_{\text{hyp}}(a, b; z) = \lim_{r \downarrow 0} \tilde{G}_{\text{ell}}(r, a, b; z),$$

(3.32)

studied in the next subsection. For further properties of the elliptic gamma function we refer to Subsection III.B in Ref. [9].

3.4. THE HYPERBOLIC GAMMA FUNCTION

The defining AΔE of the hyperbolic gamma function reads

$$\frac{G(z + ia/2)}{G(z - ia/2)} = 2\text{ch}(\pi z/b). \tag{3.33}$$

Clearly, any solution $G_{\text{hyp}}(a, b; z)$ of (3.33) can be used to solve (1.1) with $\Phi(z)$ an arbitrary hyperbolic function with period ib. Indeed, all functions $\Phi \in \mathbb{C}(\exp(2\pi z/b))$ arise via functions of the form

$$F(z) = \exp(c_0 + c_1 z + c_2 z^2) \frac{\prod_{j=1}^{M} G_{\text{hyp}}(a, b; z - \alpha_j)}{\prod_{k=1}^{N} G_{\text{hyp}}(a, b; z - \beta_k)}. \tag{3.34}$$

As before, there is a certain arbitrariness in the choice of AΔE for the building block function. Our choice $2\text{ch}(\pi z/b)$ together with the requirement that the solution be minimal will lead us to a function $G_{\text{hyp}}(a, b; z)$ with various features that would be spoiled by any other choice, however. (In particular, the constant 2 cannot be changed without losing the remarkable $(a \leftrightarrow b)$-invariance of the hyperbolic gamma function (3.35).)

Following the method to construct minimal solutions sketched in Subsection 2.1, one readily verifies that $\phi(z) \equiv \partial_z^3 \ln(2\text{ch}(\pi z/b))$ has the three properties (2.6). The Fourier transform can be done explicitly, and integrating up three times then yields our hyperbolic gamma function,

$$G_{\text{hyp}}(a, b; z) \equiv \exp\left(i \int_0^\infty \frac{dy}{y}\left(\frac{\sin(2yz)}{2\text{sh}(ay)\text{sh}(by)} - \frac{z}{aby}\right)\right), \tag{3.35}$$
$$|\text{Im}\, z| < (a + b)/2.$$

Since $G_{\text{hyp}}(a, b; z)$ has no poles and zeros in the strip $|\text{Im}\, z| < (a + b)/2$, one readily deduces from the defining AΔE (3.33) that G_{hyp} extends to a meromorphic function with poles and zeros given by

$$z = -i(k + 1/2)a - i(l + 1/2)b, \quad k, l \in \mathbb{N}, \quad \text{(poles)}, \tag{3.36}$$

$$z = i(k + 1/2)a + i(l + 1/2)b, \quad k, l \in \mathbb{N}, \quad \text{(zeros)}. \tag{3.37}$$

In Section 6 we need various other properties of $G_{\text{hyp}}(a, b; z)$. Some automorphy properties are immediate from (3.35): One has

$$G_{\text{hyp}}(a, b; -z) = 1/G_{\text{hyp}}(a, b; z), \tag{3.38}$$

$$G_{\text{hyp}}(a, b; z) = G_{\text{hyp}}(b, a; z), \tag{3.39}$$

$$G_{\text{hyp}}(\lambda a, \lambda b; \lambda z) = G_{\text{hyp}}(a, b; z), \quad \lambda > 0. \tag{3.40}$$

It is also easy to establish from the defining AΔE that the multiplicity of a pole or zero z_0 equals the number of distinct pairs $(k, l) \in \mathbb{N}^2$ giving rise to z_0, cf. (3.36), (3.37). In particular, the pole at $-i(a + b)/2$ and zero at $i(a + b)/2$ are simple, and for a/b irrational *all* poles and zeros are simple.

The remaining properties we have occasion to use are not clear by inspection, and we refer to Subsection IIIA in Ref. [9] for a detailed account. First, we need to know the residue at the simple pole $-i(a + b)/2$. It is given by

$$\text{Res}(-i(a + b)/2) = \frac{i}{2\pi}(ab)^{1/2}. \tag{3.41}$$

(Here and below, we take positive square roots of positive quantities.)

Second, we need two distinct zero step size limits of the hyperbolic gamma function. The first one reads

$$\lim_{b \downarrow 0} \frac{G_{\text{hyp}}(\pi, b; z + i\lambda b)}{G_{\text{hyp}}(\pi, b; z + i\mu b)} = \exp((\lambda - \mu)\ln(2\text{ch}z)), \tag{3.42}$$

where the limit is uniform on compact subsets of the cut plane

$$\mathbb{C} \setminus \{\pm iz \in [\pi/2, \infty)\}. \tag{3.43}$$

When $\lambda - \mu$ is an integer, this limit is immediate from the $(a \leftrightarrow b)$-invariance of G_{hyp} and the AΔE (3.33). For $\lambda - \mu \notin \mathbb{Z}$, the emergence of the logarithmic branch cuts on the imaginary axis may be viewed as a consequence of the coalescence of an infinite number of zeros and poles on the cuts.

The second zero step size limit yields the connection to the Γ-function. Consider the function

$$H(\rho; z) \equiv G_{\text{hyp}}(1, \rho; \rho z + i/2)\exp(iz\ln(2\pi\rho))/(2\pi)^{1/2}. \tag{3.44}$$

From the AΔE (3.33) and its $(a \leftrightarrow b)$-counterpart one sees that H satisfies the AΔE

$$\frac{H(\rho; z + i/2)}{H(\rho; z - i/2)} = \frac{i\text{sh}(\pi\rho z)}{\pi\rho}, \tag{3.45}$$

and reflection equation

$$H(\rho; z)H(\rho; -z) = \pi^{-1}\text{ch}\pi z. \tag{3.46}$$

(Recall also (3.38) to check (3.46).) Therefore, it should not come as a surprise that one has

$$\lim_{\rho\downarrow 0} H(\rho; z) = 1/\Gamma(iz + 1/2), \tag{3.47}$$

uniformly for z in $\mathbb{C}$-compacts.

We have made a (convenient) choice for the first parameter a of the hyperbolic gamma function, but it should be pointed out that the scale invariance (3.40) can be used to handle arbitrary a. In particular, recalling the definition (3.1), this yields the limit

$$\lim_{b\uparrow\infty} \exp\left(\frac{iz}{a}\ln(2\pi/b)\right) G_{\text{hyp}}(a, b; z - ib/2) = G_{\text{rat}}(a; z), \tag{3.48}$$

which should be compared to the trigonometric limit (3.20).

Third, we need the $|\text{Re}\,z| \to \infty$ asymptotics of G_{hyp}. To detail this, we set

$$c \equiv \max(a, b), \quad \sigma = 1 - \epsilon, \quad \epsilon > 0. \tag{3.49}$$

Then we have

$$\mp i \ln G_{\text{hyp}}(a, b; z) = -\frac{\pi z^2}{2ab} - \frac{\pi}{24}\left(\frac{a}{b} + \frac{b}{a}\right) + O(\exp(\mp 2\pi\sigma\text{Re}\,z/c)),$$
$$\text{Re}\,z \to \pm\infty. \tag{3.50}$$

Here, the bound is uniform for $\text{Im}\,z$ in $\mathbb{R}$-compacts, but it is not uniform as $\epsilon \downarrow 0$, since it is false for $\sigma = 1$ and $a = b$.

Finally, in Subsection 4.4 we need the representation

$$\partial_z \ln G_{\text{hyp}}(a, b; z) = \frac{\pi}{2ia^2} \int_{-\infty}^{\infty} du \frac{\ln(2\text{ch}(\pi u/b))}{\text{ch}^2(\pi(z - u)/a)}, \quad |\text{Im}\,z| < a/2, \tag{3.51}$$

which cannot be found in Ref. [9]. To prove its validity, we first note that when we take two more z-derivatives, then the resulting formula amounts to a special case of (2.9). Indeed, this follows upon trading the z-derivatives for u-derivatives, and then integrating by parts three times.

As a consequence, $\partial_z \ln G_{\text{hyp}}$ is given by the rhs of (3.51) plus a term of the form $Az + B$. Now since $\partial_z G_{\text{hyp}}/G_{\text{hyp}}$ is even, we must have $A = 0$. To show $B = 0$, we consider the $\text{Re}\,z \to \infty$ asymptotics of the rhs of (3.51). Changing variables $u \to z - x$, it is routine to check it reads $-i\pi z/ab + o(1)$. Comparing to the derivative of (3.50) for $\text{Re}\,z \to \infty$, we deduce $B = 0$. (By virtue of Cauchy's formula, it is legitimate to differentiate the bound (3.50).)

To conclude this subsection we add some further information on the hyperbolic gamma function. First of all, this function was introduced in another guise and from a quite different perspective in previous literature—a fact we were not aware of at the time we wrote Ref. [9]. Indeed, our hyperbolic gamma function is related to a function that is nowadays referred to as Kurokawa's double sine function [14]. The latter is usually denoted $S_2(x|\omega_1, \omega_2)$, and the relation reads

$$G_{\text{hyp}}(a, b; z) = S_2(iz + (a + b)/2|a, b). \tag{3.52}$$

The first occurrence of the double sine function is however in a series of papers by Barnes, published a century ago. He generalized the gamma function from another viewpoint to his so-called multiple gamma functions; the double sine function is then a quotient of two double gamma functions. We return to Barnes' multiple gamma functions in Subsection 5.2, where they will be tied in with the minimal solution ideas of Subsection 2.1.

Multiple gamma functions show up in particular in number theory. The double gamma and sine functions were studied from this viewpoint in a paper by Shintani [15]. He derived a product formula that can be used to tie in the hyperbolic and trigonometric gamma functions in a quite explicit way, and we continue by presenting the pertinent formulas.

First, we should mention that $G_{\text{hyp}}(a, b; z)$ admits a representation as an infinite product of Γ-functions, from which meromorphy properties in a, b and z follow by inspection. To be specific, using the scale-invariant variables

$$\lambda \equiv -iz/a, \quad \rho \equiv b/a, \tag{3.53}$$

one has

$$G_{\text{hyp}}(a, b; z + \frac{ia}{2})^2 = 2\cos(\pi\lambda/\rho)e^{2\lambda \ln 2}$$

$$\times \prod_{j=0}^{\infty} \frac{\Gamma((j + \frac{1}{2})\rho + \lambda)\,\Gamma(1 + (j + \frac{1}{2})\rho + \lambda)}{\Gamma((j + \frac{1}{2})\rho - \lambda)\,\Gamma(1 + (j + \frac{1}{2})\rho - \lambda)}$$

$$\times e^{-4\lambda \ln(j + \frac{1}{2})\rho}. \tag{3.54}$$

Here, one needs $\rho \in \mathbb{C} \setminus (-\infty, 0]$ for the infinite product to converge, cf. Prop. III.5 in Ref. [9]. (Observe that for non-real ρ the poles and zeros on the rhs are double, as should be the case.)

It follows in particular from this representation that $G_{\text{hyp}}(a, b; z)$ has an analytic continuation to $b \in i(0, \infty)$. Now for $b = i\pi/r, r > 0$, the rhs of (3.33) can be rewritten as

$$\exp(-irz)[1 - \exp(2ir[z + \pi/2r])]. \tag{3.55}$$

Comparing to the rhs of (3.11), we deduce that the quotient function

$$\exp(-rz^2/2a)G_{\text{trig}}(r, a; z + \pi/2r)/G_{\text{hyp}}(a, i\pi/r; z) \qquad (3.56)$$

is ia-periodic.

Using Shintani's formula, we can determine this ia-periodic quotient explicitly. For the case at hand his product formula amounts to

$$G_{\text{hyp}}(a, i\pi/r; z) \;=\; \exp\left(-\frac{rz^2}{2a} - \frac{1}{24}\left(ra - \frac{\pi^2}{ra}\right)\right) \qquad (3.57)$$

$$\times \; \prod_{k=1}^{\infty} \frac{1 + \exp(2i\pi a^{-1}[iz + i(k - 1/2)\pi r^{-1}])}{1 + \exp(2ir[z + i(k - 1/2)a])}.$$

From the definition (3.12) of the trigonometric gamma function we then get the remarkable relation

$$G_{\text{hyp}}(a, i\pi/r; z) \;=\; \exp\left(-\frac{rz^2}{2a} - \frac{1}{24}(ra - \frac{\pi^2}{ra})\right) \qquad (3.58)$$

$$\times \; \frac{G_{\text{trig}}(r, a; z + \pi/2r)}{G_{\text{trig}}(\pi/a, \pi/r; iz + a/2)},$$

from which the ia-periodic function (3.56) can be read off. (The reader who is familiar with modular transformation properties may find more information on this angle below Eq. (A.27) in Ref. [16], where we rederived Shintani's product formula.)

4. Generalized Zeta Functions

This section is concerned with the Hurwitz zeta function $\zeta(s, w)$, and with 'trigonometric,' 'elliptic' and 'hyperbolic' generalizations thereof. These generalizations are arrived at by combining the perspective of Subsection 2.1 with the trigonometric, elliptic and hyperbolic gamma functions of Section 3, respectively. To our knowledge, the generalized zeta functions studied in Subsections 4.2–4.4 have not appeared in previous literature. (In any event, our results reported there are hitherto unpublished.) A different generalization of the Hurwitz zeta function has been introduced and studied by Ueno and Nishizawa [17].

A suitable specialization also leads to a 'trigonometric' and 'hyperbolic' generalization of Riemann's zeta function $\zeta(s) = \zeta(s, 1)$. Via the latter we are led to some integral representations for $\zeta(s)$ that seem to be new. See also Suslov's contribution to these proceedings [18], where, among other things, different generalizations of $\zeta(s)$ are discussed.

4.1. THE RATIONAL ZETA FUNCTION

The Hurwitz zeta function is defined by

$$\zeta(s,w) \equiv \sum_{m=0}^{\infty} (w+m)^{-s}, \quad \operatorname{Re} w > 0, \quad \operatorname{Re} s > 1. \tag{4.1}$$

From this definition it is immediate that $\zeta(s,w)$ is a solution to the (additive) first order AΔE

$$\zeta(w+1) - \zeta(w) = -w^{-s}. \tag{4.2}$$

Let us now introduce

$$\phi_s(z) \equiv -(c-iz)^{-s}, \quad c > 0, \tag{4.3}$$

and consider the AΔE

$$f(z+i/2) - f(z-i/2) = \phi_s(z). \tag{4.4}$$

Using Euler's integral

$$\Gamma(x) = \int_0^{\infty} u^{x-1} e^{-u} du, \quad \operatorname{Re} x > 0, \tag{4.5}$$

we may write

$$\phi_s(z) = -\frac{1}{\Gamma(s)} \int_0^{\infty} t^{s-1} e^{-ct} e^{itz} dt, \quad \operatorname{Re} s > 0, \quad \operatorname{Im} z > -c. \tag{4.6}$$

Thus we have

$$\hat{\phi}_s(y) = \begin{cases} -(-y)^{s-1} e^{cy}/\Gamma(s), & y < 0, \\ 0, & y \geq 0, \end{cases} \tag{4.7}$$

cf. (2.7). Taking $\operatorname{Re} s \geq 2$, we deduce that $\phi_s(z)$ satisfies the conditions (2.6). The corresponding minimal solution (2.8) to the AΔE (4.4) reads

$$f_s(z) = \frac{1}{\Gamma(s)} \int_0^{\infty} dy (2y)^{s-1} e^{-2cy} e^{2iyz}/\operatorname{sh} y, \quad \operatorname{Re} s \geq 2, \tag{4.8}$$

where we may take $\operatorname{Im} z > -c - 1/2$. Writing

$$1/\operatorname{sh} y = 2e^{-y} \sum_{m=0}^{\infty} e^{-2my}, \quad y > 0, \tag{4.9}$$

this yields (using (4.5) once more)

$$f_s(z) = \sum_{m=0}^{\infty} \left(\frac{1}{2} + m + c - iz \right)^{-s} = \zeta\left(s, \frac{1}{2} + c - iz \right), \quad \mathrm{Re}\, s \geq 2.$$

$$(4.10)$$

The upshot is that for $\mathrm{Re}\, s \geq 2$ we may view the function $\zeta(s, 1/2 + c - iz)$ as the minimal solution (2.8) to the AΔE (4.4). Note that it manifestly has the properties mentioned above the alternative representation (2.9). For the case at hand the latter specializes to

$$\zeta\left(s, \frac{1}{2} + c - iz \right) = \frac{i}{2} \int_{-\infty}^{\infty} (c - iu)^{-s} \tanh \pi(z - u) du.$$

$$(4.11)$$

Next, we integrate by parts in (4.11) and change variables to obtain

$$\zeta(s, w) = \frac{\pi}{2(s-1)} \int_{-\infty}^{\infty} \frac{(w - 1/2 + ix)^{1-s}}{\mathrm{ch}^2(\pi x)} dx, \quad \mathrm{Re}\, w > 1/2.$$

$$(4.12)$$

Clearly, one can shift the contour up by $i(1 - \epsilon)/2, \epsilon > 0$, so as to handle more generally w in the right half plane. From these formulas (which we have not found in the literature) one can easily deduce some well-known features of $\zeta(s, w)$.

Specifically, one infers that for $\mathrm{Re}\, w > 0$ the function $s \mapsto \zeta(s, w)$ has a meromorphic continuation to $\mathbb{C}$, yielding a simple pole at $s = 1$ with residue 1. Moreover, one obtains

$$\zeta(0, w) = 1/2 - w, \qquad (4.13)$$

and

$$\partial_s \zeta(s, w)|_{s=0} = -\frac{1}{2} + w + \frac{\pi}{2} \int_{-\infty}^{\infty} \frac{(w - 1/2 + ix) \ln(w - 1/2 + ix)}{\mathrm{ch}^2(\pi x)} dx,$$
$$\mathrm{Re}\, w > 1/2. \qquad (4.14)$$

The latter formula gives rise to a representation for $\ln \Gamma(w)$ that seems to be new. Indeed, one also has

$$\partial_s \zeta(s, w)|_{s=0} = \ln \Gamma(w) - \frac{1}{2} \ln(2\pi), \quad \mathrm{Re}\, w > 0. \qquad (4.15)$$

This well-known relation can be more easily derived via the representation

$$\zeta(s, w) = \frac{1}{\Gamma(s)} \int_0^{\infty} \frac{t^{s-1} e^{-wt}}{1 - e^{-t}} dt, \quad \mathrm{Re}\, s > 1, \qquad (4.16)$$

which follows from (4.8) and (4.10). Indeed, noting

$$\frac{t}{1 - e^{-t}} = 1 + \frac{t}{2} + O(t^2), \quad t \to 0, \tag{4.17}$$

we can write (using (4.5))

$$\zeta(s, w) = \frac{1}{\Gamma(s)} \int_0^\infty \frac{dt}{t^2} t^s e^{-wt} \left(\frac{t}{1 - e^{-t}} - 1 - \frac{t}{2} \right) \tag{4.18}$$
$$+ w^{-s} \left(\frac{w}{s - 1} + \frac{1}{2} \right).$$

It is clear by inspection that this representation continues analytically to
$\operatorname{Re} s > -1$, and it yields (using $1/\Gamma(s) = s + O(s^2)$ for $s \to 0$)

$$\partial_s \zeta(s, w)|_{s=0} = \int_0^\infty \frac{dt}{t} e^{-wt} \left(\frac{1}{1 - e^{-t}} - \frac{1}{t} - \frac{1}{2} \right) - w + w \ln w$$
$$- \frac{\ln w}{2}. \tag{4.19}$$

Using the Γ-representation (3.7), it is now straightforward to check (4.15).
    The above properties of the Hurwitz zeta function have been known for
a long time, cf. e.g. Ref. [19]. But the minimal solution interpretation of
$\zeta(s, w)$ seems to be new. From the AΔE viewpoint it can also be understood
why the relation (4.15) between the zeta and gamma functions holds true.
Indeed, due to the (analytic continuation of the) AΔE (4.2), the lhs of
(4.15) satisfies the AΔE

$$f(w + 1) - f(w) = \ln w, \tag{4.20}$$

just as the rhs. (Of course, the constant does not follow from this reasoning.)
    We now introduce a 'rational zeta function'

$$Z_{\text{rat}}(a; s, z) \equiv a^{-s} \zeta \left(s, -\frac{iz}{a} + \frac{1}{2} \right), \quad a > 0, \quad \operatorname{Im} z > -a/2. \tag{4.21}$$

In view of (4.11) and (4.12), it admits integral representations

$$Z_{\text{rat}}(a; s, z) = \frac{i}{2a} \int_{-\infty}^\infty (-iz + ix)^{-s} \operatorname{th}(\pi x/a) dx, \quad \operatorname{Re} s > 1, \quad \operatorname{Im} z > 0, \tag{4.22}$$

$$Z_{\text{rat}}(a; s, z) = \frac{\pi}{2a^2(s - 1)} \int_{-\infty}^\infty \frac{(-iz + ix)^{1-s}}{\operatorname{ch}^2(\pi x/a)} dx, \quad \operatorname{Im} z > 0. \tag{4.23}$$

Furthermore, it satisfies the AΔE

$$Z(z + ia/2) - Z(z - ia/2) = -(-iz)^{-s}. \qquad (4.24)$$

(As before, it is understood that in (4.22)–(4.24) the logarithm branch is fixed by choosing $\ln(-iz)$ real for $z \in i(0,\infty)$.) Finally, upon combining (4.15), (4.13) and (3.1), one obtains

$$\partial_s Z_{\text{rat}}(a; s, z)|_{s=0} = \ln(G_{\text{rat}}(a; z)), \quad \text{Im}\, z > -a/2. \qquad (4.25)$$

In the following subsections we take the above state of affairs as a lead to introduce 'zeta functions' that are minimal solutions to trigonometric, elliptic and hyperbolic generalizations of (4.24). These generalizations turn out to be such that analogs of (4.25) are valid, with the rational gamma function replaced by the trigonometric, elliptic and hyperbolic gamma functions from Subsections 3.2–3.4, respectively.

4.2. THE TRIGONOMETRIC ZETA FUNCTION

Following the ideas explained at the end of the previous subsection, we start from the AΔE

$$Z(z + ia/2) - Z(z - ia/2) = -(1 - e^{2irz})^{-s}. \qquad (4.26)$$

Indeed, the s-derivative of the rhs at $s = 0$ reads $\ln(1 - \exp(2irz))$, which equals $\ln G_{\text{trig}}(z + ia/2) - \ln G_{\text{trig}}(z - ia/2)$, cf. Subsection 3.2. Thus we expect to obtain a minimal solution Z_{trig} to the (shifted) AΔE (4.26), satisfying

$$\partial_s Z_{\text{trig}}(r, a; s, z)|_{s=0} = \ln(G_{\text{trig}}(r, a; z)). \qquad (4.27)$$

We proceed by validating this expectation. Taking $z \to z + ia/2$, we obtain an AΔE

$$f(z + ia/2) - f(z - ia/2) = -(1 - qe^{2irz})^{-s}$$
$$= -\exp(s \sum_{n=1}^{\infty} n^{-1} q^n e^{2inrz}), \qquad (4.28)$$

to which the theory sketched in Subsection 2.1 applies. Indeed, for all $s \in \mathbb{C}$ the rhs is analytic in the half plane $\text{Im}\, z > -a/2$ and π/r-periodic.

As a consequence, a minimal solution to (4.28) is given by

$$\frac{iz}{a} + \frac{1}{2} \sum_{k=1}^{\infty} \frac{t_k(r, a; s)}{\text{sh}(kra)} e^{2ikrz}, \qquad (4.29)$$

with

$$t_k(r, a; s) = \frac{r}{\pi} \int_{-\pi/2r}^{\pi/2r} \exp\left(s \sum_{n=1}^{\infty} n^{-1} q^n e^{2inrx}\right) e^{-2ikrx} dx, \quad k \in \mathbb{N}^*. \tag{4.30}$$

Due to analyticity and π/r-periodicity of the integrand, the contour can be shifted down by $i(a-\epsilon)/2$ for any $\epsilon > 0$. From this we obtain a majorization

$$|t_k(r, a; s)| \leq \exp(C(\epsilon)|s|) \exp(-kr(a - \epsilon)), \quad k \in \mathbb{N}^*, \tag{4.31}$$

where $C(\epsilon) > 0$ diverges as $\epsilon \downarrow 0$. Now $t_k(s)$ is manifestly entire. Hence it readily follows that the solution (4.29) is analytic in the half plane $\mathrm{Im}\, z > -a$ for fixed $s \in \mathbb{C}$, and entire in s for z in the latter half plane.

We now define the 'trigonometric zeta function' by taking $z \to z - ia/2$ in this solution. Expanding the first exponential in (4.30), one readily obtains

$$t_k(r, a; s) = q^k p_k(s), \tag{4.32}$$

with $p_k(s)$ the polynomial

$$p_k(s) \equiv \sum_{m=1}^{k} \frac{s^m}{m!} \sum_{\substack{n_1,\ldots,n_m=1 \\ n_1+\cdots+n_m=k}}^{k} \frac{1}{n_1} \cdots \frac{1}{n_m}. \tag{4.33}$$

Thus our definition amounts to

$$Z_{\mathrm{trig}}(r, a; s, z) \equiv \frac{i(z - ia/2)}{a} + \sum_{k=1}^{\infty} \frac{p_k(s)}{2\mathrm{sh}(kra)} e^{2ikrz}, \quad \mathrm{Im}\, z > -a/2. \tag{4.34}$$

Next, we observe that (4.33) entails

$$\partial_s p_k(s)|_{s=0} = 1/k. \tag{4.35}$$

From this we deduce

$$\partial_s Z_{\mathrm{trig}}(r, a; s, z)|_{s=0} = \sum_{k=1}^{\infty} \frac{e^{2ikrz}}{2kr\mathrm{sh}(kra)}. \tag{4.36}$$

Comparing this to (3.16), we see that (4.27) holds true, as announced.

An alternative representation for Z_{trig} can be obtained by noting that when one adds 1 to the rhs of (4.26), one obtains an AΔE that admits an upward iteration solution. A uniqueness argument then yields

$$Z_{\mathrm{trig}}(r, a; s, z) = \frac{i(z - ia/2)}{a} + \sum_{l=1}^{\infty} \left((1 - q^{2l-1} e^{2irz})^{-s} - 1\right), \quad \mathrm{Im}\, z > -a/2. \tag{4.37}$$

Observe that the relation (4.27) to G_{trig} is also clear from this formula and (3.12).

The relation to the Hurwitz zeta function can be readily established from (4.37), as well. Indeed, one has

$$
\lim_{r\downarrow 0}(2r)^s Z_{\text{trig}}(r,a;s,z)
$$

$$
= \lim_{r\downarrow 0}\sum_{n=0}^{\infty}\left(\left(\frac{2r}{1-\exp(-2r[(n+1/2)a-iz])}\right)^s - (2r)^s\right)
$$

$$
= \sum_{n=0}^{\infty}[(n+1/2)a-iz]^{-s} \tag{4.38}
$$

$$
= a^{-s}\zeta(s,-iz/a+1/2), \quad \operatorname{Re}s > 1, \quad \operatorname{Im}z > -a/2.
$$

This can be abbreviated as

$$
\lim_{\alpha\downarrow 0}\alpha^s Z_{\text{trig}}(\alpha/2,a;s,z) = Z_{\text{rat}}(a;s,z), \quad \operatorname{Re}s > 1, \quad \operatorname{Im}z > -a/2, \tag{4.39}
$$

cf. (4.21).

The Riemann zeta function $\zeta(s)$ is obtained from $Z_{\text{rat}}(a;s,z)$ by choosing $a=1$ and $z=i/2$, cf. (4.21). Thus, setting

$$
\zeta_{\text{trig}}(\alpha;s) \equiv \alpha^s Z_{\text{trig}}(\alpha/2,1;s,i/2), \tag{4.40}
$$

one obtains a 'trigonometric' generalization of $\zeta(s)$. In view of (4.34) and (4.37) it admits the representations

$$
\zeta_{\text{trig}}(\alpha;s) = \alpha^s \sum_{k=1}^{\infty}\frac{e^{-k\alpha}}{1-e^{-k\alpha}}p_k(s), \tag{4.41}
$$

$$
\zeta_{\text{trig}}(\alpha;s) = \sum_{n=1}^{\infty}\left(\left(\frac{\alpha}{1-e^{-n\alpha}}\right)^s - \alpha^s\right). \tag{4.42}
$$

These formulas entail in particular that $\zeta_{\text{trig}}(\alpha;s)$ is positive for all $(\alpha,s) \in (0,\infty)^2$.

4.3. THE ELLIPTIC ZETA FUNCTION

Proceeding as before, we should start from the AΔE

$$
Z(z+ia/2) - Z(z-ia/2) = -\exp\left(s\sum_{n=1}^{\infty}\frac{\cos(2nrz)}{n\operatorname{sh}(nrb)}\right), \tag{4.43}
$$

cf. (3.21). The rhs is manifestly analytic for $|\text{Im } z| < b/2$, so we need not shift z. Thus we define Z_{ell} as the minimal solution

$$Z_{\text{ell}}(r, a, b; s, z) \equiv \frac{iz}{a} e_0(r, b; s) - \sum_{k \in \mathbb{Z}^*} \frac{e_k(r, b; s)}{2\text{sh}(kra)} e^{-2ikrz}, \tag{4.44}$$

where $e_k(s)$ are the Fourier coefficients

$$e_k(r, b; s) = \frac{r}{\pi} \int_{-\pi/2r}^{\pi/2r} \exp\left(s \sum_{n=1}^{\infty} \frac{\cos(2nrx)}{n\text{sh}(nrb)} \right) e^{2ikrx} dx, \quad k \in \mathbb{Z}. \tag{4.45}$$

Next, we note that by analyticity and π/r-periodicity of the integrand, we may shift the contour by ic, with $2c \in (-b, b)$. Therefore, choosing any $\epsilon \in (0, b]$, we obtain

$$|e_k(r, b; s)| \leq \exp(C(\epsilon)|s|) \exp(-|k|r(b - \epsilon)), \quad k \in \mathbb{Z}, \tag{4.46}$$

with $C(\epsilon) > 0$ diverging as $\epsilon \downarrow 0$. Thus Z_{ell} is analytic in the strip $|\text{Im } z| < (a + b)/2$ for fixed $s \in \mathbb{C}$, and entire in s for fixed z in the latter strip.

We proceed by demonstrating the expected relation

$$\partial_s Z_{\text{ell}}(r, a, b; s, z)|_{s=0} = \ln(G_{\text{ell}}(r, a, b; z)), \quad |\text{Im } z| < (a + b)/2. \tag{4.47}$$

To this end we observe that (4.45) entails

$$\partial_s e_k(s)|_{s=0} = \begin{cases} 0, & k = 0, \\ 1/2k\text{sh}(krb), & k \in \mathbb{Z}^*. \end{cases} \tag{4.48}$$

Hence we have

$$\partial_s Z_{\text{ell}}(r, a, b; s, z)|_{s=0} = - \sum_{k \in \mathbb{Z}^*} \frac{e^{-2ikrz}}{4k\text{sh}(krb)\text{sh}(kra)}. \tag{4.49}$$

Comparing this to (3.22), we deduce (4.47).

4.4. THE HYPERBOLIC ZETA FUNCTION

In the hyperbolic context the pertinent AΔE reads

$$Z(z + ia/2) - Z(z - ia/2) = -[2\text{ch}(\pi z/b)]^{-s}, \tag{4.50}$$

cf. (3.33). From the known Fourier transform

$$\int_{-\infty}^{\infty} \frac{e^{ixy}}{[2\mathrm{ch}(\pi x/b)]^s} dx = \frac{b}{2\pi}\Gamma(\frac{s}{2} + \frac{iby}{2\pi})\Gamma(\frac{s}{2} - \frac{iby}{2\pi})\Gamma(s)^{-1}, \quad \mathrm{Re}\, s > 0,$$
(4.51)

and the Γ-function asymptotics, we readily deduce that the conditions (2.6) apply to the function

$$\phi(z) = -\partial_z[2\mathrm{ch}(\pi z/b)]^{-s}, \quad \mathrm{Re}\, s > 0.$$
(4.52)

The minimal solution (2.9) corresponding to the rhs (4.52) can be rewritten as

$$f(z) = \frac{i\pi}{2a^2}\int_{-\infty}^{\infty} du[2\mathrm{ch}(\pi u/b)]^{-s}/\mathrm{ch}^2(\pi(z-u)/a), \quad \mathrm{Re}\, s > 0.$$
(4.53)

Integrating once with respect to z, we obtain a minimal solution

$$Z_{\mathrm{hyp}}(a,b;s,z) \equiv \frac{i}{2a}\int_{-\infty}^{\infty} du \quad [2\mathrm{ch}(\pi u/b)]^{-s}\mathrm{th}(\pi(z-u)/a), \qquad (4.54)$$
$$\mathrm{Re}\, s > 0, \qquad\qquad |\mathrm{Im}\, z| < a/2,$$

to (4.50). More precisely, (4.54) is the unique minimal solution that is odd in z. (The requirement of oddness fixes the arbitrary constant.)

Of course, we can also start from (2.8) and (4.51) to obtain a second representation

$$Z_{\mathrm{hyp}}(a,b;s,z) = \frac{ib}{2\pi^2}\int_0^{\infty} dy \quad \frac{\sin(2yz)}{\mathrm{sh}(ay)}\Gamma\left(\frac{s}{2} + \frac{iby}{\pi}\right)\Gamma\left(\frac{s}{2} - \frac{iby}{\pi}\right)\Gamma(s)^{-1},$$
$$\mathrm{Re}\, s > 0, \qquad\qquad |\mathrm{Im}\, z| < a/2. \qquad (4.55)$$

This formula appears less useful than (4.54), however.

Performing suitable contour shifts in (4.54), one easily checks that for fixed s with $\mathrm{Re}\, s > 0$ the function Z_{hyp} is analytic in the strip $|\mathrm{Im}\, z| < (a+b)/2$, and that for fixed z in the latter strip Z_{hyp} is analytic in the right half s-plane. Alternatively, these features readily follow from (4.55). But the representations (4.54) and (4.55) are ill defined already for s on the imaginary axis.

This different behavior in s (as compared to the trigonometric and elliptic cases) can be understood from the AΔE (4.50). For $\mathrm{Re}\, s < 0$ the function on the rhs is no longer polynomially bounded in the strip $|\mathrm{Im}\, z| < a/2$, so the theory summarized in Subsection 2.1 does not apply. Even so, $Z_{\mathrm{hyp}}(a,b;s,z)$ (with $|\mathrm{Im}\, z| < (a+b)/2$) admits a meromorphic continuation to the half plane $\mathrm{Re}\, s > -2b/a$.

We proceed by proving the assertion just made. To this end we begin by noting that the assertion holds true for $\partial_z^k Z_{\text{hyp}}$ with $k \geq 1$ (cf. (4.53)). Now we exploit the identity

$$\text{ch}(\pi u/b)^{-s} = \frac{b^2}{\pi^2 s^2} \partial_u^2 (\text{ch}(\pi u/b)^{-s}) + \left(1 + \frac{1}{s}\right) \text{ch}(\pi u/b)^{-s-2},$$

$$\tag{4.56}$$

which is easily checked. Inserting it in (4.54) and integrating by parts twice, we deduce the functional equation

$$Z_{\text{hyp}}(a, b; s, z) = \frac{b^2}{\pi^2 s^2} \partial_z^2 Z_{\text{hyp}}(a, b; s, z) + 4\left(1 + \frac{1}{s}\right) Z_{\text{hyp}}(a, b; s + 2, z).$$

$$\tag{4.57}$$

Consider now the two terms on the rhs of (4.57). The first one has a meromorphic extension to $\text{Re}\, s > -2b/a$, and the second one to $\text{Re}\, s > -2$. In case $b > a$, one can iterate the functional equation, so as to continue the second term further to the left. In this way one finally obtains a meromorphic continuation to the half plane $\text{Re}\, s > -2b/a$, as asserted.

At face value, (4.57) seems to entail that Z_{hyp} has a pole for $s = 0$. This is not the case, however. Indeed, we may write (4.57) as

$$
\begin{aligned}
Z_{\text{hyp}}(a, b; s, z) = \ &\frac{i}{2a} \int_{-\infty}^{\infty} du \left(-\frac{b}{as} \frac{1}{\text{ch}^2(\pi(z-u)/a)} \cdot \frac{2\text{sh}(\pi u/b)}{[2\text{ch}(\pi u/b)]^{s+1}} \right. \\
&\left. +4\left(1 + \frac{1}{s}\right) \frac{\text{th}(\pi(z-u)/a)}{[2\text{ch}(\pi u/b)]^{s+2}}\right).
\end{aligned}
$$

$$\tag{4.58}$$

From this it is clear that there can be at most a simple pole at $s = 0$. But in fact we have

$$
\begin{aligned}
\lim_{s \to 0} &s Z_{\text{hyp}}(a, b; s, z) \\
&= \frac{i}{2a} \int_{-\infty}^{\infty} du \left(-\frac{b}{a} \frac{\text{th}(\pi u/b)}{\text{ch}^2(\pi(z-u)/a)} + \frac{\text{th}(\pi(z-u)/a)}{\text{ch}^2(\pi u/b)}\right) \\
&= \frac{i}{2a} \int_{-\infty}^{\infty} du \frac{b}{\pi} \partial_u [\text{th}(\pi u/b) \text{th}(\pi(z-u)/a)] \\
&= 0.
\end{aligned}
$$

$$\tag{4.59}$$

Hence $Z_{\text{hyp}}(a, b; s, z)$ is regular at $s = 0$, as claimed. (The same reasoning shows that Z_{hyp} has no poles in the half plane $\text{Re}\, s > -2b/a$.)

Next, we study the function

$$L(a, b; z) \equiv \partial_s Z_{\text{hyp}}(a, b; s, z)|_{s=0}.$$

$$\tag{4.60}$$

Its z-derivative is given by

$$-\frac{i\pi}{2a^2}\int_{-\infty}^{\infty} du \ln[2\mathrm{ch}(\pi u/b)]/\mathrm{ch}^2(\pi(z-u)/a), \qquad (4.61)$$

cf. (4.53). Comparing this to (3.51), we infer

$$\partial_z L(a, b; z) = \partial_z \ln G_{\mathrm{hyp}}(a, b; z). \qquad (4.62)$$

Thus we have $L = G_{\mathrm{hyp}} + C$. Finally, because both L and $\ln G_{\mathrm{hyp}}$ are odd, we obtain $C = 0$. As a result, we have proved

$$\partial_s Z_{\mathrm{hyp}}(a, b; s, z)|_{s=0} = \ln(G_{\mathrm{hyp}}(a, b; z)), \quad |\mathrm{Im}\, z| < (a+b)/2. \qquad (4.63)$$

We continue by considering some special cases. First, let us observe that one has

$$Z_{\mathrm{hyp}}(a, b; 2, z) = -\frac{b^2}{4\pi^2}\partial_z^2 \ln G_{\mathrm{hyp}}(a, b; z). \qquad (4.64)$$

Indeed, when we take the z-derivative of (3.51), we can integrate by parts twice to obtain this formula, cf. (4.54) with $s = 2$. Alternatively, (4.64) follows by combining equality of the AΔEs satisfied by both functions, their minimal solution character, and oddness in z.

Second, we point out the explicit specialization

$$Z_{\mathrm{hyp}}(a, a; 1, z) = \frac{i}{4}\mathrm{th}(\pi z/2a). \qquad (4.65)$$

To check this, one need only verify that the rhs satisfies (4.50) with $b = a, s = 1$. (Indeed, (4.65) then follows from minimality and oddness.)

Third, we claim

$$Z_{\mathrm{hyp}}(a, b; 0, z) = iz/a. \qquad (4.66)$$

To show this, we use (the continuation to $s = 0$ of) (4.53) to infer that the z-derivative of the lhs is a constant. The rhs is odd and satisfies (4.50) with $s = 0$, so (4.66) follows.

Fourth, we note that $Z_{\mathrm{hyp}}(a, b; s, z)$ is not polynomially bounded for $|\mathrm{Re}\, z| \to \infty$ whenever $\mathrm{Re}\, s$ is negative. Indeed, this feature follows from the analytic continuation of the AΔE (4.50). As an illuminating special case, we use (4.53) to calculate

$$\partial_z Z_{\mathrm{hyp}}(\pi, \pi; -1, z) = \frac{i}{\pi}\int_{-\infty}^{\infty} dx \frac{\mathrm{ch}(z-x)}{\mathrm{ch}^2 x} = i\mathrm{ch}(z), \qquad (4.67)$$

so that (by oddness)

$$Z_{\text{hyp}}(\pi, \pi; -1, z) = i\text{sh}(z). \tag{4.68}$$

Next, we obtain the relation to the rational zeta function (4.21). Changing variables in (4.54), we get

$$Z_{\text{hyp}}(a, b; s, z - ib/2) = \frac{i}{2a} \int_{-\infty}^{\infty} [-2i\text{sh}(\pi(z - x)/b)]^{-s} \text{th}(\pi x/a) dx,$$
$$\text{Im}\, z \in (0, b). \tag{4.69}$$

(The logarithm implied here is real-valued for $x = 0, z \in i(0, b)$.) Recalling (4.22), we now deduce

$$\lim_{b \uparrow \infty} \left(\frac{2\pi}{b}\right)^s Z_{\text{hyp}}(a, b; s, z - ib/2) = Z_{\text{rat}}(a; s, z), \quad \text{Re}\, s > 1, \quad \text{Im}\, z > 0. \tag{4.70}$$

Via suitable contour shifts, this limiting relation can be readily extended to the half plane $\text{Im}\, z > -a/2$.

To conclude this section, we introduce and study a natural 'hyperbolic' generalization of the Riemann zeta function, viz.,

$$\zeta_{\text{hyp}}(\rho; s) \equiv (2\pi\rho)^s Z_{\text{hyp}}(1, 1/\rho; s, i/2 - i/2\rho), \quad \rho > 0. \tag{4.71}$$

Indeed, from (4.70) we have

$$\lim_{\rho \downarrow 0} \zeta_{\text{hyp}}(\rho; s) = \zeta(s), \quad \text{Re}\, s > 1. \tag{4.72}$$

From (4.71) we see that $\zeta_{\text{hyp}}(\rho; s)$ is analytic in the half plane $\text{Re}\, s > -2/\rho$ and that

$$\zeta_{\text{hyp}}(1; s) = 0. \tag{4.73}$$

Moreover, from (4.69) we deduce the representation

$$\zeta_{\text{hyp}}(\rho; s) = \frac{i}{2} \int_{-\infty}^{\infty} \text{th}(\pi x) \left(\frac{\pi\rho}{i\text{sh}(\pi\rho(x - i/2))}\right)^s dx, \quad \rho \in (0, 2), \quad \text{Re}\, s > 0. \tag{4.74}$$

We proceed by turning (4.74) into a more telling formula. To this end we first write

$$\zeta_{\text{hyp}}(\rho; s) = \frac{i}{2} \int_{-\infty}^{\infty} \text{th}(\pi x) \left(\frac{1}{ix + 1/2}\right)^s \left(\left(\frac{\pi\rho(x - i/2)}{\text{sh}(\pi\rho(x - i/2))}\right)^s \tag{4.75}\right.$$

$$- \left(\frac{\pi(x - i/2)}{\mathrm{sh}(\pi(x - i/2))} \right)^s \right) dx,$$

where the three logarithms are real for $x = 0$. The point is now that for $\mathrm{Re}\, s \in (0, 2)$ we may shift the contour up by $i/2$ and take $x \to u/\pi$ to obtain the representation

$$\zeta_{\mathrm{hyp}}(\rho; s) = \pi^{s-1} \sin(\pi s/2) \int_0^\infty \frac{du}{u^s} \coth(u) h(\rho; s, u), \quad \rho \in (0, 2),$$
$$\mathrm{Re}\, s \in (0, 2), \tag{4.76}$$

where we have introduced

$$h(\rho; s, u) \equiv \left(\frac{\rho u}{\mathrm{sh}(\rho u)} \right)^s - \left(\frac{u}{\mathrm{sh}(u)} \right)^s, \quad \rho \geq 0. \tag{4.77}$$

From the latter representation we see in particular that $\zeta_{\mathrm{hyp}}(\rho; s)$ is positive for $(\rho, s) \in (0, 1) \times (0, 2)$. It can also be exploited to obtain representations for the Riemann zeta function that have not appeared in previous literature, to our knowledge.

In order to derive the latter, we observe that (4.77) entails uniform bounds

$$|h(\rho; s, u)| \leq C_1 u^2, \quad \rho \in [0, 1), \quad \mathrm{Re}\, s \in (0, 2), \quad u \in \mathbb{R},$$
$$\tag{4.78}$$

$$|h(\rho; s, u)| \leq C_2, \quad \rho \in [0, 1), \quad \mathrm{Re}\, s \in (0, 2), \quad u \in \mathbb{R}.$$
$$\tag{4.79}$$

Therefore, choosing $\mathrm{Re}\, s \in (1, 2)$, we may interchange the $\rho \downarrow 0$ limit and the integration, yielding

$$\zeta(s) = \pi^{s-1} \sin(\pi s/2) \int_0^\infty \coth(u) \left(\frac{1}{u^s} - \frac{1}{\mathrm{sh}(u)^s} \right) du, \quad \mathrm{Re}\, s \in (1, 2), \tag{4.80}$$

cf. (4.72). Writing $u^{-s} = (1 - s)^{-1} \partial_u u^{1-s}$, we may integrate by parts to get

$$\zeta(s) = \frac{1}{s - 1} \pi^{s-1} \sin(\pi s/2) \int_0^\infty u^{1-s} \partial_u (\coth u [1 - (u/\mathrm{sh}\, u)^s]) du. \tag{4.81}$$

Evidently, the latter representation makes sense and is valid for $\mathrm{Re}\, s \in (0, 2)$. Using the functional equation, we also deduce

$$\zeta(s) = -\frac{2^{s-1}}{\Gamma(1 + s)} \int_0^\infty u^s \partial_u (\coth u [1 - (u/\mathrm{sh}\, u)^{1-s}]) du, \quad \mathrm{Re}\, s \in (-1, 1). \tag{4.82}$$

5. Barnes' Multiple Zeta and Gamma Functions

In this section we present a summary of some results from our forthcoming paper Ref. [20]. Specifically, we focus on those results that hinge on interpreting the Barnes functions from our AΔE perspective.

Barnes' multiple zeta function may be defined by the series

$$\zeta_N(s, w | a_1, \ldots, a_N) = \sum_{m_1, \ldots, m_N = 0}^{\infty} (w + m_1 a_1 + \cdots + m_N a_N)^{-s},$$

$$(5.1)$$

where we have

$$a_1, \ldots, a_N > 0, \quad \mathrm{Re}\, w > 0, \quad \mathrm{Re}\, s > N. \tag{5.2}$$

It is immediate that these functions are related by the recurrence

$$\zeta_{M+1}(s, w + a_{M+1} | a_1, \ldots, a_{M+1}) - \zeta_{M+1}(s, w | a_1, \ldots, a_{M+1}) \tag{5.3}$$
$$= -\zeta_M(s, w | a_1, \ldots, a_M),$$

with

$$\zeta_0(s, w) \equiv w^{-s}. \tag{5.4}$$

As Barnes shows [21], ζ_N has a meromorphic continuation in s, with simple poles only at $s = 1, \ldots, N$. He defined his multiple gamma function $\Gamma_N^B(w)$ in terms of the s-derivative at $s = 0$,

$$\Psi_N(w | a_1, \ldots, a_N) \equiv \partial_s \zeta_N(s, w | a_1, \ldots, a_N)|_{s=0}, \quad N \in \mathbb{N}. \tag{5.5}$$

Analytically continuing (5.3), it follows that the functions Ψ_N satisfy the recurrence

$$\Psi_{M+1}(w + a_{M+1} | a_1, \ldots, a_{M+1}) - \Psi_{M+1}(w | a_1, \ldots, a_{M+1}) \tag{5.6}$$
$$= -\Psi_M(w | a_1, \ldots, a_M).$$

Comparing (5.1) with $N = 1$ to (4.21), we infer

$$\zeta_1(s, w | a) = Z_{\mathrm{rat}}(a; s, i[w - a/2]). \tag{5.7}$$

Also, comparing (5.5) and (4.25), we get

$$\Psi_1(w | a) = \ln(G_{\mathrm{rat}}(a; i[w - a/2]))$$
$$= \ln((2\pi)^{-1/2} \exp([w/a - 1/2] \ln a) \Gamma(w/a)), \tag{5.8}$$

cf. (3.1). Now we have already seen that G_{rat} and Z_{rat} can be viewed as minimal solutions to AΔEs of the form (1.1) and (1.2), respectively (cf. Subsections 3.1 and 4.1). In Subsections 5.1 and 5.2 we sketch how, more generally, the functions ζ_N and Ψ_N can be tied in with the AΔE lore reviewed in Subsection 2.1.

In brief, ζ_{M+1} and Ψ_{M+1} may be viewed as *minimal* solutions to the equations (5.3) and (5.6), reinterpreted as AΔEs of the form (1.2), with the right-hand sides viewed as the given function $\phi(z)$. In this way we arrive at a precise version of Barnes' expression 'simplest solution' [21], and we are led to new and illuminating integral representations for the Barnes functions.

5.1. MULTIPLE ZETA FUNCTIONS

Using the identity

$$\prod_{j=1}^{N}(1 - e^{-a_j t})^{-1} = \sum_{m_1,\ldots,m_N=0}^{\infty} \exp(-t(m_1 a_1 + \cdots + m_N a_N)),$$
$$a_1,\ldots,a_N, t > 0, \tag{5.9}$$

and Euler's integral (4.5), we can rewrite $\zeta_N(s,w)$ (5.1) as

$$\zeta_N(s,w) = \frac{1}{\Gamma(s)} \int_0^\infty \frac{dt}{t} t^s e^{-wt} \prod_{j=1}^{N}(1 - e^{-a_j t})^{-1}, \quad \text{Re}\, s > N,\ \text{Re}\, w > 0. \tag{5.10}$$

Consider now the function

$$\phi_{M,s}(z) \equiv -\zeta_M(s, A_M + d - iz), \quad d > -A_M, \quad \text{Re}\, s \geq M + 2, \tag{5.11}$$

where we have introduced

$$A_N \equiv \frac{1}{2} \sum_{j=1}^{N} a_j, \quad N \in \mathbb{N}. \tag{5.12}$$

Since we choose the displacement parameter d greater than $-A_M$, we obtain a non-empty strip $|\text{Im}\, z| < A_M + d$ in which $\phi_{M,s}(z)$ is defined and analytic. Thus we can use (5.10) to write

$$\phi_{M,s}(z) = -\frac{2^{1-M}}{\Gamma(s)} \int_0^\infty dy \frac{(2y)^{s-1} e^{-2dy}}{\prod_{j=1}^{M} \text{sh}(a_j y)} \cdot e^{2iyz}, \quad \text{Re}\, s \geq M + 2,$$
$$\text{Im}\, z > -A_M - d. \tag{5.13}$$

Now the Fourier transform $\hat{\phi}_{M,s}(y)$ (2.7) can be read off from (5.13). Clearly, it belongs to $L^1(\mathbb{R})$, and it satisfies $\hat{\phi}_{M,s}(y) = O(y)$ for $y \to 0$. Moreover, using the series representation (5.1) for $\phi_{M,s}(x), x \in \mathbb{R}$, it is easily seen that $\phi_{M,s}(x)$ belongs to $L^1(\mathbb{R})$, too.

As a consequence, the conditions (2.6) are satisfied. Therefore, we obtain a minimal solution

$$
\begin{aligned}
f_{M,s}(z) &= \frac{2^{-M}}{\Gamma(s)} \int_0^\infty dy \frac{(2y)^{s-1}e^{-2dy}}{\prod_{j=1}^{M+1} \operatorname{sh}(a_j y)} \cdot e^{2iyz} \\
&= \zeta_{M+1}(s, A_{M+1} + d - iz),
\end{aligned} \tag{5.14}
$$

to the AΔE

$$
\begin{aligned}
f(z + ia_{M+1}/2) - f(z - ia_{M+1}/2) &= \phi_{M,s}(z), \\
\operatorname{Re} s \geq M + 2, \quad \operatorname{Im} z &> -A_M - d.
\end{aligned} \tag{5.15}
$$

In summary, for $\operatorname{Re} s \geq M + 2$ we may view ζ_{M+1} as the unique minimal solution (2.8).

Next, we exploit the formula (2.9) with $\phi(z)$ given by $\phi_{M,s}(z)$ (5.11). Changing variables, it yields

$$
\begin{aligned}
\zeta_{M+1}(s, A_{M+1} + d - iz) &= \frac{i}{2a_{M+1}} \int_{-\infty}^\infty dx \zeta_M(s, c - iz + ix) \\
&\qquad \times \operatorname{th} \frac{\pi}{a_{M+1}} x,
\end{aligned} \tag{5.16}
$$

where $c = A_M + d$ and where we may take $\operatorname{Im} z > -c$. This relation can now be iterated, but first we integrate by parts, using the relation

$$
\partial_w \zeta_N(s, w) = -s\zeta_N(s+1, w), \quad N \in \mathbb{N}. \tag{5.17}
$$

(This formula can be read off from (5.1).) Doing so, we obtain

$$
\begin{aligned}
\zeta_N(s, A_N + d - iz) &= \frac{\pi}{2a_N^2} \cdot \frac{1}{s-1} \int_{-\infty}^\infty dx \\
&\qquad \times \zeta_{N-1}(s-1, A_{N-1} + d - iz + ix)/\operatorname{ch}^2(\pi x/a_N) \\
&= \int_{\mathbb{R}^N} \left(\prod_{n=1}^N \frac{\pi \operatorname{ch}^{-2}(\pi x_n/a_n)}{2a_n^2(s-n)} \right) \\
&\qquad \times \left(d - iz + i \sum_{n=1}^N x_n \right)^{N-s} d^N x.
\end{aligned} \tag{5.18}
$$

(Recall (5.4) to check the last iteration step.)

Now we derived this new representation for $\operatorname{Re} s > N$ and $\operatorname{Im} z > -d$. But it is evident that it yields a meromorphic s-continuation of ζ_N to all of $\mathbb{C}$, with simple poles occurring only at $s = 1, \dots, N$. Moreover, suitable contour shifts yield analyticity in z for $\operatorname{Im} z > -A_N - d$. (Observe that we are generalizing arguments we already presented in Subsection 4.1, where we are dealing with ζ_1, cf. (5.7).)

The above integral representation (5.18) was derived for the first time in Ref. [20]. For $\operatorname{Re} s \le N+1$ one can also view $\zeta_N(s, w)$ as a minimal solution, but now in the more general sense of Theorem II.3 in Ref. [9]. (This hinges on (5.17) and its iterates.) The details can be found in Section 4 of Ref. [20].

In Ref. [20] we also approach $\zeta_N(s, w)$ from yet another, quite elementary angle. The latter does not involve AΔEs, but rather a certain class of Laplace-Mellin transforms. In this way we can easily rederive various explicit formulas and properties established in other ways by Barnes. For brevity we refrain from sketching this other approach, which is complementary to the AΔE perspective.

5.2. MULTIPLE GAMMA FUNCTIONS

As we have just shown, $\zeta_N(s, w)$ has a meromorphic continuation to $\mathbb{C}$, which is analytic in $s = 0$. From (5.18) we can also obtain a representation for the function $\Psi_N(w|a_1, \dots, a_N)$ (5.5), namely,

$$\Psi_N(A_N + d - iz) = \sum_{l=1}^{N} \frac{1}{l} \cdot \zeta_N(0, A_N + d - iz) \qquad (5.19)$$

$$+ \ (-)^{N+1} \left(\prod_{n=1}^{N} \frac{\pi}{2na_n^2} \int_{-\infty}^{\infty} \frac{dx_n}{\operatorname{ch}^2(\pi x_n/a_n)} \right) I_N(x),$$

where the integrand reads

$$I_N(x) = \left(d - iz + i \sum_{n=1}^{N} x_n \right)^N \ln \left(d - iz + i \sum_{n=1}^{N} x_n \right). \qquad (5.20)$$

The $s = 0$ value of ζ_N appearing here follows from (5.18), too, yielding an Nth degree polynomial in z. (The coefficients of the latter can be expressed in terms of the Bernoulli numbers [20].)

As before, the restriction $\operatorname{Im} z > -d$ can be relaxed to $\operatorname{Im} z > -A_N - d$ by contour shifts, revealing that $\Psi_N(w)$ is analytic for $\operatorname{Re} w > 0$. Now a second continuation of the logarithm in the shifted integrand to $z \in \mathbb{C} \setminus i(-\infty, -A_N - d]$ reveals that $\Psi_N(w)$ admits an analytic continuation to the cut plane $\mathbb{C} \setminus (-\infty, 0]$. Defining

$$\Gamma_N(w|a_1, \dots, a_N) \equiv \exp \Psi_N(w|a_1, \dots, a_N), \qquad (5.21)$$

it follows that $\Gamma_N(w)$ is analytic and zero-free in this cut plane. Furthermore, the recurrence (5.6) entails

$$\Gamma_{M+1}(w|a_1,\ldots,a_{M+1}) \tag{5.22}$$
$$= \Gamma_M(w|a_1,\ldots,a_M)\Gamma_{M+1}(w+a_{M+1}|a_1,\ldots,a_{M+1}), \quad M \in \mathbb{N}.$$

We can now determine the analytic character of $\Gamma_N(w)$ on the cut by exploiting (5.22). Specifically, taking first $M = 0$ and noting $\Gamma_0(w) = 1/w$, we can iterate (5.22) to get

$$\Gamma_1(w|a_1) = \prod_{k=0}^{l-1} \frac{1}{w+ka_1} \cdot \Gamma_1(w+la_1|a_1), \quad l \in \mathbb{N}^*. \tag{5.23}$$

From this we see that $\Gamma_1(w|a_1)$ has a meromorphic extension without zeros and with simple poles for $w \in -a_1\mathbb{N}$. Next, writing

$$\Gamma_2(w|a_1,a_2) = \prod_{k=0}^{l-1} \Gamma_1(w+ka_2|a_1) \cdot \Gamma_2(w+la_2|a_1,a_2), \quad l \in \mathbb{N}^*, \tag{5.24}$$

we deduce that $\Gamma_2(w|a_1,a_2)$ has a meromorphic extension without zeros and with poles for $w = -(k_1a_1+k_2a_2), k_1, k_2 \in \mathbb{N}$. The multiplicity of a pole w_0 equals the number of distinct pairs (k_1, k_2) such that $w_0 = -(k_1a_1+k_2a_2)$. In particular, all poles are simple when a_1/a_2 is irrational, and the pole at $w = 0$ is always simple. Proceeding recursively, it is now clear that $\Gamma_N(w)$ has a meromorphic extension, without zeros and with poles for $w = -(k_1a_1+\cdots+k_Na_N), k_1,\ldots,k_N \in \mathbb{N}$, the pole at $w = 0$ being simple.

Before concluding this section with some remarks, we would like to point out that the above account of the functions $\zeta_N(s,w)$ and $\Gamma_N(w)$ has only involved Euler's formula (4.5) and the 'minimal solution' consequences of the properties (2.6). Taking the latter for granted, the arguments in this section are self-contained, leading quickly and simply to a significant part of Barnes' results.

At this point it should be remarked that Barnes used a different normalization for his multiple gamma function $\Gamma_N^B(w)$. Specifically, the relation to $\Gamma_N(w)$ reads

$$\Gamma_N^B(w) = \rho_N\Gamma_N(w), \tag{5.25}$$

where ρ_N is Barnes' modular constant. Our use of $\Gamma_N(w)$ is in accord with most of the later literature. The constant ρ_N in (5.25) is (by definition) equal to the reciprocal residue of $\Gamma_N(w)$ at its simple pole $w = 0$. Equivalently, one has

$$w\Gamma_N^B(w) \to 1, \quad w \to 0. \tag{5.26}$$

We further remark that Kurokawa's double sine function $S_2(x|\omega_1, \omega_2)$, which we encountered in Subsection 3.4 (see (3.52)), can be defined by

$$S_2(w|a_1, a_2) \equiv \Gamma_2(a_1 + a_2 - w|a_1, a_2)/\Gamma_2(w|a_1, a_2). \qquad (5.27)$$

(Note that one may replace Γ_2 by Γ_2^B in this formula.)

Finally, the special function $\Gamma_N(w)$ can once again be interpreted as a minimal solution to an AΔE of the form (1.1). This is explained in detail in Section 4 of Ref. [20]. We also mention that the second (Laplace transform) approach alluded to at the end of Subsection 5.1 can be exploited to derive uniform large-w asymptotics away from the cut $(-\infty, 0]$, cf. Section 3 in Ref. [20].

6. A Generalized Hypergeometric Function

The subject of this section is a function $R(a_+, a_-, \mathbf{c}; v, \hat{v})$, depending on parameters $a_+, a_- \in (0, \infty)$, couplings $\mathbf{c} = (c_0, c_1, c_2, c_3) \in \mathbb{R}^4$ and variables $v, \hat{v} \in \mathbb{C}$. It generalizes both the hypergeometric function $_2F_1(a, b, c; w)$ and the Askey-Wilson polynomials $p_n(q, \alpha, \beta, \gamma, \delta; \cos v)$. (For a complete account of the features of $_2F_1$ used below we refer to Refs. [19, 23]. For information on the Askey-Wilson polynomials, see Refs. [22, 13].) The R-function was introduced in Ref. [1]. Detailed proofs of several assertions made below can be found in our paper Ref. [16].

In Subsection 6.1 we review various features of the $_2F_1$-function that admit a generalization to the R-function. The latter is defined in Subsection 6.2, where we also obtain some automorphy properties. In Subsection 6.3 we introduce the four independent hyperbolic difference operators of which the R-function is an eigenfunction. In Subsection 6.4 we derive the specialization to the Askey-Wilson polynomials. Subsection 6.5 concerns the 'nonrelativistic limit' $R \to {}_2F_1$. In Subsection 6.6 we study how the R-function is related to the Ismail-Rahman functions [24, 25], which are eigenfunctions of the trigonometric Askey-Wilson difference operator. Though we shed some light on this issue, we leave several questions open.

6.1. SOME REMINDERS ON $_2F_1$

The hypergeometric function was already known to Euler in terms of an integral representation. Our generalized hypergeometric function is defined in terms of an integral as well, but this integral representation does not generalize Euler's integral representation for $_2F_1$, but rather the much later one due to Barnes.

The latter representation can be readily understood from Gauss' series representation,

$$
{}_2F_1(a, b, c; w) = \sum_{n=0}^{\infty} \frac{\Gamma(a+n)}{\Gamma(a)} \frac{\Gamma(b+n)}{\Gamma(b)} \frac{\Gamma(c)}{\Gamma(c+n)} \frac{w^n}{n!}. \tag{6.1}
$$

Using for instance the ratio test, one sees that this power series converges for $|w| < 1$. The Barnes representation makes it possible to analytically continue ${}_2F_1$ to the cut plane $|\mathrm{Arg}(-w)| < \pi$. It reads

$$
\int_C dz \exp(-iz \ln(-w)) \cdot \frac{\Gamma(iz)\Gamma(c)}{2\pi\Gamma(c-iz)} \cdot \frac{\Gamma(a-iz)\Gamma(b-iz)}{\Gamma(a)\Gamma(b)}. \tag{6.2}
$$

Here, the logarithm branch is fixed by choosing $\ln(-w) \in \mathbb{R}$ for $w \in (-\infty, 0)$. Taking first $\mathrm{Re}\,a, \mathrm{Re}\,b > 0$, the contour C runs along the real axis from $-\infty$ to ∞, with a downward indentation at the origin to avoid the pole due to $\Gamma(iz)$. Thus it separates the downward pole sequences starting at $-ia$ and $-ib$ from the upward sequence starting at 0.

Invoking the asymptotics of the Γ-function, one sees that the integrand has exponential decay for $\mathrm{Re}\,z \to \pm\infty$, provided $|\mathrm{Arg}(-w)| < \pi$. Thus the integral yields an analytic function of w in the cut plane. After multiplication by $2\pi i$, the residues at the simple poles $z = in$ of the integrand are equal to the terms in the Gauss series (6.1). A second somewhat subtle application of the Γ-function asymptotics now shows that when one moves the contour C up across the poles $0, i, \ldots, in$, picking up $2\pi i$ times the residues in the process, then the integral over the shifted contour converges to 0 for $n \to \infty$, provided $|w| < 1$. Thus the integral (6.2) yields an analytic continuation to the cut plane $|\mathrm{Arg}(-w)| < \pi$, as advertised.

The analyticity region cannot be much improved, since the function ${}_2F_1(a, b, c; w)$ has a logarithmic branch point at $w = 1$ for generic $a, b, c \in \mathbb{C}$. In this connection we should add that the representation (6.2) can be modified to handle arbitrary $a, b \in \mathbb{C}$: For $\mathrm{Re}\,a \leq 0$ and/or $\mathrm{Re}\,b \leq 0$ one need only shift the contour C up, so that the downward pole sequences starting at $-ia$ and $-ib$ stay below it. In this way one can demonstrate that for fixed w in the cut plane one obtains a meromorphic function of a, b and c.

Next, we recall that the hypergeometric function can be used to diagonalize the two-coupling family of Schrödinger operators

$$
H(g, \tilde{g}) \equiv -\frac{d^2}{dx^2} + \frac{g(g-1)\nu^2}{\mathrm{sh}^2\nu x} - \frac{\tilde{g}(\tilde{g}-1)\nu^2}{\mathrm{ch}^2\nu x}. \tag{6.3}
$$

Specifically, one first performs the similarity transformation

$$
\tilde{H}(g, \tilde{g}) \equiv w(\nu x)^{-1/2} H(g, \tilde{g}) w(\nu x)^{1/2}, \tag{6.4}
$$

where $w(y)$ is the 'weight function'

$$w(y) \equiv \text{sh} y^{2g} \text{ch} y^{2\tilde{g}}. \tag{6.5}$$

A straightforward calculation yields

$$\tilde{H}(g, \tilde{g}) = -\frac{d^2}{dx^2} - 2\nu[g \coth(\nu x) + \tilde{g} \tanh(\nu x)]\frac{d}{dx} - \nu^2(g + \tilde{g})^2. \tag{6.6}$$

Then one has the eigenvalue equation

$$\tilde{H}(g, \tilde{g})\Psi_{\text{nr}} = p^2\Psi_{\text{nr}}, \tag{6.7}$$

where Ψ_{nr} is the nonrelativistic wave function

$$\Psi_{\text{nr}}(\nu, g, \tilde{g}; x, p) \equiv {}_2F_1\left(\frac{1}{2}\left(g + \tilde{g} - \frac{ip}{\nu}\right), \frac{1}{2}\left(g + \tilde{g} + \frac{ip}{\nu}\right), \tag{6.8}$$

$$g + \frac{1}{2}; -\text{sh}^2\nu x\right).$$

Indeed, (6.7) is simply the rational ODE satisfied by ${}_2F_1(a, b, c; w)$, transformed to hyperbolic form via the substitution $w = -\text{sh}^2\nu x$.

The wave function Ψ_{nr} (6.8) is also an eigenfunction of two analytic difference operators, one of which acts on x, while the second one acts on the spectral variable p. We will obtain this fact (which cannot be found in the textbook literature) as a corollary of the nonrelativistic limit in Subsection 6.5.

6.2. THE 'RELATIVISTIC' HYPERGEOMETRIC FUNCTION

The function $R(a_+, a_-, \mathbf{c}; v, \hat{v})$ we are about to introduce can be used in particular to diagonalize an analytic difference operator (from now on $A\Delta O$) that arises in the context of the relativistic Calogero-Moser system, cf. Ref. [2], Subsection 3.3.

But just as the nonrelativistic wave function Ψ_{nr} (6.8) serves as an eigenfunction for a 2-coupling generalization of the (reduced, two-particle) nonrelativistic Calogero-Moser Hamiltonian $H(g, 0)$ (6.3), we will find that the relativistic wave function Ψ_{rel} (6.46) we associate below with the R-function is in fact an eigenfunction of a 4-coupling generalization of the relativistic counterpart of $H(g, 0)$. Moreover, it is an eigenfunction of three more independent $A\Delta Os$ with a similar structure. (In the nonrelativistic limit two of these give rise to the $A\Delta Os$ mentioned at the end of the previous subsection.)

We have split the integrand in (6.2) in three factors to anticipate a corresponding factorization of the integrand for the R-function. Setting

$$\hat{c}_0 \equiv (c_0 + c_1 + c_2 + c_3)/2, \tag{6.9}$$

the latter reads

$$I(a_+, a_-, \mathbf{c}; v, \hat{v}, z) \equiv F(c_0; v, z)K(a_+, a_-, \mathbf{c}; z)F(\hat{c}_0; \hat{v}, z). \tag{6.10}$$

Here, the functions F and K involve the hyperbolic gamma function $G(z) \equiv G_{\text{hyp}}(a_+, a_-; z)$ from Subsection 3.4, cf. (3.35).

Specifically, F and K are defined by

$$F(d; y, z) \equiv \left(\frac{G(z + y + id - ia)}{G(y + id - ia)} \right) (y \rightarrow -y), \tag{6.11}$$

$$K(a_+, a_-, \mathbf{c}; z) \equiv \frac{1}{G(z + ia)} \prod_{j=1}^{3} \frac{G(is_j)}{G(z + is_j)}, \tag{6.12}$$

where we use the notation

$$s_1 \equiv c_0 + c_1 - a_-/2, \quad s_2 \equiv c_0 + c_2 - a_+/2, \quad s_3 \equiv c_0 + c_3,$$
$$a \equiv (a_+ + a_-)/2. \tag{6.13}$$

We have suppressed the dependence on a_+ and a_- in G and in F, since these functions are invariant under the interchange of a_+ and a_-. (Note K is not invariant, since s_1 and s_2 are not.)

Just as we first have chosen $\text{Re}\, a, \text{Re}\, b > 0$ so as to define the integration contour C in the Barnes representation (6.2), we begin by choosing

$$s_j \in (0, a), \quad j = 1, 2, 3, \quad c_0, \hat{c}_0, v, \hat{v} \in (0, \infty). \tag{6.14}$$

Then we choose once more the contour C going from $-\infty$ to ∞ in the z-plane, with a downward indentation at the origin. The choices just detailed ensure that C separates the four upward pole sequences coming from the four z-dependent G-functions in K (6.12) and the four downward pole sequences coming from the z-dependent G-functions in the two F-factors of the integrand (6.10). (At this point the reader should recall the pole-zero properties of the hyperbolic G-function, cf. (3.36), (3.37).)

Our R-function is now defined by the integral

$$R(a_+, a_-, \mathbf{c}; v, \hat{v}) \equiv \frac{1}{(a_+a_-)^{1/2}} \int_C dz I(a_+, a_-, \mathbf{c}; v, \hat{v}, z). \tag{6.15}$$

The asymptotics (3.50) of the G-function plays the same role as the Stirling formula for the Γ-function in showing that the integral converges. Indeed, using (3.50) one readily obtains

$$
I(a_+, a_-, \mathbf{c}; v, \hat{v}, z) = O\left(\exp\left(\mp 2\pi \left(\frac{1}{a_+} + \frac{1}{a_-} \right) \operatorname{Re} z \right) \right),
$$
$$
\operatorname{Re} z \to \pm\infty. \tag{6.16}
$$

Therefore, R is well defined and analytic in v and $\hat{v}$ for $\operatorname{Re} v, \operatorname{Re} \hat{v} \neq 0$.

The R-function has in fact much stronger analyticity properties, but to demonstrate these in detail is well beyond our present scope. Thus we only summarize some results, referring to Ref. [16] for proofs. Briefly, the R-function extends to a function that is meromorphic in all of its eight arguments, as long as a_+ and a_- stay in the (open) right half plane. Moreover, the pole varieties and their associated orders are explicitly known. For the case of fixed positive a_+, a_- and (generic) real c_0, c_1, c_2, c_3, the R-function is meromorphic in v and $\hat{v}$, with poles that can (but need not) occur solely for certain points on the imaginary axis. These points correspond to collisions of v- and $\hat{v}$-dependent z-poles in the integrand with z-poles in the three upward s_j-pole sequences and points that are given by the poles of the factors $1/G(\pm v + ic_0 - ia)$ and $1/G(\pm\hat{v} + i\hat{c}_0 - ia)$ in the integrand.

We conclude this subsection by listing some automorphic properties of the R-function. To this end we introduce the 'dual couplings'

$$
\hat{\mathbf{c}} \equiv J\mathbf{c}, \quad J \equiv \frac{1}{2}\begin{pmatrix} 1 & 1 & 1 & 1 \\ 1 & 1 & -1 & -1 \\ 1 & -1 & 1 & -1 \\ 1 & -1 & -1 & 1 \end{pmatrix}, \tag{6.17}
$$

whence one has (cf. (6.13))

$$
c_0 + c_j = \hat{c}_0 + \hat{c}_j, \quad s_j = \hat{s}_j, \quad j = 1, 2, 3. \tag{6.18}
$$

We also define the transposition

$$
I\mathbf{c} \equiv (c_0, c_2, c_1, c_3). \tag{6.19}
$$

Then one has the symmetries

$$
R(a_+, a_-, \mathbf{c}; v, \hat{v}) = R(a_+, a_-, \hat{\mathbf{c}}; \hat{v}, v), \quad (\text{self} - \text{duality}), \tag{6.20}
$$

$$
R(a_+, a_-, \mathbf{c}; v, \hat{v}) = R(a_-, a_+, I\mathbf{c}; v, \hat{v}), \tag{6.21}
$$

$$R(\lambda a_+, \lambda a_-, \lambda \mathbf{c}; \lambda v, \lambda \hat{v}) = R(a_+, a_-, \mathbf{c}; v, \hat{v}), \quad \lambda > 0, \text{ (scale invariance)}.$$
$$(6.22)$$

These features can be quite easily checked directly from the definition (6.15). (Use the G-function properties (3.39) and (3.40) to check (6.21) and (6.22), resp.)

6.3. EIGENFUNCTION PROPERTIES

In order to detail the four $A\Delta O$s for which our R-function is a joint eigenfunction, we introduce the quantities

$$s_\delta(y) \equiv \text{sh}(\pi y / a_\delta), \quad c_\delta(y) \equiv \text{ch}(\pi y / a_\delta), \tag{6.23}$$

$$
\begin{aligned}
C_\delta(\mathbf{c}; z) &\equiv \frac{s_\delta(z - ic_0)}{s_\delta(z)} \frac{c_\delta(z - ic_1)}{c_\delta(z)} \frac{s_\delta(z - ic_2 - ia_{-\delta}/2)}{s_\delta(z - ia_{-\delta}/2)} \\
&\quad \times \frac{c_\delta(z - ic_3 - ia_{-\delta}/2)}{c_\delta(z - ia_{-\delta}/2)},
\end{aligned}
\tag{6.24}
$$

$$
\begin{aligned}
A_\delta(\mathbf{c}; y) &\equiv C_\delta(\mathbf{c}; y)\left(T^y_{ia_{-\delta}} - 1\right) + C_\delta(\mathbf{c}; -y)\left(T^y_{-ia_{-\delta}} - 1\right) \\
&\quad + 2c_\delta(i(c_0 + c_1 + c_2 + c_3)),
\end{aligned}
\tag{6.25}
$$

where $\delta = +, -$, and where the superscript y on the shifts indicates the variable they act on. The eigenfunction properties of the R-function are now specified in the following proposition, whose proof we sketch.

Proposition 6.1 *The function $R(a_+, a_-, \mathbf{c}; v, \hat{v})$ is a joint eigenfunction of the $A\Delta O$s*

$$A_+(\mathbf{c}; v), \quad A_-(I\mathbf{c}; v), \quad A_+(\hat{\mathbf{c}}; \hat{v}), \quad A_-(I\hat{\mathbf{c}}; \hat{v}), \tag{6.26}$$

with eigenvalues

$$2c_+(2\hat{v}), \quad 2c_-(2\hat{v}), \quad 2c_+(2v), \quad 2c_-(2v). \tag{6.27}$$

Sketch of proof. In view of the symmetries (6.20) and (6.21) we need only prove the $A\Delta E$

$$A_+(\mathbf{c}; v)R(\mathbf{c}; v, \hat{v}) = 2c_+(2\hat{v})R(\mathbf{c}; v, \hat{v}). \tag{6.28}$$

Also, due to the analyticity properties of the R-function already detailed, we may restrict the parameters and imaginary parts of v and $\hat{v}$ in such a way that R is given by

$$R(\mathbf{c}; v, \hat{v}) = \frac{1}{(a_+a_-)^{1/2}} \int_C F(c_0; v, z)K(\mathbf{c}; z)F(\hat{c}_0; \hat{v}, z)dz,$$
$$(6.29)$$

and that we may let the AΔO $A_+(\mathbf{c}; v)$ act on the integrand.

A main tool in proving the second-order AΔE (6.28) is now to exploit the first-order AΔE

$$\frac{G(z + ia_-/2)}{G(z - ia_-/2)} = 2c_+(z), \tag{6.30}$$

satisfied by the G-function. Indeed, using (6.30), one readily checks that the function F (6.11) solves the two AΔEs

$$\frac{F(d; y + ia_-/2, z)}{F(d; y - ia_-/2, z)} = \frac{s_+(y + z + id - ia_-/2)}{s_+(y - z - id + ia_-/2)} \frac{s_+(y - id + ia_-/2)}{s_+(y + id - ia_-/2)}, \tag{6.31}$$

$$\frac{F(d; y, z - ia_-)}{F(d; y, z)} = \frac{1}{4s_+(y + z + id - ia_-)s_+(y - z - id + ia_-)}. \tag{6.32}$$

Using (6.31) with $d = c_0$ and $y = v$, one can calculate the quotient

$$Q(\mathbf{c}; v, z) \equiv (A_+(\mathbf{c}; v)F)(c_0; v, z)/F(c_0; v, z). \tag{6.33}$$

A key point is now that Q can be rewritten as

$$2c_+(2z + 2i\hat{c}_0) + \frac{4 \prod_{j=1}^4 c_+(z - ia_-/2 + is_j)}{s_+(v + z + ic_0 - ia_-)s_+(v - z - ic_0 + ia_-)}, \quad s_4 \equiv a. \tag{6.34}$$

This fact amounts to a functional equation that can be proved by comparing residues at simple poles and $|\mathrm{Re}\, v| \to \infty$ asymptotics. We now observe that the denominator in (6.34) appears in (6.32) with $d = c_0, y = v$. Thus we get

$$\begin{aligned}
A_+(\mathbf{c}; v)F(c_0; v, z) &= 2c_+(2z + 2i\hat{c}_0)F(c_0; v, z) + F(c_0; v, z - ia_-) \\
&\times \Pi(\mathbf{c}; z - ia_-/2),
\end{aligned} \tag{6.35}$$

where we have introduced the product

$$\Pi(\mathbf{c}; z) \equiv 16 \prod_{j=1}^4 c_+(z + is_j). \tag{6.36}$$

The upshot of these calculations is the identity

$$A_+(\mathbf{c}; v)R(\mathbf{c}; v, \hat{v}) = \frac{1}{(a_+ a_-)^{1/2}} \int_C dz [2c_+(2z + 2i\hat{c}_0)I(\mathbf{c}; v, \hat{v}, z) \tag{6.37}$$

$$+F(c_0; v, z - ia_-)\Pi(\mathbf{c}; z - ia_-/2)K(\mathbf{c}; z)F(\hat{c}_0; \hat{v}, z)].$$

To proceed, we now shift C down by ia_- in the second term and then take $z \to z + ia_-$. Then we are in the position to exploit a critical property of the function $K(\mathbf{c}; z)$: Due to (6.30) it obeys the AΔE

$$K(\mathbf{c}; z + ia_-/2) = K(\mathbf{c}; z - ia_-/2)/\Pi(\mathbf{c}; z). \qquad (6.38)$$

Therefore we obtain

$$A_+(\mathbf{c}; v)R(\mathbf{c}; v, \hat{v}) = \frac{1}{(a_+a_-)^{1/2}} \int_C dz[2c_+(2z + 2i\hat{c}_0)I(\mathbf{c}; v, \hat{v}, z)$$
$$+F(c_0; v, z)K(\mathbf{c}; z)F(\hat{c}_0; \hat{v}, z + ia_-)]. \qquad (6.39)$$

Finally, we use (6.32) with $d = \hat{c}_0, y = \hat{v}$ to get

$$F(\hat{c}_0; \hat{v}, z + ia_-) = 4s_+(\hat{v} + z + i\hat{c}_0)s_+(\hat{v} - z - i\hat{c}_0)F(\hat{c}_0; \hat{v}, z)$$
$$= 2[c_+(2\hat{v}) - c_+(2z + 2i\hat{c}_0)]F(\hat{c}_0; \hat{v}, z). \qquad (6.40)$$

Then substitution in (6.39) yields (6.28). $\square$

The joint eigenfunction property just demonstrated shows that the R-function may be viewed as a solution to a so-called bispectral problem, cf. Grünbaum's contribution to these proceedings [26]. In this respect, it has however a much more restricted character. Indeed, it solves in fact a 'quadrispectral' problem. (Note that the latter problem can be posed more generally whenever one is dealing with a pair of commuting AΔOs of the form (2.19).)

6.4. THE ASKEY-WILSON SPECIALIZATION

We continue by sketching how the Askey-Wilson polynomials arise as a specialization of the R-function. With the restriction (6.14) on the arguments in effect, we may and will use the representation (6.15). We are going to exploit the analyticity properties of the R-function in the variable $\hat{v}$ and the eigenvalue equation

$$A_+(\hat{\mathbf{c}}; \hat{v})R(\mathbf{c}; v, \hat{v}) = 2c_+(2v)R(\mathbf{c}; v, \hat{v}). \qquad (6.41)$$

To prevent nongeneric singularities, we choose $\hat{c}_0$ rationally independent of $a_+, a_-, \hat{c}_1, \hat{c}_2$ and $\hat{c}_3$. Then R has no pole at the points $\hat{v} = i\hat{c}_0 + ina, n \in \mathbb{Z}$, so we may define

$$R_n(v) \equiv R(\mathbf{c}; v, i\hat{c}_0 + ina_-), \quad n \in \mathbb{Z}. \qquad (6.42)$$

Proposition 6.2 *One has $R_n(v) = P_n(c_+(2v)), n \in \mathbb{N}$, with $P_n(u)$ a polynomial of degree n in u.*

Proof. The pole of $I(z)$ at $z = 0$ is simple and has residue $-i(a_+a_-)^{1/2}/2\pi$. (This follows from (3.41) and (3.38).) Thus we have

$$R(\mathbf{c}; v, \hat{v}) = 1 + \frac{1}{(a_+a_-)^{1/2}} \int_{\mathcal{C}^+} dz I(\mathbf{c}; v, \hat{v}, z), \qquad (6.43)$$

where $\mathcal{C}^+$ denotes the contour $\mathcal{C}$ with an upward indentation at $z = 0$ (instead of downward).

We can now let $\hat{v}$ converge to $i\hat{c}_0$ without $\hat{v}$-dependent poles crossing $\mathcal{C}^+$. The factor $1/G(-\hat{v} + i\hat{c}_0 - ia)$ in $I(z)$ has a zero for $\hat{v} = i\hat{c}_0$, whereas the factor $1/G(\hat{v} + i\hat{c}_0 - ia)$ has no pole for $\hat{v} = i\hat{c}_0$ (due to the rational independence requirement). Hence we deduce

$$R_0(v) = 1. \qquad (6.44)$$

Next, we write out the eigenvalue equation (6.41) for $\hat{v} = i\hat{c}_0 + ina_-$:

$$C_+(\hat{\mathbf{c}}; i\hat{c}_0 + ina_-)[R_{n-1}(v) - R_n(v)] + C_+(\hat{\mathbf{c}}; -i\hat{c}_0 - ina_-)$$
$$[R_{n+1}(v) - R_n(v)] + 2c_+(2ic_0)R_n(v) = 2c_+(2v)R_n(v). \qquad (6.45)$$

The rational independence assumption entails that the coefficients are well defined, and that $C_+(\hat{\mathbf{c}}; -i\hat{c}_0 - ina_-)$ does not vanish for $n \in \mathbb{N}$. Since we have $C_+(\hat{\mathbf{c}}; i\hat{c}_0) = 0$ (cf. (6.24)), we may now use (6.44) as a starting point to prove the assertion recursively. $\square$

Note that when one restricts attention to a *finite* number of the above functions $R_n(v)$, one may let $\hat{c}_0$ vary over suitable intervals without encountering singularities or zeros of the recurrence coefficients (save for $C_+(\hat{\mathbf{c}}; i\hat{c}_0)$, of course). We can now continue a_+ analytically to $-i\pi/r, r > 0$, to obtain polynomials $P_n(\cos(2rv))$ with recurrence coefficients that can be read off from (6.45). The a_+-continuation turns the hyperbolic AΔO $A_+(\mathbf{c}; v)$ into a trigonometric AΔO with eigenvalue $2\text{ch}2r(\hat{c}_0 + na_-)$ on $P_n(\cos(2rv))$. In essence, the latter AΔO is the Askey-Wilson AΔO and the recurrence is the 3-term recurrence of the Askey-Wilson polynomials. More precisely, taking $r = 1/2$, the polynomials $P_n(\cos v)$ turn into the Askey-Wilson polynomials $p_n(q, \alpha, \beta, \gamma, \delta; \cos v)$ under a suitable parameter substitution and n-dependent renormalization, cf. Ref. [16].

6.5. THE 'NONRELATIVISTIC' LIMIT $R \to {}_2F_1$

We continue by clarifying the relation between the R- and ${}_2F_1$-functions. To this end we introduce the relativistic wave function

$$\Psi_{\mathrm{rel}}(\beta, \nu, (g_0, g_1, g_2, g_3); x, p) \equiv R(\pi, \beta\nu, \beta\nu(g_0, g_1, g_2, g_3); \nu x, \beta p/2).$$
$$(6.46)$$

Now we change variables $z \to \beta\nu z$ in the integral representation (6.15), and rewrite the result as

$$\Psi_{\mathrm{rel}} = \int_C dz S_l M S_r, \qquad (6.47)$$

where

$$S_l \equiv \exp(2iz \ln 2) F(\beta\nu g_0; \nu x, \beta\nu z), \qquad (6.48)$$

$$S_r \equiv \exp(2iz \ln(2\beta\nu)) F(\beta\nu \hat{g}_0; \beta p/2, \beta\nu z), \qquad (6.49)$$

$$M \equiv \left(\frac{\beta\nu}{\pi}\right)^{1/2} \exp(-2iz \ln(4\beta\nu)) K(\pi, \beta\nu, \beta\nu(g_0, g_1, g_2, g_3); \beta\nu z).$$
$$(6.50)$$

The factorization performed here ensures that the $\beta \downarrow 0$ limit of the three factors exists. Indeed, using the two zero step size limits (3.42) and (3.47), we obtain

$$\lim_{\beta\downarrow 0} S_l = \exp(-iz \ln(\mathrm{sh}^2 \nu x)), \qquad (6.51)$$

$$\lim_{\beta\downarrow 0} S_r = \left(\frac{\Gamma(-\frac{ip}{2\nu} + \frac{1}{2}(g + \tilde{g}) - iz)}{\Gamma(-\frac{ip}{2\nu} + \frac{1}{2}(g + \tilde{g}))}\right) \quad (p \to -p), \qquad (6.52)$$
$$g \equiv g_0 + g_2, \quad \tilde{g} \equiv g_1 + g_3,$$

$$\lim_{\beta\downarrow 0} M = \frac{\Gamma(iz)\Gamma(g + \frac{1}{2})}{2\pi\Gamma(g + \frac{1}{2} - iz)}, \qquad (6.53)$$

where the limits are uniform on sufficiently small discs around any point on the contour.

When we now interchange these $\beta \downarrow 0$ limits with the contour integration, we obviously get

$$\lim_{\beta\downarrow 0} \Psi_{\mathrm{rel}}(\beta, \nu, (g_0, g_1, g_2, g_3); x, p) = \Psi_{\mathrm{nr}}(\nu, g, \tilde{g}; x, p),$$
$$(6.54)$$
$$g \equiv g_0 + g_2, \quad \tilde{g} \equiv g_1 + g_3,$$

cf. (6.8), (6.2). To date, we have no justification for this interchange. A uniform L^1 tail bound as $\beta \downarrow 0$ would suffice (by dominated convergence), but it remains to supply such a bound. In any case, we conjecture that the limit (6.54) holds true uniformly on x-compacts in $\{\text{Re}\, x > 0, |\text{Im}\, x| < \pi/2\nu\}$ and p-compacts in $\mathbb{C}$.

Let us now consider the $\beta \downarrow 0$ limits of the above four AΔOs with parameters and variables

$$a_+ = \pi, \quad a_- = \beta\nu, \quad \mathbf{c} = \beta\nu(g_0, g_1, g_2, g_3), \quad v = \nu x, \quad \hat{v} = \beta p/2. \tag{6.55}$$

Clearly, $A_-(\hat{\mathbf{c}}; \hat{v})$ and its eigenvalue $\hat{E}_- = 2\text{ch}(2\pi x/\beta)$ diverge for $\beta \downarrow 0$. For the remaining AΔOs and their eigenvalues one readily verifies the following limiting behavior:

$$A_+(\mathbf{c}; v) = 2 + \beta^2 \tilde{H}(g, \tilde{g}) + O(\beta^4), \quad \beta \downarrow 0, \tag{6.56}$$

$$E_+ = 2\text{ch}(\beta p) = 2 + \beta^2 p^2 + O(\beta^4), \quad \beta \downarrow 0, \tag{6.57}$$

$$\lim_{\beta \downarrow 0} A_-(\mathbf{c}; v) = \exp(-i\pi(g + \tilde{g})) T^x_{i\pi/\nu} + (i \to -i), \quad (\text{Re}\, x > 0), \tag{6.58}$$

$$E_- = 2\text{ch}(\pi p/\nu), \tag{6.59}$$

$$\lim_{\beta \downarrow 0} A_+(\hat{\mathbf{c}}; \hat{v}) = \frac{[p - i\nu(g + \tilde{g})]}{p} \cdot \frac{[p - i\nu - i\nu(g - \tilde{g})]}{p - i\nu}(T^p_{2i\nu} - 1)$$
$$+ (i \to -i) + 2, \tag{6.60}$$

$$\hat{E}_+ = 2\text{ch}(2\nu x). \tag{6.61}$$

It can be shown directly that the limiting operators do have the pertinent eigenvalues on Ψ_{nr} (6.8). Indeed, for the AΔO $A_+(\mathbf{c}; v)$ this amounts to (6.7). The limit AΔO on the rhs of (6.58) does have eigenvalue $2\text{ch}(\pi p/\nu)$ by virtue of the known analytic continuation of $_2F_1(a, b, c; w)$ across the logarithmic branch cut $w \in [1, \infty)$. (To appreciate the role of the cut, it may help to recall that the rhs of (6.8) reduces to $\cos(xp)$ for $g = \tilde{g} = 0$.) Finally, the eigenvalue $2\text{ch}(2\nu x)$ for the AΔO on the rhs of (6.60) can be verified by using the contiguous relations of the hypergeometric function, cf. Ref. [19].

6.6. THE RELATION TO THE ISMAIL-RAHMAN FUNCTIONS

As we mentioned above (6.17), the R-function admits a meromorphic continuation in the shift parameters a_+, a_-, provided one requires they stay in the right half plane. Equivalently, scaling out a_+ via (6.22), we retain meromorphy as long as the scale-invariant quotient $\rho = a_-/a_+$ varies over the cut plane $\mathbb{C} \setminus (-\infty, 0]$. (Just as for the hyperbolic gamma function, where the meromorphic continuation is explicitly given by (3.54), we do not know what happens when ρ converges to the cut.)

Consider now the function

$$Q(r, a, \mathbf{c}; v, \hat{v}) \equiv R(\overline{\omega}\pi r^{-1}, \omega a, \omega \mathbf{c}; \omega v, \omega \hat{v}), \quad \omega \equiv e^{i\pi/4}, \quad r, a > 0. \tag{6.62}$$

It is meromorphic in $c_0, \ldots, c_3, v$ and $\hat{v}$, and satisfies

$$AQ = 2\cos(2r\hat{v})Q, \tag{6.63}$$

$$BQ = 2\text{ch}(2\pi a^{-1}\hat{v})Q. \tag{6.64}$$

Here, A and B are the AΔOs (cf. (6.24) and (6.25))

$$A \equiv C_t(r, a, \mathbf{c}; v)[\exp(-iad/dv) - 1] + (v \to -v) + 2\text{ch}(2r\hat{c}_0), \tag{6.65}$$

$$B \equiv C_h(r, a, \mathbf{c}; v)[\exp(-\pi r^{-1}d/dv) - 1] + (v \to -v) + 2\cos(2\pi a^{-1}\hat{c}_0), \tag{6.66}$$

with

$$C_t(z) \equiv \frac{\sin r(z - ic_0)}{\sin rz} \frac{\cos r(z - ic_1)}{\cos rz} \frac{\sin r(z - ic_2 - ia/2)}{\sin r(z - ia/2)} \times \frac{\cos r(z - ic_3 - ia/2)}{\cos r(z - ia/2)}, \tag{6.67}$$

$$C_h(z) \equiv \frac{\text{sh}\pi a^{-1}(z - ic_0)}{\text{sh}\pi a^{-1}z} \frac{\text{ch}\pi a^{-1}(z - ic_1)}{\text{ch}\pi a^{-1}z} \\ \times \frac{\text{sh}\pi a^{-1}(z - ic_2 - i\pi r^{-1}/2)}{\text{sh}\pi a^{-1}(z - i\pi r^{-1}/2)} \\ \times \frac{\text{ch}\pi a^{-1}(z - ic_3 - i\pi r^{-1}/2)}{\text{ch}\pi a^{-1}(z - i\pi r^{-1}/2)}. \tag{6.68}$$

Moreover, one has

$$Q(r, a, \mathbf{c}; v, i\hat{c}_0 + ina) = P_n(r, a, \mathbf{c}; \cos(2rv)), \quad n \in \mathbb{N},$$

$$(6.69)$$

where $P_n(x)$ are the polynomials from Subsection 6.4.

It is immediate from (6.64) that we have the implication

$$Q(v + \pi/r, \hat{v}) = Q(v, \hat{v}) \Rightarrow \pm i\hat{v} = \hat{c}_0 + ka, \quad k \in \mathbb{Z}. \quad (6.70)$$

Put differently, $Q(v, \hat{v})$ is not π/r-periodic in v for generic spectral parameter $\hat{v}$. Now as we mentioned at the end of Subsection 6.4, the trigonometric A$\triangle$O A is essentially the Askey-Wilson A$\triangle$O. Ismail and Rahman [24] construct independent solutions $F_j(r, a, \mathbf{c}; v, \hat{v}), j = 1, 2$, to the Askey-Wilson A$\triangle$E (6.63) in terms of the $_8\phi_7$ basic hypergeometric function. Their solutions are manifestly π/r-periodic in v for *arbitrary* spectral parameters. But in contrast to our solution $Q(r, a, \mathbf{c}; v, \hat{v})$, the functions $F_j(r, a, \mathbf{c}; v, \hat{v})$ do not admit analytic continuation to the hyperbolic regime. This is for basically the same reason as G_{trig} (3.12) does not admit continuation to a hyperbolic gamma function: When one takes q from the open unit disc to the unit circle, one looses convergence. (Cf. the last paragraph of Subsection 3.2.)

On the other hand, the general theory sketched in Subsection 2.2 entails that we must have

$$Q(v, \hat{v}) = \mu_1(v, \hat{v})F_2(v, \hat{v}) - \mu_2(v, \hat{v})F_1(v, \hat{v}), \quad (6.71)$$

with μ_1, μ_2 ia-periodic in v, cf. (2.15). The open problem of finding these multipliers explicitly can be further illuminated by a reasoning that is of interest in its own right.

As a first step, let us note that the above A$\triangle$Os A and B commute. But in contrast to the situation considered in Subsection 2.2, the shifts in the complex plane are not in the same direction. Therefore, the space of joint eigenfunctions is left invariant by elliptic multipliers, with the period parallellogram corresponding to the two directions. But when one lets the two directions become equal, such elliptic multipliers generically will diverge. Accordingly, joint eigenfunctions in general do not converge in the pertinent limit.

The point is now that from the Ismail-Rahman solutions F_1, F_2 to (6.63) we can construct *joint* solutions J_1, J_2 to (6.63) and (6.64), as we will detail in a moment. It is quite plausible (but we could not prove) that J_1 and J_2 form a basis for the space of joint solutions, viewed as a vector space over the field of elliptic functions with periods π/r and ia. Assuming this, the problem of explicitly finding μ_1, μ_2 in (6.71) gets narrowed down to the problem of finding two elliptic functions e_j such that

$$Q(v, \hat{v}) = e_1(v, \hat{v}) J_2(v, \hat{v}) - e_2(v, \hat{v}) J_1(v, \hat{v}). \tag{6.72}$$

(In this connection it should be recalled that Q (6.62) is given in terms of a rather inaccessible integral.)

We continue by filling in the details concerning the functions J_1, J_2. They are defined by

$$J_j(r, a, \mathbf{c}; v, \hat{v}) \equiv F_j(r, a, \mathbf{c}; v, \hat{v}) F_j(\pi/a, \pi/r, ic; iv, i\hat{v}), \quad j = 1, 2. \tag{6.73}$$

(This makes sense, since F_1 and F_2 are meromorphic in $c_0, \ldots, c_3, v$ and $\hat{v}$.) To see that they solve both (6.63) and (6.64), note first that since the first F_j-factor is π/r-periodic in v, the second one is ia-periodic in v. Therefore, J_j still solves (6.63). Next, observe that the substitution

$$r, a, \mathbf{c}, v, \hat{v} \to \pi/a, \pi/r, ic, iv, i\hat{v}, \tag{6.74}$$

that turns the first factor into the second one, also turns A (and its dual) into B (and its dual). Hence the second F_j-factor solves (6.64), so that J_j solves (6.64), too.

The elliptic multiplier question is of particular interest in view of recent work by Suslov [27], and by Koelink and Stokman [28]. They study Hilbert space properties associated with an even and self-dual linear combination $\Phi(r, a, \mathbf{c}; v, \hat{v})$ of the Ismail-Rahman functions F_1, F_2. Now the above 'doubling' argument (which was suggested by the relation (3.58) between the trigonometric and hyperbolic gamma functions) can be applied to Φ as well. It entails that the function

$$\Psi(r, a, \mathbf{c}; v, \hat{v}) \equiv \Phi(r, a, \mathbf{c}; v, \hat{v}) \Phi(\pi/a, \pi/r, ic; iv, i\hat{v}), \tag{6.75}$$

is an even and self-dual solution to (6.63) and (6.64), just as $Q(r, a, \mathbf{c}; v, \hat{v})$. Thus one would be inclined to expect that the even, self-dual function μ defined by

$$Q(r, a, \mathbf{c}; v, \hat{v}) = \mu(r, a, \mathbf{c}; v, \hat{v}) \Psi(r, a, \mathbf{c}; v, \hat{v}), \tag{6.76}$$

is already elliptic with periods $\pi/r, ia$.

We cannot rule out this hunch, but it does lead to consequences that seem startling. Indeed, assuming μ is elliptic, it has doubly-periodic poles and zeros. The zeros must adjust the poles of the two Φ-factors in (6.75) (which can be seen from their series representations, cf. Ref. [28]) to those of the Q-function, whose *eventual* locations follow from Ref. [16]. But this gives rise to zero patterns for the Q- and Φ-functions that appear quite unexpected—though we cannot exclude them for the time being.

Acknowledgment

We would like to thank S. K. Suslov for the invitation to present this material and for useful discussions.

References

1. S. N. M. Ruijsenaars, *Systems of Calogero-Moser type*, in: "Proceedings of the 1994 Banff summer school Particles and fields", G. Semenoff, L. Vinet, eds., CRM Series in Mathematical Physics, Springer, New York, 1999, pp. 251–352.
2. S. N. M. Ruijsenaars, *Special functions associated with Calogero-Moser type quantum systems*, in: "Proceedings of the 1999 Montréal Séminaire de Mathématiques Supérieures, J. Harnad, G. Sabidussi, and P. Winternitz, eds., CRM Proceedings and Lecture Notes, Vol. 26, Amer. Math. Soc., Providence, 2000, pp. 189–226.
3. S. N. M. Ruijsenaars, *On relativistic Lamé functions*, in: "Proceedings of the 1997 Montréal Workshop on Calogero-Moser-Sutherland systems, J. F. van Diejen and L. Vinet, eds., CRM Series in Mathematical Physics, Springer, New York, 2000, pp. 421–440.
4. N. E. Nörlund, *Vorlesungen über Differenzenrechnung*, Springer, Berlin, 1924.
5. L.M. Milne-Thomson, *The calculus of finite differences*, Macmillan, London, 1933.
6. H. Meschkowski, *Differenzengleichungen*, Vandenhoeck & Ruprecht, Göttingen, 1959.
7. G. K. Immink, *Asymptotics of analytic difference equations*, Lect. Notes in Math. Vol. 1085, Springer, Berlin, 1984.
8. S. N. M. Ruijsenaars, *Finite-dimensional soliton systems*, in: "Integrable and superintegrable systems", B. Kupershmidt, ed., World Scientific, Singapore, 1990, pp. 165–206.
9. S. N. M. Ruijsenaars, *First order analytic difference equations and integrable quantum systems*, J. Math. Phys. **38** (1997), 1069–1146.
10. P. Schroth, *Zur Definition der Nörlundschen Hauptlösung von Differenzengleichungen*, Manuscr. Math. 4 (1978), 239–251.
11. S. N. M. Ruijsenaars, *Generalized Lamé functions. I. The elliptic case*, J. Math. Phys. **40** (1999), 1595–1626.
12. S. N. M. Ruijsenaars, *Generalized Lamé functions. II. Hyperbolic and trigonometric specializations*, J. Math. Phys. **40** (1999), 1627–1663.
13. G. Gasper and M. Rahman, *Basic hypergeometric series*, Encyclopedia of Mathematics and its Applications **35**, Cambridge Univ. Press, Cambridge, 1990.
14. N. Kurokawa, *Multiple sine functions and Selberg zeta functions*, Proc. Jap. Acad., Ser. A **67** (1991), 61–64.
15. T. Shintani, *On a Kronecker limit formula for real quadratic fields*, J. Fac. Sci. Univ. Tokyo, Sect. 1A **24** (1977), 167–199.
16. S. N. M. Ruijsenaars, *A generalized hypergeometric function satisfying four analytic difference equations of Askey-Wilson type*, Commun. Math. Phys. **206** (1999), 639–690.
17. K. Ueno, M. Nishizawa, *Quantum groups and zeta functions*, in: "Quantum groups", Karpacz 1994, (PWN, Warsaw, 1995), pp. 115–126.
18. S. K. Suslov, *Basic exponential functions on a q-quadratic grid*, this volume.

19. E. T. Whittaker and G. N. Watson, *A course of modern analysis*, Cambridge Univ. Press, Cambridge, 1973.
20. S. N. M. Ruijsenaars, *On Barnes' multiple zeta and gamma functions*, Advances in Math. **156** (2000), 107–132.
21. E. W. Barnes, *On the theory of the multiple gamma function*, Trans. Camb. Phil. Soc. **19**, 374–425 (1904).
22. R. Askey and J. Wilson, *Some basic hypergeometric orthogonal polynomials that generalize Jacobi polynomials*, Mem. Amer. Math. Soc. **54**, no. 319 (1985).
23. T. H. Koornwinder, *Jacobi functions and analysis on noncompact semisimple Lie groups*, in: "Special functions: group theoretical aspects and applications", R. A. Askey, T. H. Koornwinder, and W. Schempp, eds., Mathematics and its applications, Reidel, Dordrecht, 1984, pp. 1–85.
24. M. E. H. Ismail, M. Rahman, *The associated Askey-Wilson polynomials*, Trans. Amer. Math. Soc. **328**, 201–237 (1991).
25. M. Rahman, *Askey-Wilson functions of the first and second kinds: series and integral representations of $C_n^2(x; \beta|q) + D_n^2(x; \beta|q)$*, J. Math. Anal. Appl. **164**, 263–284 (1992).
26. F. A. Grünbaum, *The bispectral problem: an overview*, this volume.
27. S. K. Suslov, *Some orthogonal very-well-poised $_8\phi_7$-functions that generalize Askey-Wilson polynomials*, to appear in The Ramanujan Journal.
28. E. Koelink, J. V. Stokman, *The Askey-Wilson function transform*, preprint.

THE FACTORIZATION METHOD, SELF-SIMILAR POTENTIALS AND QUANTUM ALGEBRAS

V. P. SPIRIDONOV

Bogoliubov Laboratory of Theoretical Physics
JINR, Dubna, Moscow region, 141980 Russia

Abstract. The factorization method is a convenient operator language formalism for consideration of certain spectral problems. In the simplest differential operators realization it corresponds to the Darboux transformations technique for linear ODE of the second order. In this particular case the method was developed by Schrödinger and became well known to physicists due to the connections with quantum mechanics and supersymmetry. In the theory of orthogonal polynomials its origins go back to the Christoffel's theory of kernel polynomials, etc. Special functions are defined in this formalism as the functions associated with similarity reductions of the factorization chains.

We consider in this lecture in detail the Schrödinger equation case and review some recent developments in this field. In particular, a class of self-similar potentials is described whose discrete spectrum consists of a finite number of geometric progressions. Such spectra are generated by particular polynomial quantum algebras which include q-analogues of the harmonic oscillator and $su(1,1)$ algebras. Coherent states of these potentials are described by differential-delay equations of the pantograph type. Applications to infinite soliton systems, Ising chains, random matrices, and lattice Coulomb gases are briefly outlined.

Keywords. Differential-delay equations, solitons, Schrödinger equation, Darboux transformations, quantum algebras, q-special functions, random matrices, Coulomb gases.

1. Introduction

From the very beginning quantum mechanics served as a rich source of good mathematical problems. It played a major role in the development of the

J. Bustoz et al. (eds.), Special Functions 2000, 335–364.

theory of generalized functions, functional analysis, path integrals, to name a few. In the last two decades the "quantum" disease became so widespread in mathematics that it is difficult to guarantee that a randomly chosen mathematical term will not get its cousin with such an adjective in the foreseeable future (if it does not have already). The influence of quantum mechanics upon the theory of special functions is also indispensable. A bright example is given by the angular momentum theory which has lead to the Racah polynomials. An aim of this lecture is to outline another fruitful interplay between these two scientific fields inspired by the factorization method.

This method was suggested by Schrödinger as a convenient operator language tool for working with quantum mechanical spectral problems [42]. It was reformulated as a problem of searching of the factorization chain solutions by Infeld [25]. The review [26] became a basic reference in this field. A fairly recent revival of the interest to this method occurred due to the discovery of its relation to the notion of supersymmetry (see, e.g. the review [22] and references therein). The author himself turned to this subject from the quantum chromodynamics due to an idea of a generalization of the supersymmetric quantum mechanics [40]. Although it was clear that special functions play an important role in this formalism, it took some time to recognize that, heuristically, special functions are defined in this approach as the functions appearing from similarity reductions of the factorization chain (for a more precise formulation, see the Appendix).

From the mathematical side, a related techniques for solving linear ordinary differential equations of the second order was proposed by Darboux long ago [12]. Its various generalizations are referred to in the modern theory of completely integrable systems as the Laplace, Darboux, Bäcklund, dressing, etc transformations [1]; in the analysis of isomonodromic deformations one deals with the Schlesinger transformations [29]. In the theory of special functions such transformations appear as *contiguity relations*. An important step in the development of the subject was performed in a series of papers by Burchnall and Chaundy [9], which contain even some parts of the operator formulation of the approach ("the transference"). For a completeness, let us mention also the terms "shift operator" and "transmutation", which are used in some other variations of the formalism. For a rigorous mathematical treatment of some aspects of the Darboux transformations technique or the factorization method, see [11, 14, 15, 30, 41].

As far as the discrete recurrence relations are concerned, actually, it is the Christoffel's theory of kernel polynomials that provides a first constructive approach of such kind to spectral problems. This theory is based upon the simplest discrete analogue of the Darboux transformations. A complementary part to this Christoffel's transformation was found by Geronimus in [23]. First applications of the Schrödinger's factorization method to finite-difference equations is given in [37]. Recent developments in this direction are reviewed in [49]. Let us remark that the same techniques was rediscovered in the works on numerical calculations of matrix eigenvalues. More precisely, the well-known numerical LR, QR, g-algorithms, etc provide particular instances of the chains of discrete Darboux transformations. The literature on the taken subject is enormous, it is not possible to describe all its branches. The list of references given at the end of this manuscript is not complete, it contains mainly the papers encountered by the author during his own work (additional lists can be found in [47, 49]).

Despite a vast variety of existing constructions, part of which was just mentioned, we limit ourselves in this lecture to the simplest possible case based upon the stationary one-dimensional Schrödinger equation

$$L\psi(x) = -\psi_{xx}(x) + u(x)\psi(x) = \lambda\psi(x), \tag{1.1}$$

describing the motion of a non-relativistic particle on the line $x \in \mathbf{R}$ in the potential field $u(x)$, which is assumed to be bounded from below. The operator L is called the Hamiltonian or the Schrödinger operator. For convenience, the particle's mass variable and the Planck's constant $\hbar$ are removed by rescalings of the coordinate x and the energy λ. The equation (1.1) describes an eigenvalue problem for L and the eigenvalues (or the permitted bound states energies of the quantum particle) are determined from the condition that the modulus of the wave function $\psi(x)$ is square integrable, $\psi(x) \in L^2(\mathbf{R})$.

 Depending on the physical situation, the real line $\mathbf{R}$ may be replaced by an interval with an appropriate boundary conditions upon $\psi(x)$. However, if one is interested not in the spectra themselves, but in the differential operators L with some formal properties, then it is convenient to take $x, \lambda \in \mathbf{C}$.

2. The Factorization Method

Let us factorize the second order differential operator $L = -d^2/dx^2 + u(x)$ as a product of two first order ones up to some real constant λ_0:

$$L = A^+A^- + \lambda_0, \qquad A^{\pm} = \mp d/dx + f(x). \tag{2.1}$$

The function $f(x)$ is a solution of the Riccati equation $f^2(x) - f_x(x) + \lambda_0 = u(x)$. Substitution of the ansatz $f(x) = -\phi_{0,x}(x)/\phi_0(x)$ shows that $-\phi_{0,xx}(x) + u(x)\phi_0(x) = \lambda_0\phi_0(x)$, i.e. $\phi_0(x)$ is a solution of the original Schrödinger equation (1.1) for $\lambda = \lambda_0$.

If $f(x)$ is a smooth function, then A^+ is a hermitean conjugate of A^- in $L^2(\mathbf{R})$ and L is a self-adjoint operator. Under these circumstances λ_0

cannot be bigger than the smallest eigenvalue of L. Suppose that λ_0 is the smallest eigenvalue of L, and let $\psi_0(x)$ be the corresponding eigenfunction. It is well known that $\psi_0(x)$ is nodeless and may be normalized to have the unit norm, $\|\psi_0\|^2 = \int_{-\infty}^{\infty} |\psi_0(x)|^2 dx = 1$. Then,

$$\phi_0(x) = a\psi_0(x) + b\psi_0(x) \int^x \frac{dy}{\psi_0^2(y)},$$

where a, b are arbitrary constants, is the general solution of the equation $L\phi_0 = \lambda_0\phi_0$. For $b \neq 0$, the resulting $f(x)$ is singular at some point and the operators $A^\pm$ in (2.1) are not well defined. Let us exclude this situation, i.e. set $a = 1, b = 0$.

Using the lowest eigenvalue eigenfunction of L, one can always factorize L as described in (2.1) with the well-defined operators $A^\pm$. Vice versa, if one manages to find the factorization (2.1) such that the zero mode ψ_0 of the operator A^-, $A^-\psi_0 = 0$, belongs to $L^2(\mathbf{R})$ and $f(x)$ is not singular, then λ_0 is the lowest eigenvalue and

$$\psi_0(x) = \frac{e^{-\int_0^x f(y)dy}}{\left(\int_{-\infty}^{\infty} e^{-2\int_0^x f(y)dy} dx\right)^{1/2}} \tag{2.2}$$

is the corresponding normalized eigenfunction.

Let us define now a new Schrödinger operator $\tilde{L}$ by the permutation of the operator factors in (2.1)

$$\tilde{L} = A^- A^+ + \lambda_0, \tag{2.3}$$

whose potential has the form $\tilde{u}(x) = f^2(x) + f_x(x) + \lambda_0$. Evidently, one has the intertwining relations

$$LA^+ = A^+\tilde{L}, \qquad A^-L = \tilde{L}A^-, \tag{2.4}$$

playing the key role in the formalism. From (2.4) one deduces that if $\psi(x)$ satisfies (1.1), then the functions $\tilde{\psi} = A^-\psi$ provide formal eigenfunctions of $\tilde{L}$. Indeed,

$$\tilde{L}\tilde{\psi} = \tilde{L}(A^-\psi) = A^-(L\psi) = \lambda(A^-\psi). \tag{2.5}$$

Actually, this gives general solutions of the differential equation $\tilde{L}\tilde{\psi} = \lambda\tilde{\psi}$ for all λ, except of the point $\lambda = \lambda_0$, where the zero mode of A^- is located. This problem is curable since one can find the general solution of the equation $\tilde{L}\tilde{\phi}_0 = \lambda_0\tilde{\phi}_0$ separately:

$$\tilde{\phi}_0(x) = \frac{g}{\psi_0(x)} + \frac{e}{\psi_0(x)} \int^x \psi_0^2(y)dy, \tag{2.6}$$

where g, e are arbitrary constants and ψ_0 is the lowest eigenvalue eigenfunction of L (note that $A^+\psi_0^{-1} = 0$).

Denote as $\lambda_n, \tilde{\lambda}_n$ and ψ_n, $\tilde{\psi}_n = A^-\psi_n$ discrete eigenvalues and corresponding eigenfunctions of L and $\tilde{L}$ respectively. From (2.5) it follows that the spectra λ_n and $\tilde{\lambda}_n$ almost coincide $\tilde{\lambda}_n = \lambda_n, n = 1, 2, \ldots$. The only difference that may occur in these spectra concerns the zero mode of the operator A^-. In fact, the point $\lambda = \lambda_0$ does not belong to the spectrum of $\tilde{L}$. Indeed, since $\psi_0 \in L^2(\mathbf{R})$, it follows that $\psi_0^{-1} \notin L^2(\mathbf{R})$, and, as a result, for any g, e the function (2.6) cannot be normalizable.

Thus, λ_1 is the lowest eigenvalue of $\tilde{L}$ and the point λ_0 was "deleted" from the spectrum of L. Repeating the same procedure once more, i.e. taking $\tilde{L} = \tilde{A}^+\tilde{A}^- + \lambda_1$ and permuting the operator factors, one can delete the point $\lambda = \lambda_1$, etc. This procedure allows one to remove an arbitrary number of smallest eigenvalues of L. Often it is much easier to find the smallest eigenvalue of a given operator than the other ones. If the lowest eigenvalue of $\tilde{L}$ is determined separately by some means, then, by construction, it will coincide with the second eigenvalue of L, etc. This observation is the central one in the factorization method [42] since it reduces the problem of finding complete discrete spectrum of a taken operator to the problem of finding lowest eigenvalues of a sequence of operators built from L by the "factorize and permute" algorithm.

One can invert the procedure of a deletion of the smallest eigenvalue. Namely, a given $\tilde{L}$ with known lowest eigenvalue λ_1 may be factorized as $\tilde{L} = A^-A^+ + \lambda_0$, with $\lambda_0 < \lambda_1$. If the zero mode of A^- is normalizable, then the operator $L = A^+A^- + \lambda_0$ has the same spectrum as $\tilde{L}$ with an additional inserted eigenvalue at an arbitrary point $\lambda = \lambda_0$. If one factorizes L (or $\tilde{L}$) in such a way that $f(x)$ is a non-singular function, but none of the zero modes of the operators $A^\pm$ are normalizable, then the discrete spectra of L and $\tilde{L}$ coincide completely (an isospectral situation).

There are more complicated possibilities for changing spectral data of a given Schrödinger operator L. For example, if the first factorization is "bad", in the sense that $\tilde{L}$ has a singular potential and considerations given above are not valid, then one may demand that after a number of additional refactorizations one gets a well defined self-adjoint Schrödinger operator. In this way one can delete not only the lowest eigenvalues of L or to insert the new ones, but delete or insert a bunch of spectral points above the smallest one [30]. In particular, two step refactorization procedure allows one to imbed an eigenvalue into the continuous spectrum, etc [14].

Let us give now a "discrete time" formulation of the construction. Denote $L \equiv L_j$ and $\tilde{L} \equiv L_{j+1}$ and take $j \in \mathbf{Z}$ (j may be treated as a continuous parameter and we could denote $\tilde{L} \equiv L_{j-1}$ — all this is a matter of agreement). This gives an infinite sequence of Schrödinger operators

$L_j = -d^2/dx^2 + u_j(x)$ with formal factorizations

$$L_j = A_j^+ A_j^- + \lambda_j, \qquad A_j^\pm = \mp \, d/dx + f_j(x). \qquad (2.7)$$

Neighboring L_j are connected to each other via the abstract factorization chain

$$L_{j+1} = A_{j+1}^+ A_{j+1}^- + \lambda_{j+1} = A_j^- A_j^+ + \lambda_j. \qquad (2.8)$$

Intertwining relations take the form

$$A_j^- L_j = L_{j+1} A_j^-, \qquad L_j A_j^+ = A_j^+ L_{j+1}.$$

Substituting explicit forms of $A_j^\pm$ into (2.8) one gets a differential-difference equation upon $f_j(x)$:

$$(f_j(x) + f_{j+1}(x))_x + f_j^2(x) - f_{j+1}^2(x) = \mu_j \equiv \lambda_{j+1} - \lambda_j. \qquad (2.9)$$

This chain was derived in [25] and the problem of searching "exactly solvable" spectral problems was formulated as a problem of the search of solutions of (2.9) such that the points λ_j define the discrete spectrum of an operator L, say, $L \equiv L_0$. For example, one may try to find solutions of the equation (2.9) $f_j(x)$ in the form of power series in j (λ_j are considered as unknown functions of j). As shown in [25, 26] the finite term expansion occurs iff $f_j(x) = a(x)j + b(x) + c(x)/j$, where a, b, c are some elementary functions of x. This leads to the $_2F_1$ hypergeometric function and well known "old" exactly solvable potentials of quantum mechanics.

As evident from the construction, the constants $\lambda_j, j \geq 0$, determine the smallest eigenvalues of L_j under the condition that the zero modes of A_j^- are normalizable and $f_j(x)$ are not singular. In general, the $j \to j+1$ transitions may describe all three possibilities — removal or insertion of an eigenvalue and isospectral transformations (sometimes it is convenient to parameterize inserted eigenvalues as $\lambda_0 > \lambda_1 > \ldots > \lambda_n$).

For a positive integer n, let us introduce the operators

$$M_j^- = A_{j+n-1}^- \cdots A_{j+1}^- A_j^-, \qquad M_j^+ = A_j^+ A_{j+1}^+ \cdots A_{j+n-1}^+.$$

The intertwining relations

$$L_{j+n} M_j^- = M_j^- L_j, \qquad M_j^+ L_{j+n} = L_j M_j^+,$$

guarantee that for almost all λ solutions of the equations $L_j \psi^{(j)} = \lambda \psi^{(j)}$ and $L_{j+n} \psi^{(i+n)} = \lambda \psi^{(j+n)}$ are related to each other as $\psi^{(j+n)} \propto M_j^- \psi^{(j)}$ and $\psi^{(j)} \propto M_j^+ \psi^{(j+n)}$. As a result, the product $M_j^+ M_j^-$ should commute with L_j and $M_j^- M_j^+$ should commute with L_{j+n}. A simple computation leads to the equalities

$$M_j^+ M_j^- = \prod_{k=0}^{n-1} (L_j - \lambda_{j+k}), \qquad M_j^- M_j^+ = \prod_{k=0}^{n-1} (L_{j+n} - \lambda_{j+k}). \qquad (2.10)$$

Let $\psi^{(j)}(x) \in L^2(\mathbf{R})$ be an eigenfunction of L_j with the eigenvalue λ and some finite norm $\|\psi^{(j)}\|$. If we set $\psi^{(j+n)} = M_j^- \psi^{(j)}$, then

$$\|\psi^{(j+n)}\|^2 = (\lambda - \lambda_j) \cdots (\lambda - \lambda_{j+n-1}) \|\psi^{(j)}\|^2.$$

Zeroes on the r.h.s. for $\lambda = \lambda_k$ for some k indicate that the corresponding eigenvalues were deleted from the spectrum. Under the condition that zero modes of $A_j^-, j = 0, \ldots, n-1$, are normalizable and nodeless, $A_j^- \psi_0^{(j)} = 0$, $\|\psi_0^{(j)}\| = 1$, one finds that the functions

$$\psi_n^{(0)}(x) = \frac{A_0^+ \cdots A_{n-1}^+ \psi_0^{(n)}(x)}{\sqrt{(\lambda_n - \lambda_{n-1}) \cdots (\lambda_n - \lambda_0)}}$$

define the unit norm eigenfunctions of L_0 with the eigenvalues λ_n: $L_0 \psi_n^{(0)} = \lambda_n \psi_n^{(0)}$ and $\|\psi_n^{(0)}\| = 1$.

The main advantage of the factorization method consists in its pure operator language formulation. One may replace the Schrödinger operator by any other (higher order differential, finite-difference, integral, etc) operator L admitting factorizations into the well defined operator factors of a simpler nature $A^\pm$ (for some applications they are not necessarily hermitean conjugates of each other). In all cases one deals with the abstract operator factorization chain (2.8) with an appropriate interpretation of the constants λ_j. It should be noted that this method does not have straightforward generalizations to the multidimensional spectral problems, only some of its features are preservable, see e.g. [3].

3. Supersymmetry

Supersymmetry is a symmetry between bosonic and fermionic degrees of freedom of particular physical systems. The corresponding symmetry algebras are distinguished from the standard Lie algebras by the presence of anticommutator relations. The simplest superalgebra is realized upon two neighboring operators in the factorization chain. Define the matrix operators (supercharges)

$$Q^+ = \begin{pmatrix} 0 & A^+ \\ 0 & 0 \end{pmatrix}, \quad Q^- = \begin{pmatrix} 0 & 0 \\ A^- & 0 \end{pmatrix},$$

where $A^{\pm}$ are the factorization operators defined in (2.1). They form the following algebra (see, e.g. [22])

$$\{Q^+, Q^-\} = H, \qquad [H, Q^{\pm}] = (Q^{\pm})^2 = 0, \tag{3.1}$$

where the Hamiltonian

$$H = \begin{pmatrix} L - \lambda_0 & 0 \\ 0 & \tilde{L} - \lambda_0 \end{pmatrix} = -\frac{d^2}{dx^2} + f^2(x) - f_x(x)\sigma_3$$

describes a particle with spin 1/2 (a fermionic variable) on the line in an external magnetic field, σ_3 is the Pauli matrix. In terms of the hermitean supercharges $Q_1 = Q^+ + Q^-, Q_2 = (Q^+ - Q^-)/i$, the algebra takes the form $\{Q_i, Q_j\} = 2H\delta_{ij}$, $[H, Q_i] = 0$.

As a formal consequence of the algebra (3.1) the discrete spectrum of H is doubly degenerate (such degeneracies are characteristic to the supersymmetry) with possible exception of the lowest eigenvalue which cannot be negative (this is another constraint upon the supersymmetric systems).

This simple construction was generalized in [40] to a symmetry between particles with parastatistics. The corresponding symmetry algebras are polynomial in the generators. For example, in the simplest case involving a parafermion of the second order (or the spin 1 particle) H is given by a 3×3 diagonal matrix with the entries L_0, L_1, L_2, the corresponding symmetry generators being $(Q^+)_{ij} = A_i^+ \delta_{i,j-1}$ and its conjugate. For the hermitean charges $Q_{1,2}$ defined as above, one has now cubic relations

$$Q_i(\{Q_j, Q_k\} - 2H\delta_{jk}) + \text{cyclic perm. of } i, j, k = 0, \qquad [H, Q_i] = 0. \tag{3.2}$$

The discrete spectrum of H is now triply degenerate with possible exception of two smallest eigenvalues.

One can go further and propose other modifications of the algebra (3.1). For instance, one can q-deform it [46]:

$$Q^+Q^- + q^2Q^-Q^+ = H, \qquad (Q^{\pm})^2 = 0, \qquad HQ^{\pm} = q^{\pm 2}Q^{\pm}H.$$

This lifts the degeneracy of spectra, but creates a nontrivial scaling relation between spectral points. As a result, it opens a way for building various q-harmonic oscillator models. Another construction proposed in [4] refers to a polynomials generalization of (3.1). It appears after a reduction of the algebra (3.2) and similar higher order polynomial relations to a two-dimensional subspace of eigenfunctions:

$$\{Q^+, Q^-\} = P_n(H), \qquad [H, Q^{\pm}] = (Q^{\pm})^2 = 0,$$

where $P_n(H) = \prod_{k=0}^{n-1}(H - \lambda_k)$ and

$$Q^+ = \begin{pmatrix} 0 & A_0^+ \cdots A_{n-1}^+ \\ 0 & 0 \end{pmatrix}, \qquad Q^- = \begin{pmatrix} 0 & 0 \\ A_{n-1}^- \cdots A_0^- & 0 \end{pmatrix},$$

$$H = \begin{pmatrix} L_0 & 0 \\ 0 & L_n \end{pmatrix}.$$

This is a pure supersymmetry again in the sense of a symmetry between bosons and fermions. However, the consequences are quite different from the standard case (3.1). In particular, there are no such severe constraints upon the smallest eigenvalue of H (the vacuum energy).

All these general constructions are useful because they contain new concepts providing nonstandard viewpoints upon physical systems and new mathematical tools for their exploration. Special functions emerge when one starts to work with particular "exactly solvable" models with such symmetries.

4. Darboux Transformations

Let us describe the Darboux transformations technique in its modern appearance. Consider compatibility conditions of the following three linear equations

$$L_j \psi^{(j)}(x,t) = \lambda \psi^{(j)}(x,t), \qquad L_j \equiv -\partial_x^2 + u_j(x,t), \qquad (4.1)$$

$$\psi^{(j+1)}(x,t) = A_j^- \psi^{(j)}(x,t), \qquad A_j^- \equiv \partial_x + f_j(x,t), \qquad (4.2)$$

$$\psi_t^{(j)}(x,t) = B_j \psi^{(j)}(x,t), \qquad B_j \equiv -4\partial_x^3 + 6u_j(x,t)\partial_x + 3u_{j,x}(x,t), \quad (4.3)$$

where t is some additional continuous parameter ("evolution time") and $u_j(x,t)$, $f_j(x,t)$ are some free functions. The compatibility condition of (4.1) and (4.2) generates the intertwining relation $A_j^- L_j = L_{j+1} A_j^-$, the resolution of which yields the constraints $u_j = f_j^2 - f_{j,x} + \lambda_j$, $u_{j+1} = u_j + 2f_{j,x}$, where λ_j is an integration constant. Thus one arrives again to the infinite chain of nonlinear differential-difference equations (2.9) and the formal factorizations (2.7).

The compatibility condition of the equations (4.1) and (4.3) yields the celebrated Korteweg-de Vries (KdV) equation:

$$u_{j,t}(x,t) - 6u_j(x,t)u_{j,x}(x,t) + u_{j,xxx}(x,t) = 0, \qquad (4.4)$$

a completely integrable Hamiltonian system with an infinite number of degrees of freedom [1]. If we express $u_j(x,t)$ through $f_j(x,t)$ in the equation (4.4), then it takes the form

$$(-\partial_x + 2f_j) V_j(x,t) = 0, \qquad V_j(x,t) \equiv f_{j,t} - 6(f_j^2 + \lambda_j)f_{j,x} + f_{j,xxx}.$$

It appears that the compatibility conditions of this equation with the Darboux transformation (4.2) imposes an additional non-trivial constraint $V_j = 0$, which is called the modified KdV equation.

A substitution of $f_j = -\phi_{0,x}^{(j)}/\phi_0^{(j)}$ into the relation $u_j = f_j^2 - f_{j,x} + \lambda_j$ yields $-\phi_{0,xx}^{(j)} + u_j\phi_0^{(j)} = \lambda_j\phi_0^{(j)}$ and the Darboux transformation (4.2) takes the form:

$$\psi^{(j+1)} = (\phi_0^{(j)}\psi_x^{(j)} - \phi_{0,x}^{(j)}\psi^{(j)})/\phi_0^{(j)},$$

where $\phi_0^{(j)}$ is a particular solution of the j-th equation in the sequence (4.1) for $\lambda = \lambda_j$.

Let $L_j\phi_k^{(j)} = \lambda_{j+k}\phi_k^{(j)}$, i.e. let $\phi_k^{(j)}(x,t)$ are formal eigenfunctions of L_j with the eigenvalues λ_{j+k} equal to constants of integration mentioned above for all k. Denote as $W(\phi_1,\ldots,\phi_n) = \det(\partial_x^{i-1}\phi_k)$ the Wronskian of a set of functions ϕ_k. Using the identities $W(\xi(x)\phi_1, \xi(x)\phi_2) = \xi^2(x)W(\phi_1,\phi_2)$ and

$$W(W(\phi_1,\ldots,\phi_n,\xi_1), W(\phi_1,\ldots,\phi_n,\xi_2)) = W(\phi_1,\ldots,\phi_n)$$
$$\times W(\phi_1,\ldots,\phi_n,\xi_1,\xi_2),$$

one can show that [11]

$$u_{j+n}(x,t) = u_j(x,t) - 2\partial_x^2 \log W(\phi_0^{(j)},\ldots,\phi_{n-1}^{(j)}), \qquad (4.5)$$

$$f_{j+n}(x,t) = -\partial_x \log \frac{W\left(\phi_0^{(j)},\ldots,\phi_n^{(j)}\right)}{W\left(\phi_0^{(j)},\ldots,\phi_{n-1}^{(j)}\right)}, \qquad (4.6)$$

$$\psi^{(j+n+1)}(x,t) = \frac{W\left(\phi_0^{(j)},\ldots,\phi_n^{(j)},\psi^{(j)}\right)}{W\left(\phi_0^{(j)},\ldots,\phi_n^{(j)}\right)} = (\partial_x + f_{j+n})\cdots(\partial_x + f_j)\psi^{(j)}(x,t).$$
$$(4.7)$$

These formulae give an explicit representation of the transformed potentials u_{j+n} in terms of the initial Schrödinger equation solutions $\phi_k^{(j)}$.

Since the parameter t is a dummy variable during the Darboux transformations, in fact one deals simultaneously with different solutions of the KdV equation: if $u_j(x,t)$ satisfies (4.4), the same is true for $u_{j+n}(x,t)$. This gives a way for building new complicated explicit solutions of the KdV equation starting from the simple ones. For instance, one may take $u_0(x,t) = 0$ which gives $\phi_m^{(0)}(x,t) = a_m e^{\kappa_m x - 4\kappa_m^3 t} + b_m e^{-\kappa_m x + 4\kappa_m^3 t}$, and for some special choice of the signs of b_m/a_m the potential $u_n(x,t)$ becomes a nonsingular reflectionless potential with n discrete spectrum points $\lambda_m = -\kappa_m^2 < 0$. It defines the famous n-soliton solution of the KdV equation.

In the formulae given above one may permute any pair of the $\phi_k^{(j)}$ functions and the final result is not changed, i.e. such a permutation of the

Darboux transformations is a symmetry of the factorization chain (2.9). In the simplest case, it has the following explicit form [2]:

$$\tilde{f}_k = f_k - \frac{\lambda_{k+1} - \lambda_k}{f_{k+1} + f_k}, \quad \tilde{f}_{k+1} = f_{k+1} + \frac{\lambda_{k+1} - \lambda_k}{f_{k+1} + f_k}, \quad \tilde{\lambda}_k = \lambda_{k+1}, \quad \tilde{\lambda}_{k+1} = \lambda_k,$$

all other $f_j(x)$, λ_j staying intact for $j \neq k, k+1$. This discrete symmetry may be "discretized" further by passing to the discrete Schrödinger equation (or the three term recurrence relation for orthogonal polynomials) and the corresponding discrete Darboux transformations [48].

The well known KdV tau-function is introduced as

$$u_j(x, t) = -2\partial_x^2 \log \tau_j(x, t).$$

In its terms relations (4.5), (4.6) are rewritten as $\tau_{j+n} = W(\phi_0^{(j)}, \ldots, \phi_{n-1}^{(j)})\tau_j$, $f_j = -\partial_x \log \tau_{j+1}/\tau_j$. Introducing the variables $\rho_j = -\partial_x \log \tau_j$, one can write $u_j = 2\rho_{j,x}$ and $f_j = \rho_{j+1} - \rho_j$. As a result, the relation between u_j and f_j yields the equation

$$(\rho_{j+1} + \rho_j)_x - (\rho_{j+1} - \rho_j)^2 = \lambda_j,$$

which starts to play the role of the factorization chain. The function τ_j is a very convenient object since its zeros in x correspond to poles of the potential.

5. Operator Self-Similarity and Quantum Algebras

Let us turn now to the problem of searching particular solutions of the operator factorization chain (2.8). At first glance it is not clear how to proceed, but the harmonic oscillator problem — a base model for the whole quantum mechanics and quantum field theory — provides a guiding idea. One has to try to form from the factorization operators $A_j^{\pm}$ some nontrivial symmetry algebras. In fact, the relations (2.10) look already as defining relations of some algebra, but they are not closed — the relations between the eigenvalue problems for L_j and L_{j+n} are too weak (they are valid for any starting potential $u_j(x)$). In order to close the system, one has to assume that there is an additional relation between L_j and L_{j+n}, say $L_{j+n} = g(L_j)$, which would force the operators $M_j^{\pm}$ to map eigenfunctions of a taken operator L_j onto themselves.

In the simplest case one demands that the sequence of Hamiltonians L_j is periodic: $L_{j+N} = L_j$ for some period $N > 0$. As a result, $M_j^{\pm}$ commute with L_j, $[M_j^{\pm}, L_j] = 0$. This is a remarkable fact since the existence of additional conserved quantities may simplify solution of the eigenvalue problem for L_j. In the differential operator realization of $A_j^\pm$ this leads to the commuting differential operators [9]. There is a generalization of this pure periodicity condition to the periodicity up to a twist condition $L_{j+N} = UL_jU^{-1}$, where U is some invertible operator. This leads again to commuting operators $[B_j^\pm, L_j] = 0$, where $B_j^+ = M_j^+U$, $B_j^- = U^{-1}M_j^-$, but now there is an essential additional freedom in the choice of U.

Another possible "closure" or a reduction of the sequence of operators L_j consists in the requirement of their periodicity up to a constant shift and a twist, $L_{j+N} = UL_jU^{-1} + \mu$, where μ is a constant. This results in the ladder relation $[L_j, B_j^\pm] = \pm\mu B_j^\pm$ and the operator identities

$$B_j^+B_j^- = \prod_{k=0}^{N-1}(L_j - \lambda_{j+k}), \qquad B_j^-B_j^+ = \prod_{k=0}^{N-1}(L_j + \mu - \lambda_{j+k}),$$

where $B_j^\pm$ operators are defined as in the previous case. Denoting $B^\pm \equiv B_0^\pm$, $L \equiv L_0$, one can form a polynomial algebra

$$[L, B^\pm] = \pm\mu B^\pm, \qquad [B^+, B^-] = P_{N-1}(L), \qquad (5.1)$$

where $P_{N-1}(x)$ is a polynomial of the degree $N - 1$ in x. For a representation theory of such algebras, see e.g. [45]. Note that for $N = 1$ this is the Heisenberg-Weyl or the harmonic oscillator algebra, and for $N = 2$ it coincides with the $su(1,1)$ algebra.

Quantum algebras, or q-analogues of the algebras (5.1) appear from the following operator self-similarity constraint imposed upon the chain (2.8):

$$L_{j+N} = q^2UL_jU^{-1} + \mu. \qquad (5.2)$$

When the numerical factor $q^2 \neq 1$, one can remove μ by the uniform shift $L_j \to L_j + \mu/(1 - q^2)$. Therefore we assume below that $\mu = 0$. Substitute (5.2) with $\mu = 0$ into (2.10). Then the operators $L = L_0$, $B^+ = M_0^+U$, $B^- = U^{-1}M_0^-$ satisfy the following identities

$$LB^\pm = q^{\pm 2}B^\pm L, \qquad B^+B^- = \prod_{j=0}^{N-1}(L - \lambda_j), \qquad B^-B^+ = \prod_{j=0}^{N-1}(q^2L - \lambda_j).$$
$$(5.3)$$

For $N = 1$ these relations provide a realization of the q-harmonic oscillator algebra

$$B^-B^+ - q^2B^+B^- = \rho, \qquad [B^\pm, \rho] = 0, \qquad (5.4)$$

with $\rho = \lambda_0(q^2 - 1)$. One can set $\rho = 1$ by taking the normalization condition $\lambda_0 = 1/(q^2 - 1)$. Such a q-analogue of the Heisenberg-Weyl algebra was encountered in physics long ago [10, 18]. In the modern times it became quite popular due to the inspirations coming from the quantum groups, see e.g. [35]. For $N = 2$ relations (5.3) determine a particular q-analogue of the conformal algebra $su(1,1)$ admitting the Hopf algebra structure, etc [46, 47].

6. Self-Similar Potentials

Consider the Schrödinger equation realizations of the algebraic relations described in the previous section. Let us start from the closure $\tilde{L} = ULU^{-1} + \mu$, where $L = -\partial_x^2 + u(x)$ is the Schrödinger operator in the notations of Sect. 2 and U is a translation operator $Uf(x) = f(x + a)$. As a result, the operators $B^{\pm}$ take the form $B^+ = (-\partial_x + f(x))U$, $B^- = U^{-1}(\partial_x + f(x))$ and one gets the Heisenberg-Weyl algebra $[L, B^{\pm}] = \pm\mu B^{\pm}, [B^-, B^+] = \mu$. The potential entering L is defined as $u(x) = f^2(x) - f_x(x) + \lambda_0$, where $f(x)$ satisfies the following nonlinear differential-delay equation

$$(f(x) + f(x + a))_x + f^2(x) - f^2(x + a) = \mu. \qquad (6.1)$$

For $a = 0$ one gets from this equation the standard harmonic oscillator model $f(x) \propto x, u(x) \propto x^2$, which is related to the Hermite polynomials. Although for $a \neq 0$ and $\mu = 0$ the author has found a meromorphic solution of (6.1) in terms of the Weierstrass $\mathcal{P}$-function

$$f(x) = -\frac{1}{2}\frac{\mathcal{P}'(x - x_0) - \mathcal{P}'(a)}{\mathcal{P}(x - x_0) - \mathcal{P}(a)},$$

in general it is quite difficult to build its solutions analytic in some region, especially for $\mu \neq 0$. Let us remark, that the equation (6.1) may be derived also after imposing the constraint $f_{j+1}(x) = f_j(x + a)$ upon the chain (2.9).

Turn now directly to the relations (5.3). In this case one can take U as the dilation (or q-difference) operator: $Uf(x) = |q|^{1/2}f(qx)$. For real $q \neq 0$ this is a unitary operator $U^{\dagger} = U^{-1}$. Taking $L \equiv L_0$, $B^- \equiv U^{-1}(\partial_x + f_{N-1}) \cdots (\partial_x + f_0)$, $B^+ \equiv (-\partial_x + f_0) \cdots (-\partial_x + f_{N-1})U$ one realizes the identities (5.3), provided $f_j(x)$ satisfy the following system of nonlinear differential-delay equations:

$$(f_0(x) + f_1(x))_x + f_0^2(x) - f_1^2(x) = \mu_0, \quad \ldots\ldots$$

$$(f_{N-1}(x) + qf_0(qx))_x + f_{N-1}^2(x) - q^2 f_0^2(qx) = \mu_{N-1}. \qquad (6.2)$$

Equivalently, these equations appear from the chain (2.9) after imposing the following constraints

$$f_{j+N}(x) = qf_j(qx), \qquad \mu_{j+N} = q^2\mu_j, \qquad (6.3)$$

having a simple group-theoretical interpretation.

One easily arrives at (6.3) using the Lie's symmetry reduction technique (for a differential context of this theory, see e.g. [38], and for an extension to difference equations, see e.g. [31, 36]). First, one notices that if $f_j(x)$, μ_j are solutions of the chain (2.9), then their discrete scaling transformation $qf_j(qx)$, $q^2\mu_j$ gives solutions of (2.9) as well. Analogously, a shift in the numeration $f_j(x) \to f_{j+N}(x)$, $\mu_j \to \mu_{j+N}$ maps solutions of (2.9) into solutions. Let us consider a set of *self-similar* solutions of (2.9) which is invariant under a combination of these two symmetries. For example, demanding that these two transformations are equivalent to one another, we arrive to (6.3). For $N = 1$, this reduction may be rewritten as $f_j(x) = q^j f_0(q^j x)$, $\lambda_j = q^{2j}\lambda_0$ in which form it was first met in [43]. The general q-periodic reduction (6.3) was found by the author [46] in an attempt to q-deform the parastatistical supersymmetry algebras (3.2).

Solutions of the equations (6.2) have a quite complicated structure and general methods of solving such differential-delay equations give relatively weak results. For example, for $N = 1$ in [44] the existence and uniqueness of solutions analytical near the $x = 0$ point was proved under some constraints, and in [32] existence of the nonsingular for $x \in \mathbf{R}$ solutions was demonstrated (the $|x| \to \infty$ asymptotics of such solutions is not determined yet completely). Some understanding of the complexity of general solutions is obtained from considerations of special values of the parameters q, μ_j, when $f_j(x)$ can be expressed through some known functions.

Let q be arbitrary and $f_j^2(x) = \frac{1}{1-q^2}\sum_{m=0}^{N-1}\mu_m - \sum_{m=0}^{j-1}\mu_m$. This yields $L_0 = -d^2/dx^2$ or the free quantum mechanical particle, which acquires in this way a q-algebraic interpretation [47].

Suppose that $f_0(x)$ is not singular at $x = 0$, then in the crystal base limit $q \to 0$ the potential $u_0(x)$ boils down to the general KdV N-soliton potential.

Substitute now $q = 1$ into (6.2) and assume that $\sum_{m=0}^{N-1}\mu_m \neq 0$. Then for $N = 1, 2$ one easily gets the potentials $u_0(x) \propto x^2, ax^2 + b/x^2$. For $N = 3$ the corresponding system of equations for $f_j(x)$ provides a "cyclic" representation of the Painlevé-IV equation [8, 50]. The $N = 4$ case is related to the Painlevé-V function [2], etc. This gives also a commutator representation of the Painlevé equations which was noticed first in [16]. For $q = -1$ one gets similar situations with the functions $f_j(x)$ obeying certain parity symmetry [47]. If $\sum_{m=0}^{N-1}\mu_m = 0$ then for odd N and some cases of even N the potential $u_0(x)$ is expressed through the hyperelliptic functions [50, 51].

The case when q is a primitive root of unity, $q^n = 1, q \neq \pm 1$, is quite interesting [44]. For any odd n and, under certain conditions, for even n potentials are expressed through hyperelliptic functions with additional crystallographic or quasi-crystallographic symmetries. For example, for $N = 1$ and $q^3 = 1$ or $q^4 = 1$ one recovers the Lamé equation with the equianharmonic or lemniscatic Weierstrass $\mathcal{P}$-functions. Thus, this classical differential equation of the second order appears to be related to the representations of the q-harmonic oscillator algebra for q a root of unity.

The general family of self-similar potentials unifies some of the Painlevé functions with hyperelliptic functions. Due to the connections with quantum algebras, it may be taken as a new class of nonlinear q-special functions (viz. continuous q-Painlevé transcendents) defined upon the differential equations. The standard basic hypergeometric functions [21] appear in this formalism through the coherent states.

As far as the discrete spectrum of self-similar potentials is concerned, from the factorization method and (6.3) it follows that in the simplest case it is composed from N independent geometric progressions: $\lambda_{pN+k} = \lambda_k q^{2p}$, $k = 0, \ldots, N-1$, $p \in \mathbf{N}$. The same follows from the theory of unitary representations of the algebra (5.3). The only condition for the validity of this formal conclusion is that the functions $f_j(x)$ are non-singular for $x \in \mathbf{R}$ and positive for $x \to +\infty$ and negative for $x \to -\infty$.

7. Coherent States

There are many definitions of quantum mechanical coherent states [39]. In the context of spectrum generating algebras, such as (5.1) and (5.3), they are defined as eigenfunctions of the corresponding lowering operators. Such eigenfunctions play the role of generating functions for the space elements of irreducible representations of a taken algebra. Let us describe briefly coherent states for the quantum algebra (5.3) (for more details, see [47]).

Let us denote as $|\lambda\rangle$ eigenstates of an abstract operator L entering (5.3), $L|\lambda\rangle = \lambda|\lambda\rangle$. Suppose that the operators $B^\pm$ are conjugated to each other with respect to some inner product $\langle \sigma | \lambda \rangle$. We assume that the spectrum of L is not degenerate. Then the action of operators $B^\pm$ upon $|\lambda\rangle$ has the form:

$$B^-|\lambda\rangle = \prod_{k=0}^{N-1} \sqrt{\lambda - \lambda_k}\,|\lambda q^{-2}\rangle, \qquad B^+|\lambda\rangle = \prod_{k=0}^{N-1} \sqrt{\lambda q^2 - \lambda_k}\,|\lambda q^2\rangle.$$

Let N be odd, $0 < q^2 < 1$ and $\lambda_0 < \ldots < \lambda_{N-1} < q^2 \lambda_0 < 0$. Then for $\lambda < 0$ the operator L may have a discrete spectrum formed only from up to N geometric progressions corresponding to the lowest weight unitary irreducible representations of the algebra (5.3). This follows from the fact that B^- is the lowering operator for the $\lambda < 0$ states and $\prod_{k=0}^{N-1}(\lambda - \lambda_k)$ becomes negative for $\lambda < \lambda_0$. Since $B^-|\lambda_k\rangle = 0$, this problem is avoided for special values of λ, namely, for $\lambda = \lambda_{pN+k} \equiv \lambda_k q^{2p}$, $p \in \mathbf{N}$. For normalizable $|\lambda_k\rangle$, one gets series of normalizable eigenstates of the form $|\lambda_{pN+k}\rangle \propto (B^+)^p |\lambda_k\rangle$.

Coherent states of the first type are defined as eigenstates of B^-:

$$B^-|\alpha\rangle_-^{(k)} = \alpha|\alpha\rangle_-^{(k)}, \quad k = 0,\dots,N-1, \tag{7.1}$$

where α is a complex variable. Representing $|\alpha\rangle_-^{(k)}$ as a sum of the states $|\lambda_{pN+k}\rangle$ with some coefficients, one finds

$$\begin{aligned}
|\alpha\rangle_-^{(k)} &= \sum_{p=0}^{\infty} C_p^{(k)} \alpha^p |\lambda_{pN+k}\rangle \\
&= C^{(k)}(\alpha) \, {}_N\varphi_{N-1} \left(\begin{matrix} 0,\dots,0 \\ \lambda_0/\lambda_k,\dots,\lambda_{N-1}/\lambda_k \end{matrix} ; q^2, z \right) |\lambda_k\rangle,
\end{aligned}$$

where ${}_N\varphi_{N-1}$ is a standard basic hypergeometric series with the operator argument $z = (-1)^N \alpha B^+/\lambda_0 \cdots \lambda_{N-1}$ and $C^{(k)}(\alpha)$ is a normalization constant (in the set of parameters of the ${}_N\varphi_{N-1}$ function the term $\lambda_k/\lambda_k = 1$ is assumed to be absent). The superscript k simply counts the number of lowest weight irreducible representations of the algebra (5.3) each of which has its own coherent state (or a generating function). The states $|\alpha\rangle_-^{(k)}$ are normalizable for $|\alpha|^2 < |\lambda_0 \cdots \lambda_{N-1}|$. Coherent states of this type were first constructed for the q-harmonic oscillator algebra (5.4) in [5], for a particular explicit model of them, see e.g. [6].

It is not difficult to see, that zero modes of the operator L define a special degenerate representation of the algebra (5.3). Since $B^\pm$ operators commute with L in the subspace of these zero modes, one can consider them as coherent states as well.

Unusual coherent states appear from the non-highest weight representations of (5.3) corresponding to the $\lambda > 0$ eigenstates of L. Let $\lambda_+ > 0$ be a discrete spectrum point of L. Then the operators $B^\pm$ generate from $|\lambda_+\rangle$ a part of the discrete spectrum of L in the form of a bilateral geometric progression $\lambda_+ q^{2n}, n \in \mathbf{Z}$. Since $\prod_{k=0}^{N-1}(\lambda_+ - \lambda_k) > 0$ for arbitrary N and $\lambda_+ > 0$, the number of such irreducible representations in the spectrum of L is not limited. Moreover, a continuous direct sum of such representations may form a continuous spectrum of L.

An important fact is that for $\lambda > 0$ the role of lowering operator is played by B^+. Therefore one can define coherent states as eigenstates of this operator as well [47]:

$$B^+|\alpha\rangle_+ = \alpha|\alpha\rangle_+. \tag{7.2}$$

Suppose that the bilateral sequence $\lambda_+ q^{2n} > 0$ belongs to the discrete spectrum of L. Then the states $|\alpha\rangle_+$ are expanded in the series of eigenstates $|\lambda_+ q^{2n}\rangle$:

$$|\alpha\rangle_+ = \sum_{n=-\infty}^{\infty} C_n \alpha^n |\lambda_+ q^{2n}\rangle = C(\alpha) \, {}_0\psi_N \left(\begin{matrix} 0,\dots,0 \\ \lambda_0/\lambda_+,\dots,\lambda_{N-1}/\lambda_+ \end{matrix} ; q^2, z \right) |\lambda_+\rangle,$$

where $_0\psi_N$ is a bilateral q-hypergeometric series with the operator argument $z = \alpha B^-/(-\lambda_+)^N$ and $C(\alpha)$ is a normalization constant. In this case the states $|\alpha\rangle_+$ are normalizable if $|\alpha|^2 > |\lambda_0 \cdots \lambda_{N-1}|$.

Suppose that the $\lambda > 0$ region is occupied by the continuous spectrum. The corresponding states $|\lambda\rangle$ may be normalized as $\langle\sigma|\lambda\rangle = \lambda\delta(\lambda - \sigma)$. Let $N = 1$ and the states $|\lambda\rangle$ are not degenerate. Taking $\rho = 1$ in (5.4), we get the expansion

$$|\alpha\rangle_+^{(s)} = C(\alpha) \int_0^\infty \frac{\lambda^{\gamma_s}|\lambda\rangle d\lambda}{\sqrt{(-\lambda q^2(1 - q^2); q^2)_\infty}},$$

where

$$\gamma_s = \frac{2\pi i s - \ln(\alpha q^2\sqrt{1 - q^2})}{\ln q^2}, \qquad s \in \mathbf{Z}.$$

There is a countable family of such coherent states which have a unit norm for $|\alpha|^2 > 1/(1-q^2)$ under the following choice of the normalization constant $C(\alpha)$:

$$|C(\alpha)|^{-2} = \int_0^\infty \frac{\lambda^{-\nu}d\lambda}{(-\lambda q^2(1 - q^2); q^2)_\infty} = \frac{\pi}{\sin \pi\nu} \frac{(q^{2\nu}; q^2)_\infty(q^2(1 - q^2))^{\nu-1}}{(q^2; q^2)_\infty},$$

where $\nu = \ln|\alpha q\sqrt{1 - q^2}|/\ln q$. The integral under consideration is a particular subcase of a Ramanujan q-beta integral [21].

We were considering until now the abstract coherent states. In the case of Schrödinger equations with the self-similar potentials one has the doubly degenerate continuous spectrum for $\lambda > 0$. The structure of $|\alpha\rangle_\pm$ states in this case is very complicated. Even for the zero potential case (the free particle) we get a highly nontrivial situation. Let $L = -d^2/dx^2$ and

$$B^- = U^{-1}(d/dx + 1/\sqrt{1 - q^2}), \qquad B^+ = (-d/dx + 1/\sqrt{1 - q^2})U,$$

where U is the dilation operator, $Uf(x) = q^{1/2}f(qx)$, $0 < q < 1$. It can be checked directly that $B^-B^+ - q^2B^+B^- = 1$ and $L = B^+B^- - 1/(1-q^2)$. The equation $B^-\psi_\alpha^-(x) = \alpha\psi_\alpha^-(x)$ coincides now with the retarded pantograph equation, which was investigated in detail in [28]. As follows from the results of this paper, in the retarded case the pantograph equation does not admit solutions belonging to $L^2(\mathbf{R})$. However, the operator B^+ has infinitely many normalizable eigenstates. The equation $B^+\psi_\alpha^+(x) = \alpha\psi_\alpha^+(x)$ coincides now with the advanced pantograph equation

$$\frac{d}{dx}\psi_\alpha^+(x) = -\alpha q^{-3/2}\psi_\alpha^+(q^{-1}x) + \frac{q^{-1}}{\sqrt{1 - q^2}}\psi_\alpha^+(x), \qquad (7.3)$$

which has infinitely many solutions belonging to $L^2(\mathbf{R})$ [28]. For the free particle realization of the $N > 1$ symmetry algebras, one gets the generalized pantograph equations considered in [27]. As one of the open problems in this field, let us mention a need to characterize a minimal set of solutions of the equation (7.3) providing a complete basis of the Hilbert space $L^2(\mathbf{R})$.

8. Solitons, Ising Chains, and Random Matrices

Consider the KdV N-soliton solution $u_N(x,t) = -2\partial_x^2 \log \tau_N(x,t)$, where τ_N is a Wronskian of N different solutions of the free Schrödinger equation (4.5). Actually, there are many determinantal representation for this tau-function, e.g. [1]

$$\tau_N = \det \Phi, \qquad \Phi_{ij} = \delta_{ij} + \frac{2\sqrt{k_i k_j}}{k_i + k_j} e^{(\theta_i + \theta_j)/2}, \qquad (8.1)$$

$$\theta_i = k_i x - k_i^3 t + \theta_i^{(0)}, \qquad i,j = 1,2,\ldots,N.$$

The variables k_i are known to describe amplitudes of solitons (they are related to the eigenvalues of L_N as $\lambda_i = -k_i^2/4$), $\theta_i^{(0)}/k_i$ are the $t = 0$ phases of solitons and k_i^2 are their velocities.

A subclass of the self-similar potentials appearing from the q-periodic closure (6.3) can be viewed as a particular infinite soliton potential. Indeed, consider the KdV pM-soliton solution with the parameters k_j subject to the constraint $k_{j+M} = qk_j, 0 < q < 1$, and take the $p \to \infty$ limit. The limiting potential has the self-similar spectrum $k_{pM+m} = q^p k_m, m = 0,\ldots, M-1$, and an infinite number of free parameters $\theta_j^{(0)}$.

The scaled potential $\tilde{u}(x,t) = q^2 u(qx, q^3 t)$ has the same solitonic interpretation with the phases $\tilde{\theta}_j(x,t) = \theta_j(qx, q^3 t) = k_{j+M}x - k_{j+M}^3 t + \theta_j^{(0)}$. Let us demand that $\theta_{j+M}^{(0)} = \theta_j^{(0)}$, i.e. impose the additional constraint $\theta_j(qx, q^3 t) = \theta_{j+M}(x,t)$, which is seen to match with the condition (6.3). In this picture the discrete dilation $x \to qx, t \to q^3 t$ just deletes M solitons corresponding to the smallest eigenvalues of L — this feature is the simplest physical characterization of the self-similar potentials.

Let us turn now to the statistical mechanics applications. N-soliton tau-function of the KdV equation (8.1) can be represented in the following Hirota form [1, 24]:

$$\tau_N = \sum_{\sigma_i = 0,1} \exp\left(\sum_{0 \leq i < j \leq N-1} A_{ij}\sigma_i\sigma_j + \sum_{i=0}^{N-1} \theta_i\sigma_i\right). \qquad (8.2)$$

The coefficients A_{ij}, describing phase shifts of the i-th and j-th solitons after their collision, have the form

$$e^{A_{ij}} = \frac{(k_i - k_j)^2}{(k_i + k_j)^2}. \qquad (8.3)$$

As noticed in [33], for $\theta_i = \theta^{(0)}$ this tau-function coincides with the grand partition function of a lattice gas model on a line (for a two-dimensional Coulomb gas picture, see [34] and the next section). In this interpretation the discrete variables σ_i describe filling factors of the lattice sites by molecules, $\theta^{(0)}$ is a chemical potential, and A_{ij} are the interaction energies of molecules [7].

It is well known that a simple change of variables in the lattice gas partition function leads to the partition function of some Ising chains:

$$Z_N = \sum_{s_i = \pm 1} e^{-\beta E}, \qquad E = \sum_{0 \leq i < j \leq N-1} J_{ij} s_i s_j - \sum_{i=0}^{N-1} H_i s_i, \qquad (8.4)$$

where N is the number of spins $s_i = \pm 1$, J_{ij} are the exchange constants, H_i is an external magnetic field, $\beta = 1/kT$ is the inverse temperature. Indeed, a replacement of filling factors in (8.2) by the spin variables $\sigma_i = (s_i + 1)/2$ yields

$$\tau_N = e^{\varphi} Z_N, \qquad \varphi = \frac{1}{4} \sum_{i<j} A_{ij} + \frac{1}{2} \sum_{j=0}^{N-1} \theta_j, \qquad (8.5)$$

where

$$\beta J_{ij} = -\frac{1}{4} A_{ij}, \qquad \beta H_i = \frac{1}{2} \theta_i + \frac{1}{4} \sum_{j=0, i \neq j}^{N-1} A_{ij}. \qquad (8.6)$$

Similar relations with Ising chains are valid for the whole infinite Kadomtsev-Petviashvili (KP) hierarchy of equations and some other differential and finite difference nonlinear integrable evolution equations. Corresponding tau-functions have the same form (8.2) with a more complicated structure of the phase shifts A_{ij} and phases θ_i (a partial list of such equations can be found in [1, 24]).

The general relation between N-soliton solutions of integrable hierarchies and partition functions of the lattice statistical mechanics models established in [33] brings a new point of view upon the self-similar potentials as well. Namely, as shown in [33] self-similar spectra can be deduced from the condition of translational invariance of the infinite spin chains. For example, let us demand that the spin chain is invariant with respect to the shift of the lattice by one site $j \to j+1$ which assumes that $J_{i+1,j+1} = J_{ij}$. As a result, the exchange constants J_{ij} (or the soliton phase shifts A_{ij})

depend only on the distance between the sites $|i - j|$. This very simple and quite natural physical requirement forces k_i to form one geometric progression (N.B.: uniquely)

$$k_i = k_0 q^i, \qquad q = e^{-2\alpha}, \qquad A_{ij} = 2\ln|\tanh\alpha(i-j)|, \qquad (8.7)$$

where k_0 and $0 < q < 1$ are some free parameters. In a more complicated case one demands the translational invariance with respect to the shifts by M lattice sites, i.e. $J_{i+M,j+M} = J_{ij}$, and this results in the general self-similar spectra $k_{j+M} = qk_j$. The main drawback of these relations between the solitons and Ising chains is that for fixed q the temperature T (or β) is fixed too, which is clearly seen from the comparison of (8.6) and (8.7) (in the Coulomb gas interpretation one actually finds that $\beta = 2$).

For finite chains $0 \le j \le N-1$, the translational invariance is not exact. The infinite soliton limit $N \to \infty$ corresponds to the thermodynamical limit in statistical mechanics. In this picture, all coordinates of integrable hierarchies (x, t and higher hierarchy "times") are interpreted as particular parameters of the external magnetic field H_i. Since $0 < q < 1$, the x and t-dependent part of H_i decays exponentially fast for $i \to \infty$. As a result, in the limit $N \to \infty$ only the constants $\theta_i^{(0)}$ are relevant for the partition function (speaking more precisely, they determine the leading asymptotics term of the partition function at $N \to \infty$). Although in the thermodynamic limit the x, t-dependence is washed out and the Ising chain is periodic, presence of an infinite number of free parameters $\theta_i^{(0)}$ does not allow one to find a closed form expression for the leading term of Z_N. However, for the self-similar infinite soliton systems, characterized by the periodicity of $\theta_i^{(0)}$ or, equivalently, by the periodicity of the external magnetic field $H_{i+M} = H_i$, such expressions do exist.

Take $M = 1$ and let the magnetic field be homogeneous $H_i = H$. Then in the KdV equation case one gets an antiferromagnetic Ising chain: $0 < |\tanh\alpha(i-j)| < 1$ and $J_{ij} = -A_{ij}/4\beta > 0$ (a similar picture is valid for the general M). The exchange has a long distance character but its intensity falls off exponentially fast and, as a result, phase transitions are absent for nonzero temperatures. However, for $\alpha \to 0$ or $q \to 1$ the exchange constants J_{ij} are diverging. Under appropriate renormalizations one can build a model with a very small non-local exchange such that a phase transition takes place even for non-zero T.

In order to compute the partition function one can use determinantal representations for the tau-function, e.g. the Wronskian formula (4.5). As shown in [33], for the translationally invariant Ising chains in a homogeneous magnetic field tau-functions become determinants of some Toeplitz matrices and their natural M-periodic generalizations. For example, for

$M = 1$ the following result has been derived: $Z_N \to \exp(-N\beta f_I)$ for $N \to \infty$, where the free energy per site f_I has the form

$$-\beta f_I(q, H) = \ln \frac{2(q^4; q^4)_\infty \cosh \beta H}{(q^2; q^2)_\infty^{1/2}} + \frac{1}{4\pi} \int_0^{2\pi} d\nu \ln(|\rho(\nu)|^2 - q \tanh^2 \beta H),$$
(8.8)

$$|\rho(\nu)|^2 = \frac{(q^2 e^{i\nu}; q^4)_\infty^2 (q^2 e^{-i\nu}; q^4)_\infty^2}{(q^4 e^{i\nu}; q^4)_\infty^2 (q^4 e^{-i\nu}; q^4)_\infty^2} \frac{1}{4 \sin^2(\nu/2)} = q \frac{\theta_4^2(\nu/2, q^2)}{\theta_1^2(\nu/2, q^2)}$$

($\theta_{1,4}$ are the standard Jacobi theta-functions). In the derivation of this formula the Ramanujan $_1\psi_1$ sum was used at the boundary of convergence of the corresponding infinite bilateral basic hypergeometric series. A standard physicists' trick was used for dealing with that, namely, a small auxiliary parameter ϵ was introduced into the original expression for the density function $\rho(\nu)$

$$\rho(\nu) = \frac{(q^2; q^4)_\infty^2}{(q^4; q^4)_\infty^2} \sum_{k=-\infty}^{\infty} \frac{e^{i\nu k - \epsilon k}}{1 - q^{4k+2}} = \frac{(q^2 e^{i\nu - \epsilon}; q^4)_\infty (q^2 e^{-i\nu + \epsilon}; q^4)_\infty}{(e^{i\nu - \epsilon}; q^4)_\infty (q^4 e^{-i\nu + \epsilon}; q^4)_\infty},$$
(8.9)

which guarantees the absolute convergence of the series for $\epsilon > 0$. Although the limit $\epsilon \to 0$ leads to some singularity in ν on the right hand side of (8.9), it does not lead to divergences after taking the integral in (8.8) (a more precise mathematical justification of this physical "harmlessness" is desirable). The total magnetization of the lattice $m(H) = -\partial_H f_I = \lim_{N \to \infty} N^{-1} \sum_{i=0}^{N-1} \langle s_i \rangle$ has the following appealing form

$$m(H) = \left(1 - \frac{1}{\pi} \int_0^\pi \frac{\theta_1^2(\nu, q^2) d\nu}{\theta_4^2(\nu, q^2) \cosh^2 \beta H - \theta_1^2(\nu, q^2) \sinh^2 \beta H} \right) \tanh \beta H.$$
(8.10)

In order to change the fixed value of the temperature, one may try to replace (8.7) by $A_{ij} = 2n \ln|(k_i - k_j)/(k_i + k_j)|$, where n is some sequence of integers, and look for integrable equations admitting N-soliton solutions with such phase shifts. This is a highly nontrivial task, and in [33] only one more permitted value of the temperature was found. It corresponds to $n = 2$ and appears from special reduction of the N-soliton solution of the KP-equation of B-type (BKP) [13]. The general BKP τ-function generates much more complicated Ising chains than in the KdV case. The corresponding exchange constants have the form

$$\beta J_{ij} = -\frac{1}{4} A_{ij}, \quad e^{A_{ij}} = \frac{(a_i - a_j)(b_i - b_j)(a_i - b_j)(b_i - a_j)}{(a_i + a_j)(b_i + b_j)(a_i + b_j)(b_i + a_j)},$$
(8.11)

where a_i, b_i are some free parameters. Translational invariance of the spin lattice, $J_{ij} = J(i - j)$, results in the following spectral self-similarity

$$a_i = q^i, \quad b_i = bq^i, \quad q = e^{-2\alpha}, \tag{8.12}$$

where we set $a_0 = 1$ and assume that $\alpha > 0$. This gives the exchange

$$\beta J_{ij} = -\frac{1}{4} \ln \frac{\tanh^2 \alpha(i - j) - (b - 1)^2/(b + 1)^2}{\coth^2 \alpha(i - j) - (b - 1)^2/(b + 1)^2}.$$

The parameter b is restricted to the unit disk $|b| \leq 1$ due to the inversion symmetry $b \to b^{-1}$. For $-1 < b < -q$ one gets now the ferromagnetic Ising chain, i.e. $J_{ij} < 0$; for $q < b \leq 1$ and $b = e^{i\phi} \neq -1$ the chain is antiferromagnetic, i.e. $J_{ij} > 0$. In the thermodynamic limit $N \to \infty$ the partition function Z_N is given again by the determinant of a Toeplitz matrix, which is diagonalized by the discrete Fourier transformation. Similar picture holds for arbitrary M-periodic chains, see [33] for explicit expressions for the magnetization and technical details of computations.

It is well known that the n-particle Coulomb gas partition functions define probability distribution functions in the theory of random matrices. In a natural way, partition functions of the Ising models considered in the previous section may be related to the probability distribution functions of some random matrix models with a discretized set of eigenvalues. Let us describe briefly this correspondence comparing the KP N-soliton solution and Dyson's circular ensemble.

Dyson has introduced an ensemble of unitary random $n \times n$ matrices with the eigenvalues $\epsilon_j = e^{i\phi_j}$, $j = 1, \ldots, n$, such that after the integration over the auxiliary "angular variables" the probability distribution of the phases $0 \leq \phi_j < 2\pi$ takes the form $Pd\phi_1 \ldots d\phi_n \propto \prod_{i<j} |\epsilon_i - \epsilon_j|^2 d\phi_1 \ldots d\phi_n$. One may relax some of the conditions used by Dyson and work with a more general set of distribution functions. In [19] Gaudin has proposed a circular ensemble with the probability distribution law

$$Pd\phi_1 \ldots d\phi_n \propto \prod_{i<j} \left| \frac{\epsilon_i - \epsilon_j}{\epsilon_i - \omega\epsilon_j} \right|^2 d\phi_1 \ldots d\phi_n, \tag{8.13}$$

containing a free continuous parameter ω. It interpolates between the Dyson's case ($\omega = 0$) and the uniform distribution ($\omega = 1$). Let us note in passing, that the paper [19] contains in it implicitly a specific model of the q-harmonic oscillator. The model (8.13) admits also an interpretation as a gas of particles on the circle with the partition function

$$Z_n \propto \int_0^{2\pi} \ldots \int_0^{2\pi} d\phi_1 \ldots d\phi_n \exp\left(-\beta \sum_{i<j} V(\phi_i - \phi_j) \right),$$

where $\beta = 1/kT = 2$ is fixed and the potential energy has the form

$$\beta V(\phi_i - \phi_j) = \ln\left(1 + \frac{\sinh^2 \gamma}{\sin^2((\phi_i - \phi_j)/2)}\right), \qquad \omega = e^{-2\gamma}. \qquad (8.14)$$

A surprising fact is that the grand partition function of this model may be obtained in a special infinite soliton limit of the N-soliton tau-function of the KP hierarchy [33]. This means that the finite KP soliton solutions provide a discretization of the model, namely, they define a lattice gas on the circle. This leads to random matrices with a discrete set of eigenvalues. For example, one may take unitary $n \times n$ matrices with eigenvalues equal to N-th roots of unity, i.e. $\epsilon_j = \exp(2\pi i m_j/N)$, $m_j = 0, \ldots, N-1$. The probability measure is taken to be continuous in the auxiliary "angular" variables of the unitary matrices and discrete in the eigenvalue phase variables ϕ_j. More precisely, the integrals over ϕ_j are replaced by finite sums over m_j and the continuous model is recovered for $m_j, N \to \infty$ with finite m_j/N:

$$\left(\frac{2\pi}{N}\right)^n \sum_{m_1=0}^{N-1} \cdots \sum_{m_n=0}^{N-1} \xrightarrow{N\to\infty} \int_0^{2\pi} d\phi_1 \ldots \int_0^{2\pi} d\phi_n. \qquad (8.15)$$

The n-particle partition function becomes

$$Z_n(N, \omega) = \left(\frac{2\pi}{N}\right)^n \sum_{m_1=0}^{N-1} \cdots \sum_{m_n=0}^{N-1} \prod_{1 \leq i < j \leq n} \left|\frac{\epsilon_i - \epsilon_j}{\epsilon_i/\sqrt{\omega} - \sqrt{\omega}\epsilon_j}\right|^2,$$

while the grand canonical ensemble partition function takes the form

$$Z(\omega, \theta) = \sum_{n=0}^{N} \frac{Z_n(N, \omega)e^{\theta n}}{n!} \qquad (8.16)$$

$$= \sum_{\sigma_m = 0,1} \exp\left(\sum_{0 \leq m < k \leq N-1} A_{mk}\sigma_m\sigma_k + (\theta + \eta)\sum_{m=0}^{N-1} \sigma_m\right),$$

where $\eta = \ln(2\pi/N)$ is an excessive chemical potential, and

$$A_{mk} = \ln\frac{\sin^2(\pi(m-k)/N)}{\sin^2(\pi(m-k)/N) + \sinh^2 \gamma} = \ln\frac{(a_m - a_k)(b_m - b_k)}{(a_m + b_k)(b_m + a_k)},$$

are the KP solitons phase shifts for a special choice of the parameters

$$a_m = e^{2\pi i m/N}, \qquad b_m = -\omega a_m, \qquad m = 0, 1, \ldots, N-1. \qquad (8.17)$$

To conclude, the grand partition function of the Gaudin's circular ensemble coincides with the particular infinite soliton KP tau-function at zero hierarchy "times". The root of unity discretization of circular ensembles has been considered by Gaudin himself [20], where a connection with Ising chains was noticed as well, but the relation with soliton solutions of integrable equations was not established. The BKP hierarchy of equations suggests a generalization of the distribution law (8.13) to

$$Pd\phi_1 \ldots d\phi_n \propto \prod_{i<j} \left|\frac{\epsilon_i - \epsilon_j}{\epsilon_i + \epsilon_j}\right|^2 \left|\frac{\epsilon_i + \omega\epsilon_j}{\epsilon_i - \omega\epsilon_j}\right|^2 d\phi_1 \ldots d\phi_n, \qquad (8.18)$$

and its discrete analogue, which were not investigated yet appropriately.

It is well known that the n-tuple Selberg integral provides an explicit evaluation of the n particle Coulomb gas partition functions for arbitrary values of the inverse temperature β (i.e. not just for $\beta = 1, 2, 4$, corresponding to orthogonal, unitary and symplectic ensembles). It would be interesting to know whether a similar universal formula exists for the grand partition functions of lattice Coulomb gases.

9. Lattice Coulomb Gas on the Plane

Solution of the Poisson equation on the plane, $\Delta V(\mathbf{r}, \mathbf{r}') = -2\pi\delta(\mathbf{r}-\mathbf{r}')$, defines the electrostatic potential $V(z, z') = -\ln|z-z'|$, created by a charged particle placed at the point $z' = x' + iy'$. In the bounded domains with dielectric or conducting walls, the potential has a more complicated form since the normal component of the electric field $\mathbf{E} = -\nabla V$ should vanish on the surface of a dielectric, $\mathcal{E}_n = 0$, while the tangent component is zero at the metallic boundary, $\mathcal{E}_t = 0$. For simple geometric configurations of boundaries an introduction of artificial images of charges may simplify solution of the Poisson equation.

The energy of an electrostatic system ("plasma") consisting of N particles in a bounded domain of the plane is

$$E_N = \sum_{1\leq i<j\leq N} q_i q_j V(z_i, z_j) + \sum_{1\leq i\leq N} q_i^2 v(z_i) + \sum_{1\leq i\leq N} q_i \phi(z_i), \qquad (9.1)$$

where $z_j = x_j + iy_j$ and q_j are the particles' coordinates and charges. The first term is the standard Coulomb energy, the second one describes an interaction with the boundaries (or the charge-image interaction), and the last term corresponds to the external fields contribution.

Suppose that our plasma is composed from the particles of equal charges, $q_j = +1$ (in the two component case $q_j = \pm 1$), upon a discrete lattice Γ on the plane. The grand partition function of such lattice Coulomb gas is

$$G_N = \sum_{n=0}^{N} \frac{e^{\mu n}}{n!} \sum_{z_1\in\Gamma} \cdots \sum_{z_n\in\Gamma} e^{-\beta E_n},$$

where μ is an effective chemical potential. In general case one can rewrite G_N in the form [7]

$$G_N = \sum_{\{\sigma(z)\}} \exp\Big(\frac{1}{2}\sum_{z\neq z'} W(z,z')\sigma(z)\sigma(z') + \sum_{z\in\Gamma} w(z)\sigma(z)\Big), \qquad (9.2)$$

$$W(z,z') = -\beta q(z)q(z')V(z,z'), \quad w(z) = \mu(z) - \beta\Big(q^2(z)v(z) + q(z)\phi(z)\Big),$$

where $\sigma(z) = 0$ or 1 depending on whether the site with the coordinate z is empty or occupied by a particle, and the functions $q(z)$ and $\mu(z)$ characterize distribution of particles of different types. For example, in the two-component case, when $q(z_\pm) = \pm 1$ charges occupy the $\{z_\pm\}$ sublattices, one has $\mu(z_\pm) = \mu_\pm$. Note that these 2D lattice gases describe simultaneously some 2D Ising magnets.

Let us take now the N-soliton tau-function (8.2) and replace in it the soliton number j by a variable z taking N discrete values. As a result, it may be rewritten in the form

$$\tau_N = \sum_{\sigma(z)=0,1} \exp\Big(\frac{1}{2}\sum_{z\neq z'} A_{zz'}\sigma(z)\sigma(z') + \sum_{\{z\}} \theta(z)\sigma(z)\Big). \qquad (9.3)$$

But this is precisely (9.2) — one just needs to identify the phase shifts $A_{zz'}$ with the Coulomb interaction potential $W(z,z')$, and the phases $\theta(z)$ with the function $w(z)$. As a result of this observation made in [34], one can construct a number of new solvable models of Coulomb gases in addition to already known ones (see, e.g. [17, 19, 20] and references therein).

For the KP-hierarchy soliton solutions one has [13]

$$A_{zz'} = \ln\frac{(a_z - a_{z'})(b_z - b_{z'})}{(a_z + b_{z'})(b_z + a_{z'})}, \qquad \theta(z) = \theta^{(0)}(z) + \sum_{p=1}^{\infty}(a_z^p - (-b_z)^p)t_p,$$

$$(9.4)$$

where t_p is the p-th KP hierarchy "time" and a_z, b_z are some arbitrary functions of z. For the BKP-hierarchy $A_{zz'}$ are given by (8.11) and $\theta(z) = \theta^{(0)}(z) + \sum_{p=1}^{\infty}(a_z^{2p-1} + b_z^{2p-1})t_{2p-1}$, where t_{2p-1} are the BKP evolution "times".

Let us take $a_z = z = x + iy$, $b_z = -z^* = -x + iy$. Then in the KP case

$$A_{zz'} = W(z,z') = -2V(z,z') = 2\ln|z - z'| - 2\ln|z^* - z'|, \qquad (9.5)$$

where $V(z,z')$ is the potential created by a positive unit charge placed at z' over a conductor with its surface occupying the $y = 0$ line. In this case $V(z,z')$ solves the Poisson equation with the boundary condition $\mathcal{E}_t(y = $

0) = 0. The same potential is created by a positive charge placed at the point z' and its image of opposite charge located at the point $(z')^*$.

Similar to the random matrix models situation, the correspondence between 2D lattice Coulomb gases with specific boundary conditions and solitons is valid only for fixed β, which is found from the comparison of (9.5) with (9.2) to be $\beta = 2$. Since $w(z) = \theta(z)$, one finds an explicit expression for the zero time soliton phases $\theta^{(0)}(z)$ in (9.4):

$$\theta^{(0)}(z) = \mu - \beta(\ln|z^* - z| + \phi(z)), \qquad \beta = 2, \qquad (9.6)$$

where the middle term corresponds to the charge-image interaction, and $\phi(z)$ is the potential created by a neutralizing background of some density $\rho(\mathbf{r})$, $\Delta\phi(\mathbf{r}) = -2\pi\rho(\mathbf{r})$, $\phi_x(y = 0) = 0$. The harmonic term $\sum_{p=0}^{\infty}(z^p - (z^*)^p)t_p = -\beta\phi_{ext}(z)$ in (9.4) corresponds to an external electric field. One may conclude that the imaginary hierarchy times evolution of the KP soliton solutions describes the behavior of a 2D lattice Coulomb gas in a varying external electric field.

Usefulness of the conformal transformations $z \to f(z)$ is well known in the potential theory. With their help one can map plasma particles to various regions. So, the map $z \to z^n$ puts them inside a corner with the π/n angle between the conducting walls. The exponential map $z \to e^{\pi z/L}$ leads to a gas inside the strip $0 < \Im z < L$ between two parallel conductors, etc.

As to the BKP equation case, the choice $a_z = z, b_z = z^*$ leads to the Coulomb gas placed inside the upper right quarter of the plane with the dielectric and conductor walls along the x and y-axes respectively. The discrete temperature appears to be the same, $\beta = 2$. A curious set of solvable dipole gas models is generated by the self-similarity constraints imposed upon the spectral data of the corresponding multi-soliton systems. For further details, see [34].

10. Appendix. A Heuristic Guide to Special Functions of One Variable

There are many handbooks and textbooks on special functions. However, none of them contains a formal list of properties which a function should have in order to be "special". One usually talks on functions of some particular type (hypergeometric, L-functions, etc). R. Askey has suggested one universal definition that if a function is so useful that it starts to bear some name, then it is "special". Another essential, but not so universal, characteristic refers to the asymptotic behavior. Namely, for special functions one is expected to be able to deduce asymptotics at infinity from a known local behavior, i.e. the connection problems should be solvable — such an approach is advocated by people working on the Painlevé type functions.

To the author's taste these two definitions are relying upon the secondary features of special functions. One has to have already the functions in hands in order to start to investigate their properties. If one takes as a goal the search of new special functions, then it is necessary to find a definition containing a more extended set of technical tools for work. In this respect it should be stressed that even the term "classical special functions" appears to be not so stable. For example, it is by now accepted that the family of classical orthogonal polynomials consists of not just the Jacobi polynomials and their descendants, but of the essentially more complicated Askey-Wilson polynomials invented just two decades ago.

The group theory provides a number of tools for building new functions, but, unfortunately, connections with the representation theory often provide only interpretations for functions already defined by other means. Still, the symmetry approach seems to be the central one in the theory of special functions. In particular, the key "old" special functions appear from separations of variables in the very simple (and, so, useful and universal) equations [38]. For the last decade the author's research was tied to the following working definition: *special functions are the functions associated with self-similar reductions of spectral transformation chains for linear spectral problems.* Speaking differently, special functions are connected with fixed points of various continuous and discrete symmetry transformations for a taken class of eigenvalue problem. This definition works well only for special functions of one independent variable (which may, however, depend on an infinite number of parameters) and even for them it does not pretend to cover all possible special functions. On the one hand, this definition comes from the theory of completely integrable systems, where a search of self-similar solutions of nonlinear evolution equations is a standard problem [1]. On the other hand, particular examples found from this approach show that it has in its heart the *contiguous relations* — linear or nonlinear equations connecting special functions at different values of their parameters.

Schematically, a heuristic algorithm of looking for these "spectral" special functions consists of the following steps:

1. Take a linear eigenvalue problem determined by differential or difference equations.

2. Consider another linear equation involving variables entering the first equation in a different way and having the same space of solutions.

3. Resolve compatibility conditions of the taken linear problems and derive nonlinear equations for free functional coefficients entering these problems. When the second equation is differential, one gets the continuous flows associated with the KdV, KP, Toda, etc equations. When the second equation is discrete, one gets a sequence of Darboux transformations performing some discrete changes in the spectral data and providing some

discrete-time Toda, Volterra, etc chains.

4. Analyze discrete and continuous symmetries of the latter equations in the Lie group-theoretical sense, i.e. look for nontrivial continuous and discrete transformations mapping solutions of these nonlinear equations into other solutions.

5. Consider *self-similar* solutions of the derived nonlinear equations, which are invariant with respect to taken symmetry transformations. As a result, some finite sets of nonlinear differential, differential-difference, difference-difference, etc equations are emerging which define "nonlinear" special functions. Solutions of the original linear equations with coefficients determined by these self-similar functions define "linear" special functions.

The last two steps are heuristic since, despite of a big progress in the general theory of symmetry reductions (see, e.g. [31, 36]), no completely regular method has been built yet. For example, the reductions used in the derivation of the associated Askey-Wilson polynomials' recurrence coefficients [49] and in the discovery of new explicit systems of biorthogonal rational functions described in Zhedanov's lecture at this meeting did not find yet a purely group-theoretical setting.

Another important ingredient of the theory of special functions, which was not listed in this scheme, is the theory of transcendency. Painlevé functions are known to be transcendental over the differential fields built from a finite tower of Picard-Vessiot extensions of the field of rational functions. In solving differential (difference, or whatever) equations one has to determine eventually which differential field a taken solution belongs to (e.g., whether it belongs to the field of functions over which the differential equation is defined). As an open problem in this field, which was discussed partially in [47], we mention a need to find a differential (or difference) Galois theory interpretation of the self-similar reductions of factorization chains.

Acknowledgement. The author was partially supported by the RFBR grants No. 01-00299 and 01-10564.

References

1. M. J. Ablowitz and H. Segur, *Solitons and the Inverse Scattering Transform*, SIAM, Philadelphia, 1981.

2. V. E. Adler, *Recuttings of polygons*, Funkt. Anal. i ego Pril. **27** (2) (1993), 79–82.

3. A. A. Andrianov, N. V. Borisov, and M.V. Ioffe, *Factorization method and Darboux transformation for multidimensional Hamiltonians*, Theor. Math. Phys. **61** (1984), 1078–1088.

4. A. A. Andrianov, M. V. Ioffe, and V. P. Spiridonov, *Higher-derivative supersymmetry and the Witten index*, Phys. Lett. **A174** (1993), 273–279.

5. M. Arik and D. D. Coon, *Hilbert spaces of analytic functions and generalized coherent states*, J. Math. Phys. **17** (1976), 524–527.

6. R. Askey and S. K. Suslov, *The q-harmonic oscillator and the Al-Salam and Carlitz polynomials*, Lett. Math. Phys. **29** (1993), 123–132.

7. R. J. Baxter, *Exactly Solved Models in Statistical Mechanics*, Academic Press, London, 1982.

8. F. J. Bureau, *Differential equations with fixed critical points*, in: "Painlevé Transcendents", D. Levi and P. Winternitz, eds., NATO ASI Series, Series B; Vol. 278, Plenum Press, New York, 1990, pp. 103–123.

9. J. L. Burchnall and T. W. Chaundy, *Commutative ordinary differential operators I & II*, Proc. Lond. Math. Soc. **21** (1923), 420–440; Proc. Roy. Soc. **A118** (1928), 557–583.

10. D. D. Coon, S. Yu, and S. Baker, *Operator formulation of a dual multiparticle theory with nonlinear trajectories*, Phys. Rev. **D5** (1972), 1429–1433.

11. M. M. Crum, *Associated Sturm-Liouville systems*, Quart. J. Math. Oxford **6** (1955), 121–127.

12. G. Darboux, *Leçons sur la Théorie Générale des Surfaces*, Gauthier-Villars, Paris, 1889.

13. E. Date, M. Jimbo, M. Kashiwara, and T. Miwa, *Transformation groups for soliton equations*, in: "Nonlinear Integrable Systems", World Scientific, Singapore, 1983, pp. 41–119.

14. P. A. Deift, *Applications of a commutation formula*, Duke. Math. J. **45** (1978), 267–310.

15. L. D. Faddeev, *The inverse problem of the quantum scattering theory*, Uspekhi Mat. Nauk (Russ. Math. Surveys) **14** (4) (1959), 57–119.

16. H. Flaschka, *A commutator representation of Painlevé equations*, J. Math. Phys. **21** (1980), 1016–1018.

17. P. J. Forrester, B. Jancovici, and G. Téllez, *Universality in some classical Coulomb systems of restricted dimension*, J. Stat. Phys. **84** (1996), 359–378.

18. U. Frisch, and R. Bourret, *Parastochastics*, J. Math. Phys. **11** (1970), 364–390.

19. M. Gaudin, *Une famille à une paramètre d'ensembles unitaires*, Nucl. Phys. **85** (1966), 545–575.

20. M. Gaudin, *Gaz coulombien discret à une dimension*, J. Phys. (France) **34** (1973), 511–522.

21. G. Gasper and M. Rahman, *Basic Hypergeometric Series*, Cambridge University Press, Cambridge, 1990.

22. L. E. Gendenshtein and I. V. Krive, *Supersymmetry in quantum mechanics*, Uspekhi Phys. Nauk (Sov. Phys. Uspekhi) **146** (1985), 553–590.

23. Ya. L. Geronimus, ıOn the polynomials orthogonal with respect to a given number sequence and a theorem by W. Hahn, Izv. Akad. Nauk SSSR **4** (1940), 215–228.

24. R. Hirota, *Direct methods in soliton theory*, in: "Solitons", R. K. Bullough and P. J. Caudrey, eds., Springer, Berlin, 1980.

25. L. Infeld, *On a new treatment of some eigenvalue problems*, Phys. Rev. **59** (1941), 737–747.

26. L. Infeld and T.E. Hull, *The factorization method*, Rev. Mod. Phys. **23** (1951), 21–68.

27. A. Iserles, *On the generalized pantograph functional-differential equation*, Europ. J. Appl. Math. **4** (1993), 1–38.

28. T. Kato and J.B. McLeod, *The functional-differential equation* $y'(x) = ay(\lambda x) + by(x)$, Bull. Am. Math. Soc. **77** (1971), 891–937.

29. A. V. Kitaev, *Special functions of the isomonodromy type*, Acta Appl. Math. **64** (2000), 1–32 and lectures at the NATO ASI "Special functions-2000," 2000.

30. M. G. Krein, *On a continuous analogue of a Christoffel formula in the theory of orthogonal polynomials*, Doklady Akad. Nauk SSSR **19** (1957), 1095–1097.

31. D. Levi and P. Winternitz, *Continuous symmetries of discrete equations*, Phys. Lett. **A152** (1991), 335–338.

32. Yu. Liu, On functional differential equations with proportional delays, Ph.D. thesis, Cambridge University, 1996.

33. I. M. Loutsenko and V. P. Spiridonov, *Self-similar potentials and Ising models*, JETP

Lett. **66** (1997), 789–795; *Spectral self-similarity, one-dimensional Ising chains and random matrices*, Nucl. Phys. **B538** (1999), 731–758.

34. I. M. Loutsenko and V. P. Spiridonov, *Soliton solutions of integrable hierarchies and Coulomb plasmas*, J. Stat. Phys. **99** (2000), 751–767; cond-mat/9909308.

35. A. J. Macfarlane, *On q-analogues of the quantum harmonic oscillator and quantum group $SU(2)_q$*, J. Phys. A: Math. Gen. **22** (1989), 4581–4588.

36. S. Maeda, *The similarity method for difference equations*, IMA J. Appl. Math. **38** (1987), 129–134.

37. W. Miller, Jr., *Lie theory and difference equations I*, J. Math. Anal. Appl. **28** (1969), 383–399.

38. W. Miller, Jr., *Symmetry and Separation of Variables*, Addison-Wesley, Reading, 1977.

39. A. M. Perelomov, *Generalized Coherent States and Their Applications*, Springer-Verlag, Berlin, 1986.

40. V. A. Rubakov and V. P. Spiridonov, *Parasupersymmetric quantum mechanics*, Mod. Phys. Lett. **A3** (1988), 1337–1347.

41. U.-W. Schminke, *On Schrödinger's factorization method for Sturm-Liouville operators*, Proc. Roy. Soc. Edinburgh **80A** (1978), 67–84.

42. E. Schrödinger, *A method of determining quantum-mechanical eigenvalues and eigenfunctions*, Proc. Roy. Irish Acad. **A46** (1940), 9–16; *Further studies on solving eigenvalue problems by factorization*, Proc. Roy. Irish Acad. **A46**, 183–206.

43. A. B. Shabat, *The infinite dimensional dressing dynamical system*, Inverse Prob. **8** (1992), 303–308.

44. S. Skorik and V. P. Spiridonov, *Self-similar potentials and the q-oscillator algebra at roots of unity*, Lett. Math. Phys. **28** (1993), 59–74.

45. S. P. Smith, *A class of algebras similar to the enveloping algebra of sl(2)*, Trans. Amer. Math. Soc. **322** (1990), 285–314.

46. V. P. Spiridonov, *Exactly solvable potentials and quantum algebras*, Phys. Rev. Lett. **69** (1992), 398–401; *Deformation of supersymmetric and conformal quantum mechanics through affine transformations*, in: "Proceedings of the International Workshop on Harmonic Oscillators", NASA Conf. Publ. **3197**, pp. 93-108; hep-th/9208073.

47. V. P. Spiridonov, *Coherent states of the q-Weyl algebra*, Lett. Math. Phys. **35** (1995), 179–185; *Universal superpositions of coherent states and self-similar potentials*, Phys. Rev. **A52**, 1909-1935; (E) **A53**, 2903; quant-ph/9601030.

48. V. P. Spiridonov, *Symmetries of factorization chains for the discrete Schrödinger equation*, J. Phys. A: Math. Gen. **30** (1997), L15–L21.

49. V. P. Spiridonov and A. S. Zhedanov, *Discrete Darboux transformations, discrete time Toda lattice and the Askey-Wilson polynomials*, Meth. Appl. Anal. **2** (1995), 369–398.

50. A. P. Veselov and A. B. Shabat, *Dressing chain and spectral theory of Schrödinger operator*, Funk. Anal. i ego Pril. **27**(2) (1993), 1–21.

51. J. Weiss, *Periodic fixed points of Bäcklund transformations*, J. Math. Phys. **31** (1987), 2025–2039.

GENERALIZED EIGENVALUE PROBLEM AND A NEW FAMILY OF RATIONAL FUNCTIONS BIORTHOGONAL ON ELLIPTIC GRIDS

V. P. SPIRIDONOV
Bogoliubov Laboratory of Theoretical Physics
JINR, Dubna, Moscow region, 141980 Russia

AND

A. S. ZHEDANOV
Donetsk Institute for Physics and Technology
Donetsk, 83114 Ukraine

1. Introduction

Orthogonal polynomials (OP) $P_n(x)$ are known to satisfy the three-term recurrence relation

$$J P_n(x) = x P_n(x), \qquad (1.1)$$

where J is a Jacobi (i.e. a tridiagonal) matrix acting upon the polynomials as follows

$$J P_n(x) = a_n P_{n+1}(x) + b_n P_n(x) + u_n P_{n-1}(x). \qquad (1.2)$$

Currently it is commonly believed that recurrence relations of OP explicitly expressed through basic hypergeometric functions are related to the recurrence relation of Askey-Wilson polynomials [2] or of their limiting cases. The Askey-Wilson polynomials themselves are expressed in terms of the $_4\varphi_3$ basic hypergeometric function (see [1, 8] for notations).

A special subclass of these polynomials—the q-Racah polynomials —is orthogonal on the discrete "q-quadratic grid" (in the terminology of Nikiforov, Suslov, and Uvarov [18]):

$$x_s = C_1 q^s + C_2 q^{-s} + C_3, \quad s = 0, 1, 2, \dots, N, \qquad (1.3)$$

where q and $C_{1,2,3}$ are some complex numbers. The case when q is a root of unity leads to another finite-dimensional system of polynomials upon

J. Bustoz et al. (eds.), Special Functions 2000, 365–388.

discrete grids whose orthogonality relations are connected to some interesting identities of the number theory (Gauss sums) [24]. The orthogonality means that

$$\sum_{s=0}^{N} P_n(x_s)P_m(x_s)w_s = h_n\delta_{nm}$$

with some discrete weight function w_s. In the limiting case $q \to 1$, q-Racah polynomials become the Racah polynomials [14], which are orthogonal on the quadratic grid $x_s = C_1s^2 + C_1s + C_2$. It can be noted that both quadratic and q-quadratic grids are defined as solutions of the following non-linear difference equation [16, 18]

$$c(x_s^2 + x_{s+1}^2) + 2dx_sx_{s+1} + e(x_s + x_{s+1}) + f = 0 \qquad (1.4)$$

with some constants c, d, e, f.

In [34, 35] Wilson has constructed a large family of finite-dimensional *biorthogonal rational* functions (BRF). Namely, a pair of explicit rational functions $R_n(z), T_n(z)$ was presented which satisfy the relation

$$\sum_{s=0}^{N} R_n(z_s)T_m(z_s)w_s = h_n\delta_{nm} \qquad (1.5)$$

for the same grids z_s (1.3). This class of BRF was extended by Rahman in [19, 20]. The most general BRF $R_n(z), T_n(z)$ from the Rahman-Wilson family are expressed in terms of the very-well-poised balanced basic hypergeometric function $_{10}\varphi_9$. For further development of the general theory of BRF see, e.g. [10, 12, 21].

In [36] it was shown that the functions $R_n(z)$ (as well as $T_n(z)$) satisfy the three-term recurrence relations

$$J_1R_n(z) = z\, J_2R_n(z), \qquad (1.6)$$

where $J_{1,2}$ are two Jacobi matrices. Note that in contrast to (1.1) the equation (1.6) belongs to the class of so-called generalized eigenvalue problems (GEVP) [33]. It is well known that eigenfunctions of GEVP satisfy biorthogonality relation (instead of pure orthogonality for eigenfunctions of the ordinary eigenvalue problems). That is why the rational functions $R_n(z), T_n(z)$ satisfy the relation (1.5).

By analogy with the Askey-Wilson polynomials situation, it was expected that biorthogonal rational functions considered in [10, 21, 35], which are associated with the grids (1.3), are the most general ones satisfying the three-term recurrence relation (1.6) and admitting an explicit expression for $R_n(z), T_n(z)$ in terms of some series of the hypergeometric type. However, it is easy to find aruments against this belief. Indeed, rational transformations of the argument $z \to (\xi z + \eta)/(\zeta z + \sigma)$ map the space of biorthogonal rational functions onto itself and the reason for this is almost trivial. Let us introduce new Jacobi matrices $\tilde{J}_1 = \xi J_1 + \eta J_2$, $\tilde{J}_2 = \zeta J_1 + \sigma J_2$. Then we have from (1.6)

$$\tilde{J}_1 R_n(z) = \frac{\xi z + \eta}{\zeta z + \sigma} \tilde{J}_2 R_n(z), \tag{1.7}$$

i.e. GEVP (1.6) admits arbitrary rational transformation of the spectral variable z. The grids (1.3) are not invariant with respect to such transformation. Hence the class of admissible grids should be wider than the q-quadratic set (1.3). This simple observation is a key for conjecturing the existence of *elliptic grids*. Indeed, the rational q-quadratic grids

$$z_s = \kappa \frac{C_1 q^s + C_2 q^{-s} + C_3}{C_1 q^s + C_2 q^{-s} + C_4} \tag{1.8}$$

satisfy the nonlinear difference equation

$$a z_s^2 z_{s+1}^2 + b z_s z_{s+1}(z_s + z_{s+1}) + c(z_s^2 + z_{s+1}^2) + 2 d z_s z_{s+1} + e(z_s + z_{s+1}) + f = 0 \tag{1.9}$$

with some constants a, b, c, d, e, f. These constants are not arbitrary. On the one hand, there are 4 essential parameters (say, κ, C_1, C_3, C_4 for $C_2 \neq 0$) defining the grid (1.8). On the other hand, there are 5 essential parameters (say, b, c, d, e, f for $a \neq 0$) in (1.9). Hence there exists an extra condition upon the parameters a, b, c, d, e, f under which solution of the equation (1.9) is reduced to the rational q-quadratic grid. It is natural to ask what happens when we remove this constraint and demand that all the parameters a, b, c, d, e, f are arbitrary. The answer to this question in a different setting was found by Baxter [3]: the general solution of (1.9) is given by the elliptic function

$$z_s = \kappa \frac{\theta_1(h(s - y_1))\theta_1(h(s - y_2))}{\theta_1(h(s - y_3))\theta_1(h(s - y_4))}, \tag{1.10}$$

where κ, h are some scaling parameters, $\theta_1(x)$ is the Jacobi theta function (see the definition in [8] or below) and there is an extra constraint $y_1 + y_2 = y_3 + y_4$ upon the parameters y_1, y_2, y_3, y_4. Thus, we have 6 parameters in the solution (1.10): 2 intrinsic parameters of the function $\theta_1(x)$ (say, its quasi-periods one of which is related to h), three parameters y_1, y_2, y_3 and the scaling parameter κ. One of the parameters y_1, y_2, y_3 is responsible for initial conditions for z_s at $s = 0$. Hence we arrive to 5 essential parameters. Due to the special properties of the θ_1-function, the set of elliptic grids (1.10) is mapped onto itself under the rational transformations $z_s \to (\xi z_s + \eta)/(\zeta z_s + \sigma)$.

A family of explicit functions $R_n(z), T_n(z)$ which are biorthogonal on the elliptic grid (1.10) was constructed in [25, 26]. These functions are expressed in terms of elliptic analogues of the very-well-poised balanced $_{10}\varphi_9$ basic hypergeometric functions introduced by Frenkel and Turaev in [7] (for some new results on such series, see the recent work by Warnaar [31]). Thus the class of biorthogonal rational functions satisfying the three-term recurrence relation (1.6) and expressed in terms of some generalizations of the hypergeometric series is much wider than it was guessed earlier.

Main material reviewed in this lecture was published in [36, 25, 26, 37]. We present also some new results whose detailed consideration will be given elsewhere [27].

2. Biorthogonal Rational Functions and a Generalized Eigenvalue Problem

Let J_1 and J_2 be two arbitrary Jacobi (tridiagonal) matrices with real entries. Consider a GEVP

$$J_1 \mathbf{R} = z J_2 \mathbf{R}, \qquad (2.1)$$

where $\mathbf{R}$ is an infinite-dimensional vector with the components $R_0, R_1, \ldots$.

Assume that the Jacobi matrices J_1 and J_2 are not proportional to one another: $J_1 \neq \lambda J_2$ and that off-diagonal entries of both these matrices are non-zero: $J_{i,i\pm1}^{(1,2)} \neq 0$. Then the relation (2.1) is equivalent to the three-term recurrence relation

$$(z - \alpha_{n+1})R_{n+1}(z) + (v_n - \rho_n z)R_n(z) + u_n(z - \beta_{n-1})R_{n-1}(z) = 0 \quad (2.2)$$

with some coefficients $\alpha_n, \beta_n, \rho_n, v_n$. Choosing initial conditions $R_{-1} = 0, R_0 = 1$ we see that $R_n(z)$ is a rational function in z of type $[n/n]$. This means that $R_n(z)$ can be presented in the form

$$R_n(z) = \frac{P_n(z)}{(z - \alpha_1)(z - \alpha_2) \cdots (z - \alpha_n)}, \qquad (2.3)$$

where $P_n(z)$ are polynomials in z of the degree $\leq n$. Note that all poles α_i, $i = 1, 2, \ldots, n$, of the rational function $R_n(z)$ are known explicitly from the recurrence relation (2.2).

The polynomials $P_n(z)$ satisfy the three-term recurrence relation

$$P_{n+1}(z) + (v_n - \rho_n z)P_n(z) + u_n(z - \beta_{n-1})(z - \alpha_n)P_{n-1}(z) = 0, \quad (2.4)$$

which immediately follows from (2.2). Theory of polynomials satisfying recurrence relation (2.4) was considered by Ismail and Masson in [12], where they have found a number of useful properties of $P_n(z)$. In particular, an important *orthogonality condition*

$$\mathcal{L}\left\{\frac{P_n(z)z^j}{(z-\alpha_1)\cdots(z-\alpha_n)(z-\beta_1)\cdots(z-\beta_{n-1})}\right\} = 0, \quad j = 0, 1, \ldots, n-1 \tag{2.5}$$

holds, where $\mathcal{L}$ is a linear functional (it differs from the one used in [12] by a simple renormalization) which is defined on the space of all rational functions with prescribed poles α_i, β_k, $i, k = 1, 2, \ldots$.

In [36] it was shown that there exists another set of rational functions

$$T_n(z) = \frac{Q_n(z)}{(z-\beta_1)(z-\beta_2)\cdots(z-\beta_n)}, \tag{2.6}$$

where $Q_n(z)$ is a polynomial of the degree $\leq n$, such that the *biorthogonality property*

$$\mathcal{L}\{R_n(z)T_m(z)\} = \delta_{nm}, \tag{2.7}$$

holds. Note that the rational functions $T_n(z)$ have prescribed poles at β_i, $i = 1, 2, \ldots, n$. Conversely, it can be shown that the biorthogonality relation (2.7) implies the three-term recurrence relation (2.2) for the functions $R_n(z)$ [27]. Due to a symmetry between $R_n(z)$ and $T_n(z)$, a similar three-term recurrence relation holds for $T_n(z)$ as well.

There is a spectral transformation for the rational functions $R_n(z)$, which can be considered as an analogue of the Christoffel transformation for the orthogonal polynomials. Assume that λ is an arbitrary parameter such that $R_n(\lambda) \neq 0$, $T_n(\lambda) \neq 0$. It can be verified [25, 27, 36] that a pair of new rational functions

$$\tilde{R}_n(z) = \frac{z-\alpha_1}{z-\lambda}\left(R_{n+1}(z) - \frac{R_{n+1}(\lambda)}{R_n(\lambda)}R_n(z)\right), \tag{2.8}$$

$$\tilde{T}_n(z) = \frac{1}{z-\lambda}\Big((z-\beta_{n+1})T_{n+1}(z)$$
$$-\frac{(\lambda-\beta_{n+1})T_{n+1}(\lambda)}{(\lambda-\alpha_n)T_n(\lambda)}(z-\alpha_n)T_n(z)\Big)$$

satisfies the biorthogonality relation

$$\tilde{\mathcal{L}}\left\{\tilde{R}_n(z)\tilde{T}_m(z)\right\} = 0, \quad n \neq m, \tag{2.9}$$

where the new linear functional $\tilde{\mathcal{L}}$ is defined on the space of rational functions $F(z)$ by the relation

$$\tilde{\mathcal{L}}\{F(z)\} = \mathcal{L}\left\{\frac{z-\lambda}{z-\alpha_1}F(z)\right\}. \tag{2.10}$$

One can repeat this procedure choosing a chain of parameters $\lambda = \lambda_1, \lambda_2, \ldots$. This leads to the so-called integrable R_{II} discrete time chain, which provides another approach to the theory of BRF. For details concerning this approach, see [25].

Theory of BRF is related to the continued fractions of the R_{II} type [12] and to the problem of multipoint rational interpolation having a long history starting from the pioneering works by Cauchy and Jacobi (for a review see, e.g. [28]).

Thus the theory of BRF is equivalent (under some weak restrictions) to GEVP (2.1) for two arbitrary Jacobi matrices. This statement is a generalization of the well-known Chebyshev-Stieltjes equivalence between the theory of ordinary orthogonal polynomials $\mathbf{P} \equiv \{P_0, P_1(z), \ldots\}$ and the ordinary eigenvalue problem $J\mathbf{P} = z\mathbf{P}$ for tridiagonal matrices.

If a special restriction $\alpha_n = \beta_n$ is imposed, then it can be shown [36] that $T_n(z) = \kappa_n^{-1} R_n(z)$ with some coefficients $\kappa_n \neq 0$ independent on z. In this case the biorthogonality property (2.7) becomes the *pure orthogonality* property

$$\mathcal{L}\{R_n(z)R_m(z)\} = \kappa_n \delta_{nm} \qquad (2.11)$$

and we deal with the theory of orthogonal rational functions [5] (see also Njåstad's lecture at this meeting).

Assume that all poles α_i coincide with one another: $\alpha_i \equiv \alpha$, $i = 1, 2, \ldots$, and the same is true for β_i, $\beta_i \equiv \beta$. Using transformation property (1.7) one can show that if $\alpha \neq \beta$ then the theory of functions $R_n(z)$ is equivalent to the theory of Laurent biorthogonal polynomials, satisfying the three-term recurrence relation [11]

$$P_{n+1}(z) + d_n P_n(z) = z(P_n(z) + d_n P_{n-1}(z)). \qquad (2.12)$$

If, additionally, $\alpha = \beta$ we return to the theory of ordinary orthogonal polynomials. For details, see [36].

3. Elliptic Analogues of Hypergeometric Functions

In this section we describe some properties of the elliptic analogues of the very-well-poised balanced hypergeometric functions (or "modular hypergeometric functions"), which were introduced by Frenkel and Turaev in [7]. Some of these properties were derived in [7], others were announced in [26].

Let us start from the description of Jacobi theta function $\theta_1(u)$ (see, e.g. [8]):

$$\theta_1(u) = 2 \sum_{n=0}^{\infty} (-1)^n p^{(n+1/2)^2} \sin(2n+1)u$$

$$= 2p^{1/4} \sin u \prod_{n=1}^{\infty} \left(1 - 2p^{2n} \cos 2u + p^{4n}\right) \left(1 - p^{2n}\right), \qquad (3.1)$$

where p is a complex parameter, $|p| < 1$. The modular parameter τ is introduced in the standard way $p = \exp(\pi i \tau)$. This function possesses many useful properties. The most important ones are:

(i) $\theta_1(u)$ is an odd function, $\theta_1(-u) = -\theta(u)$;

(ii) $\theta_1(u)$ is quasiperiodic with respect to the shifts by π and $\pi\tau$

$$\theta_1(u + \pi) = -\theta(u), \quad \theta(u + \pi\tau) = -p^{-1} \exp(-2iu) \theta_1(u); \qquad (3.2)$$

(iii) the Riemann identity

$$\begin{aligned} &\theta_1(x + z)\theta_1(x - z)\theta_1(y + w)\theta_1(y - w) \\ &-\theta_1(x + w)\theta_1(x - w)\theta_1(y + z)\theta_1(y - z) \\ &= \theta_1(x + y)\theta_1(x - y)\theta_1(z + w)\theta_1(z - w) \end{aligned} \qquad (3.3)$$

holds for any variables x, y, z, w.

Following [7], define the "elliptic numbers" (or, simply, e-numbers):

$$[x; h, \tau] = \frac{\theta_1(\pi h x)}{\theta_1(\pi h)}, \qquad (3.4)$$

where h is an arbitrary constant. The e-numbers depend on three variables x, h and τ. In what follows the dependence on h, τ will be indicated in the notations only when necessary, i.e. we take the notations $[x] \equiv [x; h, \tau]$.

Main properties of the e-numbers are

(i) $[-x] = -[x]$;

(ii) $[x + 1/h] = -[x], \ [x + \tau/h] = -\exp(-i\pi\tau - 2\pi i h x)[x]$; (3.5)

(iii) $[x + z][x - z][y + w][y - w] - [x + w][x - w][y + z][y - z]$

$$= [x + y][x - y][z + w][z - w]; \qquad (3.6)$$

(iv) $\lim_{Im(\tau) \to +\infty} [x; h, \tau] = \frac{\sin(\pi h x)}{\sin(\pi h)}; \qquad (3.7)$

(v) $\lim_{h \to 0} [x; h, \tau] = x; \qquad (3.8)$

(vi) $[x; h, \tau] = 0$ for $x_{m_1, m_2} = (m_1 + m_2\tau)/h, \qquad (3.9)$
       $m_{1,2} = 0, \pm 1, \pm 2, \ldots.$

Properties (iv) and (v) connect the limiting cases of the e-numbers with the q-numbers $[x; q] = (q^x - q^{-x})/(q - q^{-1})$ for $q = e^{\pi i h}$ and usual numbers respectively. As shown in [24] the case when q is a root of unity require a special treatment due to some extra zeros in $[x; q]$. In order to escape similar situations in the elliptic case, we assume that

$$h \neq m_1 + m_2\tau, \tag{3.10}$$

which means that $[n] \neq 0$ for any integer $n > 0$.

We will use also the notations [7]

$$[x]_n = [x][x+1]\cdots[x+n-1], \qquad [n]! = [1]_n$$

defining natural elliptic generalizations of the Pochhammer symbol and factorial.

Elliptic analogues of the very-well-poised balanced hypergeometric functions $_{r+1}E_r$ were defined by Frenkel and Turaev as series of the following form [7] (we use slightly different notations)

$$_{r+1}E_r(a_1; a_4, a_5, \ldots, a_{r+1}; h, \tau)$$

$$= \sum_{k=0}^{\infty} \frac{[a_1]_k[a_1 + 2k]}{[k]![a_1]} \prod_{m=1}^{r-2} \frac{[a_{3+m}]_k}{[1 + a_1 - a_{3+m}]_k} \tag{3.11}$$

with the balancing restriction upon the parameters

$$\frac{r-5}{2} + \frac{r-3}{2}a_1 - \sum_{m=1}^{r-2} a_{3+m} = 0. \tag{3.12}$$

It should be noted that there are difficulties with the convergence of the infinite sum in (3.11), because there is no limit of the ratio $[a+k]/[b+k]$ for $k \to \infty$ at finite values of the modular parameter $\tau, Im\,\tau > 0$, and $h \neq 0$. This difficulty is avoided in the limit $Im(\tau) \to +\infty$ or $h \to 0$ when the infinite series (3.11) may be convergent. In the elliptic case one can resolve the problem by taking one of the parameters (say, a_{r+1}) equal to a negative integer $a_{r+1} = -n$. In this case the sum (3.11) terminates at $k = n+1$ and becomes a finite sum of elliptic functions.

The function $_{10}E_9$ is of the main importance for the following considerations. One of its most important properties derived by Frenkel and Turaev is an analogue of the Bailey's transformation formula [7]:

$$_{10}E_9(a_1; a_4, \ldots, -n; h, \tau) = {_{10}E_9}(b_1; b_4, \ldots, b_{10}; h, \tau)$$

$$\times \frac{[a_1 + 1]_n[b_1 + 1 - a_7]_n[b_1 + 1 - a_8]_n[a_1 + 1 - a_7 - a_8]_n}{[b_1 + 1]_n[a_1 + 1 - a_7]_n[a_1 + 1 - a_8]_n[b_1 + 1 - a_7 - a_8]_n}, \tag{3.13}$$

where

$$b_1 = 2a_1 + 1 - a_4 - a_5 - a_6, \quad b_4 = b_1 - a_1 + a_4,$$

$$b_5 = b_1 - a_1 + a_5, \quad b_6 = b_1 - a_1 + a_6 \tag{3.14}$$

and $\{b_7, b_8, b_9, b_{10}\}$ is an arbitrary permutation of the parameters $a_7, a_8, a_9,$ $a_{10} = -n$. A repeated use of this formula provides another useful form of the elliptic Bailey transformation

$$_{10}E_9(a_1; a_4, \ldots, a_9, -n; h, \tau) = \sigma_n \, _{10}E_9(b_1; b_4, \ldots, b_9, -n; h, \tau), \quad (3.15)$$

where

$$b_1 = a_4 - a_9 - n, \quad b_4 = a_4, \quad b_9 = a_4 - n - a_1,$$

$$b_i = 1 + a_1 - a_i - a_9, \quad i = 5, \ldots, 8 \quad (3.16)$$

and

$$\sigma_n = \frac{[a_1 + 1]_n [a_9]_n}{[1 + a_1 - a_4]_n [a_9 - a_4]_n} \prod_{i=5}^{8} \frac{[1 + a_1 - a_4 - a_i]_n}{[1 + a_1 - a_i]_n}. \quad (3.17)$$

We mention also a generalization of the terminating $_8\varphi_7$ Jackson summation formula following from (3.13) [7]

$$_8E_7(a_1; a_4, a_5, a_6, a_7, -n; h, \tau) = \frac{[a_1 + 1]_n [a_1 + 1 - a_4 - a_5]_n}{[a_1 + 1 - a_4]_n [a_1 + 1 - a_5]_n}$$

$$\times \frac{[a_1 + 1 - a_4 - a_6]_n [a_1 + 1 - a_5 - a_6]_n}{[a_1 + 1 - a_6]_n [a_1 + 1 - a_4 - a_5 - a_6]_n}. \quad (3.18)$$

The next two theorems describe generalizations of the contiguous relations for the corresponding basic hypergeometric functions $_{10}\varphi_9$ from [10]. In what follows we will use for brevity the following notations used in [10, 34]: $\Phi(a_i\pm)$ denotes the function $_{10}E_9$ with the particular parameter a_i replaced by $a_i \pm 1$, whereas other parameters being unchanged, and $\Phi_\pm$ denotes the function $_{10}E_9(a_1 \pm 2; a_4 \pm 1, \ldots, a_{10} \pm 1; h, \tau)$.

Theorem 1 *Assume that one of the parameters $a_i = -n$, where $i = 4, \ldots, 10$. Then the following identity takes place*

$$\Phi(a_9-, a_{10}+) - \Phi = \Phi_+(a_9-) \quad (3.19)$$

$$\times \frac{[a_1 + 1][a_1 + 2][a_{10} - a_9 + 1][a_{10} + a_9 - a_1 - 1]}{[1 + a_1 - a_9][2 + a_1 - a_9][a_1 - a_{10}][1 + a_1 - a_{10}]} \prod_{i=4}^{8} \frac{[a_i]}{[1 + a_1 - a_i]}.$$

Theorem 2 *Under the same assumptions as in the previous theorem, the following identity takes place*

$$\frac{[a_{10}]}{[1 + a_1 - a_9][2 + a_1 - a_9]} \prod_{i=4}^{8} [1 + a_1 - a_i - a_9] \Phi_+(a_9-)$$

$$= \frac{[a_9]}{[1 + a_1 - a_{10}][2 + a_1 - a_{10}]} \prod_{i=4}^{8} [1 + a_1 - a_i - a_{10}] \Phi_+(a_{10}-)$$

$$+ \frac{[a_{10} - a_9]}{[1 + a_1][2 + a_1]} \prod_{i=4}^{8} [1 + a_1 - a_i] \Phi. \tag{3.20}$$

4. Elliptic Rational Functions

In this section we describe a class of rational functions $R_n(z)$ expressed explicitly in terms of the elliptic analogues of hypergeometric functions $_{10}E_9$ and show that they satisfy the three-term recurrence relation (2.2).

Let d_1, d_2, d_3, d_4, d_5 and x_0, x_1, x_2 be some parameters satisfying the relations

$$x_2 = x_0 + x_1 \tag{4.1}$$

and

$$\sum_{i=1}^{5} d_i = 1 + 2(x_0 + x_2). \tag{4.2}$$

Introduce three sequences of numbers $\alpha_k, \beta_k, \lambda_k$

$$\alpha_k = \frac{[k - x_2 + e_1][k - x_2 + e_2]}{[k - x_2 + d_1][k - x_2 + d_2]},$$

$$\beta_k = \frac{[k - e_1 + 1][k - e_2 + 1]}{[k - d_1 + 1][k - d_2 + 1]},$$

$$\lambda_k = \frac{[k + x_0 - e_1][k + x_0 - e_2]}{[k + x_0 - d_1][k + x_0 - d_2]}, \tag{4.3}$$

where e_1, e_2 are arbitrary parameters with the restriction

$$e_1 + e_2 = d_1 + d_2. \tag{4.4}$$

We parametrize the argument of rational functions z in terms of an auxiliary parameter u:

$$z(u) = \frac{[u][u + e_2 - e_1]}{[u + d_2 - e_1][u + d_1 - e_1]}. \tag{4.5}$$

Using the Riemann identity (3.6) we derive the following relations

$$z(u) - \alpha_k$$
$$= \frac{[k + u + e_2 - x_2][k - u + e_1 - x_2][d_2 - e_1][e_1 - d_1]}{[u + d_2 - e_1][u + d_1 - e_1][k - x_2 + d_1][k - x_2 + d_2]}, \tag{4.6}$$

$$z(u) - \beta_{k-1} = \frac{[k + u - e_1][k - u - e_2][d_2 - e_1][e_1 - d_1]}{[u + d_2 - e_1][u + d_1 - e_1][k - d_1][k - d_2]}, \quad (4.7)$$

$$z(u) - \lambda_k$$
$$= \frac{[k + u + x_0 - e_1][k - u + x_0 - e_2][d_2 - e_1][e_1 - d_1]}{[u + d_2 - e_1][u + d_1 - e_1][k + x_0 - d_1][k + x_0 - d_2]}. \quad (4.8)$$

Introduce the functions

$$R_n(z) = {}_{10}E_9(1 - x_1; 1 + x_0 - d_3, 1 + x_0 - d_4, 1 + x_0 - d_5,$$
$$1 + u + x_0 - e_1, 1 - u + x_0 - e_2, 1 - x_2 + n, -n; h, \tau). \quad (4.9)$$

Proposition 1 *The functions $R_n(z)$ defined by (4.9) are rational functions of the type $[n/n]$ of the argument z and the poles of $R_n(z)$ are located at the points $\alpha_1, \alpha_2, \ldots, \alpha_n$.*

Let us denote as $R_n^{(+)}(z)$ the function obtained from $R_n(z)$ by shifting the parameters $x_0 \to x_0 + 1$, $x_1 \to x_1 - 2$ (and hence $x_2 \to x_2 - 1$ due to (4.1)).

Proposition 2 *The relations*

$$R_n^{(+)}(z) = \frac{z - \alpha_1}{z - \lambda_1} \epsilon_n (R_{n+1}(z) - R_n(z)), \quad (4.10)$$

$$a_n(z - \alpha_{n+1})R_n^{(+)}(z) - b_n(z - \beta_{n-1})R_{n-1}^{(+)}(z) = c_n(z - \alpha_1)R_n(z) \quad (4.11)$$

hold, where

$$\epsilon_n = \frac{[2 + n - x_1][3 + n - x_1][x_0 - n][x_0 - n + 1]}{[2 - x_1][3 - x_1][2 - x_2 + 2n][-x_0 - 1]} \prod_{i=1}^{5} \frac{[1 - x_2 + d_i]}{[1 + x_0 - d_i]},$$

$$a_n = \frac{[n + 1 - x_2]}{[2 + n - x_1][3 + n - x_1]} \prod_{i=1}^{5} [n + 1 - x_2 + d_i],$$

$$b_n = \frac{[-n]}{[1 + x_0 - n][2 + x_0 - n]} \prod_{i=1}^{5} [d_i - n],$$

$$c_n = \frac{[1 - x_2 + 2n]}{[2 - x_1][3 - x_1]} \prod_{i=1}^{5} [1 - x_2 + d_i].$$

Proof. The relation (4.10) is a direct consequence of the equation (3.20) where the relations (4.6) and (4.8) were used. Similarly, the relation (4.11) is a direct consequence of (3.20). □

As a corollary of this proposition we get the following statement.

Theorem 3 *Rational functions $R_n(z)$ satisfy the three-term recurrence relation*

$$\epsilon_n a_n (z - \alpha_{n+1})(R_{n+1}(z) - R_n(z)) - \epsilon_{n-1} b_n (z - \beta_{n-1})(R_n(z) - R_{n-1}(z))$$

$$= c_n(z - \lambda_1)R_n(z), \qquad n = 0, 1, 2, \ldots, \tag{4.12}$$

where the second term is equal to zero for $n = 0$.

The equation (4.12) belongs to the class of recurrence relations (2.1). Hence there exists a linear functional $\mathcal{L}$ providing the orthogonality property [12, 36]

$$\mathcal{L}\left\{ \frac{R_n(z)}{(z - \beta_1) \cdots (z - \beta_m)} \right\} = 0, \qquad m = 0, 1, \ldots, n - 1. \tag{4.13}$$

An explicit realization of this functional can be found for a finite-dimensional case.

Let us impose the constraint

$$x_2 - d_3 = N + 1, \tag{4.14}$$

where $N = 1, 2, \ldots$ is a positive integer. As a result, $a_N = 0$ and one may impose the constraint $R_{N+1}(z) = 0$ leading to a finite-dimensional set of BRF. Actually, $R_{N+1}(z)$ is not well defined for (4.14). For dealing with such a problem we set $d_3 = x_2 - N - 1 + \delta$ and take the limit $\delta \to 0$ in the product $a_N R_{N+1}(z)$ which does not have divergencies. Assuming such a limiting procedure, let us introduce the polynomials $P_n(z) = [n + d_3 - x_2](z - \alpha_1) \cdots (z - \alpha_n)R_n(z)$.

Proposition 3 *Zeros of the polynomial $P_{N+1}(z)$ have the following form*

$$z_s = \lambda_{s+1} = \frac{[s + x_0 - e_1 + 1][s + x_0 - e_2 + 1]}{[s + x_0 - d_1 + 1][s + x_0 - d_2 + 1]}, \qquad s = 0, 1, \ldots, N. \tag{4.15}$$

These zeros are simple provided

$$2x_0 - d_1 - d_2 \neq (m_1 + m_2 \tau)/h - 2 - M \tag{4.16}$$

for some integers $M > 0$ and $m_{1,2} = 0, \pm 1, \pm 2, \ldots$.

Knowledge of these zeros allows one to find the corresponding discrete weight function ω_s.

Proposition 4 *The rational functions $R_n(z)$ defined by (4.9) satisfy the orthogonality condition*

$$\sum_{s=0}^{N} \frac{\omega_s R_n(z_s)}{(z_s - \beta_1) \cdots (z_s - \beta_m)} = 0, \qquad m = 0, 1, \ldots, n - 1, \tag{4.17}$$

where ω_s is

$$\omega_s = \frac{[2x_0 + 2 - d_1 - d_2 + 2s][-N]_s[2x_0 + 2 - d_1 - d_2]_s}{[2x_0 + 2 - d_1 - d_2][s]![2x_0 + 3 - d_1 - d_2 + N]_s}$$
$$\times \frac{[x_0]_s[1 + d_4 - x_2]_s[1 + d_5 - x_2]_s[1 + x_0 + x_2 - d_1 - d_2]_s}{[2 - x_1]_s[3 + x_0 - d_1 - d_2]_s[-N + d_4]_s[-N + d_5]_s}. \quad (4.18)$$

From the general results concerning biorthogonal rational functions we know that the orthogonality condition (4.17) implies existence of the companion rational function $T_n(z)$ for which biorthogonality relation holds. Its explicit expression is

$$T_n(z) = {}_{10}E_9(2 + x_0 - d_1 - d_2; 2 + x_0 + x_2 - d_1 - d_2 - d_3,$$
$$2 + x_0 + x_2 - d_1 - d_2 - d_4, 2 + x_0 + x_2 - d_1 - d_2 - d_5, -n, 1 - x_2 + n,$$
$$1 + u + x_0 - e_1, 1 - u + x_0 - e_2; h, \tau). \quad (4.19)$$

The biorthogonality condition has the form

$$\sum_{s=0}^{N} \omega_s R_n(z_s) T_m(z_s) = h_n \delta_{nm}, \quad (4.20)$$

where the rational functions $R_n(z)$ and $T_n(z)$ are given by (4.9) and (4.19), and the weight function ω_s is fixed by (4.18). Similar to $R_n(z)$, $T_n(z)$ is a rational function of z with the poles located at β_k which are simple provided $d_1 + d_2 \neq (m_1 + m_2 \tau)/h + M + 2$ for any integers $m_{1,2} = 0, \pm 1, \pm 2, \ldots$ and $M = 1, 2, \ldots$.

The normalization constants have the form

$$h_n = \kappa \frac{[1 - x_2][n]![2 - x_2 + N]_n}{[1 - x_2 + 2n][-N]_n[1 - x_2]_n} \quad (4.21)$$
$$\times \frac{[2 - x_1]_n[3 + x_0 - d_1 - d_2]_n[1 - d_4]_n[1 - d_5]_n}{[-1 - x_0 - x_2 + d_1 + d_2]_n[1 + d_4 - x_2]_n[1 + d_5 - x_2]_n[-x_0]_n},$$

where κ is a constant not depending on n. For its explicit expression, see [25, 26].

5. Self-Duality and the Finite-Difference Equation

Introduce a short hand notation $d \equiv d_1 + d_2$ and consider the following transformation of the parameters

$$\tilde{x}_0 = -1 - x_0 - x_2 + d, \quad \tilde{x}_1 = x_1, \quad \tilde{x}_2 = -1 - 2x_0 + d,$$
$$\tilde{d}_3 = -2 - N - 2x_0 + d, \quad \tilde{d}_4 = -1 - 2x_0 - x_2 + d + d_4,$$
$$\tilde{d}_5 = -1 - 2x_0 - x_2 + d + d_5, \quad \tilde{d} = 2d - 1 - 2x_0 - x_2. \quad (5.1)$$

It is directly verified that

$$\tilde{R}_n(z_s) = R_s(z_n), \quad \tilde{T}_n(z_s) = T_s(z_n), \qquad 0 \le s, n \le N, \qquad (5.2)$$

where $\tilde{R}_n(z_s)$ denotes the function obtained from $R_n(z_s)$ after replacing the parameters $d, \ldots, x_2$ by $\tilde{d}, \ldots, \tilde{x}_2$. Thus the transformation (5.1) of the parameters is equivalent to the permutation of n and s. We will call (5.1) the duality transformation.

Due to the self-duality property (5.2) the finite-dimensional system of elliptic biorthogonal rational functions $R_n(z_s)$ satisfies a difference equation in the argument z_s. It can be written as a homogeneous second order difference equation in the discrete variable s parametrizing the argument z_s, which is similar to the original three-term recurrence relation (4.12).

Let us define the dual grid

$$\tilde{z}_n = \frac{[n - x_1 - \eta][n + 1 - x_0 + \eta]}{[n - x_1 - \xi][n + 1 - x_0 + \xi]},$$

where ξ, η are arbitrary parameters ($\xi \ne \eta$), and the corresponding dual spectral parameters

$$t_s = \frac{[s + 1 + \eta][s + x_0 - x_1 - \eta]}{[s + 1 + \xi][s + x_0 - x_1 - \xi]},$$

$$r_s = \frac{[s - d + 2x_0 + 1 - \eta][s - d + 2 + x_2 + \eta]}{[s - d + 2x_0 + 1 - \xi][s - d + 2 + x_2 + \xi]}.$$

Proposition 5 *Rational functions $R_n(z_s)$ satisfy the following finite-difference equation of the second order*

$$g_s(\tilde{z}_n - t_{s+1})(R_n(z_{s+1}) - R_n(z_s)) + f_s(\tilde{z}_n - r_{s-1})(R_n(z_s) - R_n(z_{s-1})) \quad (5.3)$$
$$= \kappa(\tilde{z}_n - \tilde{z}_0)R_n(z_s),$$

where $n, s = 0, 1, 2, \ldots$ and g_s, f_s are

$$g_s = \frac{[N - s][s + 1 - d + x_0 + x_2][s - d + x_0 + x_2][s + 2 - d + 2x_0]}{[3 + 2x_0 - d + 2s]}$$
$$\times \frac{[s + 2 + \xi][s + 1 + x_0 - x_1 - \xi][s + 1 - x_2 + d_4][s + 1 - x_2 + d_5]}{[2 + 2x_0 - d + 2s]},$$

$$f_s = \frac{[s][1 + s - x_1][2 + s - x_1][s - d + 2x_0 - \xi][s - d + x_2 + 1 + \xi]}{[2 + 2x_0 - d + 2s]}$$
$$\times \frac{[s - d + 2x_0 + N + 2][s - d - d_4 + 2x_0 + x_2 + 1]}{[1 + 2x_0 - d + 2s]}$$
$$\times \frac{[s - d - d_5 + 2x_0 + x_2 + 1]}{[1 + 2x_0 - d + 2s]}.$$

The constant κ is

$$\kappa = [x_0 + x_2 - d][-x_1 - \xi][x_0 - 1 - \xi][2 - x_1 + N][1 + x_0 - d_4][1 + x_0 - d_5].$$

One may equally say that $R_n(z_s)$ are eigenfunctions of a generalized eigenvalue problem for two Jacobi matrices with the eigenvalues $\tilde{z}_n$. Indeed, equation (5.3) can be rewritten in the matrix form

$$M_1 X_s^{(n)} = \lambda_n M_2 X_s^{(n)}, \tag{5.4}$$

where the tridiagonal $(N + 1) \times (N + 1)$ matrices M_1 and M_2 act on the space of $(N + 1)$-component vectors X_s, $s = 0, 1, \ldots, N$, and the vectors $X_s^{(n)} = R_n(z_s)$ are eigenvectors of the generalized eigenvalue problem (5.4).

The rational functions $R_n(z_s)$ may be interpreted also as solutions of the generalized bispectral problem. Recall that in the ordinary bispectral problem one is looking for differential or difference operators L_n and M_s such that

$$L_n F(n, s) = \mu(s) F(n, s), \qquad M_s F(n, s) = \lambda(n) F(n, s), \tag{5.5}$$

where the operator L_n acts on the space of functions of the argument n and does not depend on the variable s, whereas the operator M_s acts on the space of functions of the argument s and is independent on n (n and s may be both continuous or discrete). The function $F(n, s)$ is thus a simultaneous eigenfunction of two operators and the functions $\mu(s)$ and $\lambda(n)$ are considered as an ordinary spectral parameter and its dual. The case when both L and M are finite-dimensional tridiagonal matrices was considered by Leonard [15], who proved that the most general solution to this bispectrality problem is given by the q-Racah polynomials. An algebraic treatment of the Leonard problem is given in [30]. A detailed analysis of the case when L_n is a Schrödinger operator and M_s is an arbitrary order differential operator is given in [6]. A similar problem on the classification of all orthogonal polynomials which are eigenfunctions of some differential operators was posed by H. Krall in the 1930's.

In our case we have four tridiagonal matrices J_1, J_2, M_1, M_2 and a pair of two GEVP (2.1) and (5.4) built with the help of these matrices. The rational functions appearing as solutions of these problems are self-dual in the same spirit as the q-Racah polynomials [15] — permutation of the variables n and s is equivalent to a simple change of free parameters. In this respect one can ask what are the most general solutions of the equations (2.1), (5.4) obeying such a property when all the matrices entering them are tridiagonal. Although this problem was not solved yet, the results of the next section indicate that the self-dual elliptic BRF [25, 26] define its most general solution in the finite-dimensional setting.

Using the identity (3.4) one can simplify the expressions $\tilde{z}_n - t_{s+1}$, $\tilde{z}_n - r_{s-1}$ and $\tilde{z}_n - \tilde{z}_0$. This yields another form of the difference equation (5.3) for BRF:

$$A(n,s)(R_n(z_{s+1}) - R_n(z_s)) + B(n,s)(R_n(z_s) - R_n(z_{s-1})) = \kappa_n R_n(z_s), \quad (5.6)$$

where

$$A(n,s) = \frac{[N-s][s-n+x_0+1][s+n-x_1+2]}{[2s+2+2x_0-d]}$$
$$\times \frac{[s+1-x_2+d_4][s+1-x_2+d_5]}{[2s+2+2x_0-d]}$$
$$\times \frac{[s+2+2x_0-d][s+x_0+x_2-d][s+x_0+x_2-d+1]}{[2s+3+2x_0-d]},$$

$$B(n,s) = \frac{[s][s-n-d+x_0+x_2][s+N+2-d+2x_0]}{[2s+2+2x_0-d]}$$
$$\times \frac{[s+n+1-d+x_0]}{[2s+2+2x_0-d]}$$
$$\times \frac{[s+2-x_1][s+1-x_1][s+1-d-d_5+2x_0+x_2]}{[2s+1+2x_0-d]}$$
$$\times \frac{[s+1-d-d_4+2x_0+x_2]}{[2s+1+2x_0-d]},$$

$$\kappa_n = [n][n+1-x_2][1+x_0-d_4][1+x_0-d_5][2-x_1+N][x_0+x_2-d].$$

It is seen that the coefficients $A(n,s), B(n,s), \kappa_n$ in (5.6) do not depend on the parameters ξ and η.

6. Uniqueness of the Elliptic Grids

Let us recall how the admissible grids for polynomials can be constructed [16, 18]. Consider a $\mathbf{C} \to \mathbf{C}$ function $x(s)$, which will be called a "grid", and associate with it the following divided difference operator

$$\mathcal{D}F(x(s)) = \frac{F(x(s)) - F(x(s-1))}{x(s) - x(s-1)} \quad (6.1)$$

It is obvious that $\mathcal{D}1 = 0$ and $\mathcal{D}x(s) = 1$. One demands that the action of the operator $\mathcal{D}$ upon any polynomial in x of order n yields a polynomial of order $n-1$ in $x(s-1/2)$. For this, it is sufficient to show that

$$\mathcal{D}x^n(s) = P_{n-1}(x(s-1/2)), \quad (6.2)$$

where $P_{n-1}(x)$ is a polynomial of order $n-1$. It can be easily shown [16] that the only grids $x(s)$ satisfying such constraints are quadratic and q-quadratic grids and their limiting cases (linear and q-linear grids) [18] satisfying the condition (1.4). Askey-Wilson polynomials are the polynomials which are orthogonal on a q-quadratic grid.

One can show that under simple natural constraints upon the system of rational functions, one can recover the *elliptic grids* [27].

Consider rational functions $R_n(z)$ — ratios of two n-th degree polynomials in z. Denote as s a variable which parametrizes the grid $z(s)$. Let $\alpha_1, \alpha_2, \ldots, \alpha_n$ be n different prescribed positions of the poles of $R_n(z)$. Then $R_n(z)$ can be written as a sum of simple fractions

$$R_n(z) = t_0^{(n)} + \sum_{i=1}^{n} \frac{t_i^{(n)}}{z - \alpha_i} \qquad (6.3)$$

with the coefficients $t_i^{(n)}$, $i = 1, 2, \ldots, n$ playing the role of residues of $R_n(z)$ at the poles α_i. The coefficient $t_0^{(n)}$ can be interpreted as $\lim_{z \to \infty} R_n(z)$.

We would like to construct a lowering operator in the space of rational functions $R_n(z)$. We take as a definition of the lowering operator $\mathcal{D}$ a divided difference operator in the parametrizing variable s, which obeys the following properties:

(i) for any function $F(s)$ one has

$$\mathcal{D}F(s) = \chi(s)(F(s) - F(s-1)),$$

where $\chi(s)$ is some function to be determined;

(ii) $\mathcal{D}R_1(z(s)) = const$;

(iii) the operator $\mathcal{D}$ transforms any rational function $R_n(z(s))$ with the *prescribed* poles $\alpha_1, \ldots, \alpha_n$ to a rational function $\tilde{R}_{n-1}(z(s - 1/2))$ with some other poles $\tilde{\alpha}_1, \ldots, \tilde{\alpha}_{n-1}$.

An important restriction is the condition of independence of $\mathcal{D}$ on the order n of a rational function. From the properties (i) and (ii) one easily finds

$$\chi(s) = \left(\frac{1}{z(s) - \alpha_1} - \frac{1}{z(s-1) - \alpha_1} \right)^{-1} = \frac{(z(s) - \alpha_1)(z(s-1) - \alpha_1)}{z(s-1) - z(s)}.$$

$$(6.4)$$

Note that from (i) one has $\mathcal{D}R_0(z) = 0$.

The most non-trivial problem in the construction of the operator $\mathcal{D}$ consists in verification of the property (iii). The shift of the argument of $z(s)$ by $1/2$ as a result of the action of $\mathcal{D}$ is similar to the one taking place in the consideration of orthogonal polynomials [16, 18] and $_{10}\varphi_9$ family of BRF [21] where is comes from a convenient uniformizing parametrization of the grids.

Checking (iii) step-by-step for $n = 2, 3, \ldots$, we arrive at a necessary condition for the property (iii) to be true.

Proposition 6 *The admissible grid z_s should satisfy the Baxter's biquadratic relation*

$$A_1 u^2 v^2 + A_2 uv(u+v) + A_3(u^2 + v^2) + B_2 uv + B_3(u+v) + C_3 = 0, \quad (6.5)$$

where $u = z_{s-1/2}$, $v = z_s$ and $A_1, \ldots, C_3$ are some constants.

Equivalently, one has the conditions

$$z(s)z(s-1) = \frac{C(u)}{A(u)}, \quad (6.6)$$

$$z(s) + z(s-1) = -\frac{B(u)}{A(u)},$$

where $u = z(s - 1/2)$ and

$$
\begin{aligned}
A(u) &= A_1 u^2 + A_2 u + A_3, \\
B(u) &= A_2 u^2 + B_2 u + B_3, \\
C(u) &= A_3 u^2 + B_3 u + C_3.
\end{aligned}
\quad (6.7)
$$

As shown by Baxter [3], the general parametrization of solutions of the biquadratic equation (6.5) is given in terms of the "elliptic numbers"

$$z(s) = c \frac{[s-a][s-b]}{[s-d][s-e]}, \quad (6.8)$$

where $a, \ldots, e$ are some constants satisfying the constraint $a + b = d + e$.

The well-known quadratic and q-quadratic grids z_s are obtained under special restrictions upon the parameters $A_1, \ldots, C_3$. For example, such grids appear when all terms of the degree ≥ 3 vanish in (6.5), i.e. $A_1 = A_2 = 0$.

It can be shown that the condition (6.5) is also *sufficient* for fulfilling the property (iii) of the operator $\mathcal{D}$ if the parameters α_i satisfy the conditions $\alpha_i = z(i + s_0)$ for an arbitrary constant s_0, in which case $\tilde{\alpha}_i = \alpha_{i+3/2}$.

7. A Ramanujan-Type Continued Fraction

Ramanujan's Entry 40 terminating continued fraction [4, 22] is one of the most spectacular "explicit" results of the XXth century in the special function theory. This example is connected with the contiguous relations for a very-well-poised hypergeometric function $_9F_8$ [34]. In [32] Watson found a q-analogue of the Ramanujan's continued fraction. Further new examples — including the non-terminating q-hypergeometric series cases — were found by Gupta and Masson [9, 10, 17]. These examples are related to the contiguous relations for a very-well-poised balanced basic hypergeometric series

$_{10}\phi_9$ [8, 10]. In turn, these contiguous relations are related to biorthogonal rational functions on quadratic and q-quadratic grids [19, 20, 21, 34, 35].

Let us construct an elliptic generalization of this class of continued fractions, which is connected to the self-dual elliptic BRF described in the previous sections.

Consider two sequences U_n, V_n, $n = 0, 1, \ldots$, generated by the recurrence relations

$$U_{n+1} = \xi_n U_n + \eta_n U_{n-1}, \qquad V_{n+1} = \xi_n V_n + \eta_n V_{n-1} \qquad (7.1)$$

with the initial conditions

$$U_0 = 0, \quad U_1 = 1, \quad V_0 = 1, \quad V_1 = \xi_0. \qquad (7.2)$$

It is well known that one has the following representation of the ratio U_n/V_n as a continued fraction

$$\frac{U_n}{V_n} = \cfrac{1}{\xi_0 + \cfrac{\eta_1}{\xi_1 + \cfrac{\eta_2}{\xi_2 + \cdots + \cfrac{\eta_{n-1}}{\xi_{n-1}}}}}, \qquad n = 1, 2, \ldots. \qquad (7.3)$$

In the most often discussed examples of continued fractions, the coefficients ξ_n and η_n are linear and quadratic functions of an independent variable z respectively. In these cases $U_n(z)$ and $V_n(z)$ are polynomials in z of the degree $\leq n$ and the ratio $U_n(z)/V_n(z)$ is a rational function of the argument z. If ξ_n are linear in z and η_n do not depend on z one gets the Chebyshev-Stieltjes approach [13] to orthogonal polynomials $V_n(z)$. If $\eta_n(z)$ are quadratic in z and $\xi_n(z)$ are linear in z, one gets polynomials of the R_{II} type [12] which, in turn, give rise to the explicit biorthogonality relation for BRF [36].

More precisely, assume that

$$\xi_n(z) = \rho_n z - v_n, \qquad \eta_n(z) = -u_n(z - \alpha_n)(z - \beta_{n-1}), \qquad (7.4)$$

where $\rho_n, v_n, u_n, \alpha_n, \beta_n$ are some coefficients. Then $V_n(z) = P_n(z)$ and $U_n(z) = P_{n-1}^{(1)}(z)$, where $P_n(z)$ are n-th degree polynomials of R_{II} type satisfying the three-term recurrence relation (2.4) with the initial conditions $P_0 = 1$, $P_1(z) = \rho_0 z - v_0$. The polynomials $P_{n-1}^{(1)}(z)$ have degree $n-1$ and they are called the associated polynomials of R_{II} type. They satisfy the recurrence relation

$$P_n^{(1)}(z) + (v_n - \rho_n z)P_{n-1}^{(1)}(z) + u_n(z - \alpha_n)(z - \beta_{n-1})P_{n-2}^{(1)}(z) = 0 \qquad (7.5)$$

with the initial conditions $P_{-1}^{(1)} = 0$, $P_0^{(1)}(z) = 1$.

On the one hand, from (7.3) we have

$$\frac{P_N^{(1)}(z)}{P_{N+1}(z)} = \cfrac{1}{\rho_0 z - v_0 - \cfrac{u_1(z - \alpha_1)(z - \beta_1)}{\rho_1 z - v_1 - \cdots - \cfrac{u_N(z - \alpha_N)(z - \beta_N)}{\rho_N z - v_N}}}. \qquad (7.6)$$

On the other hand, we can expand the l.h.s. of (7.6) in partial fractions:

$$F_N(z) \equiv \frac{P_N^{(1)}(z)}{P_{N+1}(z)} = \sum_{s=0}^{N} \frac{g_s}{z - z_s}, \qquad (7.7)$$

where

$$g_s = \frac{P_N^{(1)}(z_s)}{P_{N+1}'(z_s)} \qquad (7.8)$$

and z_s, $s = 0, 1, \ldots, N$, are zeros of the polynomial $P_{N+1}(z)$. We assume that all these zeros are simple, i.e. $z_s \neq z_{s'}$ if $s \neq s'$.

For any recurrence relation of the type (7.1) one can get an analogue of the Wronskian formula

$$U_{n+1}V_n - U_n V_{n+1} = (-1)^n \eta_1 \cdots \eta_n (U_1 V_0 - U_0 V_1),$$

which yields in our case

$$P_n(z) P_n^{(1)}(z) - P_{n+1}(z) P_{n-1}^{(1)}(z) = h_n A_n(z) B_n(z), \qquad (7.9)$$

where $h_n = u_1 u_2 \cdots u_n$ and $A_n(z) = (z - \alpha_1)(z - \alpha_2) \cdots (z - \alpha_n)$, $B_n(z) = (z - \beta_0)(z - \beta_1) \cdots (z - \beta_{n-1})$. Putting $n = N$ and $z = z_s$, $s = 0, 1, \ldots, N$, in (7.9) we can express $P_N^{(1)}(z_s)$ in terms of $P_N(z_s)$. This yields an alternative expression for g_s

$$g_s = \frac{h_N A_N(z_s) B_N(z_s)}{P_{N+1}'(z_s) P_N(z_s)}, \qquad (7.10)$$

which is much more convenient for explicit evaluation.

Take now our explicit family of elliptic BRF $R_n(z(u))$ defined in (4.9) and introduce two convenient combinations of the recurrence coefficients

$$G_n = \frac{[n + 1 - x_2][n - x_0] \prod_{i=1}^{5} [n + 1 - x_2 + d_i]}{[2n + 1 - x_2][2n + 2 - x_2][n + 2 - x_1]},$$

$$D_n = \frac{[n][n + 1 - x_1] \prod_{i=1}^{5} [n - d_i]}{[2n - x_2][2n + 1 - x_2][n - x_0 - 1]}.$$

Then it is not difficult to see from (4.12), that $P_n(z) \equiv G_{n-1} \cdots G_0 A_n(z) \times R_n(z)$ are the n-th degree polynomials satisfying the three-term recurrence relation (2.4) with

$$u_n = G_{n-1}D_n, \quad \rho_n = G_n + D_n + \frac{[x_0 + 1]\prod_{i=1}^{5}[d_i - x_0 - 1]}{[n + 2 - x_1][n - 1 - x_0]},$$

$$v_n = G_n\alpha_{n+1} + D_n\beta_{n-1} + \frac{[x_0 + 1]\prod_{i=1}^{5}[d_i - x_0 - 1]}{[n + 2 - x_1][n - 1 - x_0]}\lambda_1.$$

The polynomial $P_{N+1}(z)$ has simple zeros defined in (4.15) and therefore $P'_{N+1}(z_s)$ can be calculated explicitly. The polynomial $P_N(z_s)$ is reduced to a $_8E_7$ series which can be summed with the help of the elliptic extension of the terminating Jackson sum (3.18). Omitting technical details of the computations of various entries in (7.10), we rewrite the sum (7.7) as a $_{10}E_9$ function:

$$F_N(z(u)) = K_N \frac{[u + d_2 - e_1][u + d_1 - e_1]}{[u + 1 + x_0 - e_1][u - x_0 - 1 + e_2]}$$
$$\times {}_{10}E_9(2 - x_1; 1, 1 + x_0, 1 + d_4 - x_2, 1 + d_5 - x_2, \qquad (7.11)$$
$$1 - u - x_2 + e_1, 1 + u - x_2 + e_2, -N; h, \tau),$$

where

$$K_N = \frac{[2 - x_1]}{[2 + N - x_1][1 + x_0 - d_4][1 + x_0 - d_5][d_2 - e_1][d_1 - e_1]}.$$

The formula (7.11) represents a new explicit terminating continued fraction.

It is very easy to check (7.11) for $N = 0$. For $x_2 = 1$ one gets a $_8E_7$ series and using the Frenkel-Turaev sum the fraction $F_N(z(u))$ can be written as a simple product of elliptic functions. The limit $Im(\tau) \to +\infty$ in (7.11) leads to the general class of terminating continued fractions of Gupta and Masson [10] which is known to incorporate the Ramanujan's Entry 40 and its q-analogue due to Watson.

8. Conclusions

The theory of BRF has many links to other branches of pure and applied mathematics. First of all, it is equivalent to the following theories.

(i) Theory of Cauchy-Jacobi multipoint rational interpolation. Recall (see, e.g. [28]) that within this theory one is seeking solution of the following problem: for the given function $F(z)$ and a fixed set of points $a_i, i = 1, 2, \ldots$, on the complex plane find a set of polynomials $P_n(z)$ ans $Q_n(z)$ of degrees n such that

$$F(a_i) = \frac{Q_{n-1}(a_i)}{P_n(a_i)}, \quad i = 1, 2, \ldots, 2n. \qquad (8.1)$$

It is assumed that the conditions (8.1) are fulfilled for all $n = 1, 2, \ldots$. This means that for every n we have $2n$ equations (8.1) for $2n$ unknown coefficients of the polynomials $P_n(z)$ and $Q_{n-1}(z)$. Under some natural conditions of a nondegeneracy of the problem one can show that both $P_n(z)$ and $Q_{n-1}(z)$ are determined uniquely (up to an irrelevant common constant factor) and satisfy the R_{II}-type recurrence relation (2.4). Hence one can construct the corresponding pair of BRF $R_n(z)$ and $T_n(z)$.

(ii) Theory of continued fractions of the R_{II}-type due to Ismail and Masson [12]. Indeed, the denominators $P_n(z)$ of these continued fractions satisfy (2.4) and yield a pair of BRF.

(iii) Theory of random matrices with the probability distribution of a special type. These distributions are determined by partition functions of two-dimensional Coulomb gases at a fixed temperature in the presence of an even number of external charges whose positions are fixed by the variables α_n and β_n.

As particular specializations of the BRF one can obtain the following objects.

(i) Orthogonal rational functions [5].

(ii) Polynomials of the R_I-type [12]. They correspond to the restriction $\beta_k = \beta = const$.

(iii) Laurent biorthogonal polynomials [11]. They correspond to a further specialization $\alpha_k = \alpha$, $\beta_k = \beta$, $\alpha \neq \beta$.

(iv) Polynomials orthogonal on the unit circle. These polynomials were introduced by Szegő [29] and come from a further specialization of the Laurent biorthogonal polynomials.

(v) Ordinary orthogonal polynomials.

As to the explicit classes of self-dual elliptic BRF $R_n(z)$ expressed through the modular generalizations of the hypergeometric series, they have other important features in addition to the ones mentioned above. For example, these functions are covariant with respect to the Christoffel transformation (2.9): for an appropriate choice of the parameters λ_i the function $\tilde{R}_n(z)$ is obtained from $R_n(z)$ just by a shift of its parameters. This fact enables one to consider recurrence coefficients of $R_n(z)$ as self-similar solutions of some discrete-time integrable chain [25]. For a special choice of parameters $R_n(z)$ are linked to some integrable models of statistical mechanics since they provide particular solutions of the Yang-Baxter equation [7]. Let us mention also that natural elliptic generalizations of the ordinary and q-beta integrals appear within the related set of constructions [23].

Moreover, there exists a wide class of non-self-dual elliptic BRF containing three additional continuous parameters [25]. In contrast to the elliptic

BRF described in this paper, the corresponding functions $R_n(z), T_n(z)$ have a much more complicated dependence on the argument z. In this case only the functions $R_n(z)$ are expressed through one terminating $_{10}E_9$ elliptic hypergeometric series. The companion functions $T_n(z)$ are expressed as a linear combination of two such series.

References

1. G. E. Andrews, R. Askey, and R. Roy, *Special Functions*, Cambridge University Press, Cambridge, 1999.
2. R. Askey and J. Wilson, *Some Basic Hypergeometric Orthogonal Polynomials that Generalize Jacobi Polynomials*, Mem. Amer. Math. Soc. **319** (1985).
3. R. Baxter, *Exactly Solvable Models in Statistical Mechanics*, Academic Press, London, 1982.
4. B. C. Berndt, R. L. Lamphere, and B. M. Wilson, *Chapter 12 of Ramanujan's second notebook: Continued fractions*, Rocky Mountain J. Math. **15** (1985), 235–310.
5. A. Bultheel, P. González-Vera, E. Hendriksen, and O. Njåstad, *Orthogonal Rational Functions*, Cambridge Monographs on Applied and Computational Mathematics, Vol. 5, Cambridge University Press, Cambridge, 1999.
6. J. J. Duistermaat and F. A. Grünbaum, *Differential equations in the spectral parameters*, Commun. Math. Phys. **103** (1986), 177–240.
7. I. B. Frenkel and V. G. Turaev, *Elliptic solutions of the Yang-Baxter equation and modular hypergeometric functions*, in: "The Arnold-Gelfand Mathematical Seminars", Birkhäuser, Boston, MA, 1997, pp. 171–204.
8. G. Gasper and M. Rahman, *Basic Hypergeometric Series*, Cambridge University Press, Cambridge, 1990.
9. D. P. Gupta and D. R. Masson, *Watson's basic analogue of Ramanujan's Entry 40 and its generalization*, SIAM J. Math. Anal. **25** (1994), 429–440.
10. D. P. Gupta and D. R. Masson, *Contiguous relations, continued fractions and orthogonality*, Trans. Amer. Math. Soc. **350** (1998), 769–808.
11. E. Hendriksen and H. van Rossum, *Orthogonal Laurent polynomials*, Indag. Math. (Ser. A) **48** (1986), 17–36.
12. M. E. H. Ismail and D. Masson, *Generalized orthogonality and continued fractions*, J. Approx. Theory **83** (1995), 1–40.
13. W. B. Jones and W. J. Thron, *Continued Fractions: Analytic Theory and Applications*, Addison-Wesley, Reading, MA, 1980.
14. R. Koekoek and R. F. Swarttouw, *The Askey scheme of hypergeometric orthogonal polynomials and its q-analogue*, Report 94-05, Faculty of Technical Mathematics and Informatics, Delft University of Technology, 1994.
15. D. A. Leonard, *Orthogonal polynomials, duality and association schemes*, SIAM J. Math. Anal. **13** (1982), 656–663.
16. A. Magnus, *Special non uniform lattice (SNUL) orthogonal polynomials on discrete dense sets of points*, J. Comp. Appl. Math. **65** (1995), 253–265.
17. D. R. Masson, *A generalization of Ramanujan's best theorem on continued fractions*, C.R. Math. Rep. Acad. Sci. Canada **13** (1991), 167–172.
18. A. F. Nikiforov, S. K. Suslov and V. B. Uvarov, *Classical Orthogonal Polynomials of a Discrete Variable*, Springer, Berlin, 1991.
19. M. Rahman, *An integral representation of a $_{10}\varphi_9$ and continuous biorthogonal $_{10}\varphi_9$ rational functions*, Canad. J. Math. **38** (1986), 605–618.
20. M. Rahman, *Biorthogonality of a system of rational functions with respect to a positive measure on [-1, 1]*, SIAM J. Math. Anal. **22** (1991), 1430–1441.
21. M. Rahman and S. K. Suslov, *Classical biorthogonal rational functions*, in: "Methods of Approximation Theory in Complex Analysis and Mathematical Physics", Lecture

Notes in Math. **1550**, Springer-Verlag, Berlin, 1993, pp. 131–146.

22. S. Ramanujan, *Notebooks*, Tata Institute of Fundamental Research, Bombay, 1957.
23. V. P. Spiridonov, *An elliptic beta integral*, in: *"Proceedings of the Fifth International Conference on Difference Equations and Applications"* (Temuco, Chile, January 3–7, 2000), S. Elaydi, J. Fenner-Lopez, and G. Ladas, eds, Gordon and Breach, to appear; *On an elliptic beta function*, Russian Math. Surveys **56** (2001), 181–182 (in Russian).
24. V. P. Spiridonov and A. S. Zhedanov, *Zeros and orthogonality of the Askey-Wilson polynomials for q. a root of unity*, Duke Math. J. **89** (1997), 283–305.
25. V. P. Spiridonov and A. S. Zhedanov, *Spectral transformation chains and some new biorthogonal rational functions*, Commun. Math. Phys. **210** (2000), 49–83.
26. V. P. Spiridonov and A. S. Zhedanov, *Classical biorthogonal rational functions on elliptic grids*, C.R. Math. Rep. Acad. Sci. Canada **22**, no. 2 (2000), 70–76.
27. V. P. Spiridonov and A. S. Zhedanov, *Biorthogonal rational functions on elliptic grids*, in preparation.
28. H. Stahl, *Existence and uniqueness of rational interpolants with free and prescribed poles*, in: *"Approximation Theory"*, E. B. Saff, ed., Lecture Notes in Math. **1287**, Springer-Verlag, Berlin, 1987, pp. 180–208.
29. G. Szegő, *Orthogonal Polynomials*, fourth edition, Amer. Math. Soc., 1975.
30. P. Terwilliger, *Two linear transformations each tridiagonal with respect to an eigenbasis of the other*, preprint (1999).
31. S. O. Warnaar, *Summation and transformation formulas for elliptic hypergeometric series*, preprint (2000).
32. G. N. Watson, *Ramanujan's continued fraction*, Proc. Cambridge Philos. Soc. **31** (1935), 7–17.
33. J. H. Wilkinson, *The Algebraic Eigenvalue Problem*, Clarendon Press, Oxford, 1965.
34. J. A. Wilson, Hypergeometric series, recurrence relations and some new orthogonal functions, Ph.D. thesis, University of Wisconsin, Madison, WI, 1978.
35. J. A. Wilson, *Orthogonal functions from Gram determinants*, SIAM J. Math. Anal. **22** (1991), 1147–1155.
36. A. S. Zhedanov, *Biorthogonal rational functions and generalized eigenvalue problem*, J. Approx. Theory **101** (1999), no. 2, 303–329.
37. A. S. Zhedanov and V. P. Spiridonov, *Hypergeometric biorthogonal rational functions*, Uspekhi Mat. Nauk (Russ. Math. Surveys) **54**, No. 2 (1999), 173–174 (in Russian).

ORTHOGONAL POLYNOMIALS AND COMBINATORICS

DENNIS STANTON

School of Mathematics
University of Minnesota
Minneapolis, MN 55455 USA

Abstract. An introduction is given to the use of orthogonal polynomials in distance regular graphs and enumeration. Some examples in each area are given, along with open problems.

1. Introduction

There are many ways that orthogonal polynomials can occur in combinatorics. This paper concentrates on two topics

(1) eigenvalues of distance regular graphs,
(2) generating functions for combinatorial objects.

The introduction to these two topics is well-known to the experts, and is intended for a general audience. Rather than give an exhaustive survey, specific applications which highlight the techniques of each subject are given. For (1), the application is Wilson's proof of the Erdös-Ko-Rado theorem (see §3). For (2), two applications will be given–Foata's proof of the Mehler formula for Hermite polynomials and the Kim-Zeng linearization theorem for Sheffer orthogonal polynomials. Some open problems are given along the way.

Active research topics which are not discussed include group and algebra representations, tableaux and symmetric functions.

The standard references for distance regular graphs and association schemes are [6] and [10], while [26] covers topics (1) and (2). The standard notation for hypergeometric series found in [23] is used here.

J. Bustoz et al. (eds.), Special Functions 2000, 389–409.
© 2001 *Kluwer Academic Publishers. Printed in the Netherlands.*

2. Association Schemes and Distance Regular Graphs

In classical coding theory the Krawtchouk polynomials are important for
several problems. This was generalized to polynomials in P-polynomial association schemes by Delsarte [14], and in this section we review this basic
setup.

Recall that a graph $G = (V, E)$ consists of a set V of vertices, and a set
E of unordered pairs of elements of V, called the edges of G. If the graph
G is very regular in a precise sense then associated to G is a finite sequence
of orthogonal polynomials.

The adjacency matrix A of the graph G is the $|V| \times |V|$ matrix defined
by (the rows and columns are indexed by the vertices of G)

$$A(x, y) = \begin{cases} 1 & \text{if } xy \text{ is an edge,} \\ 0 & \text{otherwise.} \end{cases}$$

We imagine the adjacency matrix as keeping track of which pairs of
vertices are distance one from each other, where the distance between two
vertices is the length of the shortest path in the graph connecting them.
This distance is finite as long as the graph is connected.

The adjacency matrix can be generalized to distance i,

$$A_i(x, y) = \begin{cases} 1 & \text{if } d(x, y) = i, \\ 0 & \text{otherwise.} \end{cases}$$

We see that $A_0 = I$, $A_1 = A$, and if the graph is connected with maximum
distance d, then $A_0 + A_1 + \cdots + A_d = J$, the all ones matrix.

We want a condition on the graph G which mimics the three-term recurrence relation

$$x p_i(x) = \alpha_i p_{i+1}(x) + \beta_i p(x) + \gamma_i p_{i-1}(x)$$

which is satisfied by any set of orthogonal polynomials. Suppose that $x, y \in V$,
with $d(x, y) = i$. If one moves distance one away from y to a vertex z,
then we have $d(x, z) = i - 1, i$ or $i + 1$.

The condition we impose on the connected graph G with maximum
distance d is distance regularity, namely the matrices A_i follow the same
rule:

$$A_i A_1 = c_{i+1} A_{i+1} + a_i A_i + b_{i-1} A_{i-1}, \quad 0 \le i \le d, \qquad (2.1)$$

for some constants c_{i+1}, a_i and b_{i-1}. If such constants exist, then they
can be found by explicit counting. For example, if we fix $x, z \in V$, with
$d(x, z) = i+1$, c_{i+1} is the number of $y \in V$ such that $d(x, y) = i, d(y, z) = 1$.

Since c_{i+1} is independent of the choice of x, z with $d(x, z) = i+1$, the graph G must be regular in a very special way, thus the term distance regular.

We now give two examples of distance regular graphs. The first is the N-dimensional cube, which is denoted $H(N, 2)$, and is referred to as the binary Hamming scheme. The vertices of $H(N, 2)$ are all N-tuples of 0's and 1's, and an edge connects two N-tuples which differ in exactly one position. The distance between two vertices is the number of positions in which they differ. In this case

$$c_{i+1} = i + 1, \quad a_i = 0, \quad b_{i-1} = N - i + 1.$$

For example to compute c_{i+1}, fix $x = (0, 0, \cdots, 0)$, and $z = (1, \cdots, 1, 0, \cdots, 0)$ where z has $i+1$ 1's so that $d(x, z) = i+1$. If $d(y, z) = 1$ and $d(x, y) = i$, then y must be obtained by switching one of the 1's in z to a 0; there are $i + 1$ choices for this switch.

The second example, $J(n, k)$, the Johnson scheme, will be considered in detail in the next section. The vertices V of $J(n, k)$ consist of all k-element subsets of the fixed n-element set $\{1, 2, \cdots, n\} = [n]$, and two subsets $\alpha, \beta \in V$ are connected by edge if $|\alpha \cap \beta| = k - 1$. In this case the distance is given by $d(\alpha, \beta) = k - |\alpha \cap \beta|$, and the constants are

$$c_{i+1} = (i + 1)^2, \quad a_i = i(n - k - i) + (k - i)i, \quad b_{i-1} = (k - i + 1)(n - i + 1).$$

This time to see that $c_{i+1} = (i + 1)^2$, fix k-subsets x and z with $|x \cap z| = k - (i+1)$. We must count how many y are distance one from z and distance i from x. To create y, we just delete a point of z from $z - x$, and add a point to z from $x - z$. This can be done in $(i + 1)^2$ ways. Note also that $|V| = \binom{n}{k}$.

These two examples have large symmetry groups–permutations of the vertices which preserve the distance in the graph–which ensure distance regularity.

It is clear from (2.1) that $A_i = p_i(A_1)$ is a polynomial of degree i in A_1, where $p_i(x)$ is an orthogonal polynomial, as long as $c_{i+1} \neq 0$ for $i + 1 \leq d$. This is the case if the graph is connected. By considering the eigenvalues of A_1 we will have a real valued polynomial instead of a matrix valued polynomial $p_i(A_1)$. The matrix A_1 is a real symmetric matrix, so A_1 has a complete set of real eigenvalues. In fact there are $d + 1$ distinct eigenvalues, $\lambda_0 > \lambda_1 > \cdots > \lambda_d$, with corresponding eigenspaces $V_0, V_1, \cdots, V_d$. Since A_i is a polynomial in A_1, the eigenvalue of A_i on V_j is $p_i(\lambda_j)$.

Note that the $d + 1$ eigenvalues are solutions to the $d + 1$ degree polynomial equation (the $i = d$ case of (2.1))

$$\lambda p_d(\lambda) = a_d p_d(\lambda) + b_{d-1} p_{d-1}(\lambda).$$

A discrete orthogonality relation can be found using the eigenspaces. Let E_j be the projection map onto the eigenspace V_j. Then we have

$$A_i = \sum_{j=0}^{d} p_i(\lambda_j) E_j.$$

If this relation is inverted, and the orthogonality of the projection maps E_j is used, the following orthogonality relations are obtained. We let v_i denote the size of a sphere of radius i.

$$\frac{1}{|V|} \sum_{n=0}^{d} p_n(\lambda_i) p_n(\lambda_j)/v_n = \delta_{ij}/\dim(V_i), \qquad (2.2a)$$

$$\frac{1}{|V|} \sum_{i=0}^{d} p_n(\lambda_i) p_m(\lambda_i) \dim(V_i) = \delta_{nm} v_n. \qquad (2.2b)$$

It is clear that (2.2b) is a discrete orthogonality relation for the finite set of orthogonal polynomials $\{p_n(x)\}_{n=0}^{d}$, while (2.2a) may or may not be a polynomial orthogonality relation.

In our two examples,

$$d = N, \quad \text{and } \lambda_j = N - 2j \quad \text{for } H(N, 2),$$

$$d = \min(k, n-k), \quad \text{and } \lambda_j = k(n-k) + j(j-n-1) \quad \text{for } J(n, k),$$

$$v_i = \binom{N}{i}, \quad \dim(V_i) = \binom{N}{i} \quad \text{for } H(N, 2),$$

$$v_i = \binom{k}{i}\binom{n-k}{i}, \quad \dim(V_i) = \binom{n}{i} - \binom{n}{i-1} \quad \text{for } J(n, k).$$

The polynomials are Krawtchouk polynomials ($p = 1/2$) and dual Hahn polynomials

$$p_i(\lambda_j) = \binom{N}{i} {}_2F_1\left(\begin{array}{cc} -i, & -j; \\ & -N \end{array} \; 2\right) \qquad (2.3a)$$

$$p_i(\lambda_j) = \binom{k}{i}\binom{n-k}{i} {}_3F_2\left(\begin{array}{ccc} -j, & j-n-1, & -i; \\ & -k, & k-n \end{array} \; 1\right). \qquad (2.3b)$$

If we put $\hat{p}_i(x) = p_i(x)/v_i$, the orthogonality relation (2.2a) becomes

$$\frac{1}{|V|} \sum_{n=0}^{d} \hat{p}_n(\lambda_i) \hat{p}_n(\lambda_j) v_n = \delta_{ij}/\dim(V_i). \qquad (2.4)$$

We see that in our two examples, $\hat{p}_i(\lambda_j)$ is a polynomial in i of degree j, so that (2.4) is also a polynomial orthogonality relation. If $\hat{p}_i(\lambda_j) = q_j(\lambda_i^*)$ for some polynomial sequence $q_j(x)$, the distance regular graph is called Q-polynomial.

The classification of distance regular graphs which are Q-polynomial is a very difficult problem. There are several infinite families (see [6], [10]) with arbitrarily large d (including $H(N,2)$ and $J(n,k)$). These families are closely related to Gelfand pairs (W, W_J) and (G, P_J), for Weyl groups W and maximal parabolic subgroups W_J, and for classical groups G over a finite field and parabolic subgroups P_J. Note that q appears here as a prime power, the order of a finite field. There is a classification theorem for orthogonal polynomials due to Leonard [6, Chap. 3], which says that any set of orthogonal polynomials whose duals (in the sense above) are orthogonal polynomials, must be special or limiting cases of the Askey-Wilson $_4\phi_3$ polynomials. Terwilliger [40] has a linear algebraic version of this theorem which hopefully will lead to an algebraic classification of the Q-polynomial distance regular graphs.

A simple application of the graph structure to the polynomials can be given by considering the multiplication

$$A_i A_j = \sum_k c_{ij}^k A_k. \tag{2.5a}$$

We know that c_{ij}^k is the number of y such that $d(x,y) = i$, $d(y,z) = j$, if x and z are fixed with $d(x,z) = k$. If each side is applied to the eigenspace V_s, we find the linearization formula

$$p_i(\lambda_s)p_j(\lambda_s) = \sum_k c_{ij}^k p_k(\lambda_s). \tag{2.5b}$$

We see from (2.5b) that the linearization coefficients are always non-negative and have a combinatorial interpretation. In example one for the Krawtchouk polynomials we have

$$c_{ij}^k = \binom{k}{(k+j-i)/2}\binom{n-k}{n-(k+i+j)/2}.$$

3. The Erdős-Ko-Rado Theorem

There are many applications of orthogonal polynomials to coding theory, see for example [15], [26]. In this section a less widely known application to extremal set theory, the Erdős-Ko-Rado theorem, will be given. It uses the eigenvalues of the $J(n,k)$ which are dual Hahn polynomials (2.3b). This proof is due to Wilson [43], with somewhat different details.

We consider the set of all k-element subsets of $[n] = \{1, 2, \cdots, n\}$; this is $J(n, k)$. We assume that $2k \leq n$, and look for subsets $\mathcal{F} \subset J(n, k)$ such that for all $A, B \in \mathcal{F}$, $|A \cap B| \geq t$. Such sets $\mathcal{F}$ are called t-intersecting.

The Erdös-Ko-Rado theorem states that

$$|\mathcal{F}| \leq \binom{n-t}{k-t} \tag{3.1}$$

as long as $n \geq n_0(k, t)$, for $n_0(k, t)$ which depends only upon k and t. Frankl [22], and then Wilson [43] found an explicit bound $n_0(k, t) = (k-t+1)(t+1)$. It is clear that the set $\mathcal{F}$ is realizable by taking all k-subsets which contain a fixed t-element subset.

If $n < n_0(k, t)$, the bound (3.1) is not correct. For values of n in this range the correct bound was conjectured by Frankl [22] and proven by Ahlswede and Khachatrian [1]. It is based upon the family of subsets

$$\mathcal{F}_i = \{A \in X_k : |A \cap [t + 2i]| \geq t + i\}$$

which clearly has the t-intersecting condition.

Theorem A [1] *If $\mathcal{F}$ is a t-intersecting family, then*

$$|\mathcal{F}| \leq \max\{|\mathcal{F}_i| : 0 \leq i \leq (n - t)/2\}.$$

If $n \geq (k - t + 1)(t + 1)$, the maximum occurs at $i = 0$. We shall give Wilson's proof of this case. We use the fact that the dual Hahn polynomials are eigenvalues of the Johnson association scheme $J(n, k)$.

Wilson's method (from Delsarte [14]) is to find an explicit matrix $\mathcal{A}$ in the span of $\{I, A, \cdots, A_d\}$ for $J(n, k)$ such that

(1) $\mathcal{A}_{\alpha\beta} = 0$ if $|\alpha \cap \beta| \geq t$,
(2) $B = I + \mathcal{A} - J/\binom{n-t}{k-t}$ is a positive semidefinite matrix.

To see that the two conditions above suffice for the proof, let $I_{\mathcal{F}}$ be the characteristic vector of the family $\mathcal{F}$. Then

$$0 \leq (I_{\mathcal{F}}, BI_{\mathcal{F}}) = |\mathcal{F}| - |\mathcal{F}|^2 / \binom{n-t}{k-t}$$

which proves that

$$|\mathcal{F}| \leq \binom{n-t}{k-t}.$$

The inequality $n \geq (k-t+1)(t+1)$ will be needed to prove that the matrix B is positive semidefinite.

To define the element $\mathcal{A}$ and thus B, we take an appropriate linear combination of the basis A_i, $0 \leq i \leq k$. Condition (1) says we should restrict to $k - t + 1 \leq i \leq k$. It is somewhat more convenient to use the basis B_j defined by

$$B_j = \sum_{s=j}^{k} \binom{s}{j} A_s. \tag{3.2}$$

Since we know the eigenvalues of each A_s as dual Hahn polynomials, we know the eigenvalues of the B_j's. However even more is true, these eigenvalues factor,

$$B_j(V_e) = (-1)^e \binom{k-e}{j-e}\binom{n-j-e}{k-e} V_e. \tag{3.3}$$

To see (3.3), just use the $_3F_2$ form of the $p_j(\lambda_e)$ given in (2.3b). The resulting double sum has two sums, each of which may be evaluated from the Chu-Vandermonde $_2F_1(1)$ evaluation.

The element $\mathcal{A}$ is defined by

$$\mathcal{A} = \sum_{i=0}^{t-1} (-1)^{t-1-i} \binom{k-1-i}{k-t}\binom{n-k-t+i}{k-t}^{-1} B_{k-i}. \tag{3.4}$$

It is clear that the non-zero matrix elements $\mathcal{A}_{\alpha\beta}$ must have $|\alpha \cap \beta| < t$, so that the first requirement for $\mathcal{A}$ is satisfied.

To prove that $B = I + \mathcal{A} - J/\binom{n-t}{k-t}$ is positive semidefinite, we check that each eigenvalue of B is non-negative. We know the eigenvalues of $\mathcal{A}$, while the eigenvalue of J is $\binom{n}{k}$ on V_0, and is 0 on V_e, $e > 0$. Thus we must show that the eigenvalue θ_e of $\mathcal{A}$ on V_e satisfies

$$\theta_e \geq -1 \quad \text{for} \quad 1 \leq e \leq k, \tag{3.5a}$$

$$\theta_0 \geq \binom{n}{k}\binom{n-t}{k-t}^{-1} - 1. \tag{3.5b}$$

Proof of (3.5). In fact much more is true; we have equality in (3.5a) for $1 \leq e \leq t$, and equality in (3.5b). We only use the inequality $n \geq (k - t + 1)(t + 1)$ to establish the inequality in (3.5a) for θ_e when $e > t$.

From (3.3) we have

$$\theta_e = (-1)^{t-1-e} \sum_{i=0}^{t-1} (-1)^i \binom{k-1-i}{k-t} \binom{k-e}{i} \binom{n-k-e+i}{k-e}$$

$$\times \binom{n-k-t+i}{k-t}^{-1} \tag{3.6}$$

$$= \frac{(-1)^e}{(t-1)!} \sum_{i=0}^{t-1} (-1)^{t-1-i} \binom{t-1}{i} (k-e-i+1)_{e-1}(n-t-k+i+1)_{t-e}.$$

We can interpret (3.6) as a $t-1$st difference $(\Delta f(x) := f(x+1) - f(x))$

$$\theta_e = \frac{(-1)^e}{(t-1)!} \Delta^{t-1} f(0), \quad f(x) = (k-e-x+1)_{e-1}(n-t-k+x+1)_{t-e}.$$

To prove (3.5a) for $1 \le e \le t$, we need only find the leading term of the polynomial $f(x)$, $(-1)^{e-1}x^{t-1}$, to see that $\theta_e = -1$.

For (3.5b), if $e = 0$,

$$f(x) = -x^{t-1} + \text{ lower order terms } + \frac{(n+1-t)_t}{k-x},$$

so that

$$\Delta^{t-1}\left(\frac{1}{k-x}\right) = \frac{(t-1)!}{(k-x-t+1)_t}$$

implies

$$\theta_0 = -1 + \frac{(n+1-t)_t}{(k-t+1)_t} = -1 + \frac{\binom{n}{k}}{\binom{n-t}{k-t}}$$

as required.

We also need the inequalities of $\theta_e \ge -1$ for $e > t$ in (3.5a). For example if we take $e = t+1$, then clearly

$$(-1)^{t+1}f(x) = -x^{t-1} + \text{ lower order terms } + \frac{(-1)^{t+1}(n-2t)_{t-1}}{n-t-k+x},$$

which again implies that θ_{t+1} is a sum of two terms just as θ_0 was

$$\theta_{t+1} = -1 + \frac{(n-2t)_t}{(n-k-t)_t} > -1$$

since $n - 2t > 0$ and $n - k - t > 0$.

If $e = t+2$,

$$(-1)^{t+2}f(x) = -x^{t-1} + \text{ lower order terms}$$

$$-\frac{(-1)^{t+2}(n-2t-1)_{t+1}}{n-t-k+x}+\frac{(-1)^{t+2}(n-2t-2)_{t+1}}{n-t-k+x-1},$$

which implies that θ_{t+2} is a sum of three terms

$$\theta_{t+2}=-1+\frac{(n-2t-1)_{t+1}}{(n-k-t)_t}-\frac{(n-2t-2)_{t+1}}{(n-k-t-1)_t}.$$

Thus

$$\begin{aligned}\theta_{t+2}&\geq-1 \quad\text{if } n\geq(k-t+1)(t+1),\\ \theta_{t+2}&=-1 \quad\text{if } n=(k-t+1)(t+1).\end{aligned}$$

The remaining inequality $\theta_e>-1$ for $e>t+2$, can be established using a $_3F_2(1)$ transformation

$$_3F_2\left(\begin{array}{ccc}a, & b, & -n;\\ & c, & 1-d-n\end{array}\,1\right)=\,_3F_2\left(\begin{array}{ccc}a, & c-b, & -n;\\ & c, & a+d\end{array}\,1\right). \tag{3.7}$$

One may write θ_e as a terminating $_3F_2$,

$$\theta_e=(-1)^{t-1-e}\binom{k-1}{k-t}\binom{n-k-e}{k-e}\binom{n-k-t}{k-t}^{-1} \tag{3.8}$$

$$\times\,_3F_2\left(\begin{array}{ccc}1-t, & e-k, & n-k-e+1;\\ & 1-k, & n-k-t+1\end{array}\,1\right)\quad\text{if } e\geq t+2.$$

Then applying (3.7) with $a=e-k$, $b=n-k-e+1$, $c=n-k-t+1$, $d=k-t+1$, we find

$$\theta_e=(-1)^{t-1-e}\frac{\binom{n-k-e}{k-e}\binom{e-1}{t-1}}{\binom{n-k-t}{k-t}}\,_3F_2\left(\begin{array}{ccc}1-t, & e-k, & e-t;\\ & e-t+1, & n-k-t+1\end{array}\,1\right)$$

$$=\frac{(e-1)!(k-t)!}{(t-1)!(e-t)!(k-e)!}\sum_{s=0}^{t-1}\frac{(1-t)_s(e-k)_s}{s!(n-k-e+1)_{e-t+s}}\frac{e-t}{e-t+s}. \tag{3.9}$$

Since we are assuming $t+2\leq e\leq k$, each term in the sum is positive, and each term is a decreasing function of n. Thus $(-1)^{t-1-e}\theta_e>0$, and $\theta_{t+2}=-1$ for $n=(k-t+1)(t+1)$ implies $0>\theta_{t+2}>-1$ for $n>(k-t+1)(t+1)$. It is enough to prove that $|\theta_{e+2+j}|<1$ if $j>0$. This can be done by induction by comparing the absolute values of the s term in (3.9) for θ_{e+j+2} and θ_{e+j+3} and concluding that the absolute values decrease as a function of j. $\qquad\Box$

The equalities in (3.5a) and (3.5b) can easily be done without appealing to the finite difference operator Δ. (3.8) is terminating zero-balanced $_3F_2$, which can be summed if $1 \leq e \leq t$, using [23, (1.9.3)]

$$_3F_2 \left(\begin{array}{ccc} -m_1 - m_2, & b_1 + m_1, & b_2 + m_2; \\ & b_1, & b_2 \end{array} \; 1 \right) = (-1)^{m_1 + m_2} \frac{(m_1 + m_2)!}{(b_1)_{m_1} (b_2)_{m_2}}$$

where m_1 and m_2 are non-negative integers. The choices $b_1 = 1 - k$, $b_2 = n - k - t + 1$, $m_1 = e - 1$, and $m_2 = t - e$ verify that $\theta_e = -1$ for $1 \leq e \leq t$ in (4.5a). A $_3F_2$ transformation also establishes the case (3.5b).

The Ahlswede-Khachatrian version of the Erdös-Ko-Rado theorem which considers $n < (k - t + 1)(t + 1)$ is below. The extremal family $\mathcal{F}_r$ that is chosen depends upon which multiple of $(k - t + 1)$ bounds n.

Theorem B [1]. *If*

$$(k - t + 1)\left(2 + \frac{t-1}{r+1}\right) < n < (k - t + 1)\left(2 + \frac{t-1}{r}\right),$$

then $\mathcal{F}_r$ is the largest t-intersecting family, up to permutations.

Note that $r = 0$ is the case that Wilson did. Their proof uses compressing techniques from extremal set theory. An eigenvalue proof along the lines of Wilson's proof is unknown.

A t-intersecting family $\mathcal{F}$ is a subset of $J(n, k)$ such that $d(A, B) \leq k - t$ for all $A, B \in \mathcal{F}$. The largest t-intersecting family can be sought for any distance regular graph. Or one could specify a set of distances, and ask for the largest possible subset which avoids these distances. Delsarte [14] did this and restated it as a linear programming problem using the orthogonal polynomials $p_i(\lambda_j)$. Wilson solves this linear programming problem in the case $n \geq (k - t + 1)(t + 1)$.

Let M be the allowed set of distances for a subset $\mathcal{F}$ of a distance regular graph. Let a_i, $i \in M$, be the average number of points in $\mathcal{F}$ at distance i,

$$a_i = \frac{1}{|\mathcal{F}|} |\{(x, y) \in \mathcal{F} \times \mathcal{F} : d(x, y) = i\}|,$$

so $a_0 = 1$, and $\sum_{i \in M} a_i = |\mathcal{F}|$.

Theorem C [14]. *For any $\mathcal{F}$,*

$$\sum_{i \in M} a_i p_i(\lambda_k) \geq 0 \text{ for } k = 1, 2, \cdots, d. \tag{3.10}$$

One then sets up a linear program using the inequalities in (3.10), the objective function is $\sum_{i \in M} a_i$, subject to $a_0 = 1$, $a_i \geq 0, i \in M - \{0\}$. For e-error correcting codes one considers the choice $M = \{0, 2e + 1, \cdots, d\}$; there is a large literature [15] on this problem in $H(N, q)$ using Krawtchouk polynomials. Many such questions remain open for the other known infinite families of distance regular graphs.

There is a beautiful theorem of Delsarte [14] relating the location of zeros of the kernel polynomials for $p_n(\lambda)$ to the existence of certain subsets of the graph. Also there is very interesting recent work of Curtin and Nomura [11] classifying spin models motivated from knot theory. Several conjectures are stated there concerning the classification of distance regular graphs with specific orthogonal polynomials.

4. Enumeration and Classical Orthogonal Polynomials

The second aspect of orthogonal polynomials in combinatorics to be discussed is enumeration. There are many ways polynomials appear, including rook theory [24] and matching theory [26]. In this section we consider examples related to the exponential formula.

One may consider a set of objects S_n, which has a natural decomposition into smaller disjoint sets $S_n = \cup_{k=0}^{n} S_{nk}$. It is natural to consider the generating function

$$p_n(x) = \sum_{k=0}^{n} |S_{nk}| x^k.$$

One may hope that the analysis related to the polynomials is closely related to the combinatorics of the finite sets S_n.

One example of this setup is to take S_n to be the set of all matchings on an n-element set, or equivalently, the set of all involutions in the symmetric group on n letters. We let S_{nk} be the set of involutions with exactly $n - 2k$ fixed points. Here we have

$$p_n(x) = \sum_{k=0}^{\lfloor n/2 \rfloor} \binom{n}{2k} ((2k - 1)(2k - 3) \cdots 1) \, x^{n-2k} = \tilde{H}_n(x),$$

which is a version of a Hermite polynomial.

Foata [18] showed that non-trivial results could result in this way, using combinatorial techniques. One of his most beautiful results, which has been very influential, is his proof [19] of the Mehler formula for Hermite polynomials. It uses the idea of an exponential generating function, which we now explain.

We consider graphs on n vertices whose vertices are labeled with the integers $1, 2, \cdots, n$. Any such graph uniquely decomposes into connected components, whose vertices are also labeled. If $\text{All}_n = 2^{\binom{n}{2}}$ is the total number of labeled graphs on n vertices, and Conn_n is the total number of connected labeled graphs on n vertices then we have

$$\text{ALL}(t) = \sum_{n=0}^{\infty} \text{All}_n \frac{t^n}{n!} = \exp(\text{CONN}(t)) = \exp\left(\sum_{k=1}^{\infty} \text{Conn}_k \frac{t^k}{k!}\right). \tag{4.1}$$

The basic reason (4.1) holds is the following: if the coefficient of t^n is found on the right side, a sum of multinomial coefficients appears. Each multinomial coefficient counts how many ways there are to shuffle the labels amongst the vertices of the connected components.

Moreover the relation $\text{ALL}(t) = \exp(\text{CONN}(t))$ holds in a weighted version, weights may be attached to the connected components. For example, if we take involutions, the connected components are singleton points and single edges. If we weight each singleton by x and each edge by 1, the exponential generating function of the connected components is $xt + t^2/2$, and we have the generating function for a rescaled version of the Hermite polynomials

$$\text{ALLINV}(t) = \exp(xt + t^2/2) = \sum_{n=0}^{\infty} \hat{H}_n(x) \frac{t^n}{n!}.$$

We next give Foata's proof [19] of the Mehler formula

$$\sum_{n=0}^{\infty} \hat{H}_n(x)\hat{H}_n(y)\frac{t^n}{n!} = \frac{1}{\sqrt{1-t^2}} \exp\left(\frac{2txy + t^2(x^2+y^2)}{2(1-t^2)}\right). \tag{4.2}$$

Proof. We consider the left side as $\text{ALL}(t)$, where we have pairs of involutions (σ, τ) each on n letters. We think of the edges of σ as colored blue and those of τ as colored red. Vertices that are blue fixed points have weight x, while red fixed points have weight y. Any edge has weight 1. For example if $n = 11$,

$$\sigma = (18)(2\ 11)(34)(5)(6)(7\ 10)(9),$$
$$\tau = (13)(2)(48)(59)(6\ 10)(7)(11),$$

then $\text{wt}(\sigma, \tau) = x^3 y^3$.

Suppose that we draw σ and τ on a single diagram. What are the resulting connected components? There are four possibilities:

(1) an even cycle with edges alternating red-blue,
(2) a path of even length, with edges alternating red-blue,

(3) a path of odd length, with edges alternating red-blue, red fixed points at both ends,

(4) a path of odd length, with edges alternating red-blue, blue fixed points at both ends.

In our example these four possibilities all occur:

$$3 \xrightarrow{\text{blue}} 4 \xrightarrow{\text{red}} 8 \xrightarrow{\text{blue}} 1 \xrightarrow{\text{red}} 3,$$

$$7 \xrightarrow{\text{blue}} 10 \xrightarrow{\text{red}} 6,$$

$$2 \xrightarrow{\text{blue}} 11,$$

$$5 \xrightarrow{\text{red}} 9.$$

If the labeled cycle in (1) has $2n$ vertices, then it has weight t^{2n}. The number of such cycles is $(2n-1)!$, so in this case the exponential generating function is

$$1 + \sum_{n=1}^{\infty} (2n-1)! \frac{t^{2n}}{(2n)!} = 1 + \sum_{n=1}^{\infty} \frac{t^{2n}}{2n} = -\frac{1}{2} \log(1 - t^2).$$

Upon exponentiating we arrive at the factor $1/\sqrt{1 - t^2}$ in (4.2).

There are $(2n)!$ labeled paths of length $2n$ of type (2) each with weight xyt^{2n+1}. So the exponential generating function is

$$xy \sum_{n=1}^{\infty} (2n)! \frac{t^{2n+1}}{(2n)!} = xyt/(1 - t^2).$$

This is the first term in the exponential of (4.2).

There are $(2n)!/2$ labeled paths of length $2n - 1$ of type (3) each with weight $x^2 t^{2n}$. So the exponential generating function is

$$x^2 \sum_{n=1}^{\infty} (2n)! \frac{t^{2n}}{2(2n)!} = t^2 x^2 / 2(1 - t^2).$$

This is the second term in the exponential of (4.2).

There are $(2n)!/2$ labeled paths of length $2n - 1$ of type (4) each with weight $y^2 t^{2n}$. So the exponential generating function is

$$y^2 \sum_{n=1}^{\infty} (2n)! \frac{t^{2n}}{2(2n)!} = t^2 y^2 / 2(1 - t^2).$$

This is the third and last term in the exponential of (4.2).   □

This proof led to a combinatorial study of other classical polynomials: for example Laguerre, Jacobi, Meixner, [7], [12], [35], [36], [41], and to generalized versions of the exponential formula [8]. Foata and Garsia [20] generalized this proof to multilinear generating functions for Hermite polynomials.

The Mehler formula is an example of a bilinear generating function, which naturally gives an integral evaluation. For a general set of orthogonal polynomials $p_n(x)$ with measure $d\mu(x)$, if $h_n = 1/\|p_n\|^2$, and the bilinear generating function is

$$P(x, y, t) = \sum_{n=0}^{\infty} h_n p_n(x) p_n(y) t^n,$$

then

$$\int_{-\infty}^{\infty} P(x, y, t) P(x, z, w) d\mu(x) = P(y, z, tw). \qquad (4.3)$$

This was noted by Bowman [9], who realized that for the q-Hermite polynomials (4.3) becomes the Askey-Wilson integral!

The q-Hermite polynomials may be defined by the generating function

$$(q\text{-Hermite GF}) \quad \sum_{n=0}^{\infty} H_n(x|q) \frac{t^n}{(q; q)_n} = \frac{1}{(te^{i\theta}, te^{-i\theta}; q)_\infty}, \quad x = \cos\theta,$$

and satisfy the orthogonality relation

$$\frac{(q; q)_\infty}{2\pi} \int_0^\pi H_n(\cos\theta|q) H(\cos\theta|q)(e^{2i\theta}, e^{-2i\theta}; q)_\infty d\theta = \delta_{mn}(q; q)_n.$$

The q-Mehler formula is ($x = \cos\theta, y = \cos\phi$)

$$(4.4)$$

$$\sum_{n=0}^{\infty} H_n(x|q) H_n(y|q) \frac{t^n}{(q; q)_n} = \frac{(t^2; q)_\infty}{(te^{i(\theta+\phi)}, te^{-i(\theta+\phi)}, te^{i(\theta-\phi)}, te^{i(-\theta+\phi)}; q)_\infty}.$$

In this case (4.3) becomes, if $z = \cos\gamma$, $a = te^{i\phi}$, $b = te^{-i\phi}$, $c = te^{i\gamma}$, $d = te^{-i\gamma}$

$$\frac{(q; q)_\infty}{2\pi} \int_0^\pi \frac{(e^{2i\theta}, e^{-2i\theta}; q)_\infty}{(ae^{i\theta}, ae^{-i\theta}, be^{i\theta}, be^{-i\theta}, ce^{i\theta}, ce^{-i\theta}, de^{i\theta}, de^{-i\theta}; q)_\infty} \qquad (4.5)$$

$$= \frac{(abcd; q)_\infty}{(ab, ac, ad, bc, bd, cd; q)_\infty},$$

the Askey-Wilson integral.

One may ask if a q-version of Foata's proof of the Mehler's proves (4.4). One can interpret the q-Hermite polynomials as generating functions of involutions with a q-statistic (see [25] and [32]). A combinatorial version of the q-exponential formula [25] establishes the q-Hermite generating function, although it does not prove the q-Mehler formula (4.4). Hung Ngo [38] found a Foata-style proof, although the q-analogue of the "overlay" map is more complicated. New multilinear versions of the q-Mehler formula (see [31] for what is known) should be found; they presumably lead to a multivariable Askey-Wilson integral, perhaps one due to Gustafson [28]. This may be related to Rogers-Ramanujan identities of higher rank [2]. A combinatorial proof of the Askey-Wilson integral (4.5) exists (see [32]).

A combinatorial proof of an equivalent form of (4.4) by counting subspaces of a finite field of order q exists, we review it here.

Let

$$h_n(x) = \sum_{k=0}^{n} \begin{bmatrix} n \\ k \end{bmatrix}_q x^k$$

be the generating function for the number of k-dimensional subspaces in an n-dimensional vector space over a finite field of order q. The q-Mehler formula (4.4) may be reformulated as

$$\sum_{n=0}^{\infty} h_n(x) h_n(y) \frac{r^n}{(q;q)_n} = \frac{(xyr^2;q)_\infty}{(r, xr, yr, xyr; q)_\infty}. \tag{4.6}$$

Clearly $h_n(x) h_n(y)$ is the generating function for pairs of subspaces, according to their respective dimensions. By choosing first the intersection of these two spaces to be k-dimensional, and then extending the intersection to each space we have

$$h_n(x) h_n(y) = \sum_k \begin{bmatrix} n \\ k \end{bmatrix}_q (xy)^k \sum_{j,l} \begin{bmatrix} n-k \\ j \end{bmatrix}_q x^j \begin{bmatrix} n-k-j \\ l \end{bmatrix}_q q^{lj} y^l. \tag{4.7}$$

Then three applications of the q-binomial theorem show that (4.7) implies (4.6).

Another fundamental generating function whose combinatorics is not understood is (see [30] for references)

$$\sum_{n=0}^{\infty} q^{n^2/4} H_n(x|q) \frac{t^n}{(q;q)_n} = (qt^2;q^2)_\infty \mathcal{E}_q(x;t), \tag{4.8}$$

where $\mathcal{E}_q(x;t)$ is the quadratic q-exponential function,

$$\mathcal{E}_q(x;t) = \frac{(t^2;q^2)_\infty}{(qt^2;q^2)_\infty} \sum_{n=0}^{\infty} \left(\prod_{j=0}^{n-1} (1 + 2iq^{(1-n)/2+j}x - q^{1-n+2j}) \right) q^{n^2/4} \frac{(-it)^n}{(q;q)_n},$$

which satisfies

$$\lim_{q \to 1} \mathcal{E}_q(x; t(1-q)/2) = e^{xt}.$$

Equation (4.8) is another q-version of the Hermite generating function $exp(xt - t^2/2)$. $\mathcal{E}_q(x; t)$ appears in Suslov's addition theorem [39] which generalizes $e^{tx}e^{ty} = e^{t(x+y)}$ and deserves further study (see also [30]).

5. Enumeration and General Orthogonal Polynomials

A combinatorial model for general orthogonal polynomials was given by Viennot [41]. The key combinatorial structure is a weighted lattice path, and such paths have been extensively studied. In this section we state the initial results and give some examples of the Viennot theory.

We assume that the orthogonal polynomial sequence $p_n(x)$ is monic, so the fundamental three-term recurrence relation takes the form

$$p_{n+1}(x) = (x - b_n)p_n(x) - \lambda_n p_{n-1}(x), \quad p_{-1}(x) = 0, \quad p_0(x) = 1. \tag{5.1}$$

(5.1) allows for a simple inductive interpretation for $p_n(x)$. We consider all lattice paths starting at the origin $(0,0)$, with three types of steps

(1) northeast $(=(1,1))$ blue edges,
(2) north $(=(1,0))$ red edges,
(3) north-north $(=(2,0))$ black edges.

An example of such a lattice path is

$$(0,0) \xrightarrow{\text{red}} (1,0) \xrightarrow{\text{blue}} (2,1) \xrightarrow{\text{black}} (4,1) \xrightarrow{\text{black}} (6,1) \xrightarrow{\text{red}} (7,1).$$

Basically we see that there are three ways for a path to terminate at y-coordinate $n + 1$; these are the three terms on the right side of (5.1). If we weight a blue northeast edge by x, a red north edge from y-coordinate n to y coordinate $n + 1$ by $-b_n$, and a black north-north edge from y-coordinate $n - 1$ to y coordinate $n + 1$ by $-\lambda_n$, then (5.1) shows that $p_{n+1}(x)$ is the generating function for all paths from $(0,0)$ to the line $y = n + 1$. The weight of the lattice path in the example is $(-b_0)(x)(-\lambda_3)(-\lambda_5)(-b_6)$. This tautological combinatorial interpretation for $p_n(x)$ always applies, but it does not necessarily give the "leanest" combinatorial interpretation.

An orthogonality relation is replaced by a moment functional $L(x^n) = \mu_n$. Since a measure may not be uniquely determined, but the moments are determined, this is appropriate. The moments must be given in terms of the recurrence coefficients b_n and λ_n, for example,

$$\mu_0 = 1, \quad \mu_1 = b_0, \quad \mu_2 = b_0^2 + \lambda_1, \quad \mu_3 = b_0^3 + 2b_0\lambda_1 + b_1\lambda_1.$$

In fact as a polynomial in $\vec{b}$ and $\vec{\lambda}$, each μ_n has coefficients which are always positive, and the precise terms which appear can be identified. To do so requires a special type of lattice path in the plane.

A Motzkin path P is a lattice path in the plane starting at $(0,0)$, and ending at $(n,0)$ which stays at or above the x-axis, whose individual steps are either $(1,0)$ (east), $(1,1)$ (northeast), or $(1,-1)$ (southeast). The weight of a Motzkin path is the product of the weights of the individual steps:

$$wt((i,k) \to (i+1,k)) = b_k,$$
$$wt((i,k) \to (i+1,k+1)) = 1,$$
$$wt((i,k) \to (i+1,k-1)) = \lambda_k.$$

Theorem D Viennot [41].

$$\mu_n = \sum_{P:(0,0) \overset{\text{Motzkin}}{\longrightarrow} (n,0)} wt(P).$$

As an example, for μ_3 there are four Motzkin paths:

$$P_1 \; : \; (0,0) \to (1,0) \to (2,0) \to (3,0), \quad wt(P_1) = b_0^3,$$
$$P_2 \; : \; (0,0) \to (1,0) \to (2,1) \to (3,0), \quad wt(P_1) = b_0\lambda_1,$$
$$P_3 \; : \; (0,0) \to (1,1) \to (2,1) \to (3,0), \quad wt(P_1) = b_1\lambda_1,$$
$$P_4 \; : \; (0,0) \to (1,1) \to (2,0) \to (3,0), \quad wt(P_1) = b_0\lambda_1,$$

which gives $\mu_3 = b_0^3 + 2b_0\lambda_1 + b_1\lambda_1$.

Theorem D allows one to combinatorially evaluate integrals if one happens to know a representing measure $d\mu(x)$ for the polynomials. For example, the Askey-Wilson integral (4.5) can be evaluated [32] using these lattice paths for the q-Hermite polynomials. The idea is that the combinatorics of the paths replaces the analysis of integration. Viennot's theory was heavily influenced by Flajolet [16],[17].

The linearization coefficients $a(m,n,k)$ are defined by

$$p_n(x)p_m(x) = \sum_{k=|m-n|}^{m+n} a(m,n,k)p_k(x),$$

which is equivalent to

$$a(m,n,k) = \frac{L(p_m p_n p_k)}{L(p_k p_k)}.$$

There has been much combinatorial work on the interpretation of $L(\prod_{i=1}^{k} \times p_{n_i}(x))$, which thereby includes $a(m, n, k)$. For example, for the Hermite polynomials, by completing the square in the integral of the product of k generating functions, one finds that

$$\sum_{n_1,\cdots,n_k \geq 0} L\left(\prod_{i=1}^{k} \hat{H}_{n_i}(x)\right) \frac{t_1^{n_1}}{n_1!} \cdots \frac{t_k^{n_k}}{n_k!} = \exp(e_2(t_1, \cdots, t_k))$$

where $e_2(t_1, \cdots, t_k)$ is the elementary symmetric function of degree 2. These are connected components for the complete matchings of a graph with k types of vertices, where the edges must connect vertices of different types. We call these matchings "inhomogeneous." Thus $L(\prod_{i=1}^{k} \hat{H}_{n_i}(x))$ is the number of complete inhomogeneous matchings of the complete graph $K_{n_1+\cdots+n_k}$, where any element of $\{n_1 + \cdots + n_{i-1} + 1, \cdots, n_1 + \cdots + n_i\}$ is a vertex of type i. This is a theorem of Azor, Gillis, and Victor [4]; it was proven using the Viennot machinery in [13]. Other versions exist, for Legendre [27], Laguerre [3], [21], Meixner [12], Krawtchouk, Charlier, and Meixner-Pollaczek [44] polynomials (conspicuously missing from this list are the ultraspherical polynomials, and Rahman's result for the Jacobi polynomials), but recently a unified theorem has been given by Kim and Zeng [33].

They consider Sheffer orthogonal polynomials, those that satisfy the recurrence relation

$$p_{n+1}(x) = (x - (ab + u_3 n + u_4 n))p_n(x) - n(b + n - 1)u_1 u_2 p_{n-1}(x).$$
$$(5.2)$$

Note that if $a = b = u_1 = u_2 = u_3 = u_4 = 1$, these are just the Laguerre polynomials $L_n(x)$, whose measure is $e^{-x}dx$ on $[0, \infty)$ and whose nth moment $\mu_n = n!$. For the polynomials in (5.2), μ_n is a polynomial in the parameters a, b, u_1, u_2, u_3, u_4 whose coefficients sum to $n!$. μ_n can be given [33] as a generating function of permutations according to fixed points, cycles, double ascents, double descents, peaks, and valleys,

$$\mu_n = \sum_{\pi \in S_n} u_1^{\#\text{peak}(\pi)} u_2^{\#\text{valley}(\pi)} u_3^{\#\text{double ascent}(\pi)} u_4^{\#\text{double descent}(\pi)}$$
$$\times a^{\#\text{fix point}(\pi)} b^{\#\text{cycles}(\pi)}.$$

The Kim-Zeng theorem combinatorially interprets the integral of a product of these polynomials. They define a permutation analogue of the inhomogeneous matchings which appear in the Hermite result. Let $[i, j]$ denote the closed interval of integers from i to j. Let $N = n_1 + \cdots + n_k$,

$$A_1 = [1, n_1], \quad A_i = [n_1 + \cdots + n_{i-1} + 1, n_1 + \cdots + n_i], \quad \text{for } 2 \leq i \leq k.$$

We consider the set D_N of all $N!$ permutations of $A_1 \cup A_2 \cup \cdots \cup A_k$ such that

$$\pi(A_i) \cap A_i = \emptyset$$

for all i. The set D_N can be considered as a set of generalized derangements.

Theorem E Kim-Zeng [33]. *If $p_n(x)$ are the Sheffer polynomials given by (5.2), then*

$$L\left(\prod_{i=1}^{k} p_{n_i}(x)\right) = \sum_{\pi \in D_N} u_1^{\#\text{peak}(\pi)+\text{mat}(\pi)} u_2^{\#\text{valley}(\pi)+\text{mat}(\pi)}$$

$$\times u_3^{\#\text{double ascent}(\pi)-\text{mat}(\pi)} u_4^{\#\text{double descent}(\pi)-\text{mat}(\pi)} b^{\#\text{cycles}(\pi)},$$

where $\text{mat}(\pi)$ is the number of color matches of π (see [33]).

Theorem E specializes to the known results for the classical orthogonal polynomials, and establishes the positivity of the coefficients.

By and large, the exact q-analogues of the classical results for $L(\prod_{i=1}^{k} \times p_{n_i}(x))$ are not understood. One would hope for just a weighted version of the Kim-Zeng result, but for the natural choices, this is not the case [33]. Only the q-Hermite polynomials are known to have such results [32].

The orthogonal polynomial framework can suggest new results in enumeration. Here is one example involving the Stirling numbers of the second kind $S(n,k)$, which count the number of set partitions of $[n]$ into k blocks. The q-Stirling numbers of the second kind $S_q(n,k)$ are polynomials in q, with positive integer coefficients which sum to $S(n,k)$. Thus it is natural to consider $S_q(n,k)$ as a generating function for a statistic on set partitions of $[n]$ into k blocks. This was done by Milne [37]. However, since $S_q(n,k)$ appear naturally in the moments of the q-Charlier polynomials

$$\mu_n = \sum_{k=1}^{n} S_q(n,k)a^k,$$

the Viennot theory offers its own statistic, which is not obviously q-Stirling distributed. Considering these two statistics led Wachs and White [42] to new bivariate equidistribution theorems for these two statistics.

There are also many relations to continued fractions [16], [17] and determinants, via the moment generating function [29] and Hankel determinants [34]. Because the entries of the determinants are weighted paths, and such determinants are combinatorially understood from the Gessel-Viennot theory, this has led to widespread work.

References

1. Ahlswede and L. Khachatrian, *The complete intersection theorem for systems of finite sets*, Eur. J. Comb. **18** (1997), 125–136

2. G. Andrews, A. Schilling, and O. Warnaar, *An A_2 Bailey lemma and Rogers-Ramanujan-type identities*, J. Amer. Math. Soc. **12** (1999), 677–702.

3. R. Askey, M. E. H. Ismail, and T. Koornwinder, *Weighted permutation problems and Laguerre polynomials*, J. Comb. Th. A **25** (1978), 277–287.

4. R. Azor, J. Gillis, and J. Victor, *Combinatorial applications of Hermite polynomials*, SIAM J. Math. Anal. **13** (1982), 879–890.

5. E. Bannai, *Algebraic combinatorics—recent topics in association schemes*, Sugaku **45** (1993), 55–75.

6. E. Bannai and T. Ito, *Algebraic combinatorics I. Association schemes*, Benjamin/Cummings, 1984.

7. F. Bergeron, *Combinatoire des polynômes orthogonaux classiques: une approche unifiée*, Eur. J. Comb. **11** (1990), 393–401.

8. F. Bergeron, G. Labelle, and P. Leroux, *Combinatorial species and tree-like structures*, Encyclopedia of Mathematics and its Applications, Vol. 67, Cambridge Univ. Press, 1998.

9. D. Bowman, *An easy proof of the Askey-Wilson integral and applications of the method*, Math. Anal. Appl. **245** (2000), 560–569.

10. A. Brouwer, A. Cohen, and A. Neumaier, *Distance-regular graphs*, Ergebnisse der Mathematik und ihrer Grenzgebiete (3), vol. 18, Springer, 1989.

11. B. Curtin and K. Nomura, *Some formulas for spin models on distance-regular graphs*, J. Comb. Th. B **75** (1999), 206–236.

12. A. de Medicis, *Combinatorics of Meixner polynomials: linearization coefficients*, Eur. J. Comb. **19** (1998), 355–367.

13. M. de Sainte-Catherine and X. Viennot, *Combinatorial interpretation of integrals of products of Hermite, Laguerre and Tchebycheff polynomials*, Lecture Notes in Math., vol. 1171, Springer, 1985, 120–128.

14. Ph. Delsarte, *An algebraic approach to the association schemes of coding theory*, Philips Res. Rep. Suppl. No. 10, (1973), 1–97.

15. Ph. Delsarte and V. Levenshtein, *Association schemes and coding theory*, IEEE Trans. Inform. Th. **44** (1998), 2477–2504.

16. P. Flajolet, *Combinatorial aspects of continued fractions*, Disc. Math. **32** (1980), 125–161.

17. P. Flajolet, *Combinatorial aspects of continued fractions, part II*, Ann. Disc. Math. **9** (1980), 217–222.

18. D. Foata, *Combinatoire des identités sur les polynômes orthogonaux*, in: "Proceedings of the International Congress of Mathematicians", PWN, Warsaw, 1984, pp. 1541–1553.

19. D. Foata, *A combinatorial proof of the Mehler formula*, J. Comb. Th. Ser. A **24** (1978), 367–376.

20. D. Foata and A. Garsia, *A combinatorial approach to the Mehler formulas for Hermite polynomials*, Proc. Sympos. Pure Math. **34** (1979), 163–179.

21. D. Foata and D. Zeilberger, *Laguerre polynomials, weighted derangements and positivity*, SIAM J. Disc. Math. **1** (1988), 425–433.

22. P. Frankl, *The Erdös-Ko-Rado theorem is true for $n = ckt$*, Colloq. Math. Soc. Janos Bolyai **18** (1976), 365–375.

23. G. Gasper and M. Rahman, *Basic Hypergeometric Series*, Cambridge University Press, Cambridge, 1990.

24. I. Gessel, *Generalized rook polynomials and orthogonal polynomials*, in: *q-Series and partitions*, IMA Vol., in: "Mathematics and Its Applications", Vol. 18, 1989.

25. I. Gessel, *A q-analog of the exponential formula*, Disc. Math. **40** (1982), 69–80.

26. C. Godsil, *Algebraic Combinatorics*, Chapman and Hall, New York, 1993.

27. J. Gillis, J. Jedwab, and D. Zeilberger, *A combinatorial interpretation of the integral of the product of Legendre polynomials*, SIAM J. Math. Anal. **19** (1988), 1455–1461.
28. R. Gustafson, *Some q-beta integrals on SU(n) and Sp(n) that generalize the Askey-Wilson and Nasrallah-Rahman integrals*, SIAM J. Math. Anal. **25** (1994), 441–449.
29. G. Han, A. Randrianarivony, and J. Zeng, *Un autre q-analogue des nombres d'Euler*, Sem. Lothar. Combin. **B42e** (1999), 22 pp.
30. M. Ismail and D. Stanton, *Addition theorems for the q-exponential function*, Contemp. Math. **254** (2000), 235–245.
31. M. Ismail and D. Stanton, *On the Askey-Wilson and Rogers polynomials*, Can. J. Math. **40** (1988), 1025–1045.
32. M. Ismail, D. Stanton, and G. Viennot, *The combinatorics of q-Hermite polynomials and the Askey-Wilson integral*, Eur. J. Comb. **8** (1987), 379–392.
33. D. Kim and J. Zeng, *A combinatorial formula for the linearization coefficients of general Sheffer polynomials*, Eur. J. Comb., to appear.
34. C. Krattenthaler, *Advanced Determinant Calculus*, Séminaire Lotharingien de Combinatoire **B42q** (1999), 67 pp.
35. J. Labelle and Y. Yeh, *Combinatorial proofs of some limit formulas involving orthogonal polynomials*, Disc. Math. **79** (1989), 77–93.
36. P. Leroux and V. Strehl, *Jacobi polynomials: combinatorics of the basic identities*, Disc. Math. **57** (1985), 167–187.
37. S. Milne, *Restricted growth functions, rank row matchings of partition lattices, and q-Stirling numbers*, Advances in Math. **43** (1982), 173–196.
38. Hung Ngo, personal communication, 2000
39. S. Suslov, *Addition theorems for some q-exponential and trigonometric functions*, Methods Appl. Anal. **4** (1997), 11–32.
40. P. Terwilliger, *Two linear transformations each tridiagonal with respect to an eigenbasis of the other*, preprint 1999.
41. G. Viennot, *Une theorie combinatoire des polynomes orthogonaux generaux*, Notes de conference donnes a l'UQAM, 1983.
42. M. Wachs and D. White, *p, q-Stirling numbers and set partition statistics*, J. Comb. Th. A **56** (1991), 27–46.
43. R. Wilson, *The exact bound in the Erdös-Ko-Rado theorem*, Combinatorica 4 (1984), 247–257.
44. J. Zeng, *Linearisation de produits de polynômes de Meixner, Krawtchouk, et Charlier*, SIAM J. Math. Anal. **21** (1990), 1349–1368.

BASIC EXPONENTIAL FUNCTIONS ON A q-QUADRATIC GRID

SERGEI K. SUSLOV
Department of Mathematics
Arizona State University
Tempe, AZ 85287-1804 USA

Dedicated to Joaquin Bustoz on his 60th birthday

Abstract. A review of basic exponential functions, basic trigonometric functions, and basic Fourier series on a q-quadratic grid is given.

1. Introduction

In this paper a review of basic exponential functions and basic trigonometric functions on a q-quadratic grid will be given. A few q-analogs of the exponential and trigonometric functions are known. The old, F. H. Jackson q-analogs of the exponential function,

$$e_q(x) := \sum_{n=0}^{\infty} \frac{x^n}{(q;q)_n} = \frac{1}{(x;q)_\infty}, \quad |x| < 1, \tag{1.1}$$

$$E_q(x) := \sum_{n=0}^{\infty} \frac{q^{n(n-1)/2}\, x^n}{(q;q)_n} = (-x;q)_\infty, \tag{1.2}$$

were introduced at the beginning of the last century. (Summation formulas are due to Euler.) They were studied in detail and have been found useful in many applications. See papers by Koornwinder and Swarttouw [43], Kalnins and Miller [36]–[37], Floreanini and Vinet [17], Koornwinder [42], and references therein.

J. Bustoz et al. (eds.), Special Functions 2000, 411–456.

Jackson's q-exponential functions are very special cases of the basic hypergeometric series defined by

$$
\begin{aligned}
& {}_r\varphi_s\left(\begin{array}{c} a_1, a_2, ..., a_r \\ b_1, b_2, ..., b_s \end{array}; q, t\right) \\
& = \sum_{n=0}^{\infty} \frac{(a_1, a_2, \ldots, a_r; q)_n}{(q, b_1, b_2, \ldots, b_s; q)_n} \left((-1)^n q^{n(n-1)/2}\right)^{1+s-r} t^n,
\end{aligned}
\tag{1.3}
$$

where the standard notations for the q-shifted factorials are

$$
(a; q)_n := \prod_{k=0}^{n-1}(1 - aq^k), \qquad (a_1, a_2, \ldots, a_r; q)_n := \prod_{k=1}^{r}(a_k; q)_n,
\tag{1.4}
$$

$$
(a; q)_\infty := \lim_{n\to\infty}(a; q)_n, \qquad (a_1, a_2, \ldots, a_r; q)_\infty := \prod_{k=1}^{r}(a_k; q)_\infty,
\tag{1.5}
$$

$$
(a; q)_\alpha := \frac{(a; q)_\infty}{(aq^\alpha; q)_\infty}, \qquad (a_1, a_2, \ldots, a_r; q)_\alpha := \prod_{k=1}^{r}(a_k; q)_\alpha,
\tag{1.6}
$$

provided $|q| < 1$. For an excellent account on the theory of basic hypergeometric series see [22]. See also Ismail's lectures at this Advanced Study Institute [28].

Some new q-exponential and q-trigonometric functions have been introduced in [7] and [35]. They have been under intense investigation during the last few years and appear from the different contexts. Basic analogs of the expansion formulas of the plane wave were found by Ismail and Zhang [35], Ismail, Rahman and Zhang [33], Floreanini, LeTourneux and Vinet [20]–[21], and Ismail, Rahman and Stanton [32].

"Addition" theorems for the basic exponential and trigonometric functions were established by Suslov [57], [61] and by Ismail and Stanton [34] who also evaluated several integrals involving these basic exponential functions. Floreanini, LeTourneux and Vinet gave an algebraic proof of Suslov's addition formula [20].

Bustoz and Suslov [10] have established the orthogonality property with respect to an absolutely continuous measure and introduced the corresponding basic analog of Fourier series. Numerical investigation of these series have been recently started by Gosper and Suslov [23] with the help of the Macsyma computer algebra system. Explicit expansions of many elementary and q-functions in basic Fourier series are found in [60].

In this paper we intend to lay a sound foundation for this study. In Section 2, we introduce basic exponential functions on a q-quadratic grid.

"Addition" theorems are discussed in Section 3. Section 4 deals with the q-trigonometric functions. Some expansions and integrals due to Ismail and Stanton are discussed in Section 5. In Section 6, we give an elementary proof of the orthogonality property of the q-trigonometric system using a special case of the integral evaluated by Ismail and Stanton. An important q-analog of the expansion of the plane wave in the ultraspherical polynomials, which is due to Ismail and Zhang, is discussed in Section 7. A proof of the completeness of q-trigonomenric systems on the basis of the Ismail–Zhang formula and the q-Lommel polynomials is given in Section 8. A short review of basic Fourier series is presented in Section 9. Extensions of the Bernoulli polynomials and Riemann zeta function are introduced in Sections 10 and 11. Some interesting open problems and further extensions of the theory are discussed in Section 12.

2. Basic Exponential Functions on a q-Quadratic Grid

The basic exponential function on a q-quadratic grid can be introduced as

$$
\begin{aligned}
\mathcal{E}_q\left(x, y; \alpha\right) &= \frac{\left(\alpha^2; q^2\right)_\infty}{\left(q\alpha^2; q^2\right)_\infty} \\
&\times \sum_{n=0}^{\infty} \frac{q^{n^2/4}\, \alpha^n}{(q; q)_n} e^{-in\varphi}\left(-q^{(1-n)/2}e^{i\theta+i\varphi}, -q^{(1-n)/2}e^{i\varphi-i\theta}; q\right)_n \\
&= \frac{\left(\alpha^2; q^2\right)_\infty}{\left(q\alpha^2; q^2\right)_\infty} \\
&\times \left[{}_4\varphi_3\left(\begin{array}{c} -q^{1/2}e^{i\theta+i\varphi}, -q^{1/2}e^{i\theta-i\varphi}, -q^{1/2}e^{i\varphi-i\theta}, -q^{1/2}e^{-i\theta-i\varphi} \\ -q, \quad q^{1/2}, \quad -q^{1/2} \end{array}; q, \alpha^2\right) \right. \\
&\quad + \frac{2q^{1/4}}{1-q}\alpha\left(\cos\theta + \cos\varphi\right) \\
&\quad \left. \times\ {}_4\varphi_3\left(\begin{array}{c} -qe^{i\theta+i\varphi}, -qe^{i\theta-i\varphi}, -qe^{i\varphi-i\theta}, -qe^{-i\theta-i\varphi} \\ -q, \quad q^{3/2}, \quad -q^{3/2} \end{array}; q, \alpha^2\right)\right].
\end{aligned}
\tag{2.1}
$$

Here $x = \cos\theta$, $y = \cos\varphi$ and $|\alpha| < 1$. Analytic continuation in a larger domain will be discussed in the next section. The function $\mathcal{E}_q\left(x, y; \alpha\right)$ is an analog of $\exp\alpha\left(x + y\right)$,

$$
\lim_{q\to 1^-} \mathcal{E}_q\left(x, y; \alpha\left(1 - q\right)/2\right) = \exp\alpha\left(x + y\right).
\tag{2.2}
$$

We also introduce

$$
\mathcal{E}_q\left(x; \alpha\right) = \mathcal{E}_q\left(x, 0; \alpha\right)
\tag{2.3}
$$

$$
= \frac{(\alpha^2; q^2)_\infty}{(q\alpha^2; q^2)_\infty} \sum_{n=0}^\infty \frac{q^{n^2/4} \alpha^n}{(q;q)_n} (-i)^n \left(-iq^{(1-n)/2} e^{i\theta}, -iq^{(1-n)/2} e^{-i\theta}; q \right)_n
$$

$$
= \frac{(\alpha^2; q^2)_\infty}{(q\alpha^2; q^2)_\infty} \left[{}_2\varphi_1 \left(\begin{array}{c} -qe^{2i\theta}, -qe^{-2i\theta} \\ q \end{array} ; q^2, \alpha^2 \right) \right.
$$

$$
\left. + \frac{2q^{1/4}}{1-q} \alpha \cos\theta \; {}_2\varphi_1 \left(\begin{array}{c} -q^2 e^{2i\theta}, -q^2 e^{-2i\theta} \\ q^3 \end{array} ; q^2, \alpha^2 \right) \right]
$$

as the q-analog of $\exp(\alpha x)$. The following simple properties hold

$$
\mathcal{E}_q(0,0;\alpha) = \mathcal{E}_q(0;\alpha) = 1, \tag{2.4}
$$

$$
\mathcal{E}_q(x,y;\alpha) = \mathcal{E}_q(y,x;\alpha). \tag{2.5}
$$

The functions $\mathcal{E}_q(x,y;\alpha)$ and $\mathcal{E}_q(x;\alpha)$ satisfy the following equation

$$
\frac{\delta u}{\delta x} = \frac{2q^{1/4}}{1-q} \alpha \, u, \tag{2.6}
$$

which is a q-version of

$$
\frac{d}{dx} e^{\alpha(x+y)} = \alpha \, e^{\alpha(x+y)}
$$

on a q-quadratic grid. The operator $\delta/\delta x$ is the standard Askey–Wilson divided difference operator

$$
\frac{\delta u(z)}{\delta x(z)} = \frac{u(z+1/2) - u(z-1/2)}{x(z+1/2) - x(z-1/2)} \tag{2.7}
$$

with $x(z) = (q^z + q^{-z})/2 = \cos\theta$, $q^z = e^{i\theta}$. Applying this operator to (2.6) once again one obtains the second order difference equation

$$
\frac{\delta^2 u}{\delta x^2} = \left(\frac{2q^{1/4}}{1-q} \alpha \right)^2 u. \tag{2.8}
$$

A double series extension of the q-exponential function (2.1) was recently introduced in [61] as

$$
\mathcal{E}_q(x,y;\alpha,\beta) \tag{2.9}
$$

$$
= \frac{(\beta^2; q^2)_\infty}{(q\alpha^2; q^2)_\infty} \sum_{n=0}^\infty \frac{q^{n^2/4} \beta^n}{(q;q)_n}
$$

$$
\times e^{-in\varphi} \left(-q^{(1-n)/2} e^{i\theta+i\varphi} \alpha/\beta, \; -q^{(1-n)/2} e^{i\varphi-i\theta} \alpha/\beta; q \right)_n
$$

$$\times {}_2\varphi_2 \left(\begin{array}{c} q^{-n}, \quad \alpha^2/\beta^2 \\ -q^{(1-n)/2}e^{i\theta+i\varphi}\alpha/\beta, \ -q^{(1-n)/2}e^{i\varphi-i\theta}\alpha/\beta \end{array} ; q, qe^{2i\varphi} \right).$$

This function appears, in a natural way, in an "addition" theorem for the q-exponetial functions discussed in the next section. If $\beta = \alpha$, the second series terminates and we obtain (2.1). Changing the order of summation gives an alternate form

$$\mathcal{E}_q \left(x, y; \alpha, \beta \right) \qquad\qquad (2.10)$$

$$= \frac{(\beta^2; q^2)_\infty}{(q\alpha^2; q^2)_\infty} \sum_{k=0}^{\infty} \frac{(\alpha^2/\beta^2; q)_k}{(q;q)_k} q^{k^2/4} \left(\beta e^{i\varphi} \right)^k$$

$$\times \sum_{n=0}^{\infty} \frac{q^{n(n-2k)/4}}{(q;q)_k} \beta^n e^{-in\varphi}$$

$$\times \left(-q^{(1-n+k)/2}e^{i\theta+i\varphi}\alpha/\beta, -q^{(1-n+k)/2}e^{i\varphi-i\theta}\alpha/\beta; q \right)_n.$$

When $\beta = \alpha$ the first sum terminates and we obtain (2.1) once again. The second sum can be reduced to the sum of two ${}_4\varphi_3$-series similar to ones in (2.1); this expression is too lengthy to present here.

The function $\mathcal{E}_q \left(x, y; \alpha, \beta \right)$ given by (2.9) is an analog of $\exp \left(\alpha x + \beta y \right)$. Indeed, in the limit $q \to 1^-$ one gets

$$\lim_{q \to 1^-} \mathcal{E}_q \left(x, y; (1-q)\alpha/2, (1-q)\beta/2 \right) \qquad\qquad (2.11)$$

$$= \sum_{n=0}^{\infty} \frac{(\beta/2)^n}{n!} e^{-in\varphi} \left(1 + e^{i\theta+i\varphi}\alpha/\beta \right)^n \left(1 + e^{i\varphi-i\theta}\alpha/\beta \right)^n$$

$$\times \sum_{k=0}^{n} \frac{(-n)_k}{k!} \left(-\frac{(1-\alpha^2/\beta^2) e^{2i\varphi}}{(1 + e^{i\theta+i\varphi}\alpha/\beta)(1 + e^{i\varphi-i\theta}\alpha/\beta)} \right)^k$$

$$= \sum_{n=0}^{\infty} \frac{(\beta/2)^n}{n!} e^{-in\varphi} \left(1 + e^{i\theta+i\varphi}\alpha/\beta \right)^n \left(1 + e^{i\varphi-i\theta}\alpha/\beta \right)^n$$

$$\times \left(1 + \frac{(1-\alpha^2/\beta^2) e^{2i\varphi}}{(1 + e^{i\theta+i\varphi}\alpha/\beta)(1 + e^{i\varphi-i\theta}\alpha/\beta)} \right)^n$$

$$= \sum_{n=0}^{\infty} \frac{(\alpha x + \beta y)^n}{n!} = \exp \left(\alpha x + \beta y \right)$$

by the binomial theorem. We shall also see in the next section that function $u = \mathcal{E}_q \left(x, y; \alpha, \beta \right)$ in (2.9)–(2.10) is a double series solution of the difference equations (2.6) and (2.8). Methods of solving of these equations are discussed in [7], [56] and [57]. They do not involve this double series solution.

The function $\mathcal{E}_q(x, y; \alpha, \beta)$ satisfies the following simple properties

$$\mathcal{E}_q(x, y; \alpha, \alpha) = \mathcal{E}_q(x, y; \alpha), \quad \mathcal{E}_q(x, y; \alpha, \beta) = \mathcal{E}_q(y, x; \beta, \alpha),$$
(2.12)

$$\mathcal{E}_q(x, 0; \alpha, \beta) = \mathcal{E}_q(x; \alpha), \quad \mathcal{E}_q(0, y; \alpha, \beta) = \mathcal{E}_q(y; \beta).$$
(2.13)

These properties are closely related to the q-addition theorems.

3. Addition Theorems

The addition theorem for the exponential function can be written as

$$\exp \alpha(x + y) = \exp \alpha x \, \exp \alpha y.$$
(3.1)

It is also well-known that $\exp(\alpha x)$ is the only measurable solution of the functional equation $f(x + y) = f(x) f(y)$. Jackson's q-exponential functions (1.1)–(1.2) satisfy the following q-addition formulas

$$e_q(x + y) = e_q(y) \, e_q(x), \quad E_q(x + y) = E_q(x) \, E_q(y),$$
(3.2)

provided that $xy = qyx$. See [42] for a nice review on special functions of q-commuting variables and the bibliography. A commutative q-analog of the addition theorem (3.1) involving the q-exponential functions (2.1) and (2.3) has been established in [57].

Theorem 1

$$\mathcal{E}_q(x, y; \alpha) = \mathcal{E}_q(x; \alpha) \, \mathcal{E}_q(y; \alpha).$$
(3.3)

Proof. We shall follow here the idea of one of the original proofs [57] using arguments of the theory of analytic functions. The functions $\mathcal{E}_q(x, y; \alpha)$, $\mathcal{E}_q(x; \alpha)$ and $\mathcal{E}_q(x; -\alpha)$ satisfy the second order difference equation (2.8). Also, the q-exponential functions $\mathcal{E}_q(x; \pm\alpha)$ are linearly independent when $\alpha \neq 0$. Therefore, see for example Ruijsenaar's contribution to this volume [52], section 2.2, or the Appendix of this paper,

$$\mathcal{E}_q(x, y; \alpha) = A \, \mathcal{E}_q(x; \alpha) + B \, \mathcal{E}_q(x; -\alpha),$$
(3.4)

where $A = A(z, s)$ and $B = B(z, s)$ are, generally speaking, some functions of period 1 in z. Here $x = \cos\theta$, $e^{i\theta} = q^z$ and $y = \cos\varphi$, $e^{i\varphi} = q^s$. Applying $\delta/\delta x$ to both sides of (3.4), with the help of (2.6)

$$\mathcal{E}_q(x, y; \alpha) = A \, \mathcal{E}_q(x; \alpha) - B \, \mathcal{E}_q(x; -\alpha),$$

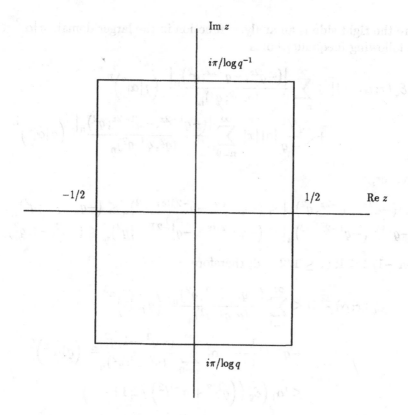

Figure 1. The parallelogram of the double periodicity in the complex z-plane.

or

$$\mathcal{E}_q\left(x, y; \alpha\right) = A\, \mathcal{E}_q\left(x; \alpha\right). \tag{3.5}$$

The functions $\mathcal{E}_q\left(x, y; \alpha\right)$ and $\mathcal{E}_q\left(x; \alpha\right)$, which are analytic when $|\alpha| < 1$ by their definitions (2.1) and (2.3), have the natural purely imaginary period $T = 2\pi i / \log q^{-1}$ in z. Hence, the coefficient

$$A(z, s) = \mathcal{E}_q\left(x, y; \alpha\right) / \mathcal{E}_q\left(x; \alpha\right) \tag{3.6}$$

is a doubly periodic analytic function in z. Using (III.3) of [22], one can rewrite (2.3) as

$$\mathcal{E}_q\left(x; \alpha\right) = {}_2\varphi_1\left(\begin{array}{c} -e^{2i\theta}, -e^{-2i\theta} \\ q \end{array}; q^2, q\alpha^2\right) \tag{3.7}$$

$$+ \frac{2q^{1/4}}{1-q}\alpha\cos\theta\; {}_2\varphi_1\left(\begin{array}{c} -qe^{2i\theta}, -qe^{-2i\theta} \\ q^3 \end{array}; q^2, q\alpha^2\right),$$

where the right side is an analytic function in the larger domain $q |\alpha|^2 < 1$. The following inequality holds

$$|\mathcal{E}_q (x; \alpha) - 1| \le \sum_{n=1}^{\infty} \frac{\left|(-q^{2z}, -q^{-2z}; q^2)_n\right|}{(q, q^2; q^2)_n} \left(q |\alpha|^2\right)^n$$
$$+ \frac{2q^{1/4}}{1 - q} |\alpha| |x| \sum_{n=0}^{\infty} \frac{\left|(-q^{1+2z}, -q^{1-2z}; q^2)_n\right|}{(q^2, q^3; q^2)_n} \left(q |\alpha|^2\right)^n.$$

Moreover,

$$\left|(-q^{2z}, -q^{-2z}; q^2)_n\right| \le \left(-q^{2\,\mathrm{Re}\,z}, -q^{-2\,\mathrm{Re}\,z}; q^2\right)_n \le \left(-q, -q^{-1}; q^2\right)_n,$$
$$\left|(-q^{1+2z}, -q^{1-2z}; q^2)_n\right| \le \left(-q^{1+2\,\mathrm{Re}\,z}, -q^{1-2\,\mathrm{Re}\,z}; q^2\right)_n \le \left(-q^2, -1; q^2\right)_n,$$

when $-1/2 \le \mathrm{Re}\,z \le 1/2$ and, therefore,

$$|\mathcal{E}_q (x; \alpha) - 1| \le \sum_{n=1}^{\infty} \frac{(-q, -q^{-1}; q^2)_n}{(q, q^2; q^2)_n} \left(q |\alpha|^2\right)^n$$
$$+ q^{-1/4} \frac{1 + q}{1 - q} |\alpha| \sum_{n=0}^{\infty} \frac{(-q^2, -1; q^2)_n}{(q^2, q^3; q^2)_n} \left(q |\alpha|^2\right)^n$$
$$\le |\alpha| \left(\mathcal{E}_q \left(\left(q^{1/2} + q^{-1/2}\right)/2; 1\right) - 1\right)$$

by (3.7) when $|\alpha| < 1$. As a result,

$$|\mathcal{E}_q (x; \alpha) - 1| \le C |\alpha|$$

for all $x = (q^z + q^{-z})/2$, $-1/2 \le \mathrm{Re}\,z \le 1/2$ and $|\alpha| < 1$. Here, the positive constant $C = \mathcal{E}_q \left(\left(q^{1/2} + q^{-1/2}\right)/2; 1\right) - 1$ depends on q only. Hence,

$$|\mathcal{E}_q (x; \alpha)| \ge 1 - C |\alpha| > 0$$

for all $|\alpha| < \delta = \min\left(C^{-1}, 1\right)$ and for all z in the parallelogram with the vertices at $-1/2 + i\pi/\log q$, $1/2 + i\pi/\log q$, $1/2 + i\pi/\log q^{-1}$ and $-1/2 + i\pi/\log q^{-1}$ in Figure 1. The last inequality shows that for sufficiently small $|\alpha|$ the $\mathcal{E}_q (x; \alpha)$ does not have zeros in z in the parallelogram.

Thus, $A(z, s)$ given by (3.6) is an entire doubly periodic function in z when $|\alpha| < \delta$ that is a constant by Liouville's theorem. One can find this constant choosing $x = 0$ (or $\theta = \pi/2$) in (3.5) which results in the addition formula (3.3) in the open disk $|\alpha| < \delta$. Analytic continuation of (3.3) in the parameter α to the entire complex α-plane completes the proof. $\square$

Formula (3.3) has attracted some attention and different proofs of this relation were given in [20], [34], and [57]. Floreanini, LeTourneux, and Vinet

[20] have used group theoretical methods. Ismail and Stanton [34] gave two different proofs on the basis of the connection coefficients for the q-Hermite polynomials. See also [57] for the original proof by the direct series manipulations and the Appendix at the end of this paper for another independent proof.

Although $\mathcal{E}_q(x;\alpha)$ is an analog of $\exp \alpha x$, the function $\mathcal{E}_q(x;\alpha)$ is not symmetric in x and α, so one would expect $\mathcal{E}_q(x;\alpha)$ to have two different addition theorems. Equation (3.3) gives the addition theorem in the variable x. Recently Ismail and Stanton [34] have found an important expansion formula (5.1) and called it the addition theorem in the variable α because it becomes $\exp \alpha x \, \exp \beta x = \exp(\alpha+\beta)x$ when $q \to 1^-$. In [61] the author was able to find another version of the "addition" formula with respect to both variables x and α which extends Theorem 1.

Theorem 2

$$\mathcal{E}_q(x,y;\alpha,\beta) = \mathcal{E}_q(x;\alpha)\ \mathcal{E}_q(y;\beta). \tag{3.8}$$

This formula can be thought of as a general analog of the relation

$$\exp(\alpha x + \beta y) = \exp \alpha x \, \exp \beta y.$$

Clearly, Theorem 2 gives the addition formula (3.3) when $\beta = \alpha$ due to (2.9). The case of the addition theorem in the variable α, raised by Ismail and Stanton, arises when $y = x$. Theorem 1 simplifies the product of two single series to a similar single series, while Theorem 2 allows to factor the double series into the product of two single series. It is worth mentioning also that the "addition" formulas (3.3) and (3.8) give an analytic continuation of the q-exponential functions $\mathcal{E}_q(x,y;\alpha)$ and $\mathcal{E}_q(x,y;\alpha,\beta)$ in the entire complex α and β-planes because the analytic continuation of the $2\varphi_1$-functions in (2.3) is well-known (see, for example, [22], [10], and [57]).

Proof. Our proof of (3.8) uses the connection relation (10.2) of [10] which we rewrite here as

$$\frac{(q\alpha^2;q^2)_\infty}{(q\beta^2;q^2)_\infty}\, \mathcal{E}_q(\cos\theta;\alpha) \tag{3.9}$$

$$= \frac{1}{2\pi} \int_0^\pi \frac{\left(q, \alpha^2/\beta^2, e^{2i\psi},\ e^{-2i\psi};\ q\right)_\infty \mathcal{E}_q(\cos\psi;\beta)\ d\psi}{\left(e^{i\theta+i\psi}\alpha/\beta, e^{i\theta-i\psi}\alpha/\beta, e^{-i\theta+i\psi}\alpha/\beta, e^{-i\theta-i\psi}\alpha/\beta; q\right)_\infty}$$

provided that $\alpha < \beta$. Multiplying both sides of (3.9) by $\mathcal{E}_q(\cos\varphi;\beta)$ and then using the addition formula (3.3), the symmetry relation (2.5) and the definition (2.1) one gets

$$\frac{(q\alpha^2;q^2)_\infty}{(q\beta^2;q^2)_\infty}\, \mathcal{E}_q(\cos\theta;\alpha)\ \mathcal{E}_q(\cos\varphi;\beta) \tag{3.10}$$

$$= \frac{1}{2\pi} \int_0^\pi \frac{\left(q, \alpha^2/\beta^2, e^{2i\psi}, e^{-2i\psi}; q\right)_\infty \mathcal{E}_q\left(\cos\varphi, \cos\psi; \beta\right)}{\left(e^{i\theta+i\psi}\alpha/\beta, e^{i\theta-i\psi}\alpha/\beta, e^{-i\theta+i\psi}\alpha/\beta, e^{-i\theta-i\psi}\alpha/\beta; q\right)_\infty} \, d\psi$$

$$= \frac{(\beta^2; q^2)_\infty}{(q\beta^2; q^2)_\infty} \left(q, \alpha^2/\beta^2; q\right)_\infty \sum_{n=0}^\infty \frac{q^{n^2/4}}{(q; q)_n} \left(\beta e^{-i\varphi}\right)^n$$

$$\times \frac{1}{2\pi} \int_0^\pi \frac{\left(e^{2i\psi}, e^{-2i\psi}, -q^{(1-n)/2}e^{i\varphi+i\psi}, -q^{(1-n)/2}e^{i\varphi-i\psi}; q\right)_\infty}{\left(-q^{(1+n)/2}e^{i\varphi+i\psi}, -q^{(1+n)/2}e^{i\varphi-i\psi}; q\right)_\infty}$$

$$\times \frac{d\psi}{\left(e^{i\theta+i\psi}\alpha/\beta, e^{i\theta-i\psi}\alpha/\beta, e^{-i\theta+i\psi}\alpha/\beta, e^{-i\theta-i\psi}\alpha/\beta; q\right)_\infty}.$$

The last integral can be evaluated as the special case $a = b = 0$ of the Nassrallah and Rahman integral,

$$\frac{1}{2\pi} \int_0^\pi \frac{\left(e^{2i\psi}, e^{-2i\psi}, ge^{i\psi}, ge^{-i\psi}; q\right)_\infty \, d\psi}{\left(ce^{i\psi}, ce^{-i\psi}, de^{i\psi}, de^{-i\psi}, fe^{i\psi}, fe^{-i\psi}; q\right)_\infty} \tag{3.11}$$

$$= \frac{(cg; q)_\infty}{(q, cd, cf; q)_\infty} \, {}_2\varphi_1\left(\begin{array}{c} g/d, \, g/f \\ cg \end{array}; q, \, df\right),$$

see (6.3.2) and (6.3.8) of [22]. Therefore

$$\mathcal{E}_q\left(\cos\theta; \alpha\right) \mathcal{E}_q\left(\cos\varphi; \beta\right) \tag{3.12}$$

$$= \frac{(\beta^2; q^2)_\infty}{(q\alpha^2; q^2)_\infty} \sum_{n=0}^\infty \frac{q^{n^2/4}}{(q; q)_n} \beta^n e^{-in\varphi} \left(-q^{(1-n)/2}e^{i\theta+i\varphi}\alpha/\beta; q\right)_n$$

$$\times {}_2\varphi_1\left(\begin{array}{c} -q^{(1-n)/2}e^{i\theta+i\varphi}\beta/\alpha, \, q^{-n} \\ -q^{(1-n)/2}e^{i\theta+i\varphi}\alpha/\beta \end{array}; q, \, -q^{(1+n)/2}e^{i\varphi-i\theta}\alpha/\beta\right).$$

Use of the transformation (III.3) of [22] completes the proof. □

The function $\mathcal{E}_q\left(x, y; \alpha, \beta\right)$ on the left side of (3.8) is an analog of the exponential function $\exp\left(\alpha x + \beta y\right)$, see (2.11). Due to the addition theorem (3.8) this function $u = \mathcal{E}_q\left(x, y; \alpha, \beta\right)$ in (2.9)–(2.10) is a double series solution of the difference equations (2.6) and (2.8). Equation (3.8) leads also to the product formula

$$\mathcal{E}_q\left(x, y; \alpha, \beta\right) \mathcal{E}_q\left(z, w; \gamma, \delta\right) = \mathcal{E}_q\left(x, z; \alpha, \gamma\right) \mathcal{E}_q\left(y, w; \beta, \delta\right), \tag{3.13}$$

which is, obviously, a q-analog of

$$\exp\left(\alpha x + \beta y\right) \exp\left(\gamma z + \delta w\right) = \exp\left(\alpha x + \gamma z\right) \exp\left(\beta y + \delta w\right).$$

Equation (3.13) is an extension of the product formula (7.7) of [57] in the case of the q-quadratic lattice under consideration.

Two limiting cases of (3.12) are of interest. When $\beta \to 0$, we obtain the following generating function

$$\left(q\alpha^2; q^2\right)_\infty \mathcal{E}_q\left(x; \alpha\right) = \sum_{n=0}^\infty \frac{q^{n^2/4}}{(q; q)_n} \, \alpha^n \, H_n\left(x|q\right) \qquad (3.14)$$

for the continuous q-Hermite polynomials

$$H_n\left(\cos\theta|q\right) = \sum_{k=0}^n \frac{(q; q)_n}{(q; q)_k \, (q; q)_{n-k}} \, e^{i(n-2k)\theta}, \qquad (3.15)$$

see, for example, [22]. One needs this generating function in order to derive the connecting formula (3.9) [10]. Different proofs of (3.14) are given in [35], [18], [20], [32], [34], and [57]; we give an independent proof in the Appendix. It is worth noting that the series in (3.14) analytically continues the left side to an entire function in x and α.

Another limiting case, $\alpha \to 0$, results in the generating relation found in [34]

$$\mathcal{E}_q\left(x; \beta\right) = \left(\beta^2; q^2\right)_\infty \sum_{n=0}^\infty \frac{q^{n^2/4}}{(q; q)_n} \, \beta^n \, H_n\left(x|q^{-1}\right), \qquad (3.16)$$

where

$$H_n\left(\cos\theta|q^{-1}\right) = \sum_{k=0}^n q^{k^2-kn} \frac{(q; q)_n}{(q; q)_k \, (q; q)_{n-k}} \, e^{i(n-2k)\theta}. \qquad (3.17)$$

4. Basic Trigonometric Functions

The basic cosine $C_q\left(x, y; \omega\right)$ and sine $S_q\left(x, y; \omega\right)$ functions can be introduced by the following analog of the Euler formula

$$\mathcal{E}_q\left(x, y; i\omega\right) = C_q\left(x, y; \omega\right) + i S_q\left(x, y; \omega\right). \qquad (4.1)$$

Hence,

$$C_q\left(x, y; \alpha\right) = \frac{\left(-\omega^2; q^2\right)_\infty}{\left(-q\omega^2; q^2\right)_\infty} \qquad (4.2)$$

$$\times \, {}_4\varphi_3\left(\begin{array}{c} -q^{1/2}e^{i\theta+i\varphi}, -q^{1/2}e^{i\theta-i\varphi}, -q^{1/2}e^{i\varphi-i\theta}, -q^{1/2}e^{-i\theta-i\varphi} \\ -q, \quad q^{1/2}, \quad -q^{1/2} \end{array}; q, -\omega^2\right)$$

and

$$S_q(x, y; \alpha) = \frac{(-\omega^2; q^2)_\infty}{(-q\omega^2; q^2)_\infty} \frac{2q^{1/4}}{1-q} \omega (\cos\theta + \cos\varphi) \tag{4.3}$$
$$\times {}_4\varphi_3 \left(\begin{array}{c} -qe^{i\theta+i\varphi}, -qe^{i\theta-i\varphi}, -qe^{i\varphi-i\theta}, -qe^{-i\theta-i\varphi} \\ -q, \quad q^{3/2}, \quad -q^{3/2} \end{array} ; q, -\omega^2 \right).$$

They are q-analogs of $\cos\omega(x+y)$ and $\sin\omega(x+y)$, respectively. Indeed, by (2.2) and (4.1),

$$\lim_{q \to 1^-} C_q(x, y; \omega(1-q)/2) = \cos\omega(x+y), \tag{4.4}$$

$$\lim_{q \to 1^-} S_q(x, y; \omega(1-q)/2) = \sin\omega(x+y). \tag{4.5}$$

The special cases $y = 0$ ($\varphi = \pi/2$) of (4.2)–(4.3) are

$$C_q(x; \omega) = \frac{(-\omega^2; q^2)_\infty}{(-q\omega^2; q^2)_\infty} \; {}_2\varphi_1 \left(\begin{array}{c} -qe^{2i\theta}, -qe^{-2i\theta} \\ q \end{array} ; q^2, -\omega^2 \right) \tag{4.6}$$

and

$$S_q(x; \omega) = \frac{(-\omega^2; q^2)_\infty}{(-q\omega^2; q^2)_\infty} \frac{2q^{1/4}\omega}{1-q} \cos\theta \tag{4.7}$$
$$\times {}_2\varphi_1 \left(\begin{array}{c} -q^2e^{2i\theta}, -q^2e^{-2i\theta} \\ q^3 \end{array} ; q^2, -\omega^2 \right).$$

The following q-addition formulas have been established in [57].

Theorem 3

$$C_q(x, y; \omega) = C_q(x; \omega) C_q(y; \omega) - S_q(x; \omega) S_q(y; \omega), \tag{4.8}$$
$$S_q(x, y; \omega) = S_q(x; \omega) C_q(y; \omega) + C_q(x; \omega) S_q(y; \omega). \tag{4.9}$$

Proof. Use

$$\mathcal{E}_q(x, y; i\omega) = \mathcal{E}_q(x; i\omega) \; \mathcal{E}_q(y; i\omega)$$

and the analog of Euler formula (4.1). $\qquad\qquad\square$

Further extension of Theorem 3 can be given. Let us introduce the basic trigonometric functions

$$C_q(x, y; \omega, \varkappa) = \frac{1}{2} \left(\mathcal{E}_q(x, y; i\omega, i\varkappa) + \mathcal{E}_q(x, y; -i\omega, -i\varkappa) \right), \tag{4.10}$$

$$S_q(x, y; \omega, \varkappa) = \frac{1}{2i} \left(\mathcal{E}_q(x, y; i\omega, i\varkappa) - \mathcal{E}_q(x, y; -i\omega, -i\varkappa) \right) \tag{4.11}$$

as analogs of $\cos(\omega x + \varkappa y)$ and $\sin(\omega x + \varkappa y)$, respectively. The following addition formulas hold [61].

Theorem 4

$$C_q\left(x,y;\omega,\varkappa\right)=C_q\left(x;\omega\right)C_q\left(y;\varkappa\right)-S_q\left(x;\omega\right)S_q\left(y;\varkappa\right),$$

(4.12)

$$S_q\left(x,y;\omega,\varkappa\right)=S_q\left(x;\omega\right)C_q\left(y;\varkappa\right)+C_q\left(x;\omega\right)S_q\left(y;\varkappa\right).$$

(4.13)

These formulas are, obviously, q-analogs of

$$\cos\left(\omega x+\varkappa y\right)=\cos\omega x\cos\varkappa y-\sin\omega x\sin\varkappa y,$$ (4.14)

$$\sin\left(\omega x+\varkappa y\right)=\sin\omega x\cos\varkappa y+\cos\omega x\sin\varkappa y.$$ (4.15)

Clearly, our Theorem 3 is the special case $\varkappa=\omega$ of Theorem 4.

5. Some Expansions and Integrals

5.1. MAIN RESULTS

Ismail and Stanton [34] have found the following expansion formula

$$\mathcal{E}_q\left(x;\alpha\right)\mathcal{E}_q\left(x;\beta\right)$$ (5.1)

$$=\sum_{m=0}^{\infty}q^{m^2/4}\alpha^m\,H_m\left(x|q\right)\frac{\left(-\alpha\beta q^{(m+1)/2};q\right)_\infty}{\left(q\alpha^2,q\beta^2;q^2\right)_\infty}\frac{\left(-q^{(1-m)/2}\beta/\alpha;q\right)_m}{\left(q;q\right)_m},$$

where $H_m\left(x|q\right)$ are the continuous q-Hermite polynomials (3.15). They call expansion (5.1) the "addition" theorem with respect to the parameter α because it becomes $\exp\alpha x\;\exp\beta x=\exp\left(\alpha+\beta\right)x$ in the limit $q\to1^-$.

The expansion (5.1) leads to the following integral evaluation

$$\int_0^\pi\mathcal{E}_q\left(\cos\theta;\alpha\right)\mathcal{E}_q\left(\cos\theta;\beta\right)\,H_m\left(\cos\theta|q\right)\,\left(e^{2i\theta},e^{-2i\theta};q\right)_\infty\,d\theta$$

$$=2\pi q^{m^2/4}\alpha^m\,\frac{\left(-\alpha\beta q^{(m+1)/2};q\right)_\infty\left(-q^{(1-m)/2}\beta/\alpha;q\right)_m}{\left(q;q\right)_\infty\left(q\alpha^2,q\beta^2;q^2\right)_\infty}$$

(5.2)

due to the orthogonality relation for the continuous q-Hermite polynomials [22]

$$\int_0^\pi H_m\left(\cos\theta|q\right)H_n\left(\cos\theta|q\right)\,\left(e^{2i\theta},e^{-2i\theta};q\right)_\infty\,d\theta=2\pi\frac{\left(q;q\right)_n}{\left(q;q\right)_\infty}\delta_{mn}.$$

(5.3)

An important extension of the integral (5.2), namely,

$$\int_0^\pi\mathcal{E}_q\left(\cos\theta;\alpha\right)\mathcal{E}_q\left(\cos\theta;\beta\right)\,C_m\left(\cos\theta;\gamma|q\right)\,\frac{\left(e^{2i\theta},\;e^{-2i\theta};\;q\right)_\infty}{\left(\gamma e^{2i\theta},\;\gamma e^{-2i\theta};\;q\right)_\infty}\,d\theta$$

(5.4)

$$= 2\pi \frac{\left(\gamma, \ \gamma q^{m+1}, -\alpha\beta q^{(m+1)/2}; q\right)_\infty}{\left(q, \ \gamma^2 q^m; q\right)_\infty \left(q\alpha^2, \ q\beta^2; q^2\right)_\infty} \alpha^m q^{m^2/4} \frac{\left(-q^{(1-m)/2}\beta/\alpha; q\right)_m}{\left(q; q\right)_m}$$

$$\times \, {}_2\varphi_2 \left(\begin{array}{c} -q^{(m+1)/2}\alpha/\beta, \, -q^{(m+1)/2}\beta/\alpha \\ \gamma q^{m+1}, \, -q^{(m+1)/2}\alpha\beta \end{array} ; q, \, -q^{(m+1)/2}\alpha\beta\gamma \right),$$

has also been found in [34]. Here $C_m\,(x; \gamma|q)$ are the continuous q-ultraspherical polynomials defined by

$$C_m\,(\cos\theta; \beta|q) = \sum_{k=0}^m \frac{(\beta; q)_k\,(\beta; q)_{m-k}}{(q; q)_k\,(q; q)_{m-k}} \, e^{i(m-2k)\theta}, \tag{5.5}$$

see, for example, [22]; the continuous q-Hermite polynomials are their special case

$$H_m\,(x|q) = (q; q)_m\,C_m\,(x; 0|q) \tag{5.6}$$

by (3.15) and (5.5).

The expansion formula associated with (5.4) is

$$\mathcal{E}_q\,(\cos\theta; \alpha)\,\mathcal{E}_q\,(\cos\theta; \beta) \tag{5.7}$$

$$= \sum_{m=0}^\infty q^{m^2/4} \alpha^m \frac{\left(-\alpha\beta q^{(m+1)/2}; q\right)_\infty \left(-q^{(1-m)/2}\beta/\alpha; q\right)_m}{(\gamma; q)_m\,(q\alpha^2, \ q\beta^2; q^2)_\infty}$$

$$\times \, C_m\,(\cos\theta; \gamma|q)$$

$$\times \, {}_2\varphi_2 \left(\begin{array}{c} -q^{(m+1)/2}\alpha/\beta, \, -q^{(m+1)/2}\beta/\alpha \\ \gamma q^{m+1}, \, -q^{(m+1)/2}\alpha\beta \end{array} ; q, \, -q^{(m+1)/2}\alpha\beta\gamma \right).$$

The expansions formulas and integrals found by Ismail and Stanton in [34] are important contributions in the growing area of q-series. For example, the special case $m = 0$ and $\gamma = q^{1/2}$ of the integral (5.4) leads to the orthogonality property of the basic trigonometric system originally found in [10] by a different method. We shall give this elementary proof of the orthogonality relation in the next section. See also [60] and [64] for further applications of the Ismail and Stanton expansions and integrals in the theory of basic Fourier series.

5.2. PROOF OF (5.1)

In this section we intend to give two proofs of the expansion (5.1). Our first proof is, essentially, the original proof of this result due to Ismail and Stanton [34] on the basis of the generating function (3.14) and the linearization formula [22],

$$\frac{H_m\,(x|q)\,H_n\,(x|q)}{(q; q)_m\,(q; q)_n} = \sum_{k=0}^{\min(m,n)} \frac{H_{m+n-2k}\,(x|q)}{(q; q)_k\,(q; q)_{m-k}\,(q; q)_{n-k}}, \tag{5.8}$$

for the continuous q-Hermite polynomials.

Use of (3.14) and (5.8) gives

$$
\left(q\alpha^2, q\beta^2; q^2\right)_\infty \mathcal{E}_q\left(x; \alpha\right) \mathcal{E}_q\left(x; \beta\right) \tag{5.9}
$$

$$
= \sum_{m,n=0}^\infty \frac{q^{(m^2+n^2)/4} \alpha^m \beta^n}{(q;q)_m (q;q)_n} H_m\left(x|q\right) H_n\left(x|q\right)
$$

$$
= \sum_{m,n,k} q^{(m^2+n^2)/4} \alpha^m \beta^n \frac{H_{m+n-2k}\left(x|q\right)}{(q;q)_k (q;q)_{m-k} (q;q)_{n-k}}
$$

$$
= \sum_{M,N} q^{(M^2+N^2)/4} \alpha^M \beta^N \frac{H_{M+N}\left(x|q\right)}{(q;q)_M (q;q)_N}
$$

$$
\times \sum_{k=0}^\infty \frac{q^{k(k-1)/2}}{(q;q)_k} \left(\alpha\beta q^{(M+N+1)/2}\right)^k
$$

$$
= \sum_{M,N} q^{(M^2+N^2)/4} \alpha^M \beta^N \frac{H_{M+N}\left(x|q\right)}{(q;q)_M (q;q)_N} \left(-\alpha\beta q^{(M+N+1)/2}; q\right)_\infty .
$$

We substitute $m = M + k$ and $n = N + k$ in the third line here and use the Euler summation formula (1.2) in the last but one line. Letting $M + N = m$ in the last expression and using (I.12) of [22] one can write

$$
\sum_{m=0}^\infty q^{m^2/4} \alpha^m H_m\left(x|q\right) \left(-\alpha\beta q^{(m+1)/2}; q\right)_\infty
$$

$$
\times \sum_{N=0}^\infty \frac{(q^{-m}; q)_N}{(q;q)_N} \left(-\frac{\beta}{\alpha} q^{(m+1)/2}\right)^N
$$

$$
= \sum_{m=0}^\infty q^{m^2/4} \alpha^m H_m\left(x|q\right) \left(-\alpha\beta q^{(m+1)/2}; q\right)_\infty \left(-\frac{\beta}{\alpha} q^{(1-m)/2}; q\right)_m
$$

by the q-binomial theorem and the first proof is complete.

Our second proof of (5.1) uses also (3.14) and (5.8), but we evaluate the integral (5.2). Obviously, this integral evaluation is equivalent to (5.1). Let us start from the evaluation of the following "matrix elements",

$$
\int_0^\pi H_m\left(\cos\theta|q\right) \mathcal{E}_q\left(\cos\theta; \alpha\right) H_n\left(\cos\theta|q\right) \left(e^{2i\theta}, e^{-2i\theta}; q\right)_\infty d\theta
$$

$$
= 2\pi \frac{\alpha^{m+n} q^{(m+n)^2/4}}{(q\alpha^2; q^2)_\infty (q;q)_\infty} \, {}_2\varphi_1\left(\begin{matrix} q^{-m}, q^{-n} \\ 0 \end{matrix} ; q, \frac{q}{\alpha^2}\right), \tag{5.10}
$$

which is of independent interest. Using the linearization formula (5.8) in the integrand, one gets

$$\sum_k \frac{(q;q)_m (q;q)_n}{(q;q)_k (q;q)_{m-k} (q;q)_{n-k}}$$

$$\times \int_0^\pi \mathcal{E}_q (\cos\theta;\alpha) H_{m+n-2k} (\cos\theta|q) \left(e^{2i\theta}, e^{-2i\theta}; q\right)_\infty d\theta$$

$$= 2\pi \frac{(q;q)_m (q;q)_n}{(q\alpha^2;q^2)_\infty (q;q)_\infty} \sum_k \frac{q^{(m+n-2k)^2/4} \alpha^{m+n-2k}}{(q;q)_k (q;q)_{m-k} (q;q)_{n-k}}$$

with the help of (3.14) and (5.3). Transformation of the sum of the q-shifted factorials in the last line by (I.12) of [22] results in (5.10).

In order to complete the proof, let us denote the integral in the left side of (5.2) as I. Applying (3.14) and (5.10), one gets

$$I = \frac{1}{(q\alpha^2;q^2)_\infty} \sum_{n=0}^\infty \frac{q^{n^2/4} \beta^n}{(q;q)_n} \tag{5.11}$$

$$\times \int_0^\pi H_m (\cos\theta|q) \mathcal{E}_q (\cos\theta;\alpha) H_n (\cos\theta|q) \left(e^{2i\theta}, e^{-2i\theta}; q\right)_\infty d\theta$$

$$= 2\pi \frac{q^{m^2/4} \alpha^m}{(q\alpha^2, q\beta^2; q^2)_\infty (q;q)_\infty}$$

$$\times \sum_{n=0}^\infty q^{n(n+m)/2} \frac{(\alpha\beta)^n}{(q;q)_n} \sum_k \frac{(q^{-m}, q^{-n}; q)_k}{(q;q)_k} \left(\frac{q}{\alpha^2}\right)^k.$$

The change the order of summation and the substitution $n = k + l$ with the help of (I.12) of [22] give

$$I = 2\pi \frac{q^{m^2/4} \alpha^m}{(q\alpha^2, q\beta^2; q^2)_\infty (q;q)_\infty}$$

$$\times \sum_k \frac{(q^{-m}; q)_k}{(q;q)_k} \left(-\frac{\alpha}{\beta} q^{(m+1)/2}\right)^k \sum_l \frac{q^{l(l-1)/2}}{(q;q)_l} \left(\alpha\beta q^{(m+1)/2}\right)^l.$$

These series can be summed by the q-binomial theorem and by the Euler summation formula (1.2), respectively. This results in (5.2) and our second proof is complete.

5.3. PROOF OF (5.4)

This proof is due to Ismail and Stanton [34], with somewhat different details. Let us denote the integral in the left side of (5.4) as I_m. Using expansion (5.1),

$$
I_m = \sum_{n=0}^{\infty} q^{n^2/4} \alpha^n \frac{\left(-\alpha\beta q^{(n+1)/2}; q\right)_{\infty} \left(-q^{(1-n)/2}\beta/\alpha; q\right)_n}{(q\alpha^2, q\beta^2; q^2)_{\infty}}
$$
$$
\times \int_0^{\pi} \frac{H_n\left(\cos\theta|q\right) C_m\left(\cos\theta; \gamma|q\right)}{(q; q)_n} \frac{\left(e^{2i\theta}, e^{-2i\theta}; q\right)_{\infty}}{\left(\gamma e^{2i\theta}, \gamma e^{-2i\theta}; q\right)_{\infty}} d\theta.
$$
(5.12)

In order to evaluate this integral, Ismail and Stanton have used the special case $\gamma = 0$ of the Rogers connection coefficient formula [22],

$$
C_m\left(x; \gamma|q\right) \tag{5.13}
$$
$$
= \sum_{k=0}^{[n/2]} \frac{\beta^k \left(\gamma/\beta; q\right)_k \left(\gamma; q\right)_{n-k} \left(1 - \beta q^{n-2k}\right)}{(q; q)_k \left(q\beta; q\right)_{n-k} (1 - \beta)} C_{n-2k}\left(x; \beta|q\right),
$$

namely,

$$
\frac{H_n\left(\cos\theta|q\right)}{(q; q)_n} = \sum_{k=0}^{[n/2]} \frac{\gamma^k \left(1 - \gamma q^{n-2k}\right)}{(q; q)_k \left(q\gamma; q\right)_{n-k} (1 - \gamma)} C_{n-2k}\left(\cos\theta; \gamma|q\right). \tag{5.14}
$$

This gives

$$
\int_0^{\pi} \frac{H_n\left(\cos\theta|q\right) C_m\left(\cos\theta; \gamma|q\right)}{(q; q)_n} \frac{\left(e^{2i\theta}, e^{-2i\theta}; q\right)_{\infty}}{\left(\gamma e^{2i\theta}, \gamma e^{-2i\theta}; q\right)_{\infty}} d\theta \tag{5.15}
$$
$$
= \sum_{k=0}^{[n/2]} \frac{\gamma^k \left(1 - \gamma q^{n-2k}\right)}{(q; q)_k \left(q\gamma; q\right)_{n-k} (1 - \gamma)}
$$
$$
\times \int_0^{\pi} C_m\left(\cos\theta; \gamma|q\right) C_{n-2k}\left(\cos\theta; \gamma|q\right) \frac{\left(e^{2i\theta}, e^{-2i\theta}; q\right)_{\infty}}{\left(\gamma e^{2i\theta}, \gamma e^{-2i\theta}; q\right)_{\infty}} d\theta
$$
$$
= 2\pi \frac{(\gamma, q\gamma; q)_{\infty}}{(q, \gamma^2; q)_{\infty}} \sum_{k=0}^{[n/2]} \frac{\gamma^k \left(\gamma^2; q\right)_m}{(q; q)_k \left(q\gamma; q\right)_{n-k} (q; q)_m} \delta_{m,n-2k}
$$

by the orthogonality relation for the continuous q-ultraspherical polynomials [22]

$$
\int_0^{\pi} C_m\left(\cos\theta; \beta|q\right) C_n\left(\cos\theta; \beta|q\right) \frac{\left(e^{2i\theta}, e^{-2i\theta}; q\right)_{\infty}}{\left(\beta e^{2i\theta}, \beta e^{-2i\theta}; q\right)_{\infty}} d\theta
$$
$$
= 2\pi \frac{(\beta, q\beta; q)_{\infty} \left(\beta^2; q\right)_n (1 - \beta)}{(q, \beta^2; q)_{\infty} (q; q)_n (1 - \beta q^n)} \delta_{mn}. \tag{5.16}
$$

Therefore

$$I_m = 2\pi \frac{(\gamma, q\gamma; q)_\infty}{(q, \gamma^2; q)_\infty (q\alpha^2, q\beta^2; q^2)_\infty} \frac{(\gamma^2; q)_m}{(q; q)_m} \tag{5.17}$$

$$\times \sum_{k=0}^\infty \frac{q^{(m+2k)^2/4} \alpha^{m+2k} \gamma^k}{(q; q)_k (q\gamma; q)_{m+k}}$$

$$\times \left(-\alpha\beta q^{(1+m)/2+k}; q\right)_\infty \left(-q^{(1-m)2-k}\beta/\alpha; q\right)_{m+2k}.$$

This series can be transformed to the $_2\varphi_2$ in the right side of (5.4) with the help of the following relations

$$(q\gamma; q)_{m+k} = (q\gamma; q)_m \left(\gamma q^{m+1}; q\right)_k,$$

$$\left(-\alpha\beta q^{(1+m)/2+k}; q\right)_\infty = \frac{\left(-\alpha\beta q^{(1+m)/2}; q\right)_\infty}{\left(-\alpha\beta q^{(1+m)/2}; q\right)_k},$$

$$\left(-q^{(1-m)/2-k}\beta/\alpha; q\right)_{m+2k}$$
$$= \left(-q^{(1-m)/2-k}\beta/\alpha; q\right)_{m+k} \left(-q^{(1-m)/2}\beta/\alpha; q\right)_k,$$

$$\left(-q^{(1-m)/2-k}\beta/\alpha; q\right)_{m+k}$$
$$= \left(-q^{(1-m)/2-k}\beta/\alpha; q\right)_k \left(-q^{(1-m)/2}\beta/\alpha; q\right)_m,$$

$$\left(-q^{(1-m)/2-k}\beta/\alpha; q\right)_k$$
$$= \left(-q^{(1+m)/2}\alpha/\beta; q\right)_k \left(q^{-(m+1)/2}\beta/\alpha\right)^k q^{-k(k-1)/2},$$

see Appendix I of [22] for the convenient formulas of transformations for the q-shifted factorials. This completes the proof.

6. Orthogonality Property

The orthogonality relation for the trigonometric functions on the interval $(-l, l)$ is

$$\frac{1}{2l} \int_{-l}^l \exp\left(i\frac{\pi m}{l}x\right) \exp\left(-i\frac{\pi n}{l}x\right) dx = \delta_{mn}. \tag{6.1}$$

Bustoz and Suslov [10] have established an analog of this important property for the q-exponential function $\mathcal{E}_q(x; i\omega)$ using the arguments of the Sturm–Liouville theory for the second order divided-difference Askey–Wilson operator; see, for example, [46] and [56]. A special case of the integral (5.4) gives rise to an elementary proof of this property which plays a key role in the theory of basic Fourier series.

Theorem 5 *The following orthogonality property holds*

$$\int_0^\pi \mathcal{E}_q(\cos\theta; i\omega_m)\, \mathcal{E}_q(\cos\theta; -i\omega_n)\, \left(e^{2i\theta},\, e^{-2i\theta};\, q\right)_{1/2} d\theta \tag{6.2}$$

$$= 2k(\omega_n)\, \delta_{mn}.$$

Here

$$k(\omega) = \pi \frac{\left(q^{1/2}; q\right)_\infty^2}{(q; q)_\infty^2} \frac{\left(-\omega^2; q^2\right)_\infty}{\left(-q\omega^2; q^2\right)_\infty} \sum_{k=0}^\infty \frac{q^{k/2}}{1 + \omega^2 q^k}; \tag{6.3}$$

$\omega_m, \omega_n = 0, \pm\omega_1, \pm\omega_2, \pm\omega_3, \dots$ *and* $\omega_0 = 0, \omega_1, \omega_2, \omega_3, \dots$, *are nonnegative solutions of the transcendental equation*

$$\left(-i\omega; q^{1/2}\right)_\infty = \left(i\omega; q^{1/2}\right)_\infty \tag{6.4}$$

arranged in ascending order of magnitude.

Proof. Using (III.4) of [22], let us transform the $_2\varphi_2$-function in the right side of (5.4) with $m = 0$ and $\gamma = q^{1/2}$ to a $_2\varphi_1$ which can be summed by a consequence of the q-binomial theorem,

$$\frac{(\alpha^2; q)_\infty}{\left(-q^{1/2}\alpha\beta; q\right)_\infty} \, {}_2\varphi_1\left(\begin{array}{c} -q^{1/2}\beta/\alpha, -q\beta/\alpha \\ q^{3/2} \end{array}; q, \alpha^2\right)$$

$$= \frac{(\alpha^2; q)_\infty (1 - q^{1/2})}{\left(-q^{1/2}\alpha\beta; q\right)_\infty (\alpha + \beta)} \sum_{k=0}^\infty \frac{\left(-\beta/\alpha; q^{1/2}\right)_{2k+1}}{\left(q^{1/2}; q^{1/2}\right)_{2k+1}} \, \alpha^{2k+1}$$

$$= \frac{(\alpha^2; q)_\infty (1 - q^{1/2})}{2\left(-q^{1/2}\alpha\beta; q\right)_\infty (\alpha + \beta)} \left[\frac{\left(-\beta; q^{1/2}\right)_\infty}{\left(\alpha; q^{1/2}\right)_\infty} - \frac{\left(\beta; q^{1/2}\right)_\infty}{\left(-\alpha; q^{1/2}\right)_\infty}\right].$$

Therefore

$$\int_0^\pi \mathcal{E}_q(\cos\theta; \alpha)\, \mathcal{E}_q(\cos\theta; \beta)\, \left(e^{2i\theta},\, e^{-2i\theta};\, q\right)_{1/2} d\theta \tag{6.5}$$

$$= \pi \frac{\left(q^{1/2}; q\right)_\infty^2}{(q; q)_\infty^2} \frac{\left(-\alpha,\, -\beta; q^{1/2}\right)_\infty - \left(\alpha,\, \beta; q^{1/2}\right)_\infty}{(\alpha + \beta)\left(q\alpha^2,\, q\beta^2; q^2\right)_\infty}.$$

This basic integral is, obviously, a q-extension of the elementary integral

$$\int_{-1}^{1} e^{\alpha x} \, e^{\beta x} \, dx = \frac{e^{\alpha+\beta} - e^{-\alpha-\beta}}{\alpha+\beta}. \tag{6.6}$$

Since,

$$\left(-\alpha, -\beta; q^{1/2}\right)_{\infty} - \left(\alpha, \beta; q^{1/2}\right)_{\infty} \tag{6.7}$$

$$= \left[\left(-\alpha; q^{1/2}\right)_{\infty} - \left(\alpha; q^{1/2}\right)_{\infty}\right] \left(-\beta; q^{1/2}\right)_{\infty}$$

$$+ \left[\left(-\beta; q^{1/2}\right)_{\infty} - \left(\beta; q^{1/2}\right)_{\infty}\right] \left(\alpha; q^{1/2}\right)_{\infty},$$

the orthogonality relation (6.1) follows immediately from (6.5) if $\alpha = i\omega_m$, $\beta = -i\omega_n$ and $\omega_n \neq \omega_m$ are the roots of the transcendental equation (6.4).

In order to obtain the expression (6.3) for the normalization constant $k(\omega)$ one can take the limit $\beta \to -\alpha$ in the right side of (6.5),

$$\int_0^{\pi} \mathcal{E}_q (\cos\theta; \alpha) \; \mathcal{E}_q (\cos\theta; -\alpha) \left(e^{2i\theta}, e^{-2i\theta}; q\right)_{1/2} d\theta \tag{6.8}$$

$$= \pi \frac{\left(q^{1/2}; q\right)_{\infty}^2}{\left(q; q\right)_{\infty}^2} \frac{\left(\alpha^2; q^2\right)_{\infty}}{\left(q\alpha^2; q^2\right)_{\infty}} \frac{d}{d\alpha} \log \frac{\left(-\alpha; q^{1/2}\right)_{\infty}}{\left(\alpha; q^{1/2}\right)_{\infty}},$$

and then substitute $\alpha = i\omega$. This results in (6.3) after evaluation of the derivative and our proof is complete. $\square$

Equation (6.4) can be rewritten in terms of the basic sine function $S_q(\eta; \omega)$ as

$$S_q \left(\frac{1}{2} \left(q^{1/4} + q^{-1/4}\right); \omega\right) = \frac{(-i\omega; q^{1/2})_{\infty} - (i\omega; q^{1/2})_{\infty}}{2i(-q\omega^2; q^2)_{\infty}} = 0, \tag{6.9}$$

see [10] for more details. We shall use the notation $\eta = \left(q^{1/4} + q^{-1/4}\right)/2$ throughout the paper. One can also see that the orthogonality relation (6.2) holds at the zeros of the basic cosine function

$$C_q \left(\frac{1}{2} \left(q^{1/4} + q^{-1/4}\right); \omega\right) = \frac{(-i\omega; q^{1/2})_{\infty} + (i\omega; q^{1/2})_{\infty}}{2(-q\omega^2; q^2)_{\infty}} = 0. \tag{6.10}$$

We leave the details to the reader.

7. Ismail and Zhang Formula

The classical expansion of the plane wave in the ultraspherical polynomials is

$$e^{irx} = \left(\frac{2}{r}\right)^{\nu} \Gamma(\nu) \sum_{m=0}^{\infty} i^m \left(\nu + m\right) J_{\nu+m}(r) C_m^{\nu}(x). \tag{7.1}$$

Ismail and Zhang [35] had found the following q-analog of this formula

$$\mathcal{E}_q\left(x;i\omega\right) = \frac{(q;q)_\infty\,\omega^{-\nu}}{(q^\nu;q)_\infty\,(-q\omega^2;q^2)_\infty} \tag{7.2}$$

$$\times \sum_{m=0}^\infty i^m\left(1-q^{\nu+m}\right)\,q^{m^2/4}\,J^{(2)}_{\nu+m}\left(2\omega;q\right)\,C_m\left(x;q^\nu|q\right),$$

where $J^{(2)}_{\nu+m}\left(2\omega;q\right)$ is Jackson's q-Bessel function defined as

$$J^{(2)}_\nu\left(x;q\right) = \frac{(q^{\nu+1};q)_\infty}{(q;q)_\infty}\sum_{n=0}^\infty q^{(\nu+n)n}\frac{(-1)^n\,(x/2)^{\nu+2n}}{(q;q)_n\,(q^{\nu+1};q)_n}. \tag{7.3}$$

Different proofs of this important formula were given in [18], [33], [32], and [34]. One can obtain (7.2) as the limiting case $\beta \to 0$ of the Ismail-Stanton expansion (5.7). Ismail and Stanton [34] have found the following simple proof of (7.2) using the generation relation (3.14) and the connection coefficient formula (5.14):

$$\left(q\alpha^2;q^2\right)_\infty\mathcal{E}_q\left(x;\alpha\right)$$

$$= \sum_{n=0}^\infty \frac{q^{n^2/4}}{(q;q)_n}\,\alpha^n\,H_n\left(x|q\right)$$

$$= \sum_{n,k} q^{n^2/4}\alpha^n\,\frac{\beta^k\left(1-\beta q^{n-2k}\right)}{(q;q)_k\,(q\beta;q)_{n-k}\,(1-\beta)}\,C_{n-2k}\left(x;\beta|q\right)$$

$$= \sum_m \frac{1-\beta q^m}{1-\beta}q^{m^2/4}\alpha^m\,C_m\left(x;\beta|q\right)$$

$$\times \sum_k \frac{\alpha^{2k}\beta^k}{(q;q)_k\,(q\beta;q)_{m+k}}q^{k(m+k)}.$$

We have substituted here $n = m + 2k$ in the second line. The sum over k gives the q-Bessel function and the infinite products in (7.2) when $\alpha = i\omega$ and $\beta = q^\nu$.

8. Completeness and Basic Lommel Polynomials

The completeness of the trigonometric system $\left\{e^{i\pi kx}\right\}_{k=-\infty}^\infty$ on the interval $(-1,1)$ is one of the fundamental facts in the theory of trigonometric series. Bustoz and Suslov [10] have proved a similar result for the system of the basic trigonometric functions $\left\{\mathcal{E}_q\left(x;i\omega_k\right)\right\}_{k=-\infty}^\infty$ using the methods of the theory of entire functions [9], [44] and [45]. We shall discuss here a direct, more "computational" proof of an extension of this property using the Ismail–Zhang formula (7.2) and the q-Lommel polynomials introduced in [26]. The following theorem states the completeness of the basic trigonometric system in the weighted $\mathcal{L}_\rho^2$-space related to the continuous q-ultraspherical polynomials $C_m(x; q^\nu | q)$.

Theorem 6 Let $j_{\mu,k}(q)$ be zeros of Jackson's q-Bessel function $J_\mu^{(2)}(x; q)$. The q-trigonometric system $\{\mathcal{E}_q(x; i\omega_k)\}_{k=-\infty}^\infty$ is complete in the weighted $\mathcal{L}_\rho^2(-1, 1)$ space, where $\rho(x)$ is the weight function of the continuous q-ultraspherical polynomials $C_m(x; q^\nu | q)$ for $0 < \nu < 1/2$ if: (a) $\omega_k = \frac{1}{2} j_{\nu-1,k}(q)$, $k = \pm 1, \pm 2, ...$; (b) $\omega_k = \frac{1}{2} j_{\nu,k}(q)$, $k = 0, \pm 1, \pm 2, ...$.

Proof. Consider the first case. The q-Lommel polynomials are generated by the recurrence relation

$$q^{n\nu + n(n-1)/2} J_{\nu+n}^{(2)}(x; q) \tag{8.1}$$
$$= h_n^\nu \left(\frac{1}{x}; q\right) J_\nu^{(2)}(x; q) - h_{n-1}^{\nu+1}\left(\frac{1}{x}; q\right) J_{\nu-1}^{(2)}(x; q).$$

We shall use the notation $h_n^\nu(y; q) := h_{n,\nu}(y; q)$ instead of the original one in [26]. Letting $x = 2\omega_k = j_{\nu-1,k}(q)$ in (8.1), one gets

$$q^{n\nu + n(n-1)/2} J_{\nu+n}^{(2)}(j_{\nu-1,k}(q); q) \tag{8.2}$$
$$= h_n^\nu \left(\frac{1}{j_{\nu-1,k}(q)}; q\right) J_\nu^{(2)}(j_{\nu-1,k}(q); q)$$

and (7.2) takes a simpler form

$$\mathcal{E}_q(x; i\omega_k) \left(-q\omega_k^2; q^2\right)_\infty \omega_k^\nu \tag{8.3}$$
$$= \frac{(q; q)_\infty}{(q^\nu; q)_\infty} \sum_{n=0}^\infty i^n \left(1 - q^{\nu+n}\right) q^{n(2-4\nu-n)/4} C_n(x; q^\nu | q)$$
$$\times h_n^\nu \left(\frac{1}{j_{\nu-1,k}(q)}; q\right) J_\nu^{(2)}(j_{\nu-1,k}(q); q).$$

The orthogonality relation for the q-Lommel polynomials had been established in [26],

$$(1 + (-1)^{m+n}) \sum_{k=1}^\infty \frac{A_k(\nu)}{j_{\nu-1,k}^2(q)} h_n^\nu \left(\frac{1}{j_{\nu-1,k}(q)}; q\right) h_m^\nu \left(\frac{1}{j_{\nu-1,k}(q)}; q\right)$$
$$= \frac{q^{n\nu + n(n-1)/2}}{1 - q^{\nu+n}} \delta_{mn}, \tag{8.4}$$

where the jumps of the step function $A_k(\nu)$ can be found from the partial fraction decomposition

$$\sum_{k=1}^{\infty} A_k(\nu) \frac{z}{j_{\nu-1,k}^2(q) - z^2} = \frac{J_{\nu}^{(2)}(z;q)}{J_{\nu-1}^{(2)}(z;q)}. \qquad (8.5)$$

The q-trigonometric system $\{\mathcal{E}_q(x; i\omega_k)\}_{k=-\infty}^{\infty}$ is complete on $(-1,1)$ if it is closed [1]. Suppose that this system is not closed. Thus, there exists at least one function $f(x)$, not identically zero, that

$$\int_{-1}^{1} f(x) \; \mathcal{E}_q(x; i\omega_k) \; \rho(x) \; dx = 0, \qquad k = \pm 1, \pm 2, ..., \qquad (8.6)$$

where $\rho(x)$ is the weight function in the orthogonality relation for the continuous q-ultraspherical polynomials (5.16). It can be shown that the series in (7.2) and (8.3) are uniformly covergent in x on compacts. Substituting (8.3) in (8.6) one gets

$$\sum_{n=0}^{\infty} i^n \left(1 - q^{\nu+n}\right) \; q^{n(2-4\nu-n)/4} \; h_n^{\nu}\left(\frac{1}{j_{\nu-1,k}(q)}; q\right) \qquad (8.7)$$

$$\times \int_{-1}^{1} f(x) \; C_n(x; q^{\nu}|q) \; \rho(x) \; dx = 0.$$

Multiplying this equation by $h_m^{\nu}(y; q)$ and integrating with respect to the purely discrete measure $d\mu(y)$ with a bounded support defined by (8.4)–(8.5) we obtain, formally, that

$$\int_{-1}^{1} f(x) \; C_n(x; q^{\nu}|q) \; \rho(x) \; dx = 0, \qquad n = 0, 1, 2, ... \qquad (8.8)$$

in view of the orthogonality relation (8.4). The system of the continuous q-ultraspherical polynomials $\{C_n(x; q^{\nu}|q)\}_{n=0}^{\infty}$ is closed on $(-1,1)$ when $\nu > 0$. Thus, $f(x)$ is identically zero and the q-trigonometric system $\{\mathcal{E}_q(x; i\omega_k)\}_{k=-\infty}^{\infty}$ is complete in the functional space under consideration, if we can justify the change of the order of summation in the following double sum

$$\sum_{n=0}^{\infty} \left(\sum_{k=-\infty}^{\infty} \mu_k h_n(y_k) h_m(y_k) \; c_n f_n \right), \qquad (8.9)$$

where

$$c_n = i^n \left(1 - q^{\nu+n}\right) \; q^{n(2-4\nu-n)/4}, \qquad \mu_k = \frac{A_k(\nu)}{j_{\nu-1,k}^2(q)},$$

$$h_p\left(y_k\right) = h_p^\nu\left(\frac{1}{j_{\nu-1,k}\left(q\right)};q\right), \quad p = m, n;$$

$$f_n = \int_{-1}^{1} f\left(x\right)\, C_n\left(x;q^\nu|q\right)\, \rho\left(x\right)\, dx.$$

We also introduce

$$B_n := \sum_{k=-\infty}^{\infty} \left|\mu_k h_n\left(y_k\right) h_m\left(y_k\right)\right|\, \left|c_n\right|\left|f_n\right|. \tag{8.10}$$

In view of Theorem 8.2 of [48], see also [71] and [74], one can change the order of summation in the double series (8.9) if $B_n < \infty$ for all $n = 0, 1, 2, \ldots$ and the sum $\sum_{n=0}^{\infty} B_n$ converges. In fact,

$$
\begin{aligned}
B_n &\leq \left(\sum_{k=-\infty}^{\infty} \mu_k \left|h_n\left(y_k\right)\right|^2\right)^{1/2}\left(\sum_{k=-\infty}^{\infty} \mu_k \left|h_m\left(y_k\right)\right|^2\right)^{1/2}\left|c_n\right|\left|f_n\right| \\
&= \left(\frac{q^{m\nu+m(m-1)/2}}{1-q^{\nu+m}}\right)^{1/2}\left(1-q^{\nu+n}\right)^{1/2}\, q^{n(1-2\nu)/4}\left|f_n\right|
\end{aligned}
$$

by the Cauchy–Schwarz inequality and (8.4). In the same fashion,

$$
\begin{aligned}
\sum_{n=0}^{\infty} B_n &\leq \left(2\pi q^{m\nu+m(m-1)/2}\frac{1-q^\nu}{1-q^{\nu+m}}\right)^{1/2} \\
&\quad \times \left(\sum_{n=0}^{\infty} \frac{\left(q^\nu, q^{\nu+1};q\right)_\infty \left(q^{2\nu};q\right)_n}{\left(q, q^{2\nu};q\right)_\infty \left(q;q\right)_n} q^{n(1-2\nu)/2}\right)^{1/2} \\
&\quad \times \left(\sum_{n=0}^{\infty} \frac{\left(q, q^{2\nu};q\right)_\infty \left(q;q\right)_n \left(1-q^{\nu+n}\right)}{2\pi \left(q^\nu, q^{\nu+1};q\right)_\infty \left(q^{2\nu};q\right)_n \left(1-q^\nu\right)}\left|f_n\right|^2\right)^{1/2} \\
&= \left(2\pi \frac{q^{m\nu+m(m-1)/2}}{1-q^{\nu+m}}\frac{\left(q^\nu, q^\nu;q\right)_\infty}{\left(q, q^{2\nu};q\right)_\infty}\right)^{1/2} \\
&\quad \times \left(\frac{\left(q^{1/2+\nu};q\right)_\infty}{\left(q^{1/2-\nu};q\right)_\infty}\right)^{1/2}\left(\int_{-1}^{1}\left|f\left(x\right)\right|^2 \rho\left(x\right)\, dx\right)^{1/2}
\end{aligned}
\tag{8.11}
$$

by Parseval's formula for the system of the continuous q-ultraspherical polynomials and by the q-binomial theorem when $0 < \nu < 1/2$. This completes the proof of the first part of our theorem. The second case is similar and we leave the details to the reader. $\qquad\square$

Remark 1 The system of basic exponential functions $\{\mathcal{E}_q(x; i\omega_k)\}_{k=-\infty}^{\infty}$ in Theorem 6 is not orthogonal in the weighted $\mathcal{L}_\rho^2(-1, 1)$ space, where $\rho(x)$ is the weight function of the continuous q-ultraspherical polynomials $C_m(x; q^\nu | q)$ for $0 < \nu < 1/2$. Several results on the completeness of the basic trigonometric system in $\mathcal{L}^p$-spaces for $p \geq 1$ have been established in [63] using the theory of entire functions. In particular, Theorem 6 holds for $0 < \nu < 1$ that includes the orthogonal case $\nu = 1/2$; see equations (9.12)–(9.13) below for the relations between the q-Bessel functions and the q-trigonometric functions. There is an overlap of the results of the two methods of the investigation of the completeness when $0 < \nu < 1/2$. An independent proof of the completeness in the most important practical case $\nu = 1/2$, when the system is orthogonal, is presented in [65].

Remark 2 As it has been pointed out in [60], in the special case $\nu = 3/2$ the orthogonality relation

$$(1 + (-1)^{m+n}) \sum_{k=1}^{\infty} \frac{1}{\omega_k^2 \, \kappa(\omega_k)} \, h_n^{3/2}\left(\frac{1}{2\omega_k}; q\right) h_m^{3/2}\left(\frac{1}{2\omega_k}; q\right)$$
$$= \frac{q^{(n+1)^2/2}}{1 - q^{n+3/2}} \, \delta_{mn} \tag{8.12}$$

and the completeness relation

$$\sum_{n=0}^{\infty} \frac{1 - q^{n+3/2}}{q^{(n+1)^2/2}} \, h_n^{3/2}\left(\frac{1}{2\omega_k}; q\right) h_n^{3/2}\left(\frac{1}{2\omega_l}; q\right) = \omega_k^2 \, \kappa(\omega_k) \, \delta_{kl} \tag{8.13}$$

for the corresponding q-Lommel polynomials can be written explicitly with the help of

$$\kappa(\omega) = \sum_{k=0}^{\infty} \frac{q^{k/2}}{1 + \omega^2 q^k}. \tag{8.14}$$

This function coincides with the normalization constant $k(\omega)$ given by (6.3) up to a factor. Similar result holds for $\nu = 1/2$ [64].

9. Basic Fourier Series

9.1. DEFINITION

A periodic function $f(x)$ with period $2l$ can be represented, under certain conditions, as the Fourier series,

$$f(x) = a_0 + \sum_{n=1}^{\infty} \left(a_n \cos\frac{\pi n}{l}x + b_n \sin\frac{\pi n}{l}x\right), \tag{9.1}$$

where

$$a_0 = \frac{1}{2l} \int_{-l}^{l} f(x) \, dx, \qquad (9.2)$$

$$a_n = \frac{1}{l} \int_{-l}^{l} f(x) \cos \frac{\pi n}{l} x \, dx, \qquad (9.3)$$

$$b_n = \frac{1}{l} \int_{-l}^{l} f(x) \sin \frac{\pi n}{l} x \, dx. \qquad (9.4)$$

For the extensive theory of these series see [1], [8], [66], [74], and [76].

Basic Fourier series have been recently introduced in [10] as the following q-analog of (9.1)

$$f(\cos \theta) = a_0 + \sum_{n=1}^{\infty} \left(a_n C_q(\cos \theta; \omega_n) + b_n S_q(\cos \theta; \omega_n) \right), \qquad (9.5)$$

where $\omega_0 = 0, \omega_1, \omega_2, \omega_3, \ldots$ are nonnegative roots of the transcendental equation (6.4) arranged in ascending order of magnitude; and the q-Fourier coefficients are given by

$$a_0 = \frac{1}{2k(0)} \int_0^{\pi} f(\cos \theta) \left(e^{2i\theta}, e^{-2i\theta}; q \right)_{1/2} d\theta, \qquad (9.6)$$

$$a_n = \frac{1}{k(\omega_n)} \int_0^{\pi} f(\cos \theta) \, C_q(\cos \theta; \omega_n) \qquad (9.7)$$
$$\times \left(e^{2i\theta}, e^{-2i\theta}; q \right)_{1/2} d\theta,$$

$$b_n = \frac{1}{k(\omega_n)} \int_0^{\pi} f(\cos \theta) \, S_q(\cos \theta; \omega_n) \qquad (9.8)$$
$$\times \left(e^{2i\theta}, e^{-2i\theta}; q \right)_{1/2} d\theta.$$

The complex form of the basic Fourier series (9.1) is

$$f(\cos \theta) = \sum_{n=-\infty}^{\infty} c_n \, \mathcal{E}_q(\cos \theta; i\omega_n) \qquad (9.9)$$

with

$$c_n = \frac{1}{2k(\omega_n)} \int_0^{\pi} f(\cos \theta) \, \mathcal{E}_q(\cos \theta; -i\omega_n) \qquad (9.10)$$
$$\times \left(e^{2i\theta}, e^{-2i\theta}; q \right)_{1/2} d\theta,$$

where $\omega_n = 0, \pm\omega_1, \pm\omega_2, \pm\omega_3, \ldots$ and $\omega_0 = 0 < \omega_1 < \omega_2 < \omega_3 < \ldots$ are nonnegative roots of (6.4); the normalization constants $k(\omega_n)$ are defined by (6.3). These expressions, of course, merely indicate how the coefficients of our basic Fourier series are to be determined on the hypothesis that the expansion exists and is uniformly convergent. We shall discuss the convergence of series (9.5) and (9.9) later.

Although classical Fourier series have a long and distinguished history, not much is known about the basic Fourier series. Bustoz and Suslov [10] had established the completeness of the basic trigonometric system $\{\mathcal{E}_q(x; i\omega_n)\}_{n=-\infty}^{\infty}$ which leads to some elementary facts about convergence of these series. Numerical investigation of the basic Fourier series had been started by Gosper and Suslov [23] with the help of the Macsyma computer algebra system. The special Macsyma program "namesum" was written by Bill Gosper for numerical evaluation of the infinite sums and infinite products. Explicit expansions in the basic Fourier series lead, in a natural way, to a new class of formulas never investigated before from the analytical or numerical point of view. Bustoz and Suslov [10] have discussed a few examples; more expansions have been found in [60]. It is worth mentioning also, that the method of basic Fourier series leads naturally to extensions of the classical Bernoulli and Euler polynomials and the zeta function [60], [64]. Our main objective in the next sections is to give a short overview of these results. See the original papers [10], [23], [60] and [64] for the proofs of the theorems and more details.

9.2. EIGENVALUES

In order to study basic Fourier series (9.5) and (9.9) one needs to investigate properties of ω-zeros of the basic trigonometric functions,

$$C_q(\eta; \omega), \qquad S_q(\eta; \omega), \qquad \eta = \left(q^{1/4} + q^{-1/4}\right)/2.$$
$$(9.11)$$

These zeros cannot be found explicitly which makes the theory of the basic Fourier series somewhat similar to the theory of the Fourier–Bessel series [73]. Properties of zeros of the functions (9.11) have been established in [10] on the basis of the Sturm–Liouville theory for the difference equation (2.8). It can also be shown that the basic sine $S_q(\eta; \omega)$ and cosine $C_q(\eta; \omega)$ functions are just multiples of the q-Bessel functions $J_{1/2}^{(2)}(2\omega; q)$ and $J_{-1/2}^{(2)}(2\omega; q)$, namely,

$$S_q(\eta; \omega) = \frac{(q; q)_\infty}{(q^{1/2}; q)_\infty} \frac{\omega^{1/2}}{(-q\omega^2; q^2)_\infty} J_{1/2}^{(2)}(2\omega; q),$$
$$(9.12)$$

$$C_q\left(\eta;\omega\right) = \frac{(q;q)_\infty}{\left(q^{1/2};q\right)_\infty} \frac{\omega^{1/2}}{\left(-q\omega^2;q^2\right)_\infty} J^{(2)}_{-1/2}\left(2\omega;q\right), \qquad (9.13)$$

see [10], [32] and [47]. The properties of zeros of the q-Bessel functions $J^{(2)}_\nu\left(r;q\right)$ were established in Ismail's papers [25] and [26] by a different method. This gives two independent approaches to the following theorems.

Theorem 7 *The basic sine $S_q\left(\eta;\omega\right)$ and basic cosine $C_q\left(\eta;\omega\right)$ functions have an infinity of real ω-zeros when $0 < q < 1$.*

Theorem 8 *The basic sine $S_q\left(\eta;\omega\right)$ and basic cosine $C_q\left(\eta;\omega\right)$ functions have only real ω-zeros when $0 < q < 1$.*

Theorem 9 *If $0 < q < 1$, then the real ω-zeros of the basic sine $S\left(\eta;\omega\right)$ and basic cosine $C\left(\eta;\omega\right)$ functions are simple.*

Theorem 10 *If $\omega_1, \omega_2, \omega_3, \ldots$ are the positive zeros of $S_q\left(\eta;\omega\right)$ arranged in ascending order of magnitude, and $\varpi_1, \varpi_2, \varpi_3, \ldots$ are those of $C_q\left(\eta;\omega\right)$, then*

$$0 = \omega_0 < \varpi_1 < \omega_1 < \varpi_2 < \omega_3 < \varpi_3 < \ldots, \qquad (9.14)$$

if $0 < q < 1$.

The tables of the zeros of the basic sine and cosine functions (9.11) are presented in [23] for different values of parameter q. Although these zeros can be found only numerically, there are simple asymptotic formulas

$$\omega_n = q^{1/4-n} - c_1\left(q\right) + o\left(1\right), \qquad \varpi_n = q^{3/4-n} - c_1\left(q\right) + o\left(1\right) \tag{9.15}$$

as $n \to \infty$. Here,

$$c_1\left(q\right) = \frac{1}{2} \frac{q^{1/4}}{1-q^{1/2}} \frac{\left(q;q^2\right)^2_\infty}{\left(q^2;q^2\right)^2_\infty} = \frac{1}{2} \frac{q^{1/4}\left(1+q^{1/2}\right)\left(1+q\right)}{\Gamma^2_{q^2}\left(1/2\right)} \tag{9.16}$$

and $\Gamma_{q^2}\left(z\right)$ is the q-gamma function [22]. The function $c_1\left(q\right)$ is a nonnegative, monotonically increasing function on $[0,1]$ (see [23] for the graph of this function) and, therefore, its maximum value on this interval is

$$\lim_{q\to 1^-} c_1\left(q\right) = 2/\pi \approx 0.63661977236758.$$

These formulas have been found numerically in [23], a rigorous proof is given in [62]. See also [23] for graphing of the basic trigonometric functions related to (9.11). Some monotonisity properties of zeros of (9.12)–(9.13) have been discussed in [23] and [30].

The sequences of the increasing lower bounds and the decreasing upper bounds for the first zeros ω_1 and ϖ_1 of the $S_q(\eta;\omega)$ and $C_q(\eta;\omega)$, respectively, have been found by Gosper and Suslov with the help of the Euler–Rayleigh method [31]. For example, the following inequalities hold

$$q^{-3/4}\left((1-q)\left(1-q^{3/2}\right)\right)^{1/2}$$
$$< \omega_1 < q^{-3/4}\left(\frac{\left(1-q^{3/2}\right)\left(1-q^{5/2}\right)(1+q)}{1+2q-q^{5/2}}\right)^{1/2}$$

and

$$q^{-1/4}\left(\left(1-q^{1/2}\right)(1-q)\right)^{1/2}$$
$$< \varpi_1 < q^{-1/4}\left(\frac{\left(1-q^{1/2}\right)\left(1-q^{3/2}\right)(1+q)}{1+2q-q^{3/2}}\right)^{1/2}.$$

See [23] for more details. In [62] the convergent sequences of the increasing lower bounds and the decreasing upper bounds were constructed for all zeros ω_n and ϖ_n of the $S_q(\eta;\omega)$ and $C_q(\eta;\omega)$.

9.3. CONVERGENCE

Elementary facts about the convergence of the q-Fourier series follow from the completeness of the basic trigonometric system $\{\mathcal{E}_q(x;i\omega_n)\}_{n=-\infty}^{\infty}$ [10].

Theorem 11 *If $f(x)$ and $g(x)$ have the same q-Fourier series, then $f \equiv g$.*

Theorem 12 *If $f(x)$ is continuous and $S[f]$, the q-Fourier series of function f, converges uniformly on $[-1,1]$, then its sum is $f(x)$.*

These theorems are the only tools available for the study of the explicit q-Fourier expansions at the current stage of the investigation.

Parseval's formula

$$2k(0)a_0^2 + \sum_{n=1}^{\infty} k(\omega_n)\left(a_n^2 + b_n^2\right) \qquad (9.17)$$
$$= \sum_{n=-\infty}^{\infty} 2k(\omega_n)|c_n|^2 = \int_{-1}^{1}|f(x)|^2\,\rho(x)\,dx$$

holds for the basic Fourier series due to the completeness of the q-trigonometric system $\{\mathcal{E}_q(x;i\omega_n)\}_{n=-\infty}^{\infty}$. Here c_n are the q-Fourier coefficients of $f(x)$, $2k(\omega_n)$ and $\rho(x)$ are the L^2-norm and the weight function, respectively.

9.4. EXAMPLES

Explicit expansions in basic Fourier series lead to lots of new interesting formulas. Bustoz and Suslov [10] have considered some of those; more q-Fourier expansions have been recently found in [60]. We can discuss here only a few examples.

9.4.1. *Linear function*

A periodic function $p_1(x)$ which is defined in the interval $(-1, 1)$ by $p_1(x) = x$ can be represented as

$$x = \sum_{n=-\infty}^{\infty} {}' \frac{(-1)^{n-1}}{i\pi n}\, e^{i\pi n x} \tag{9.18}$$

$$= 2 \sum_{n=1}^{\infty} (-1)^{n-1}\, \frac{\sin \pi n x}{\pi n}.$$

The following q-extension of this formula has been found in [10]

$$x = \frac{1}{2} \left(q^{1/4} + q^{-1/4} \right) \tag{9.19}$$

$$\times \sum_{n=-\infty}^{\infty} {}' \frac{(-1)^{n-1}}{i\,\kappa(\omega_n)\,\omega_n} \sqrt{\frac{(-q\omega_n^2; q^2)_\infty}{(-\omega_n^2; q^2)_\infty}}\, \mathcal{E}_q(x; i\omega_n)$$

$$= \left(q^{1/4} + q^{-1/4} \right)$$

$$\times \sum_{n=1}^{\infty} \frac{(-1)^{n-1}}{\kappa(\omega_n)\,\omega_n} \sqrt{\frac{(-q\omega_n^2; q^2)_\infty}{(-\omega_n^2; q^2)_\infty}}\, S_q(x; \omega_n),$$

where ω_n are roots of (6.9). This series converges uniformly on $[-1, 1]$, the details including the convenient uniform bounds of the basic exponential and trigonometric functions are given in [60]; see also [23] for numerical investigation of the convergence.

9.4.2. *Quadratic function*

Expansion of the quadratic function in the basic Fourier series is

$$x^2 = \frac{\left(1 + q^{1/2}\right)^2}{4\left(1 + q^{1/2} + q\right)} + \frac{\left(1 + q^{1/2}\right)\left(1 - q^2\right)}{4q} \tag{9.20}$$

$$\times \sum_{n=-\infty}^{\infty} {}' \frac{(-1)^n}{i\,\kappa(\omega_n)\,\omega_n^2} \sqrt{\frac{(-q\omega_n^2; q^2)_\infty}{(-\omega_n^2; q^2)_\infty}}\, \mathcal{E}_q(x; i\omega_n)$$

$$= \frac{\left(1+q^{1/2}\right)^2}{4\left(1+q^{1/2}+q\right)} - \frac{\left(1+q^{1/2}\right)\left(1-q^2\right)}{2q}$$

$$\times \sum_{n=1}^{\infty} \frac{(-1)^{n-1}}{\kappa\left(\omega_n\right)\omega_n^2} \sqrt{\frac{\left(-q\omega_n^2; q^2\right)_\infty}{\left(-\omega_n^2; q^2\right)_\infty}} \, C_q\left(x; \omega_n\right),$$

which is clearly a q-analog of

$$x^2 = \frac{1}{3} - 4\sum_{n=1}^{\infty} (-1)^{n-1} \frac{\cos \pi n x}{(\pi n)^2}. \tag{9.21}$$

See [23] for numerical investigation of convergence in (9.20).

9.4.3. Generalized power function
The q-Fourier cosine and sine expansion for even and odd "generalized power functions" are

$$\prod_{k=0}^{m-1} \left(\left(1+q^{k+1/2}\right)^2 - 4q^{k+1/2}x^2\right) \tag{9.22}$$

$$= \frac{(q;q)_{2m}}{\left(q^{1/2},\, q^{3/2}; q\right)_m} + 2\frac{(q;q)_{2m}}{\left(q^{1/2}; q\right)_m} \, q^{-m^2/2}$$

$$\times \sum_{n=1}^{\infty} \frac{(-1)^{n-1}}{\kappa\left(\omega_n\right)\omega_n^{m+1}} \, h_{m-1}\left(\frac{1}{2\omega_n}; q\right) \sqrt{\frac{\left(-q\omega_n^2; q^2\right)_\infty}{\left(-\omega_n^2; q^2\right)_\infty}} \, C_q\left(x; \omega_n\right),$$

$m > 1$, and

$$x \prod_{k=0}^{m-2} \left(\left(1+q^{k+1/2}\right)^2 - 4q^{k+1/2}x^2\right) \tag{9.23}$$

$$= \frac{(q;q)_{2m-1}}{\left(q^{1/2}; q\right)_m} \, q^{-m^2/2+1/4}$$

$$\times \sum_{n=1}^{\infty} \frac{(-1)^{n-1}}{\kappa\left(\omega_n\right)\omega_n^m} \, h_{m-1}\left(\frac{1}{2\omega_n}; q\right) \sqrt{\frac{\left(-q\omega_n^2; q^2\right)_\infty}{\left(-\omega_n^2; q^2\right)_\infty}} \, S_q\left(x; \omega_n\right),$$

$m > 0$, respectively. The special q-Lommel polynomials $h_n\left(x; q\right)$ are defined by

$$h_n\left(x; q\right) = h_n^{3/2}\left(x; q\right). \tag{9.24}$$

These formulas allow us to expand the elementary powers of the higher degrees in the basic Fourier series. One can also obtain relations with the q-Bernoulli polynomials discussed in the next section.

The q-Fourier series for the continuous q-ultraspherical polynomials and some other polynomial examples can be found in [60].

9.4.4. *Basic cosine and sine functions*

The Fourier series of the cosine function in the interval $(-\pi, \pi)$ has the form

$$\cos \omega x = \frac{2}{\pi} \sin \omega \pi \left[\frac{1}{2\omega} + \sum_{n=1}^{\infty} (-1)^n \frac{\omega \cos nx}{\omega^2 - n^2} \right]. \tag{9.25}$$

The basic Fourier expansion for the q-cosine function $C_q(x; \omega)$ is found in [60],

$$C_q(x; \omega) = 2S_q(\eta; \omega) \tag{9.26}$$

$$\times \left[\frac{1}{2\kappa(0)\omega} + \sum_{n=1}^{\infty} \frac{(-1)^n \omega}{\kappa(\omega_n)(\omega^2 - \omega_n^2)} \sqrt{\frac{(-q\omega_n^2; q^2)_\infty}{(-\omega_n^2; q^2)_\infty}} C_q(x; \omega_n) \right].$$

In a similar fashion, the classical Fourier series for the sine function in the interval $(-\pi, \pi)$ has the form

$$\sin \omega x = \frac{2}{\pi} \sin \omega \pi \sum_{n=1}^{\infty} (-1)^n \frac{n \sin nx}{\omega^2 - n^2}. \tag{9.27}$$

An analog of this formula for the q-sine function $S_q(x; \omega)$ is

$$S_q(x; \omega) = 2S_q(\eta; \omega) \tag{9.28}$$

$$\times \sum_{n=1}^{\infty} \frac{(-1)^n \omega_n}{\kappa(\omega_n)(\omega^2 - \omega_n^2)} \sqrt{\frac{(-q\omega_n^2; q^2)_\infty}{(-\omega_n^2; q^2)_\infty}} S_q(x; \omega_n).$$

See [60] for the proof of these expansions.

9.4.5. *Basic exponential function*

Expansion of the exponential function $\exp(\alpha x)$ in Fourier series on $(-\pi, \pi)$ is given by

$$e^{\alpha x} = \frac{e^{\alpha \pi} - e^{-\alpha \pi}}{\pi} \left[\frac{1}{2\alpha} + \sum_{n=1}^{\infty} \frac{(-1)^n}{\alpha^2 + n^2} (\alpha \cos nx - n \sin nx) \right]. \tag{9.29}$$

The expansion of $\mathcal{E}_q(x; \alpha)$ in the basic Fourier series has a similar form,

$$\mathcal{E}_q(x; \alpha) = \frac{(-\alpha; q^{1/2})_\infty - (\alpha; q^{1/2})_\infty}{(q\alpha^2; q^2)_\infty} \tag{9.30}$$

$$\times \left[-\frac{1}{2\kappa(0)\,\alpha} + \sum_{n=1}^{\infty} \frac{(-1)^n}{\kappa(\omega_n)\,(\alpha^2 + \omega_n^2)} \right.$$

$$\left. \times \sqrt{\frac{(-q\omega_n^2; q^2)_\infty}{(-\omega_n^2; q^2)_\infty}} \left(\alpha C_q(x; \omega_n) - \omega_n S_q(x; \omega_n)\right) \right].$$

This follows directly from the basic Fourier series for the $C_q(x; \omega)$ and $S_q(x; \omega)$ and the q-Euler formula (4.1).

Formula (9.30) admits further generalization, namely,

$$\mathcal{E}_q(\cos\theta; \alpha)\, \frac{\left(q^{1/2}e^{2i\theta},\, q^{1/2}e^{-2i\theta};\, q\right)_\infty}{\left(\gamma e^{2i\theta},\, \gamma e^{-2i\theta};\, q\right)_\infty}\, C_m(\cos\theta; \gamma|q) \qquad (9.31)$$

$$= \frac{\pi\left(\gamma,\, \gamma q^{m+1}; q\right)_\infty}{(q; q)_m\, \left(q,\, \gamma^2 q^m; q\right)_\infty\, (q\alpha^2; q^2)_\infty}$$

$$\times \sum_{n=-\infty}^{\infty} \frac{\left(i\alpha\omega_n q^{(m+1)/2}; q\right)_\infty}{k(\omega_n)\, (-q\omega_n^2; q^2)_\infty}\, \alpha^m q^{m^2/4}$$

$$\times \left(iq^{(1-m)/2}\omega_n/\alpha; q\right)_m\, \mathcal{E}_q(\cos\theta; i\omega_n)$$

$$\times {}_2\varphi_2\left(\begin{matrix} iq^{(m+1)/2}\omega_n/\alpha, & -iq^{(m+1)/2}\alpha/\omega_n \\ \gamma q^{m+1}, & iq^{(m+1)/2}\alpha\omega_n \end{matrix}\; ;\; q,\; iq^{(m+1)/2}\alpha\gamma\omega_n \right).$$

Here $C_m(\cos\theta; \gamma|q)$ are the continuous q-ultraspherical polynomials (5.5). This expansion follows from the integral (5.4). It has many interesting special and limiting cases including expansions for the "generalized powers" (9.22)–(9.23); see [60] for more details.

9.4.6. Basic cosecant and cotangent functions

The partial fractions decompositions for the cosecant and cotangent functions are

$$\frac{1}{\sin\omega} = \frac{1}{\omega} + \sum_{n=1}^{\infty} (-1)^n \left(\frac{1}{\omega - \pi n} + \frac{1}{\omega + \pi n}\right) \qquad (9.32)$$

and

$$\cot\omega = \frac{1}{\omega} + \sum_{n=1}^{\infty} \left(\frac{1}{\omega - \pi n} + \frac{1}{\omega + \pi n}\right). \qquad (9.33)$$

Special cases $x = 0$ and $x = \eta$ of the expansion (9.26) result in the following q-extensions

$$\frac{1}{S_q(\eta; \omega)} \qquad (9.34)$$

$$= \frac{1}{2\kappa(0)\,\omega} + \sum_{n=1}^{\infty} \frac{(-1)^n}{\kappa(\omega_n)} \left(\frac{1}{\omega - \omega_n} + \frac{1}{\omega + \omega_n} \right) \sqrt{\frac{(-q\omega_n^2; q^2)_\infty}{(-\omega_n^2; q^2)_\infty}}$$

and

$$\mathrm{Cot}_q (\omega) := \frac{C_q(\eta; \omega)}{S_q(\eta; \omega)} = \frac{1}{\kappa(0)\,\omega} + \sum_{n=1}^{\infty} \frac{1}{\kappa(\omega_n)} \left(\frac{1}{\omega - \omega_n} + \frac{1}{\omega + \omega_n} \right), \tag{9.35}$$

respectively.

The following partial fraction decomposition

$$\sum_{n=-\infty}^{\infty} \frac{1}{\kappa(\omega_n)(\omega - \omega_n)^2} = \frac{(-\omega^2; q^2)_\infty}{(-q\omega^2; q^2)_\infty} \frac{\kappa(\omega)}{S_q^2(\eta; \omega)} \tag{9.36}$$

has been found in [60] as a q-analog of another elementary formula

$$\sum_{n=-\infty}^{\infty} \frac{1}{(\omega - \pi n)^2} = \frac{1}{\sin^2 \omega}. \tag{9.37}$$

10. Bernoulli Polynomials, Numbers and Their q-Extensions

The classical Bernoulli polynomials can be introduced by means of the generating function

$$\frac{\alpha e^{\alpha x}}{e^\alpha - 1} = 1 + \sum_{m=1}^{\infty} B_m(x) \frac{\alpha^m}{m!}, \qquad |\alpha| < 2\pi. \tag{10.1}$$

They have the following Fourier expansions

$$B_{2m-1}(x) = (-1)^m \frac{2(2m-1)!}{(2\pi)^{2m-1}} \sum_{n=1}^{\infty} \frac{\sin 2\pi n x}{n^{2m-1}}, \tag{10.2}$$

$$B_{2m}(x) = (-1)^{m-1} \frac{2(2m)!}{(2\pi)^{2m}} \sum_{n=1}^{\infty} \frac{\cos 2\pi n x}{n^{2m}} \tag{10.3}$$

with $m = 1, 2, 3, \dots$. The Bernoulli numbers are defined as

$$B_m = B_m(0). \tag{10.4}$$

One can see that $B_0 = 1, B_1 = -1/2, B_2 = 1/6, B_4 = -1/30, B_6 = 1/42$; $B_{2m+1} = 0$ when $m = 1, 2, \dots$. These numbers appear as the coefficients in Taylor's expansion

$$\frac{\omega}{2} \cot \frac{\omega}{2} = 1 + \sum_{m=1}^{\infty} (-1)^m \frac{B_{2m}}{(2m)!} \omega^{2m} \tag{10.5}$$

$$= 1 - \frac{1}{12}\,\omega^2 - \frac{1}{720}\,\omega^4 - \frac{1}{30240}\,\omega^6 - \dots .$$

The special case $x = 0$ of (10.1) is the generating function

$$\frac{\alpha}{e^\alpha - 1} = 1 + \sum_{m=1}^{\infty} B_m \, \frac{\alpha^m}{m!}. \tag{10.6}$$

See, for example, [4] and [74] for other properties of the Bernoulli polynomials and numbers.

The method of the basic Fourier series discussed in the previous section allows us to introduce q-analogs of the Bernoulli polynomials in a similar manner [60]. Expansion (9.30) for the q-exponential function $\mathcal{E}_q(x; \alpha)$ gives rise to the following generating relation

$$\frac{\alpha(q\alpha^2; q^2)_\infty \, \mathcal{E}_q(x; \alpha)}{(-\alpha; q^{1/2})_\infty - (\alpha; q^{1/2})_\infty} \tag{10.7}$$

$$= \frac{1}{2}\left(1 - q^{1/2}\right) + \sum_{m=1}^{\infty} \alpha^{2m-1} \, \mathcal{B}_{2m-1}(x; q) + \sum_{m=1}^{\infty} \alpha^{2m} \, \mathcal{B}_{2m}(x; q),$$

where

$$\mathcal{B}_{2m-1}(x; q) \tag{10.8}$$

$$= (-1)^{m-1} \sum_{n=1}^{\infty} \frac{(-1)^{n-1}}{\kappa(\omega_n)\,\omega_n{}^{2m-1}} \sqrt{\frac{(-q\omega_n^2; q^2)_\infty}{(-\omega_n^2; q^2)_\infty}} \, S_q(x; \omega_n),$$

$$\mathcal{B}_{2m}(x; q) = (-1)^m \sum_{n=1}^{\infty} \frac{(-1)^{n-1}}{\kappa(\omega_n)\,\omega_n{}^{2m}} \sqrt{\frac{(-q\omega_n^2; q^2)_\infty}{(-\omega_n^2; q^2)_\infty}} \, C_q(x; \omega_n) \tag{10.9}$$

are q-analogs of the Bernoulli polynomials. It can be shown that functions $\mathcal{B}_m(x; q)$ given by (10.8) and (10.9) are, indeed, polynomials of the exact degrees m in x [60].

We call $\mathcal{B}_m(x; q)$ the q-Bernoulli polynomials. One can also introduce the q-Bernoulli numbers as follows

$$\mathcal{B}_m(q) := \mathcal{B}_m(-\eta; q) \tag{10.10}$$

in accordance with the classical case (10.4). The first five q-Bernoulli polynomials and numbers are explicitly given in [60]. As in the classical case,

$\mathcal{B}_{2m+1}(q) = 0$, $m = 1, 2, 3, \dots$. The q-analog of Taylor's expansion (10.5) has the form

$$\omega \operatorname{Cot}_q(\omega) = 1 - q^{1/2} + 2 \sum_{m=1}^{\infty} (-1)^m B_{2m}(q) \; \omega^{2m}.$$

(10.11)

Several properties of the q-Bernoulli polynomials and numbers are similar to those for the classical ones. See [60] for more details on the investigation of the q-Bernoulli polynomials and numbers. Other q-analogs of the Bernoulli polynomials and numbers were studied in [2], [11]–[13], [38], [54], [68], and [72]. Analogs of the Euler polynomials and numbers are introduced in a similar fashion in our forthcoming paper [64].

11. Extension of Riemann's Zeta Function

The Riemann zeta function is, usually, introduced as the Dirichlet series of the form

$$\zeta(z) = \sum_{n=1}^{\infty} n^{-z}.$$

(11.1)

This series converges uniformly and absolutely for Re $z > 1$ and, therefore, defines a holomorphic function in the half-plane Re $z > 1$. There exists an analytic continuation of this function in the entire complex plane; see, for example, [4] and [74]. In view of the relation,

$$\zeta(2m) = (-1)^{m-1} \frac{(2\pi)^{2m}}{2(2m)!} B_{2m}, \qquad m > 0,$$

(11.2)

Riemann's zeta function can be viewed as an extension of Bernoulli's numbers to an arbitrary complex index.

An extension of the zeta function can be introduced as [60]

$$\zeta_q(z) = \sum_{n=1}^{\infty} \frac{1}{\kappa(\omega_n) \; \omega_n{}^z}.$$

(11.3)

It can be shown that the right side here is a uniformly and absolutely convergent series of analytic functions in any domain Re $z > 1$ and, consequently, the series is an analytic function in such a domain. It can also be shown that, as in the classical case, the $\zeta_q(z)$ has a simple pole at $z = 1$ and it has no other singularities for Re $z > 0$. An analytic continuation in the entire complex plane is an open problem.

The following expression

$$\zeta_q(2m) = (-1)^{m-1} \mathcal{B}_{2m}(q) \tag{11.4}$$

gives the q-analog of the classical relation (11.2) between the zeta function and the Bernoulli numbers. Two important special cases are

$$\zeta_q(2) = \frac{q^{1/2}}{2\left(1 - q^{3/2}\right)}, \qquad \zeta_q(4) = \frac{q^2}{2\left(1 - q^{3/2}\right)^2 \left(1 - q^{5/2}\right)}, \tag{11.5}$$

which are q-analogs of the classical Euler results,

$$\zeta(2) = \sum_{n=1}^{\infty} \frac{1}{n^2} = \frac{\pi^2}{6}, \qquad \zeta(4) = \sum_{n=1}^{\infty} \frac{1}{n^4} = \frac{\pi^4}{90}. \tag{11.6}$$

Euler considered also the following series

$$\phi(z) = \sum_{n=1}^{\infty} \frac{(-1)^{n-1}}{n^z} \tag{11.7}$$

related to the zeta function as

$$\phi(z) = \left(1 - 2^{1-z}\right) \zeta(z). \tag{11.8}$$

Special values are

$$\phi(2) = \frac{\pi^2}{12}, \qquad \phi(4) = \frac{7\pi^4}{720} \tag{11.9}$$

in view of (11.6) and (11.8).

We can introduce a q-extension of $\phi(z)$ as

$$\phi_q(z) = \sum_{n=1}^{\infty} \frac{(-1)^{n-1}}{\kappa(\omega_n)\,\omega_n{}^z} \sqrt{\frac{(-q\omega_n^2; q^2)_\infty}{(-\omega_n^2; q^2)_\infty}} \tag{11.10}$$

and evaluate the following sums

$$\phi_q(2m) = (-1)^m \, B_{2m}(0; q) \tag{11.11}$$

in terms of values of the q-Bernoulli polynomials $B_{2m}(x; q)$ at $x = 0$. These values can also be found in terms of the q-Bernoulli numbers [60].

Extensions of Euler's results (11.9) are

$$\phi_q(2) = \frac{q}{2(1 + q)\left(1 - q^{3/2}\right)}, \tag{11.12}$$

$$\phi_q(4) = \frac{q^{5/2}\left(1 + 2q + q^{3/2} + 2q^2 + q^3\right)}{2\left(1 + q\right)^2\left(1 + q^2\right)\left(1 - q^{3/2}\right)^2\left(1 - q^{5/2}\right)}. \tag{11.13}$$

One can see that

$$\phi_q(2)/\zeta_q(2) = \frac{q^{1/2}}{1 + q} \to \frac{1}{2}, \tag{11.14}$$

$$\phi_q(4)/\zeta_q(4) = q^{1/2}\frac{1 + 2q + q^{3/2} + 2q^2 + q^3}{(1 + q)^2(1 + q^2)} \to \frac{7}{8} \tag{11.15}$$

as $q \to 1^-$ (compare with (11.8) for $z = 2$ and $z = 4$, respectively).

We hope that this observation will lead, eventually, to a q-extension of the theory of the zeta and related functions. Other q-analogs of the zeta function were discussed by Al-Salam [2], Satoh [53], Tsumura [67], [68], [69], [70], and in the recent papers by Cherednik [14]–[16]. See also Ruijsenaars' contribution to these proceedings [52], where – among other things – his novel results on trigonometric, elliptic and hyperbolic generalizations of the Hurwitz zeta function are presented for the first time. See also Ruijsenaars' original papers [49]–[51] and references therein.

12. Concluding Remarks

In conclusion, I would like to outline several directions for future investigation.

1. Positivity. The q-exponential function $\mathcal{E}_q(x; \alpha)$ is a positive function for all $x \geq 0$ when $0 \leq \alpha < q^{-1/2}$ by (3.7). A simple estimate, similar to one in the proof of Theorem 1, shows that $\mathcal{E}_q(x; \alpha) > 0$ when $x > -a$, $a > 0$ and $\mathcal{E}_q(a; |\alpha|) < 2$, $q\alpha^2 < 1$. A similar consideration on the basis of (3.14) instead of (3.7) leads to the inequality $e(x, \alpha) \geq 2 - e(a, |\alpha|)$, $|x| < a$ for the entire function $e(x, \alpha) = (q\alpha^2; q^2) \mathcal{E}_q(x; \alpha)$. Therefore $e(x, \alpha) > 0$ when $x > -a$ and $e(a, |\alpha|) < 2$. Almost nothing is known about the positivity of the q-exponential function and/or about its zeros in larger domains.

2. More explicit q-Fourier expansions – joy of discovering of new formulas! This, obviously, requires more explicit integral evaluations involving the q-exponential functions. The fundamental Askey–Wilson integral [6] and its extension, the Nassrallah–Rahman integral [22], are expected to play the central role in this study.

3. Investigation of the summability of the basic Fourier series. Methods of summation and expansion theorems somewhat similar to the classical ones should be found. An analog of Abel's summation method

has already been discussed in [10]. A systematic study has not been undertaken.

4. Extension of the classical results on the completeness of the trigonometric systems $\left\{e^{i\lambda_n x}\right\}_{n=-\infty}^{\infty}$ established in [44]–[45] to the q-exponential functions. See [63] for the 'first step' in this direction.

5. The q-Fourier transform. See [5] and references therein for analogs of the Fourier integral related to the Poisson kernels of the Askey–Wilson polynomials and their special cases. This was developed as an extension of Wiener's apporoach [75] to the classical Fourier transform. Another version, related to the q-exponential functions, is to be investigated.

6. Group theoretical interpretation of the q-addition theorems.

7. Further investigation of the q-analogs of the Bernoulli and Euler polynomials and numbers. The q-Bernoulli polynomials and numbers have been briefly discussed here on the basis of the q-Fourier series; see also [60]. Analogs of the Euler polynomials and numbers can be introduced in a similar fashion with the help of another version of the q-Fourier series which uses the zeros of (6.10) instead of (6.9) as eigenvalues. This has been examined in [64].

8. Thorough investigation of the q-zeta function. Analytic continuation and a "functional equation". Zeros and numerical analysis in the complex plane.

9. Theory of the q-Fourier–Bessel series and the corresponding integral transforms. See [29] for the new orthogonality relation of the q-Bessel functions on a q-quadratic grid. The theory of these functions somewhat similar to [73] has not been developed.

10. Theory of the orthogonal basic hypergeometric functions. It is well known that the q-exponential and q-trigonometric functions discussed in this paper are solutions of a very special case of the general difference equation of hypergeometric type on nonuniform lattices [7], [46] and [56]. The Askey–Wilson polynomials [6] – and more generally the classical orthogonal polynomials [3], [39], [46] – are well-known as the most important orthogonal solutions of this equation. In the last few years Ismail, Masson and Suslov [29], and Suslov [56], [58], [59] have found another type of orthogonal solutions of this difference equation with respect to continuous weight functions at the higher levels of the nonterminating $2\varphi_1$ and $8\varphi_7$-functions, correspondingly. Recently Koelink and Stokman [40] have established an orthogonality relation of the $8\varphi_7$-function under consideration when the measure includes an absolutely continuous part and an infinite series of mass points. See also Koelink and Stokman's contribution [41] to these proceedings for more details. All these results have to be generalized to the highest level of the nonterminating $10\varphi_9$-biorthogonal rational functions and, possibly, to the

elliptic hypergeometric functions. See Spiridonov and Zhedanov's contribution to these proceedings [55] for the recent progress in the theory of these functions and bibliography.

13. Appendix

Here we shall give independent proofs of the addition formula (3.3) and the generating relation (3.14) on the basis of a general approach discussed by Ruijsenaars [52]. Consider the difference equation of hypergeometric type in a self-adjoint form

$$\frac{\Delta}{\nabla x_1(z)}\left(\sigma(z)\,\rho(z)\,\frac{\nabla y(z)}{\nabla x(z)}\right) + \lambda\rho(z)\,y(z) = 0, \qquad (13.1)$$

where $\Delta f(z) = \nabla f(z+1) = f(z+1) - f(z)$, $x_1(z) = x(z+1/2)$ and $x(z) = (q^z + q^{-z})/2$. Suppose that y_1 and y_2 are two solutions of (13.1) corresponding to the same eigenvalue λ. Then

$$\Delta\left[\sigma(z)\,\rho(z)\,W(y_1, y_2)\right] = 0, \qquad (13.2)$$

where

$$W(y_1, y_2) = \begin{vmatrix} y_1(z) & y_2(z) \\ \dfrac{\nabla y_1(z)}{\nabla x(z)} & \dfrac{\nabla y_2(z)}{\nabla x(z)} \end{vmatrix} = \frac{y_2(z)\,y_1(z-1) - y_1(z)\,y_2(z-1)}{x(z) - x(z-1)} \qquad (13.3)$$

is the difference analog of the Wronskian. See, for example, [46], [52], [56], and [58]–[59] for more details. If y_3 is another solution of (13.1) corresponding to the same eigenvalue, then a straighforward calculation shows

$$y_3(z) = -\frac{W(y_2, y_3)}{W(y_1, y_2)}\,y_1(z) + \frac{W(y_1, y_3)}{W(y_1, y_2)}\,y_2(z). \qquad (13.4)$$

It follows also from (13.2) that the ratios $W(y_k, y_3)/W(y_1, y_2)$, $k = 1, 2$ are periodic functions in z of period 1. The last equation gives us the possibility to find relations between different solutions of (13.1). We apply this method here only at the level of the q-exponential functions to derive the addition formula (3.3) and the generating function (3.14).

For the q-exponential functions equation (13.1) reduces to (2.8) and we can choose $\sigma = \rho = 1$. Consider three different solutions discussed in Section 3,

$$y_1(z) = \mathcal{E}_q(x; \alpha), \quad y_2(z) = \mathcal{E}_q(x; -\alpha), \quad y_3(z) = \mathcal{E}_q(x, y; \alpha). \qquad (13.5)$$

Then, by (13.2), $\Delta\left[W\left(y_k, y_l\right)\right] = 0$ and for $|\alpha| < 1$ all the q-Wronskians are constants as doubly periodic entire functions. One can find these constants using special values of x. Equations (2.6) and (13.3) for $W\left(y_1, y_3\right)$ give

$$\mathcal{E}_q\left(x; \alpha\right)\ \mathcal{E}_q\left(x\left(z - 1/2\right), y; \alpha\right) - \mathcal{E}_q\left(x, y; \alpha\right)\ \mathcal{E}_q\left(x\left(z - 1/2\right); \alpha\right) = A. \tag{13.6}$$

Letting $x = \eta = x\left(\pm 1/4\right) = \left(q^{1/4} + q^{-1/4}\right)/2$,

$$A = \mathcal{E}_q\left(\eta; \alpha\right)\ \mathcal{E}_q\left(\eta, y; \alpha\right) - \mathcal{E}_q\left(\eta, y; \alpha\right)\ \mathcal{E}_q\left(\eta; \alpha\right) = 0 \tag{13.7}$$

and

$$W\left(y_1, y_3\right) = 0. \tag{13.8}$$

In the case of $W\left(y_1, y_2\right)$, in the same vain,

$$\mathcal{E}_q\left(x; \alpha\right)\ \mathcal{E}_q\left(x\left(z - 1/2\right); -\alpha\right) + \mathcal{E}_q\left(x; -\alpha\right)\ \mathcal{E}_q\left(x\left(z - 1/2\right); \alpha\right) = B, \tag{13.9}$$

and for $x = \eta$,

$$B = 2\,\mathcal{E}_q\left(\eta; \alpha\right)\ \mathcal{E}_q\left(\eta; -\alpha\right) = 2\frac{\left(\alpha^2; q^2\right)_\infty}{\left(q\alpha^2; q^2\right)_\infty}, \tag{13.10}$$

in view of

$$\mathcal{E}_q\left(\eta; \alpha\right) = \frac{\left(-\alpha; q^{1/2}\right)_\infty}{\left(q\alpha^2; q^2\right)_\infty}. \tag{13.11}$$

This follows directly from (2.3) by a consequence of the q-binomial theorem, or from (5.37)–(5.38) and (7.4) of [10]. As a result,

$$W\left(y_1, y_2\right) = -\frac{4q^{1/4}\alpha}{1 - q}\ \frac{\left(\alpha^2; q^2\right)_\infty}{\left(q\alpha^2; q^2\right)_\infty}. \tag{13.12}$$

Finally, substituting (13.8) in (13.4) and letting $x = 0$ we get

$$\frac{W\left(y_2, y_3\right)}{W\left(y_1, y_2\right)} = -\mathcal{E}_q\left(y; \alpha\right) \tag{13.13}$$

and our proof of the first addition formula (3.3) is complete. We have also evaluated the q-Wronskian

$$W\left(y_2, y_3\right) = \frac{4q^{1/4}\alpha}{1 - q}\ \frac{\left(\alpha^2; q^2\right)_\infty}{\left(q\alpha^2; q^2\right)_\infty}\ \mathcal{E}_q\left(y; \alpha\right) \tag{13.14}$$

which is of independent interest.

Our next goal is to derive the generating function (3.14). Consider the series

$$F(x, \alpha) := \sum_{n=0}^{\infty} \frac{q^{n^2/4}}{(q; q)_n} \alpha^n H_n(x|q). \qquad (13.15)$$

Using the difference-differentiation formula

$$\frac{\delta}{\delta x} H_n(x|q) = 2q^{(1-n)/2} \frac{1 - q^n}{1 - q} H_{n-1}(x|q) \qquad (13.16)$$

for the continuous q-Hermite polynomials – see, for example, [57] – one can show that the function

$$y_4(z) = F(x, \alpha) \qquad (13.17)$$

is a solution of (2.6) and (2.8). Thus,

$$y_4(z) = -\frac{W(y_2, y_4)}{W(y_1, y_2)} y_1(z) + \frac{W(y_1, y_4)}{W(y_1, y_2)} y_2(z). \qquad (13.18)$$

Evaluation of $W(y_1, y_4)$ gives

$$\mathcal{E}_q(x; \alpha) \ F(x(z - 1/2), \alpha) - F(x, \alpha) \ \mathcal{E}_q(x(z - 1/2); \alpha) = C \qquad (13.19)$$

and letting $x = \eta$ we obtain $C = 0$, or

$$W(y_1, y_4) = 0. \qquad (13.20)$$

Substituting this in (13.18) and choosing $x = 0$ one gets

$$F(0, \alpha) = -\frac{W(y_2, y_4)}{W(y_1, y_2)}. \qquad (13.21)$$

So, we need to evaluate

$$F(0, \alpha) = \sum_{n=0}^{\infty} \frac{q^{n^2/4}}{(q; q)_n} \alpha^n H_n(0|q), \qquad (13.22)$$

where

$$H_{2k+1}(0|q) = 0, \qquad H_{2k}(0|q) = (-1)^k (q; q^2)_k. \qquad (13.23)$$

These special values can be found from (3.15), see also Ex 7.17 of [22]; or from the Rogers generating function given, for example, in Ex 1.28 of the same book. Therefore

$$F(0, \alpha) = \sum_{n=0}^{\infty} (-1)^k \frac{q^{k^2}}{(q^2; q^2)_k} \alpha^{2k} = (q\alpha^2; q^2)_\infty \qquad (13.24)$$

by (1.2) and

$$F(x, \alpha) = (q\alpha^2; q^2)_\infty \, \mathcal{E}_q(x; \alpha) \qquad (13.25)$$

which is the generating relation (3.14). In a similar fashion one can derive (3.16).

This method can be used to derive relations between different solutions of the difference equation of the hypergeometric type (13.1) up to the $_8\varphi_7$-level.

Acknowledgments. The author thanks Joaquin Bustoz, John McDonald and Simon Ruijsenaars for valuable discussions and help. He was supported in part by NSF grant # DMS 9803443.

References

1. N. I. Akhiezer, *Theory of Approximation*, Frederick Ungar Publishing Co., New York, 1956.
2. W. A. Al-Salam, *q-Bernoulli numbers and polynomials*, Math. Nachr. **17** (1959), 239–260.
3. G. E. Andrews and R. A. Askey, *Classical orthogonal polynomials*, in: "Polynômes orthogonaux et applications", Lecture Notes in Math. **1171**, Springer-Verlag, 1985, pp. 36–62.
4. G. E. Andrews, R. A. Askey, and R. Roy, *Special Functions*, Cambridge University Press, Cambridge, 1999.
5. R. A. Askey, M. Rahman, and S. K. Suslov, *On a general q-Fourier transformation with nonsymmetric kernels*, J. Comp. Appl. Math. **68** (1996), 25–55.
6. R. A. Askey and J. A. Wilson, *Some basic hypergeometric orthogonal polynomials that generalize Jacobi polynomials*, Memoirs Amer. Math. Soc., Number **319** (1985).
7. N. M. Atakishiyev and S. K. Suslov, *Difference hypergeometric functions*, in: "Progress in Approximation Theory: An International Perspective", A. A. Gonchar and E. B. Saff, eds, Springer Series in Computational Mathematics, Vol. 19, Springer-Verlag, 1992, pp. 1–35.
8. N. K. Bary, *A Treatise on Trigonometric Series*, in two volumes, Macmillan, New York, 1964.
9. R. P. Boas, *Entire Functions*, Academic Press, New York, 1954.
10. J. Bustoz and S. K. Suslov, *Basic analog of Fourier series on a q-quadratic grid*, Methods Appl. Anal. **5** (1998), 1–38.
11. L. Carlitz, *q-Bernoulli numbers and polynomials*, Duke Math. J. **15** (1948), 987–1000.
12. L. Carlitz, *q-Bernoulli and Eulerian numbers*, Trans. Amer. Math. Soc. **76** (1954), 332–350.
13. L. Carlitz, *Expansions of q-Bernoulli numbers*, Duke Math. J. **25** (1958), 355–364.

14. I. Cherednik, *From double Hecke algebra to analysis*, Documenta Mathematica, Extra Volume ICM 1998; Preprint math.QA/9806097, v2, 13 Jul 1998.
15. I. Cherednik, *On q-analogues of Riemann's zeta*, Preprint math.QA/9804099, v3, 8 Oct 1998.
16. I. Cherednik, *Calculations of zeros of a q-zeta function numerically*, Preprint math.QA/9804100, v2, 12 May 1998.
17. R. Floreanini and L. Vinet, *On the quantum group and quantum algebra approach to q-special functions*, Lett. Math. Phys. **27** (1993), 179–190.
18. R. Floreanini and L. Vinet, *A model for the continuous q-ultraspherical polynomials*, J. Math. Phys. **36** (1995), 3800–3813.
19. R. Floreanini and L. Vinet, *More on the q-oscillator algebra and q-orthogonal polynomials*, J. Phys. **A28** (1995), L287–L293.
20. R. Floreanini, J. LeTourneux, and L. Vinet, *Symmetry techniques for the Al-Salam-Chihara polynomials*, J. Phys. **A30** (1997), 3107–3114.
21. R. Floreanini, J. LeTourneux, and L. Vinet, *Symmetries and continuous q-orthogonal polynomials*, in: "Algebraic Methods and q-Special Functions", J. F. van Diejen and L. Vinet, eds., CRM Proceedings & Lecture Notes, Vol. 22, American Mathematical Society, 1999, pp. 135–144.
22. G. Gasper and M. Rahman, *Basic Hypergeometric Series*, Cambridge University Press, Cambridge, 1990.
23. R. W. Gosper, Jr. and S. K. Suslov, *Numerical investigation of basic Fourier series*, in "q-Series from a Contemporary Perspective", M. E. H. Ismail, D. R. Stanton, eds., Contemporary Mathematics, Vol. 254, Amer. Math. Soc., 2000, pp. 199–227.
24. Y. Chen, M. E. H. Ismail, and K. A. Muttalib, *Asymptotics of basic Bessel functions and q-Laguerre polynomials*, J. Comp. Appl. Math. **54** (1994), 263–272.
25. M. E. H. Ismail, *The basic Bessel functions and polynomials*, SIAM J. Math. Anal. **12** (1981), 454–468.
26. M. E. H. Ismail, *The zeros of basic Bessel functions, the functions $J_{\nu+ax}(x)$, and associated orthogonal polynomials*, J. Math. Anal. Appl. **86** (1982), 1–19.
27. M. E. H. Ismail, *Some properties of Jackson's third q-Bessel functions*, to appear.
28. M. E. H. Ismail, *Lectures on q-orthogonal polynomials*, this volume.
29. M. E. H. Ismail, D. R. Masson, and S. K. Suslov, *The q-Bessel functions on a q-quadratic grid*, in: "Algebraic Methods and q-Special Functions", J. F. van Diejen and L. Vinet, eds., CRM Proceedings & Lecture Notes, Vol. 22, Amer. Math. Soc., 1999, pp. 183–200.
30. M. E. H. Ismail and M.E. Muldoon, *On the variation with respect to a parameter of zeros of Bessel and q-Bessel functions*, J. Math. Anal. Appl. **135** (1988), 187–207.
31. M. E. H. Ismail and M.E. Muldoon, *Bounds for the small real and purely imaginary zeros of Bessel and related functions*, Methods Appl. Anal. **2** (1995), 1–21.
32. M. E. H. Ismail, M. Rahman, and D. Stanton, *Quadratic q-exponentials and connection coefficient problems*, Proc. Amer. Math. Soc. **127** (1999) #10, 2931–2941.
33. M. E. H. Ismail, M. Rahman, and R. Zhang, *Diagonalization of certain integral operators II*, J. Comp. Appl. Math. **68** (1996), 163–196.
34. M. E. H. Ismail and D. Stanton, *Addition theorems for the q-exponential functions*, in "q-Series from a Contemporary Perspective", M. E. H. Ismail, D. R. Stanton, eds., Contemporary Mathematics, Vol. 254, Amer. Math. Soc., 2000, pp. 235–245.
35. M. E. H. Ismail and R. Zhang, *Diagonalization of certain integral operators*, Advances in Math. **108** (1994), 1–33.
36. E. G. Kalnins and W. Miller, *Models of q-algebra representations: q-integral transforms and "addition theorems"*, J. Math. Phys. **35** (1994), 1951–1975.
37. E. G. Kalnins and W. Miller, *q-Algebra representations of the Euclidean, Pseudo-Euclidean and oscillator algebras, and their tensor products*, in: "Symmetries and Integrability of Difference Equations", D. Levi, L. Vinet, and P. Winternitz, eds., CRM Proceedings & Lecture Notes, Vol. 9, Amer. Math. Soc., 1996, pp. 173–183.
38. N. Koblitz, *On Carlitz's q-Bernoulli numbers*, J. Number Theory **14** (1982), 332–

339.
39. R. Koekoek and R. F. Swarttouw, *The Askey scheme of hypergeometric orthogonal polynomials and its q-analogues*, Report 94–05, Delft University of Technology, 1994.
40. E. Koelink and J. V. Stokman, *The Askey–Wilson function transform*, preprint (2000), 19 pp.
41. E. Koelink and J. V. Stokman, *The Askey–Wilson function transform scheme*, this volume.
42. T. Koornwinder, *Special functions and q-commuting variables*, in: "Special Functions, q-Series and Related Topics", M. E. H. Ismail, D. R. Masson, and M. Rahman, eds., Fields Institute Communications, Vol. 14, Amer. Math. Soc., 1997, pp. 131–166.
43. T. Koornwinder and R. F. Swarttouw, *On q-analogues of the Fourier and Hankel transforms*, Trans. Amer. Math. Soc. **333** (1992), 445–461.
44. B. Ya. Levin, *Distribution of Zeros of Entire Functions*, Translations of Mathematical Monographs, Vol. 5, Amer. Math. Soc., Providence, Rhode Island, 1980.
45. N. Levinson, *Gap and Density Theorems*, Amer. Math. Soc. Colloq. Publ., Vol. 36, New York, 1940.
46. A. F. Nikiforov, S. K. Suslov, and V. B. Uvarov, *Classical Orthogonal Polynomials of a Discrete Variable*, Nauka, Moscow, 1985 [in Russian]; English translation, Springer–Verlag, Berlin, 1991.
47. M. Rahman, *An integral representation and some transformation properties of q-Bessel functions*, J. Math. Anal. Appl. **125** (1987), 58–71.
48. W. Rudin, *Principles of Mathematical Analysis*, third edition, McGraw–Hill, New York, 1976.
49. S. N. M. Ruijsenaars, *First order analytic difference equations and integrable quantum systems*, J. Math. Phys. **38** (1997), 1069–1146.
50. S. N. M. Ruijsenaars, *Special functions associated with Calogero–Moser type quantum systems*, in: "Proceedings of the 1999 Montréal Séminaire de Mathématiques Supérieures", J. Harnad, G. Sabidussi, and P. Winternitz, eds., CRM Proceedings & Lecture Notes, Amer. Math. Soc., 2000, pp. 189–226.
51. S. N. M. Ruijsenaars, *On Barnes' multiple zeta and gamma functions*, Advances in Math. **156** (2000), 107–132.
52. S. N. M. Ruijsenaars, *Special functions defined by analytic difference equations*, this volume.
53. J. Satoh, *q-Analogue of Riemann's ζ-function and q-Euler numbers*, J. Number Theory **31** (1989), 346–362.
54. A. Sharma, *q-Bernoulli and Euler numbers of higher order*, Duke Math. J. **25** (1958), 343–353.
55. V. P. Spiridonov and A. S. Zhedanov, *Generalized eigenvalue problem and a new family of rational functions biorthogonal on elliptic grids*, this volume.
56. S. K. Suslov, *The theory of difference analogues of special functions of hypergeometric type*, Russian Math. Surveys **44** (1989), 227–278.
57. S. K. Suslov, *"Addition" theorems for some q-exponential and q-trigonometric functions*, Methods Appl. Anal. **4** (1997), 11–32.
58. S. K. Suslov, *Some orthogonal very-well-poised $_8\varphi_7$-functions*, J. Phys. A: Math. Gen. **30** (1997), 5877–5885.
59. S. K. Suslov, *Some orthogonal very-well-poised $_8\varphi_7$-functions that generalize Askey–Wilson polynomials*, The Ramanujan Journal **5** (2001), 183–218.
60. S. K. Suslov, *Some expansions in basic Fourier series and related topics*, J. Approx. Theory, to appear.
61. S. K. Suslov, *Another addition theorem for the q-exponential function*, J. Phys. A: Math. Gen. **33** (2000), L375–380.
62. S. K. Suslov, *Asymptotics of zeros of basic sine and cosine functions*, submitted.
63. S. K. Suslov, *Completeness of basic trigonometric system in $\mathcal{L}^p$*, submitted.
64. S. K. Suslov, *Basic Fourier series and q-Euler polynomials*, submitted to The Ra-

manujan Journal.

65. S. K. Suslov, *A note on completeness of basic trigonometric system in* $\mathcal{L}^2$, submitted to Rocky Mountain J. Math.

66. G. P. Tolstov, *Fourier Series*, Dover, New York, 1962.

67. H. Tsumura, *On the values of a q-analogue of the q-adic L-functions*, Mem. Fac. Sci. Kyushu Univ. **44** (1990), 49–60.

68. H. Tsumura, *A note on q-analogues of the Dirichlet series and q-Bernoulli numbers*, J. Number Theory **39** (1991), 251–256.

69. H. Tsumura, *On evaluation of the Dirichlet series at positive integers by q-calculation*, J. Number Theory **48** (1994), 383–391.

70. H. Tsumura, *A note on q-analogues of the Dirichlet series*, Proc. Japan Acad. **75** (1999), 23–25.

71. W. R. Wade, *An Introduction to Analysis*, Prentice–Hall, Upper Saddle River, New Jersey, 1995.

72. M. Ward, *A calculus of sequences*, Amer. J. Math. **58** (1936), 255–266.

73. G. N. Watson, *A Treatise on the Theory of Bessel Functions*, second edition, Cambridge University Press, Cambridge, 1944.

74. E. T. Whittaker and G. N. Watson, *A Course of Modern Analysis*, fourth edition, Cambridge University Press, Cambridge, 1952.

75. N. Wiener, *The Fourier Integral and Certain of Its Applications*, Cambridge University Press, Cambridge, 1933; Dover edition published in 1948.

76. A. Zygmund, *Trigonometric Series*, second edition, Cambridge University Press, Cambridge, 1968.

PROJECTION OPERATOR METHOD FOR QUANTUM GROUPS

V. N. TOLSTOY

Institute of Nuclear Physics
Moscow State University
Moscow, 119899 Russia

1. Introduction

At present there can be no doubt to say that the quantum group theory has been one of the most important, modern and rapidly developing directions of mathematics and mathematical physics in the end of the twentieth century. Although initially (in the early 1980's) the quantum group theory was formulated for solving problems in the theory of integrable systems and statistical physics, later the surprising connection of this theory with many branches of mathematics, and the theoretical and mathematical physics was discovered. Today the quantum group theory is connected with such mathematical fields as special functions (especially, with q-orthogonal polynomials and basic hypergeometric series), the theory of difference and differential equations, combinatorial analysis and representation theory, matrix and operator algebras, noncommutative geometry, knot theory, topology, category theory and so on. From point of view of the mathematical physics there exists interconnection of the quantum group theory with the quantum inverse scattering method, conformal and quantum fields theory and so on. It is expected that the quantum groups will provide deeper understanding of concept of symmetry in physics.

In the broad sense the notation "quantum groups" [7] involves different deformations of the universal enveloping algebras $U(g)$ of Lie algebras and superalgebras g, such as: q-deformations, Yangians, elliptic and dynamic quantum groups, mixed deformations (for example, two-parameter deformations) and so on. In the restricted sense the quantum groups mainly mean the q-deformations of the universal enveloping algebras $U_q(g)$ of Lie algebras and superalgebras g (sometimes the q-deformations of the groups

J. Bustoz et al. (eds.), Special Functions 2000, 457–488.
© 2001 *Kluwer Academic Publishers. Printed in the Netherlands.*

and supergroups are also included here). In what follows we use the notation of the quantum groups in this restricted sense.

It is well known that the method of projection operators for usual (non-quantized) Lie algebras and superalgebras is a powerful and universal method for solution of many problems in the representation theory. For example, the method allows to classify irreducible modules, to decompose modules on submodules (e.g. to analyze structure of Verma modules), to describe reduced (super)algebras (which are connected with reduction of a (super)algebra to (super)subalgebra), to construct bases of modules (e.g. the Gelfand-Tsetlin's type), to develop the detailed theory of Clebsch-Gordan coefficients and another elements of Wigner-Racah calculus (including compact analytic formulas of these elements and their symmetry properties) and so on. It is evident that the projection operators of quantum groups play the same role in their representation theory.

In these lectures we develop the projection operator method for quantum groups. Here the term "quantum groups" means q-deformed universal enveloping algebras of contragredient Lie (super)algebras of finite growth (these (super)algebras include all finite-dimensional simple Lie algebras and classical superalgebras, infinite-dimensional affine (Kac-Moody) algebras and superalgebras). Conventionally, contains of the lectures can be divided on two parts. Basis fragments of the first part are: a combinatorial structure of root systems, the q-analog of the Cartan-Weyl basis, the extremal projector and the universal R-matrix for any contragredient Lie (super)algebra of finite growth. It should be noted that the explicit expressions for the extremal projectors and the universal R-matrices are ordered products of special q-series depending on noncommutative Cartan-Weyl generators.

In the second part (Sects. 9-12) we consider some applications of the extremal projectors. Here we use the projector operator method to develop the theory of the Clebsch-Gordan coefficients for the quantum algebras $U_q(su(2))$ and $U_q(su(3))$. In particular, we give a very compact general formula for the canonical $U_q(su(3)) \supset U_q(su(2))$ Clebsch-Gordan coefficients in terms of the $U_q(su(2))$ Wigner $3nj$-symbols which are connected with the basic hyperheometric series. Then we apply the projection operator method for the construction of the q-analog of the Gelfand-Tsetlin basis for $U_q(su(n))$. Finally using analogy between the extremal projector $p(U_q(sl(2)))$ of the quantum algebra $U_q(sl(2))$ and the $\delta(x)$-function we introduce 'adjoint extremal projectors' $p^{(n)}(U_q(sl(2))$ $(n=1,2,\dots)$ which are some generalizations of the extremal projector $p(U_q(sl(2)))$, and which are analogies of the derivatives of the $\delta(x)$-function, $\delta^{(n)}(x)$ $(n=1,2,\dots)$. The elements $p^{(n)}(U_q(sl(2))$ can be applied to construction and description of decomposable representations of quantum algebra $U_q(sl(2))$ (for details see [6]).

2. Preliminary Information

Let $g(A, \tau)$ be a contragredient Lie (super)algebra of finite growth[1] with a symmetrizable Cartan matrix A (i.e. $A = DA^{sym}$, where $A^{sym} = (a_{ij}^{sym})_{i,j \in I}$ is a symmetrical matrix, and D is an invertible diagonal matrix, $D = \operatorname{diag}(d_1, d_2, \ldots, d_r)$), $\tau \subset I$, $I := \{1, 2, \ldots, r\}$, and let $\Pi := \{\alpha_1, \ldots, \alpha_r\}$ be a system of simple roots for $g(A, \tau)$.

The Lie (super)algebra $g := g(A, \tau)$ and its universal enveloping algebra $U(g)$ are completely determined by the Chevalley generators $e'_{\pm \alpha_i}$, h'_{α_i} ($i = 1, 2, \ldots, r$) with the defining relations [9]:

$$[h'_{\alpha_i}, h'_{\alpha_j}] = 0, \qquad\qquad [h'_{\alpha_i}, e'_{\pm \alpha_j}] = \pm a_{ij}^{sym} e'_{\pm \alpha_i}, \qquad (2.1)$$

$$[e'_{\alpha_i}, e'_{-\alpha_j}] = \delta_{ij} h'_{\alpha_i}, \qquad (\operatorname{ad} e'_{\pm \alpha_i})^{n_{ij}} e'_{\pm \alpha_j} = 0 \quad \text{for } i \neq j, \qquad (2.2)$$

where the positive integers n_{ij} are given as follows: $n_{ij} = 1$ if $a_{ii}^{sym} = a_{ij}^{sym} = 0$, $n_{ij} = 2$ if $a_{ii}^{sym} = 0$, $a_{ij}^{sym} \neq 0$, and $n_{ij} = -2a_{ij}^{sym}/a_{ii}^{sym} + 1$ if $a_{ii}^{sym} \neq 0$. Moreover it is necessary also to add all nontrivial relations of the form:

$$[[e'_{\pm \alpha_i}, e'_{\pm \alpha_j}], [e'_{\pm \alpha_i}, e'_{\pm \alpha_k}]] = 0 \qquad (2.3)$$

for each triple of simple roots α_i, α_j, α_k if they satisfy the condition:

$$a_{ii}^{sym} = a_{jk}^{sym} = 0, \qquad a_{ij}^{sym} = -a_{ik}^{sym} \neq 0. \qquad (2.4)$$

Throughout the paper the brackets $[\cdot, \cdot]$ and the symbol "ad" denote the supercommutator in $U(g)$, i.e.

$$(\operatorname{ad} a)b \equiv [a, b] = ab - (-1)^{\deg(a)\deg(b)} ba \qquad (2.5)$$

for all homogeneous elements a, $b \in g$, where

$$\deg(h'_{\alpha_i}) = 0 \text{ for } i \in I, \quad \deg(e'_{\pm \alpha_i}) = 0 \text{ for } i \notin \tau, \quad \deg(e'_{\pm \alpha_i}) = 1 \text{ for } i \in \tau. \qquad (2.6)$$

Remarks (i) The triple relation (2.3) may appear only in the supercase for the following situation in the Dynkin diagram:

$$\overset{\displaystyle \alpha_j}{\bullet} \rule[0.5ex]{3em}{0.4pt} \overset{\displaystyle \alpha_i}{\otimes} \rule[0.5ex]{3em}{0.4pt} \overset{\displaystyle \alpha_k}{\bullet} \qquad (2.7)$$

Here α_i is an odd gray root, and α_j, a_k are not connected and they can be of any color (degree): white, gray or dark, and moreover the lines connected

---

[1]These (super)algebras include all finite-dimensional simple Lie algebras and classical superalgebras, infinite-dimensional affine (Kac-Moody) algebras and superalgebras.

the node α_i with the nodes α_j and a_k can be non-single. The second equality (2.2) is ordinary called the Serre relation therefore the equality (2.3) may be called the triple Serre relation because it connects three root vectors e_{α_i}, e_{α_j} and e_{α_k}.

(ii) Besides the relations of type (2.3) the additional Serre relations of higher order can also occur but we don't give them here because these relations appear only for the Dynkin diagrams of special type. A total list of such diagrams and corresponding additional Serre relations can be found in the paper [30].

Let Δ_+ be a system of all positive roots of the (super)algebra $g(A, \tau)$. Any root γ of Δ_+ has the form: $\gamma = \sum_i^r n_i \alpha_i$, where all n_i are nonnegative integers. The total system of all roots, Δ, has the form: $\Delta = \Delta_+ \bigcup (-\Delta_+)$. On the system Δ there is a bilinear form $(\cdot, \cdot)$ such that $(\alpha_i, \alpha_j) = a_{ij}^{sym}$. The form is positive definite for all simple finite-dimensional Lie algebras and it is nondegenerate for all finite-dimensional contragredient Lie superalgebras. With respect to this form the simple roots $\alpha_i \in \Pi$ are classified (colored) as follows:

- A simple root α_i is called even (white) if $(\alpha_i, \alpha_i) \neq 0$ and $2\alpha_i \notin \Delta_+$. (In this case $i \notin \tau$).
- A simple root α_i is called odd, dark if $(\alpha_i, \alpha_i) \neq 0$ and $2\alpha_i \in \Delta_+$. (In this case $i \in \tau$).
- A simple root α_i is called odd, grey if $(\alpha_i, \alpha_i) = 0$. One can show that doubled grey roots don't exist, $2\alpha_i \notin \Delta_+$. (In this case $i \in \tau$).

The grey and dark roots occur only in the supercase.

Let γ be any root of Δ_+, this root is called odd if in its decomposition on the simple roots, $\gamma = \sum_i^r n_i \alpha_i$, the sum of the coefficients n_i for all odd roots α_i is odd. Otherwise the root γ is called even. The parity of the negative root $-\gamma$ coincides with the parity of the positive root γ.

Coloring of the roots is extended on all system Δ as follows:

- All even roots are white. A white root is pictured by the white node $\circ$.
- An odd root γ is called grey if 2γ is not any root. This root is pictured by the grey node $\otimes$.
- An odd root γ is called dark if 2γ is a root. This root is pictured by the dark node $\bullet$.

In the case of the affine Kac-Moody (super)algebras all roots are also divided into real and imaginary. Every imaginary root $n\delta$ satisfies the condition $(n\delta, \gamma) = 0$ for all $\gamma \in \Delta$. For the real roots this condition is not valid.

3. Combinatorial Structure of Root Systems

At first we remind the definition of the reduced system of the positive root system Δ_+ for any contragredient (super)algebras of finite growth.

Definition 3.1 *The system $\underline{\Delta}_+$ is called the reduced system if it is defined by the following way: $\underline{\Delta}_+ = \Delta_+\backslash\{2\gamma \in \Delta_+\,|\,\gamma$ is odd$\}$. That is the reduced system $\underline{\Delta}_+$ is obtained from the total system Δ_+ by removing of all doubled roots 2γ where γ is a dark odd root.*

Combinatorial structure of root system of the contragredient Lie (super) algebra of finite growth is connected with notation of the normal ordering in the reduced system of positive roots.

Definition 3.2 *We say that the system $\underline{\Delta}_+$ is in normal ordering if each composite (not simple) root $\gamma = \alpha + \beta$ $(\alpha,\beta,\gamma \in \underline{\Delta}_+)$, where α and β are not proportional roots $(\alpha \neq \lambda\beta)$, is written between its components α and β. It means that in the normal ordering system $\underline{\Delta}_+$ we have either*

$$\ldots,\alpha,\ldots,\alpha+\beta,\ldots,\beta,\ldots\,, \tag{3.1}$$

or

$$\ldots,\beta,\ldots,\alpha+\beta,\ldots,\alpha,\ldots\,. \tag{3.2}$$

We say also that $\alpha \prec \beta$ if α is located on the left side of β in the normal ordering system $\underline{\Delta}_+$, i.e. this corresponds to the case (3.1).

The normal ordering system $\underline{\Delta}_+$ is denoted by the symbol $\vec{\underline{\Delta}}_+$. It is evident that boundary (end) roots in $\vec{\underline{\Delta}}_+$ are simple. The combinatorial structure of the root system $\underline{\Delta}_+$ is described the following theorem.

Theorem 3.1 *(i) Normal ordering in the system $\underline{\Delta}_+$ exists for any mutual location of the simple roots α_i, $i = 1, 2, \ldots, r$.*
(ii) Any two normal orderings $\vec{\underline{\Delta}}_+$ and $\vec{\underline{\Delta}}'_+$ can be obtained one from another by compositions of the following elementary inversions:

$$\alpha,\beta \leftrightarrow \beta,\alpha\,, \tag{3.3}$$

$$\alpha,\alpha+\beta,\beta \leftrightarrow \beta,\alpha+\beta,\alpha\,, \tag{3.4}$$

$$\alpha,\alpha+\beta,\alpha+2\beta,\beta \leftrightarrow \beta,\alpha+2\beta,\alpha+\beta,\alpha\,, \tag{3.5}$$

$$\alpha,\alpha+\beta,2\alpha+3\beta,\alpha+2\beta,\alpha+3\beta,\beta \leftrightarrow \beta,\alpha+3\beta,\alpha+2\beta,2\alpha+3\beta,\alpha+\beta,\alpha, \tag{3.6}$$

$$\alpha, \delta + \alpha, 2\delta + \alpha, \ldots, \infty\delta + \alpha, \delta, 2\delta, 3\delta, \ldots, \infty\delta, \infty\delta - \alpha, \ldots, 2\delta - \alpha, \delta - \alpha \leftrightarrow$$

$$\tag{3.7}$$

$$\leftrightarrow \delta - \alpha, 2\delta - \alpha, \ldots, \infty\delta - \alpha, \delta, 2\delta, 3\delta, \ldots, \infty\delta, \infty\delta + \alpha, \ldots, 2\delta + \alpha, \delta + \alpha, \alpha ,$$

$$\alpha, \delta + 2\alpha, \delta + \alpha, 3\delta + 2\alpha, 2\delta + \alpha, \ldots, \infty\delta + \alpha, (2\infty + 1)\delta + 2\alpha, (\infty + 1)\delta + \alpha, \delta, 2\delta, \ldots,$$

$$\infty\delta, (\infty + 1)\delta - \alpha, (2\infty + 1)\delta - 2\alpha, \infty\delta - \alpha, \ldots, 2\delta - \alpha, 3\delta - 2\alpha, \delta - \alpha, \delta - 2\alpha, \leftrightarrow$$

$$\tag{3.8}$$

$$\leftrightarrow \delta - 2\alpha, \delta - \alpha, 3\delta - 2\alpha, 2\delta - \alpha, \ldots, \infty\delta - \alpha, (2\infty + 1)\delta - 2\alpha, (\infty + 1)\delta - \alpha, \delta, 2\delta, \ldots,$$

$$\infty\delta, (\infty + 1)\delta + \alpha, (2\infty + 1)\delta + 2\alpha, \infty\delta + \alpha, \ldots, 2\delta + \alpha, 3\delta + 2\alpha, \delta + \alpha, \delta + 2\alpha, \alpha,$$

where $\alpha - \beta$ is not any root.

A proof of the second part *(ii)* for the case of the finite-dimensional simple Lie algebras can be found in [2]. The full proof of the theorem is given in the outgoing paper [28].

The root systems in (3.3)–(3.8) belong to the (super)algebras of rank 2. The combinatorial theorem permits to construct a q-analog of the Cartan-Weyl basis and to reduce the proof of basic theorems for extremal projector and the universal R-matrix for the quantum (super)algebra of arbitrary rank to the proof of such theorems for the quantum (super)algebras of rank 2.

4. Quantized Lie (Super)Algebras

A quantum (q-deformed) universal enveloping (super)algebra $U_q(g)^2$, where g is a contragredient Lie (super)algebra of finite growth, may be consider as a deformation f (reserving the grading) of the universal enveloping algebra $U(g): U(g) \overset{f}{\mapsto} U_q(g)$ ($e'_{\pm\alpha_i} \overset{f}{\mapsto} e_{\pm\alpha_i}$, $h'_{\alpha_i} \overset{f}{\mapsto} h_{\alpha_i}$), which modifies the relations (2.2) (2.3). More precisely we have the following definition [26, 11].

Definition 4.1 *The quantum (super)algebra $U_q(g)$ (where $g := g(A, \tau)$ is a contragredient Lie (super)algebra of finite growth), is an associative (super)algebra over $\mathbb{C}[q, q^{-1}]$ with Chevalley generators $e_{\pm\alpha_i}$, $k_{\alpha_i}^{\pm 1} := q^{\pm h_{\alpha_i}}$, ($i \in I := \{1, 2, \ldots, r\}$), and the defining relations:*

$$k_{\alpha_i} k_{\alpha_i}^{-1} = k_{\alpha_i}^{-1} k_{\alpha_i} = 1 , \qquad k_{\alpha_i} k_{\alpha_j} = k_{\alpha_j} k_{\alpha_i} , \tag{4.1}$$

$$k_{\alpha_i} e_{\pm\alpha_j} k_{\alpha_i}^{-1} = q^{\pm(\alpha_i, \alpha_j)} e_{\pm\alpha_j} , \qquad [e_{\alpha_i}, e_{-\alpha_j}] = \delta_{ij} \frac{k_{\alpha_i} - k_{\alpha_i}^{-1}}{q - q^{-1}} , \tag{4.2}$$

---

[2]We shall also use the term "the quantized Lie (super)algebra g".

$$(\mathrm{ad}_q\, e_{\pm\alpha_i})^{n_{ij}} e_{\pm\alpha_j} = 0 \qquad\qquad \text{for } i \neq j , \qquad\qquad (4.3)$$

where the positive integers n_{ij} are the same as in the relations (2.2). Moreover, if any three simple roots α_i, α_j, α_k satisfy the condition (2.4) then there are the additional triple relations of the form:

$$[[e_{\pm\alpha_i}, e_{\pm\alpha_j}]_q, [e_{\pm\alpha_i}, e_{\pm\alpha_k}]_q]_q = 0 . \qquad\qquad (4.4)$$

Here in (4.1)–(4.4) the brackets $[\cdot, \cdot]$ is the usual supercommutator (2.5), and $[\cdot, \cdot]_q$ and ad_q denote the q-deformed supercommutator (q-supercommutator) in $U_q(g)$:

$$(\mathrm{ad}_q\, e_\alpha)e_\beta \equiv [e_\alpha, e_\beta]_q = e_\alpha e_\beta - (-1)^{\deg(e_\alpha)\deg(e_\beta)} q^{(\alpha,\beta)} e_\beta e_\alpha ,$$
$$(4.5)$$

where (α, β) is a scalar product of the roots α and β, and the parity function $\deg(\cdot)$ is given by

$$\deg(k_{\alpha_i}) = 0 \text{ for } i \in I, \quad \deg(e_{\pm\alpha_i}) = 0 \text{ for } i \notin \tau, \quad \deg(e_{\pm\alpha_i}) = 1 \text{ for } i \in \tau.$$
$$(4.6)$$

Below we shall use the following short notation:

$$\vartheta(\gamma) := \vartheta(e_\gamma) = \deg(e_\gamma) . \qquad\qquad (4.7)$$

Remarks (i) It is not hard to verify that the the relations (4.1)–(4.4) are invariant with respect to the replacement of q by q^{-1}.
(ii) The outer q-supercommutator in (4.4) is really the usual supercommutator since $(\alpha_i + \alpha_j, \alpha_i + \alpha_k) = 0$.
(iii) The remark (ii) after the formula (2.7) is also valid.

Clearly, the quantum (super)algebra $U_q(g)$ reduces to the usual universal enveloping (super)algebra $U(g)$ if $q \to 1$.

By direct calculations we can show that quantum (super)algebra $U_q(g)$ is a Hopf (super)algebra with respect to a comultiplication Δ_q, an antipode S_q and a counit ε defined as

$$\Delta_q(k_{\alpha_i}^{\pm 1}) = k_{\alpha_i}^{\pm 1} \otimes k_{\alpha_i}^{\pm 1} , \qquad\qquad S_q(k_{\alpha_i}^{\pm 1}) = k_{\alpha_i}^{\mp 1} ,$$
$$\Delta_q(e_{\alpha_i}) = e_{\alpha_i} \otimes 1 + k_{\alpha_i}^{-1} \otimes e_{\alpha_i} , \qquad S_q(e_{\alpha_i}) = -k_{\alpha_i} e_{\alpha_i} ,$$
$$\Delta_q(e_{-\alpha_i}) = e_{-\alpha_i} \otimes k_{\alpha_i} + 1 \otimes e_{-\alpha_i} , \qquad S_q(e_{-\alpha_i}) = -e_{-\alpha_i} k_{\alpha_i}^{-1} , \quad (4.8)$$

$$\varepsilon(e_{\pm\alpha_i}) = 0 , \qquad \varepsilon(k_{\alpha_i}) = \varepsilon(1) = 1 , \qquad\qquad (4.9)$$

Both in the quantum and non-quantum case we can directly use the Chevalley generators for construction of a 'monomial' basis in all universal enveloping (super)algebra $U_q(g)$ $(U(g))$. Bases of such kind were proposed by Verma for the non-quantized case and by Lusztig [17] for the general case. The Lusztig basis is an universal one and it is called the canonical basis. Both the Verma basis and the Lusztig basis have rather complicated algebraic structure and therefore they were not used in a broad fashion until now. It is well known that a monomial basis constructed of Cartan-Weyl generators is a more algebraically simple basis. Therefore a natural problem is to construct a q-analog of the Cartan-Weyl basis (the quantum Cartan-Weyl basis) for the quantum (super)algebra $U_q(g)$.

Our method for construction of the q-analog of the Cartan-Weyl basis and its general properties and also properties of the extremal projector and the universal R-matrix are closely connected with the combinatorial structure for the root system of the Lie (super)algebra g.

5. Quantum Cartan-Weyl Basis

The q-analog of the Cartan-Weyl basis for $U_q(g)$ is constructed by using the following inductive algorithm [26], [11]–[15].

We fix some normal ordering $\vec{\underline{\Delta}}_+$ and put by induction

$$e_\gamma := [e_\alpha, e_\beta]_q, \qquad e_{-\gamma} := [e_{-\beta}, e_{-\alpha}]_{q^{-1}} \qquad (5.1)$$

if $\gamma = \alpha + \beta$, $\alpha \prec \gamma \prec \beta$ $(\alpha, \beta, \gamma \in \underline{\Delta}_+)$, and the segment $[\alpha; \beta] \subseteq \vec{\underline{\Delta}}_+$ is minimal one including the root γ, i.e. the segment has not another roots α' and β' such that $\alpha' + \beta' = \gamma$. Moreover we put

$$k_\gamma := \prod_{i=1}^{r} k_{\alpha_i}^{l_i} , \qquad (5.2)$$

if $\gamma = \sum_{i=1}^{r} l_i \alpha_i$ $(\gamma \in \underline{\Delta}_+,\ \alpha_i \in \Pi)$.

By this procedure one can construct the total quantum Cartan-Weyl basis for all quantized finite-dimensional simple contragredient Lie (super) algebras. In the case of the quantized infinite-dimensional affine Kac-Moody (super)algebras we have to apply one more additional condition. Namely, first we construct all root vectors e_γ $(\gamma \in \Delta)$ by means of the given procedure, and then we overdeterminate the generators $e_{n\delta}$ of the imaginary roots $n\delta \in \Delta$ in a way that the new generators $e'_{n\delta}$ are mutually commutative if they are not conjugate generators. Because of the fact that we do not have a sufficient place here to describe the overdetermination of imaginary root generators in details, we are restricted to a consideration of finite-dimensional case, i.e. when g is a finite-dimensional simple contragredient Lie (super)algebra.

The quantum Cartan-Weyl basis is characterized by the following properties [11]–[15].

Proposition 5.1 *The root vectors $\{e_{\pm\gamma}\}$ ($\gamma \in \underline{\Delta}_+$) satisfy the following relations:*

$$k_\alpha^{\pm 1} e_\gamma = q^{\pm(\alpha,\gamma)} e_\gamma k_\alpha^{\pm 1} , \qquad (5.3)$$

$$[e_\gamma, e_{-\gamma}] = a(\gamma) \frac{k_\gamma - k_\gamma^{-1}}{q - q^{-1}} , \qquad (5.4)$$

$$[e_\alpha, e_\beta]_q = \sum_{\alpha \prec \gamma_1 \prec \ldots \prec \gamma_n \prec \beta} C_{m_i,\gamma_i} e_{\gamma_1}^{m_1} e_{\gamma_2}^{m_2} \cdots e_{\gamma_n}^{m_n} , \qquad (5.5)$$

where $\sum_i^n m_i \gamma_i = \alpha + \beta$, and the coefficients $C...$ are rational functions of q and they do not depend on the Cartan elements k_{α_i}, $i = 1, 2, \ldots n$, and also

$$[e_\beta, e_{-\alpha}] = \sum C'_{m_i,\gamma_i;m'_j,\gamma'_j} e_{-\gamma_1}^{m_1} e_{-\gamma_2}^{m_2} \cdots e_{-\gamma_p}^{m_p} e_{\gamma'_1}^{m'_1} e_{\gamma'_2}^{m'_2} \cdots e_{\gamma'_s}^{m'_s} \qquad (5.6)$$

where the sum is taken on $\gamma_1, \ldots, \gamma_p, \gamma'_1, \ldots, \gamma'_s$ and $m_1, \ldots, m_p, m'_1, \ldots, m'_s$ such that

$$\gamma_1 \prec \ldots \prec \gamma_p \prec \beta \prec \gamma'_1 \prec \ldots \prec \gamma'_s , \quad \sum_l (m'_l \gamma'_l - m_l \gamma_l) = \beta - \alpha$$

and the coefficients $C'...$ are rational functions of q and k_α or k_β. The monomials $e_{\gamma_1}^{n_1} e_{\gamma_2}^{n_2} \cdots e_{\gamma_p}^{n_p}$ and $e_{-\gamma_1}^{n_1} e_{-\gamma_2}^{n_2} \cdots e_{-\gamma_p}^{n_p}$, ($\gamma_1 \prec \gamma_2 \prec \cdots \prec \gamma_p$), generate (as a linear space over $U_q(\mathcal{H})$) subalgebras $U_q(b_+)$ and $U_q(b_-)$ correspondingly. The monomials

$$e_{-\gamma_1}^{n_1} e_{-\gamma_2}^{n_2} \cdots e_{-\gamma_p}^{n_p} e_{\gamma'_1}^{n'_1} e_{\gamma'_2}^{n'_2} \cdots e_{\gamma'_s}^{n'_s} ,$$

where $\gamma_1 \prec \gamma_2 \prec \cdots \prec \gamma_p$ and $\gamma'_1 \prec \gamma'_2 \prec \cdots \prec \gamma'_s$), generate $U_q(g)$ over $U_q(\mathcal{H})$.

Here the algebra $U_q(\mathcal{H})$ is generated by the Cartan elements k_{α_i} ($i = 1, 2, \ldots, r$).

Now we consider some extensions of $U_q(g)$, $U_q(b_+) \otimes U_q(b_-)$ and $U_q(g) \otimes U_q(g)$ since the extremal projector and the universal R-matrix are elements of these extensions.

6. Taylor Extensions of $U_q(g)$, $U_q(b_+) \otimes U_q(b_-)$ and $U_q(g) \otimes U_q(g)$

Let Fract $(U_q(K))$ be a field of fractions over $U_q(K)$, i.e. Fract $(U_q(K))$ is an associative algebra of rational functions of the elements $k_{\alpha_i}^{\pm 1}$, ($i = 1, 2, \ldots, r$). We put

$$\tilde{U}_q(g) = \text{Fract} (U_q(K)) \otimes_{U_q(K)} U_q(g) . \qquad (6.1)$$

Evidently, the extension $\tilde{U}_q(g)$ is an associative algebra. The algebra $\tilde{U}_q(g)$ is called the Cartan extension of the quantum algebra $U_q(g)$.

Let $\{e_{\pm\gamma}\}$, $\gamma \in \underline{\Delta}_+$, be the root vectors of the quantum Cartan-Weyl basis built in accordance with some fixed normal ordering in $\underline{\Delta}_+$. Let us construct a formal Taylor series on the following monomials

$$e_{-\beta}^{n_\beta} \cdots e_{-\gamma}^{n_\gamma} e_{-\alpha}^{n_\alpha} \, e_{\alpha}^{m_\alpha} e_{\gamma}^{m_\gamma} \cdots e_{\beta}^{m_\beta} \qquad (6.2)$$

with coefficients from Fract $(U_q(K))$, where $\alpha \prec \gamma \prec \cdots \prec \beta$ in a sense of the fixed normal ordering in $\underline{\Delta}_+$ and nonnegative integers $n_\beta, n_\gamma \ldots, n_\alpha, m_\alpha, m_\gamma \ldots, m_\beta$ are subjected to the constraints

$$\left| \sum_{\gamma \in \underline{\Delta}_+} (n_\gamma - m_\gamma) c_i^{(\gamma)} \right| \leq \text{const} , \qquad i = 1, 2, \cdots, r , \qquad (6.3)$$

where $c_i^{(\gamma)}$ are coefficients in a decomposition of the root γ with respect to the system of simple roots Π. Let $T_q(g)$ be a linear space of all such formal series. We have the following simple proposition.

Proposition 6.1 *The linear space $T_q(g)$ is an associative algebra with respect to a multiplication of formal series.*

The algebra $T_q(g)$ is called the Taylor extension of $U_q(g)$.

Let Fract $(U_q(K \otimes K))$ be a field of fractions generated by the following elements: $1 \otimes k_{\alpha_i}$, $k_{\alpha_i} \otimes 1$ and $q^{h_{\alpha_i} \otimes h_{\alpha_j}}$, $(i, j = 1, 2, \ldots, r)$. Let us consider a formal Taylor series of the following monomials

$$e_{\alpha}^{n_\alpha} e_{\gamma}^{n_\gamma} \cdots e_{\beta}^{n_\beta} \otimes e_{-\beta}^{m_\beta} \cdots e_{-\gamma}^{m_\gamma} e_{-\alpha}^{m_\alpha} \qquad (6.4)$$

with coefficients from Fract $(U_q(K \otimes K))$, where $\alpha \prec \gamma \prec \cdots \prec \beta$ in a sense of the fixed normal ordering in $\underline{\Delta}_+$ and nonnegative integers $n_\beta, \ldots, n_\alpha$, $m_\alpha, \ldots, m_\beta$ are subjected to the constraint (6.3). Let $T_q(b_+ \otimes b_-)$ be a linear space of all such formal series. The following proposition holds.

Proposition 6.2 *The linear space $T_q(b_+ \otimes b_-)$ is an associative algebra with respect to a multiplication of formal series.*

The algebra $T_q(b_+ \otimes b_-)$ will be called the Taylor extension of $U_q(b_+) \otimes U_q(b_-)$.

At least we consider a formal Taylor series of the following monomials

$$e_{-\beta}^{m_\beta} \cdots e_{-\gamma}^{m_\gamma} e_{-\alpha}^{m_\alpha} e_{\alpha}^{n_\alpha} e_{\gamma}^{n_\gamma} \cdots e_{\beta}^{n_\beta} \otimes e_{-\beta}^{m'_\beta} \cdots e_{-\gamma}^{m'_\gamma} e_{-\alpha}^{m'_\alpha} e_{\alpha}^{n'_\alpha} e_{\gamma}^{n'_\gamma} \cdots e_{\beta}^{n'_\beta}$$

$$(6.5)$$

with coefficients from Fract $(U_q(K \otimes K))$, where $\alpha \prec \gamma \prec \cdots \prec \beta$ in a sense of the fixed normal ordering in Δ_+ and nonnegative integers $n_\beta, \ldots, n_\alpha, m_\alpha,$

$\ldots, m_\beta$ and $n'_\beta, \ldots, n'_\alpha, m'_\alpha, \ldots, m'_\beta$ are subjected to the constraints

$$\left| \sum_{\gamma \in \Delta_+} (n_\gamma + n'_\gamma - m_\gamma - m'_\gamma) c_i^{(\gamma)} \right| \le \text{const} , \qquad i = 1, 2, \cdots, r .$$
(6.6)

Let $T_q(g \otimes g)$ be a linear space of all such formal series. The following simple proposition holds.

Proposition 6.3 *The linear space $T_q(g \otimes g)$ is an associative algebra with respect to a multiplication of formal series.*

The algebra $T_q(g \otimes g)$ will be called the Taylor extension of $U_q(g) \otimes U_q(g)$. Evidently the following embedding hold

$$T_q(g \otimes g) \supset T_q(b_+ \otimes b_-) ,$$
$$T_q(g \otimes g) \supset T_q(g) \otimes T_q(g) \supset \Delta_q(T_q(g)) .$$
(6.7)

7. Extremal Projector

By definition, the extremal projector for $U_q(g)$ is a nonzero element $p :=$ $p(U_q(g))$ of the Taylor extension $T_q(g)$, satisfying the equations

$$e_{\alpha_i} p = p e_{-\alpha_i} = 0 \quad (\forall\, \alpha_i \in \Pi) , \qquad p^2 = p .$$
(7.1)

Acting by the extremal projector p on any highest weight $U_q(g)$-module M we obtain a space $M^0 = pM$ of highest weight vectors for M (if pM has no singularities).

Fix some normal ordering $\vec{\Delta}_+$ and let $\{e_{\pm\gamma}\}$ $(\gamma \in \Delta_+)$ be the corresponding Cartan-Weyl generators. The following statement holds for any quantized finite-dimensional contragredient Lie (super)algebra[3] g [26, 12].

Theorem 7.1 *The equations (7.1) have a unique nonzero solution in the space of the Taylor extension $T_q(g)$ and this solution has the form*

$$p = \prod_{\gamma \in \vec{\Delta}_+} p_\gamma ,$$
(7.2)

where the order in the product coincides with the chosen normal ordering of Δ_+ and the elements p_γ are defined by the formulae

$$p_\gamma = \sum_{m \ge 0} \frac{(-1)^m}{(m)_{\bar{q}_\gamma}!} \varphi_{\gamma,m} e_{-\gamma}^m e_\gamma^m ,$$
(7.3)

[3]The theorem is also valid for the quantized infinite-dimensional affine Kac-Moody (super)algebras, but in this case the formulas (7.3) and (7.4) for the imaginary roots $\gamma = n\delta$ should be more detailed (see [10, 28] as examples).

$$\varphi_{\gamma,m} = \frac{(q-q^{-1})^m q^{-\frac{1}{4}m(m-3)(\gamma,\gamma)} q^{-m(\rho,\gamma)}}{(a(\gamma))^m \prod\limits_{l=1}^{m} \left(k_\gamma q^{(\rho,\gamma)+\frac{1}{2}(\gamma,\gamma)} - (-1)^{(l-1)\theta(\gamma)} k_\gamma^{-1} q^{-(\rho,\gamma)-\frac{1}{2}(\gamma,\gamma)} \right)} \; . \quad (7.4)$$

Here ρ is a linear function such that $(\rho, \alpha_i) = \frac{1}{2}(\alpha_i, \alpha_i)$ for all simple roots $\alpha_i \in \Pi$; $a(\gamma)$ is a factor in the relation (5.4); $\bar{q}_\gamma := (-1)^{\theta(\gamma)} q^{-(\gamma,\gamma)}$; the symbol $(m)_q$ is given by the formula:

$$(m)_q := \frac{q^m - 1}{q - 1} \; . \quad (7.5)$$

In the limit $q \to 1$ we obtain the extremal projector for the (super)algebra g: $\lim\limits_{q \to 1} p(U_q(g)) = p(g)$ [1, 2, 24, 25]. A proof of the theorem actually reduces to the proof for the case of the quantized (super)algebras of rank 2, and it is similar to the case of non-deformed finite-dimensional simple Lie algebras [2].

8. Universal R-Matrix

By definition, the universal R-matrix for the Hopf (super)algebra $U_q(g)$ is an invertible element of the Taylor extension $T_q(b_+ \otimes b_-)$, satisfying the equations

$$\tilde{\Delta}_q(x) = R\Delta(x)R^{-1}, \qquad \forall \, x \in U_q(g), \quad (8.1)$$

$$(\Delta_q \otimes \mathrm{id})R = R^{13}R^{23} \, , \qquad (\mathrm{id} \otimes \Delta_q)R = R^{13}R^{12}, \quad (8.2)$$

where $\tilde{\Delta}_q$ is an opposite comultiplication:

$$\tilde{\Delta}_q = \sigma\Delta_q, \quad \sigma(x \otimes y) = (-1)^{\deg x \deg y} y \otimes x$$

for all homogeneous elements $x, y \in U_q(g)$. In (8.2) we use standard notation $R^{12} = \sum a_i \otimes b_i \otimes 1$, $R^{13} = \sum a_i \otimes 1 \otimes b_i$, $R^{23} = \sum 1 \otimes a_i \otimes b_i$ if R has a form $R = \sum a_i \otimes b_i$.

We employ the following standard notation for the q-exponential:

$$\exp_q(x) := 1 + x + \frac{x^2}{(2)_q!} + \ldots + \frac{x^n}{(n)_q!} + \ldots = \sum_{n \geq 0} \frac{x^n}{(n)_q!} \, , \quad (8.3)$$

where $(n)_q$ is defined by the formulas (7.5).

Fix some normal ordering $\vec{\Delta}_+$ and let $\{e_{\pm\gamma}\}$ ($\gamma \in \underline{\Delta}_+$) be the corresponding Cartan-Weyl generators. The following statement holds for any quantized finite-dimensional contragredient Lie (super)algebra[4] g [11, 12, 13].

---

[4]The theorem is also valid for the quantized infinite-dimensional affine Kac-Moody (super)algebras, but in this case the formula (8.5) for the imaginary roots $\gamma = n\delta$ should be more detailed (see [29, 14, 15].

Theorem 8.1 *The equation (8.1) has a unique (up to a multiplicative constant) invertible solution in the space of the Taylor extension $T_q(b_+ \otimes b_-)$ and this solution has the form*

$$R = \left(\prod_{\gamma \in \vec{\Delta}_+} R_\gamma \right) \cdot K, \tag{8.4}$$

where the order in the product coincides with the chosen normal ordering $\vec{\Delta}_+$ and the elements R_γ and K are defined by the formulas

$$R_\gamma = \exp_{\bar{q}_\gamma} \left((-1)^{\theta(\gamma)} (q - q^{-1})(a(\gamma))^{-1}(e_\gamma \otimes e_{-\gamma}) \right), \tag{8.5}$$

$$K = q^{\sum_{i,j} d_{ij}(h_{\alpha_i} \otimes h_{\alpha_j})} \tag{8.6}$$

where $a(\gamma)$ is a factor from the relation (5.4), and d_{ij} is an inverse matrix for a symmetrical Cartan matrix (a_{ij}^{sym}) if (a_{ij}^{sym}) is not degenerated. (In a case of a degenerated (a_{ij}^{sym}) we extend it up to a non-degenerated matrix $(\tilde{a}_{ij}^{sym})$ and take an inverse to this extended matrix). Moreover the solution (8.4) is the universal R-matrix, i.e. it satisfies the equations (8.2) too.

A proof of the theorem actually reduces to the proof for the case of the (super)algebras of rank 2 (see [11]).

In the rest sections we consider some applications of the extremal projectors.

9. Clebsch-Gordan and Racah Coefficients for the Quantum Algebra $U_q(su(2))$

Let $J_\pm$, $q^{\pm J_0}$ be generators of the quantum algebra $U_q(su(2))$. These generators satisfy the standard relations:

$$q^{J_0} J_\pm = q^{\pm 1} J_\pm q^{J_0}, \qquad [J_+, J_-] = \frac{q^{2J_0} - q^{-2J_0}}{q - q^{-1}} \equiv [2J_0],$$

$$J_\pm^* = J_\mp, \qquad J_0^* = J_0, \qquad q^* = q \ (\text{or } q^{-1}). \tag{9.1}$$

Here and in what follows we use the notation: $[x] = (q^x - q^{-x})/(q - q^{-1})$. The Hopf structure of $U_q(su(2))$ is given by the following formulas for the comultiplication Δ_q, and the antipode S_q:

$$\Delta_q(J_0) = J_0 \otimes 1 + 1 \otimes J_0, \qquad S_q(J_0) = -J_0,$$

$$\Delta_q(J_\pm) = J_\pm \otimes q^{J_0} + q^{-J_0} \otimes J_\pm, \qquad S_q(J_\pm) = -q^{\pm 1} J_\pm. \tag{9.2}$$

Let $\{|jm\rangle\}$ be a canonical basis of the $U_q(su(2))$-irreducible representation (IR) with the spin j. These basis vectors satisfy the relations:

$$q^{J_0}|jm\rangle = q^m|jm\rangle ,$$
$$J_\pm|jm\rangle = \sqrt{[j \mp m][j \pm m + 1]}|jm \pm 1\rangle . \qquad (9.3)$$

The vector $|jm\rangle$ can be represented in the following form

$$|jm\rangle = F^j_{m;j}|jj\rangle , \qquad (9.4)$$

where

$$F^j_{m;j} = \sqrt{\frac{[j+m]!}{[2j]![j-m]!}}\, J_-^{j-m} , \qquad (9.5)$$

and $|jj\rangle$ is the highest weight vector, i.e.

$$J_+|jj\rangle = 0 . \qquad (9.6)$$

The operator $F^j_{m;j}$ is called the lowering operator. We can also introduce the rising operator

$$F^j_{j;m} = \sqrt{\frac{[j+m]!}{[2j]![j-m]!}}\, J_+^{j-m} , \qquad (9.7)$$

which has the property

$$|jj\rangle = F^j_{j;m}|jm\rangle . \qquad (9.8)$$

The extremal projector p for $U_q(su(2))$ can be represented in the form

$$p = \sum_{n=0}^{\infty} \frac{(-1)^n \bar{\Gamma}_q(2J_0+2)}{[n]!\bar{\Gamma}_q(2J_0+n+2)}\, J_-^n J_+^n , \qquad (9.9)$$

where $\bar{\Gamma}_q(x)$ is the modified q-gamma function

$$\bar{\Gamma}_q(x + 1) = [x]\Gamma_q(x) . \qquad (9.10)$$

This function is connected with the standard Heine-Thomae q-gamma function $\Gamma_q(x)$ by the relation $\bar{\Gamma}_q(x) = q^{x(x-1)/4}\Gamma_q(x)$. The extremal projector p satisfies the relations:

$$J_+p = pJ_- = 0 , \qquad p^2 = p . \qquad (9.11)$$

We multiply the extremal projector p by the lowering an rising operators as follows

$$P^j_{m;m'} := F^j_{m;j}\, p\, F^j_{j;m'}\,. \tag{9.12}$$

Below we assume that the operator $P^j_{m;m'}$ acts in a vector space of the weight m'. The operator $P^j_{m;m'}$ is called the general projection operator.

Let $\{|j_i m_i\rangle\}$ be canonical bases of two IRs j_i $(i = 1, 2)$. Then $\{|j_1 m_1\rangle| \times j_2 m_2\rangle\}$ be an 'uncoupled' bases in the representation $j_1 \otimes j_2$ of $U_q(su(2)) \otimes U_q(su(2))$. In this representation there is another basis $|j_1 j_2 : j_3 m_3\rangle_q$ which is called a 'coupled' basis with respect to $\Delta_q(U_q(su(2)))$. We can expand the coupled basis in terms of the uncoupled basis $\{|j_1 m_1\rangle|j_2 m_2\rangle\}$:

$$|j_1 j_2 : j_3 m_3\rangle_q = \sum_{m_1, m_2} \left(j_1 m_1\, j_2 m_2 | j_3 m_3\right)_q |j_1 m_1\rangle |j_2 m_2\rangle\,, \tag{9.13}$$

where the matrix element $\left(j_1 m_1\, j_2 m_2 | j_3 m_3\right)_q$ is called the Clebsch-Gordan coefficient (CGC). After some manipulations we can show that CGC is presented by

$$\left(j_1 m_1\, j_2 m_2 | j_3 m_3\right)_q = \frac{\langle j_1 m_1 | \langle j_2 m_2 | \Delta_q(P^{j_3}_{m_3; j_3}) | j_1 j_1\rangle | j_2 j_3 - j_1\rangle}{\sqrt{\langle j_1 j_1 | \langle j_2 j_3 - j_1 | \Delta_q(P^{j_3}_{j_3; j_3}) | j_1 j_1\rangle | j_2 j_3 - j_1\rangle}}\,. \tag{9.14}$$

This is a formula for calculation of CGCs. Using the explicit expression (9.12) for the general projection operator $P^{j_3}_{m_3; j_3}$, the formulas (9.2) for the comultiplication Δ_q and the actions (9.3) for the generators of $U_q(su(2))$ on the canonical basis vectors $|j_i m_i\rangle$ $(i = 1, 2)$ it is not hard to calculate the numerator and the denominator of the right side of (9.14). As result we obtain the following expression for CGC of the quantum algebra $U_q(su(2))$:

$$\left(j_1 m_1\, j_2 m_2 | j_3 m_3\right)_q = \delta_{m_1 + m_2, m_3} q^{-\frac{1}{2}(j_1 + j_2 - j_3)(j_1 + j_2 + j_3 + 1) + j_1 m_2 - j_2 m_1}$$

$$\times \sqrt{\frac{[2j_3 + 1][j_1 + j_2 - j_3]![j_1 - j_2 + j_3]![j_1 + j_2 + j_3 + 1]![j_2 - m_2]![j_3 + m_3]!}{[-j_1 + j_2 + j_3]![j_1 + m_1]![j_1 - m_1]![j_2 + m_2]![j_3 - m_3]!}}$$

$$\times \sum_n \frac{(-1)^{j_1 + j_2 - j_3 - n} q^{n(j_1 + m_1)} [2j_2 - n]![j_1 + j_2 - m_3 - n]!}{[n]![j_1 + j_2 - j_3 - n]![j_2 - m_2 - n]![j_1 + j_2 + j_3 + 1 - n]!}\,. \tag{9.15}$$

A total list of different explicit expressions and symmetry properties for the q-CGCs can be found, for example, in [16, 20, 21, 22].

The general formula (9.15) can be expressed in terms of the basic hypergeometric series

$$
\begin{aligned}
\left(j_1 m_1\, j_2 m_2 \middle| j_3 m_3\right)_q = {} & \delta_{m_1+m_2, m_3}\, q^{-\frac{1}{2}(j_1+j_2-j_3)(j_1+j_2+j_3+1)+j_1 m_2-j_2 m_1} \\
& \times (-1)^{j_1+j_2-j_3} \sqrt{\frac{[2j_3+1][j_1-j_2+j_3]!([2j_2]!)^2}{[j_1+j_2-j_3]![-j_1+j_2+j_3]![j_1+j_2+j_3+1]!}} \\
& \times \sqrt{\frac{[j_3+m_3]!([j_1+j_2-m_3]!)^2}{[j_1+m_1]![j_1-m_1]![j_2+m_2]![j_2-m_2]![j_3-m_3]!}} \\
& \times {}_3\Phi_2\left(\begin{matrix} -j_1-j_2+j_3,\ -j_1-j_2-j_3-1,\ -j_2-m_2 \\ -2j_2,\qquad -j_1-j_2+m_3 \end{matrix}\middle| q^2, q^{2(j_1+m_1+1)}\right).
\end{aligned}
\tag{9.16}
$$

We can also obtain the explicit expression of the Racah coefficients or the 6j-symbols for $U_q(su(2))$ using the extremal projector. Let $\{|j_i m_i\rangle\}$ be canonical bases of three IRs j_i $(i = 1, 2, 3)$. The Racah coefficients for $U_q(su(2))$ (or the q-Racah coefficients) are matrix elements of the transformation between two couplings of these representations: $j_1 \otimes (j_2 \otimes j_3)$ and $(j_1 \otimes j_2) \otimes j_3$. i.e.

$$
\left|j_1, j_2 j_3(j_{23}) : jm\right\rangle_q = \sum_{j_{12}} U(j_1 j_2 j j_3; j_{12} j_{23})_q \left|j_1 j_2(j_{12}), j_3 : jm\right\rangle_q ,
\tag{9.17}
$$

where, for example, the vector $\left|j_1, j_2 j_3(j_{23}) : jm\right\rangle_q$ corresponds to the first coupling scheme $j_1 \otimes (j_2 \otimes j_3)$ and it has the form:

$$
\begin{aligned}
\left|j_1, j_2 j_3(j_{23}) : jm\right\rangle_q = {} & \sum_{\substack{m_2 m_3 \\ m_1 m_{23}}} \left(j_1 m_1\, j_{23} m_{23} \middle| jm\right)_q \left(j_2 m_2\, j_3 m_3 \middle| j_{23} m_{23}\right)_q \\
& \times \left|j_1 m_1\right\rangle \left|j_2 m_2\right\rangle \left|j_3 m_3\right\rangle .
\end{aligned}
\tag{9.18}
$$

The q-Racah coefficient $U(j_1 j_2 j j_3; j_{12} j_{23})_q$ is connected with the q-6j-symbol $\{ ::: \}_q$ by the standard relation

$$
U(j_1 j_2 j j_3; j_{12} j_{23})_q = (-1)^{j_1+j_2+j_3+j} \sqrt{[2j_{12}+1]![2j_{23}+1]!} \left\{ \begin{matrix} j_1 & j_2 & j_{12} \\ j_3 & j & j_{23} \end{matrix} \right\}_q .
\tag{9.19}
$$

It is not hard to obtain the formula for calculation of the q-6j-symbols in terms of projection operators:

$$
\begin{aligned}
\left\{ \begin{matrix} j_1 & j_2 & j_{12} \\ j_3 & j & j_{23} \end{matrix} \right\}_q = {} & \frac{(-1)^{j_1+j_2+j_3+j}}{\sqrt{[2j_{12}+1]![2j_{23}+1]!}} \\
& \times \frac{\langle j_1 j - j_{23} | \langle j_2 j_{23} - j_2 | \langle j_2 j_2 | P^{j_{23}}(23) P^j(123) P^{j_{12}}(12) | j_1 j_{12} - j_2 \rangle | j_2 j_2 \rangle | j_3 j - j_{12} \rangle}{(j_1 j_{12} - j_2, j_2 j_2 | j_{12} j_{12})_q (j_{12} j_{12}, j_3 j - j_{12} | jj)_q (j_1 j - j_{23}, j_{23} j_{23} | jj)_q (j_2 j_2, j_{23} - j_2 | j_{23} j_{23})_q},
\end{aligned}
\tag{9.20}
$$

where the notations are used: $P^j := P^j_{jj}$, $P^{j_{23}}(23) := \mathrm{id} \otimes \Delta_q(P^{j_{23}})$, $P^j(123) := (\Delta_q \otimes \mathrm{id})\Delta_q(P^j)$, $P^{j_{12}}(12) := \Delta_q(P^{j_{12}}) \otimes \mathrm{id}$.

We substitute the explicit expressions for all special CGCs in the denominator of the right side of (9.20) and then we use the actions of the generators of $U_q(su(2))$ on the canonical basis vectors $|j_i m_i\rangle$ $(i = 1, 2, 3)$, and as result we obtain the explicit expression for the q-$6j$-symbol [21, 22]:

$$
\begin{Bmatrix} j_1 & j_2 & j_{12} \\ j_3 & j & j_{23} \end{Bmatrix}_q = (-1)^{j_1+j_{12}+j_{23}+j_3} \frac{\nabla(j_1 j_2 j_{12})_q \nabla(j_2 j_3 j_{23})_q}{[j_1-j_2+j_{12}]![-j_1+j_2+j_{12}]!}
$$
$$
\times \frac{\nabla(j_{12} j_3 j)_q \nabla(j_1 j_{23} j)_q [j_{12}+j_3+j+1]![j_1+j_{23}+j+1]!}{[j_2-j_3+j_{23}]![-j_2+j_3+j_{23}]![j_1-j_{23}+j]![-j_{12}+j_3+j]!}
$$
$$
\times \sum_z \frac{(-1)^z [j_1+j-j_{23}+z]![j_3+j-j_{12}+z]![j_{12}+j_{23}+j_2-j-z]!}{[z]![j_1+j_{23}-j-z]![j_{12}+j_3-j-z]![j_2+j-j_{12}-j_{23}+z]![2j+1+z]!},
$$
(9.21)

where we use the notation:

$$
\nabla(j_1 j_2 j_3)_q = \sqrt{\frac{[j_1+j_2-j_3]![j_1-j_2+j_3]![-j_1+j_2+j_3]!}{[j_1+j_2+j_3+1]!}}.
$$
(9.22)

A total list of different explicit expressions and symmetry properties for the q-$6j$-symbols (or the q-Racah coefficients) can be found in the papers [4, 16, 21, 22].

The general formula (9.21) of the q-$6j$-symbol can be expressed in terms of the following basic hypergeometric series

$$
\begin{Bmatrix} j_1 & j_2 & j_{12} \\ j_3 & j & j_{23} \end{Bmatrix}_q = (-1)^{j_1+j_{12}+j_{23}+j_3} \frac{\nabla(j_1 j_2 j_{12})_q \nabla(j_2 j_3 j_{23})_q \nabla(j_{12} j_3 j)_q}{[j_1-j_2+j_{12}]![-j_1+j_2+j_{12}]![j_2-j_3+j_{23}]!}
$$
$$
\times \frac{\nabla(j_1 j_{23} j)_q [j_{12}+j_3+j+1]![j_1+j_{23}+j+1]![j_2+j_{12}+j_{23}-j]!}{[-j_2+j_3+j_{23}]![j_1+j_{23}-j]![j_{12}+j_3-j]![j_2+j-j_{12}-j_{23}]![2j+1]!}
$$
$$
\times {}_4\Phi_3 \left(\begin{matrix} -j_1-j_{23}+j, -j_{12}-j_3+j, j_1-j_{23}+j+1, j_3+j-j_{12}+1 \\ -j_2-j_{12}-j_{23}+j, j_2-j_{12}-j_{23}+j+1, 2j+2 \end{matrix} \middle| q^2, q^2 \right).
$$
(9.23)

If we set $j_{12} = j_1 + j_2$ in (9.21) we obtain simple explicit expression for the special (so called 'stretched') q-$6j$-symbol:

$$
\begin{Bmatrix} j_1 & j_2 & j_1+j_2 \\ j_3 & j & j_{23} \end{Bmatrix}_q = (-1)^{j_1+j_2+j_3+j} \left[\frac{[2j_1]![2j_2]![j_1+j_2+j_3+j+1]!}{[2j_1+2j_2+1]![j_1+j+j_{23}+1]![j_2+j_3+j_{23}+1]!} \right]^{\frac{1}{2}}
$$
(9.24)

$$
\times \left[\frac{[j_1+j_2-j_3+j]![j_1+j_2+j_3-j]![-j_1+j+j_{23}]![-j_2+j_3+j_{23}]!}{[-j_1-j_2+j_3+j]![j_1+j-j_{23}]![j_1-j+j_{23}]![j_2+j_3-j_{23}]![j_2-j_3+j_{23}]!} \right]^{\frac{1}{2}}.
$$

In the next section we consider a more complicated example of application of the projection operator method for calculation of a general expression for CGCs of the quantum algebra $U_q(su(3))$.

10. Clebsch-Gordan Coefficients for the Quantum Algebra $U_q(su(3))$

Let $\Pi := \{\alpha_1, \alpha_2\}$ be a system of simple roots of the Lie algebra $sl(3)$ ($sl(3) := sl(3, \mathbb{C}) \simeq A_2$), endowed with the following scalar product: $(\alpha_1, \alpha_1) = (\alpha_2, \alpha_2) = 2$, $\alpha_1, \alpha_2) = (\alpha_2, \alpha_1) = -1$. The root system Δ_+ of $sl(3)$ consists of the roots $\alpha_1, \alpha_1 + \alpha_2, \alpha_2$.

The quantum Hopf algebra $U_q(sl(3))$ is generated by the Chevalley elements $q^{\pm h_{\alpha_i}}$, $e_{\pm\alpha_i}$ $(i = 1, 2)$ with the relations (11.1), (11.2) where $i, j = 1, 2$.

For construction of the composite root vectors $e_{\pm(\alpha_1 + \alpha_2)}$ we fix the following normal ordering in Δ_+:

$$\alpha_1, \ \alpha_1 + \alpha_2, \ \alpha_2. \tag{10.1}$$

According to this ordering we set

$$e_{\alpha_1 + \alpha_2} := [e_{\alpha_1}, e_{\alpha_2}]_{q^{-1}}, \qquad e_{-\alpha_1 - \alpha_2} := [e_{-\alpha_2}, e_{-\alpha_1}]_q. \tag{10.2}$$

Let us introduce another standard notations for the Cartan-Weyl generators:

$$
\begin{aligned}
e_{12} &:= e_{\alpha_1}, & e_{21} &:= e_{-\alpha_1}, & e_{11} - e_{22} &:= h_{\alpha_1}, \\
e_{23} &:= e_{\alpha_2}, & e_{32} &:= e_{-\alpha_2}, & e_{22} - e_{33} &:= h_{\alpha_2}, \\
e_{13} &:= e_{\alpha_1 + \alpha_2}, & e_{31} &:= e_{-\alpha_1 - \alpha_2}, & e_{11} - e_{33} &:= h_{\alpha_1} + h_{\alpha_2}.
\end{aligned} \tag{10.3}
$$

The explicit formula for the extremal projector (7.2) specialized to the case of $U_q(sl(3))$ has the form

$$p = p_{12} p_{13} p_{23}, \tag{10.4}$$

where the elements p_{ij} $(1 \leq i < j \leq 3)$ are given by

$$
\begin{aligned}
p_{ij} &= \sum_{n=0}^{\infty} \frac{(-1)^n}{[n]!} \varphi_{ij,n} e_{ij}^n e_{ji}^n, \\
\varphi_{ij,n} &= q^{-(j-i-1)n} \left\{ \prod_{s=1}^{n} [e_{ii} - e_{jj} + j - i + s] \right\}^{-1}.
\end{aligned} \tag{10.5}
$$

The extremal projector p satisfies the relations:

$$e_{ij} p = p e_{ji} = 0 \quad (i < j), \qquad p^2 = p. \tag{10.6}$$

The quantum algebra $U_q(su(3))$ can be considered as the quantum algebra $U_q(sl(3))$ endowed with the additional Cartan involution *:

$$h_{\alpha_i}^* = h_{\alpha_i}, \qquad e_{\pm\alpha_i}^* = e_{\mp\alpha_i}, \qquad q^* = q \ (\text{or } q^{-1}). \tag{10.7}$$

Let $(\lambda\mu)$ be a finite-dimensional IR of $U_q(su(3))$ with the highest weight $(\lambda\mu)$ (λ and μ are nonnegative integers). The vector of the highest weight, denoted by the symbol $|(\lambda\mu)h\rangle$, satisfy the relations

$$h_{\alpha_1}|(\lambda\mu)h\rangle = \lambda|(\lambda\mu)h\rangle , \qquad h_{\alpha_2}|(\lambda\mu)h\rangle = \mu|(\lambda\mu)h\rangle ,$$

$$e_{ij}|(\lambda\mu)h\rangle = 0 \qquad (i < j) . \tag{10.8}$$

Labeling of another basis vectors in the IR $(\lambda\mu)$ depends on the choice of subalgebras of $U_q(u(3))$ (in other words which reduction chain from $U_q(u(3))$ to subalgebras is chosen). Here we use the Gelfand-Tsetlin reduction chain:

$$U_q(su(3)) \supset U_q(u_Y(1)) \otimes U_q(su_T(2)) \supset U_q(u_{T_0}(1)) , \tag{10.9}$$

where the subalgebra $U_q(su_T(2))$ is generated by the elements

$$T_+ := e_{23} , \qquad T_- := e_{32} , \qquad T_0 := \tfrac{1}{2}(e_{22} - e_{33}) , \tag{10.10}$$

the subalgebra $U_q(u_{T_0}(1))$ is generated by q^{T_0}, and the subalgebra $U_q(u_Y(1))$ is generated by q^Y where[5]:

$$Y = -\tfrac{1}{3}\left(2h_{\alpha_1} + h_{\alpha_2}\right) . \tag{10.11}$$

In the case of the reduction chain (10.9) the basis vectors of IR $(\lambda\mu)$ are denoted by

$$|(\lambda\mu)jtt_z\rangle . \tag{10.12}$$

Here the quantum number set jtt_z characterize the hypercharge y and the T-spin t and its projection t_z:

$$q^{T_0}|(\lambda\mu)jtt_z\rangle = q^{t_z}|(\lambda\mu)jtt_z\rangle ,$$

$$T_\pm|(\lambda\mu)jtt_z\rangle = \sqrt{[t \mp t_z][t \pm t_z + 1]}|(\lambda\mu)jtt_z \pm 1\rangle ,$$

$$q^Y|(\lambda\mu)jtt_z\rangle = q^y|(\lambda\mu)jtt_z\rangle , \tag{10.13}$$

where the parameter j is connected with the eigenvalue y of the "hypercharge" operator Y as follows

$$y = -\tfrac{1}{3}(2\lambda + \mu) + 2j . \tag{10.14}$$

---

[5]In the classical non-deformed case in the elementary particle theory the subalgebra $su_T(2)$ is called the T-spin algebra and the element Y is the hypercharge operator.

We can show (see [18, 19]) that the quantum numbers jt are taken all nonnegative integers and half-integers such that the sum $\frac{1}{2}\mu + j + t$ is an integer and they are subjected to the constraints:

$$\begin{cases} \frac{1}{2}\mu + j - t \geq 0 \,, \\ \frac{1}{2}\mu - j + t \geq 0 \,, \\ -\frac{1}{2}\mu + j + t \geq 0 \,, \\ \frac{1}{2}\mu + j + t \geq \lambda + \mu \,. \end{cases} \tag{10.15}$$

For every fixed t the projection t_z runs the values $t_z = -t, -t+1, \ldots, t-1, t$. It is not hard to show that the orthonormalized vectors (10.12) can be represented in the following form

$$\left\{ |(\lambda\mu)jtt_z\rangle := N_{jt}^{(\lambda\mu)} P_{t_z;t}^t \mathcal{R}_{\frac{1}{2}\mu-t}^j |(\lambda\mu)h\rangle \right\} \,, \tag{10.16}$$

where $P_{t_z;t_z'}^t$ is the general projection operator of the type (9.12) for the quantum algebra $U_q(su_T(2))$, the element $\mathcal{R}_{\frac{1}{2}\mu-t}^j$ is a component of the irreducible tensor operator of rank j, $(-j \leq j_z \leq j)$:

$$\mathcal{R}_{j_z}^j = \sqrt{\frac{[2j]!}{[j-j_z]![j+j_z]!}} (-1)^{j_z} q^{2j^2-j+j_z} e_{21}^{j-j_z} e_{31}^{j+j_z} q^{-jh_{\alpha_1}-(j+j_z)T_0} \,, \tag{10.17}$$

the normalizing factor $N_{jt}^{(\lambda\mu)}$ has the form

$$N_{jt}^{(\lambda\mu)} = (-1)^{t-\frac{1}{2}\mu} q^{-(j+\frac{1}{2}\mu-t)(j-\mu+t)-2j^2+j\lambda-\mu+2t}$$
$$\times \sqrt{\frac{[\lambda+\frac{1}{2}\mu-j+t+1]![\lambda+\frac{1}{2}\mu-j-t]![\frac{1}{2}\mu+j+t+1]![\frac{1}{2}\mu-j+t]!}{[\lambda]![\mu]![\lambda+\mu+1]![2j]![2t+1]!}} \,. \tag{10.18}$$

The operator

$$F_{jtt_z;h}^{(\lambda\mu)} = N_{jt}^{(\lambda\mu)} P_{t_z;t}^t \mathcal{R}_{\frac{1}{2}\mu-t}^j \tag{10.19}$$

is called the lowering operator and the conjugate operator

$$F_{h;jtt_z}^{(\lambda\mu)} = (F_{jtt_z;h}^{(\lambda\mu)})^* \tag{10.20}$$

is called the rising operator of $U_q(su(3))$.

If the extremal operator p acts on a vector of the weight $(\lambda\mu)$ with respect to the operators h_{α_1} and h_{α_2} then we write $p^{(\lambda\mu)}$. We multiply the extremal projector $p^{(\lambda\mu)}$ by the lowering and rising operators as follows

$$P_{jtt_z;j't't_z'}^{(\lambda\mu)} = F_{jtt_z;h}^{(\lambda\mu)} p^{(\lambda\mu)} F_{h;j't't_z'}^{(\lambda\mu)} \,, \tag{10.21}$$

and we assume that the resulting operator acts on a vector space of the fixed weight $(-\frac{1}{3}(2\lambda+\mu)+2j', t'_z)$ with respect to the operators Y and T_0. The operator $P^{(\lambda\mu)}_{jtt_z;j't't'_z}$ is called the general projection operator of $U_q(su(3))$.

For convenience we introduce the short notations: $\Lambda := (\lambda\mu)$ and $\gamma := jtt_z$ and therefore the basis vector (10.12) will be denoted by $|\Lambda\gamma\rangle$. Let $\{|\Lambda_i\gamma_i\rangle\}$ be Gelfand-Tsetlin bases of two IRs Λ_i ($i = 1, 2$). Then $\{|\Lambda_1\gamma_1\rangle| \times \Lambda_2\gamma_2\rangle\}$ be a uncoupled bases in the representation $\Lambda_1 \otimes \Lambda_2$ of $U_q(su(3)) \otimes U_q(su(3))$. In this representation there is another coupled basis $|\Lambda_1\Lambda_2 : s\Lambda_3\gamma_3\rangle_q$ with respect to $\Delta_q(U_q(su(3)))$ where the index s classifies multiple representations Λ. We can expand the coupled basis in terms of the uncoupled basis $\{|\Lambda_1\gamma_1\rangle|\Lambda_2\gamma_2\rangle\}$:

$$|\Lambda_1\Lambda_2 : s\Lambda_3\gamma_3\rangle_q = \sum_{\gamma_1,\gamma_2} (\Lambda_1\gamma_1\,\Lambda_2\gamma_2|s\Lambda_3\gamma_3)_q|\Lambda_1\gamma_1\rangle|\Lambda_2\gamma_2\rangle,$$
(10.22)

where the matrix element $(\Lambda_1\gamma_1\,\Lambda_2\gamma_2|\Lambda_3\gamma_3)_q$ is the Clebsch-Gordan coefficient of $U_q(su(3))$. In just the same way as for the non-quantized Lie algebra $su(3)$ (see [18, 19]) we can show that any CGC of $U_q(su(3))$ can be represented in terms of the linear combination of the matrix elements of the projection operator (10.21)

$$(\Lambda_1\gamma_1\,\Lambda_2\gamma_2|s\Lambda_3\gamma_3)_q = \sum_{\gamma'_2} C(\gamma'_2)\langle\Lambda_1\gamma_1|\langle\Lambda_2\gamma_2|\Delta_q(P^{\Lambda_3}_{\gamma_3;h})|\Lambda_1 h \rangle|\Lambda_2\gamma'_2\rangle.$$
(10.23)

Classification of multiple representations Λ_3 in the representation $\Lambda_1 \otimes \Lambda_2$ is special problem and we shall not discuss it here. For the non-deformed algebra $su(3)$ this problem was considered in details in [18, 19]. Concerning the matrix elements in the right-side of (10.23) we give here an explicit expression for the more general matrix element:

$$\langle\Lambda_1\gamma_1|\langle\Lambda_2\gamma_2|\Delta_q(P^{\Lambda_3}_{\gamma_3,\gamma'_3})|\Lambda_1\gamma'_1 \rangle|\Lambda_2\gamma'_2\rangle.$$
(10.24)

Using a tensor form of the projection operator (10.21) and the Wigner-Racah calculus for the subalgebra $U_q(su(2))$ [20] it is not hard to obtain the following result (see [5]):

$$\langle\Lambda_1\gamma_1|\langle\Lambda_2\gamma_2|\Delta_q(P^{\Lambda}_{\gamma_3,\gamma'_3})|\Lambda_1\gamma'_1\rangle|\Lambda_2\gamma'_2\rangle$$

$$= (t_1t_{1z}\,t_2t_{2z}|t_3t_{3z})_q\,(t_1t'_{1z}\,t_2t'_{2z}|t_3t'_{3z})_q\,A \sum_{j''_1j''_2t''_1t''_2t''_3} C_{j''_1j''_2t''_1t''_2t''_3}$$

$$\times \left\{\begin{matrix} j_1-j''_1 & j_2-j''_2 & j_1+j_2-j''_1-j''_2 \\ t''_1 & t''_2 & t''_3 \\ t_1 & t_2 & t_3 \end{matrix}\right\}_q \left\{\begin{matrix} j'_1-j''_1 & j'_2-j''_2 & j'_1+j'_2-j''_1-j''_2 \\ t''_1 & t''_2 & t''_3 \\ t'_1 & t'_2 & t'_3 \end{matrix}\right\}_q.$$
(10.25)

Here

$$A = [\lambda+1][\mu+1][\lambda+\mu+2]$$

$$\times \left[\frac{[2t_1+1][2t_2+1][2j_1+1]![2j_2+1]![\lambda_3+\frac{1}{2}\mu_3-j_3+t_3+1]![\lambda_3+\frac{1}{2}\mu_3-j_3-t_3]!}{[\lambda_1+\frac{1}{2}\mu_1-j_1+t_1+1]![\lambda_1+\frac{1}{2}\mu_1-j_1-t_1]![\lambda_2+\frac{1}{2}\mu_2-j_2+t_2+1]![\lambda_2+\frac{1}{2}\mu_2-j_2-t_2]![2j_3]!} \right]^{\frac{1}{2}}$$

$$\times \left[\frac{[2t_1'+1][2t_2'+1][2j_1'+1]![2j_2'+1]![\lambda_3+\frac{1}{2}\mu_3-j_3'+t_3'+1]![\lambda_3+\frac{1}{2}\mu_3-j_3'-t_3']!}{[\lambda_1+\frac{1}{2}\mu_1-j_1'+t_1'+1]![\lambda_1+\frac{1}{2}\mu_1-j_1'-t_1']![\lambda_2+\frac{1}{2}\mu_2-j_2'+t_2'+1]![\lambda_2+\frac{1}{2}\mu_2-j_2'-t_2']!} \right]^{\frac{1}{2}},$$

$$(10.26)$$

the coefficient $C_{j_1''j_2''t_1''t_2''t_3''}$ does not contain any 'inner' summation (without summation!) and has the form

$$C_{j_1''j_2''t_1''t_2''t_3''} = (-1)^{2(j_1+j_2+j_3'-j_1''-j_2'')} q^\psi$$

$$\times \frac{[2j_1+2j_2-2j_1''-2j_2''+1]![2j_1'+2j_2'-2j_1''-2j_2''+1]![2t_2''+1][2t_3''+1]}{[2j_1'']![2j_2'']![2j_1-2j_1'']![2j_2-2j_2'']![2j_1'-2j_1'']![2j_2'-2j_2'']![2j_1+2j_2-2j_3-2j_1''-2j_2'']!}$$

$$\times \frac{[\lambda_1+\frac{1}{2}\mu_1-j_1''+t_1''+1]![\lambda_1+\frac{1}{2}\mu_1-j_1''-t_1'']![\lambda_2+\frac{1}{2}\mu_2-j_2''+t_2''+1]![\lambda_2+\frac{1}{2}\mu_2-j_2''-t_2'']!}{[\lambda_3+\frac{1}{2}\mu_3+j_1+j_2-j_3-j_1''-j_2''+t_3''+2]![\lambda_3+\frac{1}{2}\mu_3+j_1+j_2-j_3'-j_1''-j_2''-t_3''+1]!}$$

$$\times \left\{ \begin{matrix} j_1-j_1'' & j_1'' & j_1 \\ \frac{1}{2}\mu_1 & t_1 & t_1'' \end{matrix} \right\}_q \left\{ \begin{matrix} j_2-j_2'' & j_2'' & j_2 \\ \frac{1}{2}\mu_2 & t_2 & t_2'' \end{matrix} \right\}_q \left\{ \begin{matrix} j_3 & j_1+j_2-j_3-j_1''-j_2'' & j_1+j_2-j_1''-j_2'' \\ t_3'' & t_3 & \frac{1}{2}\mu_3 \end{matrix} \right\}_q$$

$$\times \left\{ \begin{matrix} j_1'-j_1'' & j_1'' & j_1' \\ \frac{1}{2}\mu_1 & t_1' & t_1'' \end{matrix} \right\}_q \left\{ \begin{matrix} j_2'-j_2'' & j_2'' & j_2' \\ \frac{1}{2}\mu_2 & t_2' & t_2'' \end{matrix} \right\}_q \left\{ \begin{matrix} j_3' & j_1'+j_2'-j_3'-j_1''-j_2'' & j_1'+j_2'-j_1''-j_2'' \\ t_3'' & t_3' & \frac{1}{2}\mu_3 \end{matrix} \right\}_q,$$

$$(10.27)$$

where

$$\psi = \sum_{i=1}^{2} \left(2\varphi(\lambda_i,\mu_i,j_i'',t_i'') - \varphi(\lambda_i,\mu_i,j_i,t_i) - \varphi(\lambda_i,\mu_i,j_i',t_i') \right.$$

$$\left. -t_i(t_i+1) - t_i'(t_i'+1) \right) - 2\varphi(\lambda_3,\mu_3,j_3'',t_3'') + \varphi(\lambda_3,\mu_3,j_3,t_3)$$

$$+ \varphi(\lambda_3,\mu_3,j_3',t_3') + j_3''(4\lambda_3+2\mu_3+2) - 2t_3''(t_3''-1) - 2\mu_3$$

$$- (j_2+j_2'-2j_2'')(2\lambda_1+\mu_1-6j_1'') - (j_3+j_3'')(j_3+j_3''+1)$$

$$- (j_3'+j_3'')(j_3'+j_3''+1) + 4(j_1-j_1'')(j_2-j_2'') + 4(j_1'-j_1'')(j_2'-j_2''),$$

$$(10.28)$$

$$\varphi(\lambda,\mu,j,t) = \frac{1}{2}(\frac{1}{2}\mu+j-t)(\frac{1}{2}\mu+j+t-3) + j(\lambda-2j+1),$$

$$j_3'' = j_1+j_2-j_3-j_1''-j_2'' = j_1'+j_2'-j_3'-j_1''-j_2''. \qquad (10.29)$$

A detailed proof of the formulas (10.25)-(10.29) is given in [5].

The q-$9j$-symbol of $U_q(su(2))$ in (10.25) can be expressed in terms of q-$6j$-symbols. This expression has the form:

$$
\begin{Bmatrix} j_1 & j_2 & j_{12} \\ j_3 & j_4 & j_{34} \\ j_{13} & j_{24} & j \end{Bmatrix}_q = \sum_z (-1)^{2z} q^{c(z)+c(j_{24})+c(j_{34})+c(j)} [2z+1]
$$

$$
\times \begin{Bmatrix} j_1 & j_2 & j_{12} \\ z & j_3 & j_{13} \end{Bmatrix}_q \begin{Bmatrix} j_3 & j_4 & j_{34} \\ j & j_{12} & z \end{Bmatrix}_q \begin{Bmatrix} j_{13} & j_{24} & j \\ j_4 & z & j_2 \end{Bmatrix}_q \quad (10.30)
$$

where $c(j) := j(j+1)$. In our case $j_{12} = j_1 + j_2$ and we have in the right side of (10.30) one particular ('stretched') q-$6j$-symbol (9) which does not contain a summation. Therefore each q-$9j$-symbol in the right-side of (10.25) is expressed in terms of a linear combination of products of the general q-$6j$-symbols. Since the general q-$6j$-symbol can be expressed in terms of the basic hypergeometric series $_4\Phi_3\left(\genfrac{}{}{0pt}{}{\cdots}{\cdots}; q, q\right)$ therefore the q-$9j$-symbol is expressed in terms of a linear combination of products of two basic hypergeometric series $_4\Phi_3\left(\genfrac{}{}{0pt}{}{\cdots}{\cdots}; q, q\right)$.

Thus, *the general matrix element (10.25) can be expressed in terms of a linear combination of products of four basic hypergeometric series* $_4\Phi_3\left(\genfrac{}{}{0pt}{}{\cdots}{\cdots}; q, q\right)$.

11. Gelfand-Tsetlin Basis for $U_q(u(n))$

Let $\Pi := \{\alpha_1, \dots, \alpha_{n-1}\}$ be a system of simple roots of the Lie algebra $sl(n)$ $(sl(n) := sl(n, \mathbb{C}) \simeq A_{n-1})$ endowed with the following scalar product: $(\alpha_i, \alpha_j) = (\alpha_j, \alpha_i)$, $(\alpha_i, \alpha_i) = 2$, $(\alpha_i, \alpha_{i+1}) = -1$, $(\alpha_i, \alpha_j) = 0$ $((|i - j| > 1)$.

The quantum Hopf algebra $U_q(sl(n))$ is generated by the Chevalley elements $q^{\pm h_{\alpha_i}}$, $e_{\pm \alpha_i}$ $(i = 1, 2, \dots, n-1)$ with the defining relations:

$$
q^{h_{\alpha_i}} q^{-h_{\alpha_i}} = q^{-h_{\alpha_i}} q^{h_{\alpha_i}} = 1, \qquad q^{h_{\alpha_i}} q^{h_{\alpha_j}} = q^{h_{\alpha_j}} q^{h_{\alpha_i}},
$$

$$
q^{h_{\alpha_i}} e_{\pm \alpha_j} q^{-h_{\alpha_i}} = q^{\pm(\alpha_i, \alpha_j)} e_{\pm \alpha_j}, \qquad [e_{\alpha_i}, e_{-\alpha_j}] = \delta_{ij} [h_{\alpha_i}],
$$

$$
[e_{\pm \alpha_i}, e_{\pm \alpha_j}] = 0 \qquad (|i - j| \geq 2),
$$

$$
[[e_{\pm \alpha_i} e_{\pm \alpha_j}]_q e_{\pm \alpha_j}]_q = 0 \qquad (|i - j| = 1), \qquad (11.1)
$$

$$
\Delta_q(h_{\alpha_i}) = h_{\alpha_i} \otimes 1 + 1 \otimes h_{\alpha_i}, \qquad S_q(h_{\alpha_i}) = -h_{\alpha_i},
$$

$$
\Delta_q(e_{\pm \alpha_i}) = e_{\pm \alpha_i} \otimes q^{\frac{h_{\alpha_i}}{2}} + q^{-\frac{h_{\alpha_i}}{2}} \otimes e_{\pm \alpha_i}, \qquad S_q(e_{\alpha_i}) = -q^{\pm 1} e_{\alpha_i}.
$$

$$
(11.2)
$$

Below we shall use another basis in the Cartan subalgebra of the algebra
$sl(n)$ $(U_q(sl(n))$. Namely we set

$$
\begin{aligned}
e_{11} &= \tfrac{1}{n}\left((n-1)h_{\alpha_1}+(n-2)h_{\alpha_2}+\cdots+2h_{\alpha_{n-2}}+h_{\alpha_{n-1}}+N\right), \\
e_{22} &= \tfrac{1}{n}\left((n-1)h_{\alpha_1}+(n-2)h_{\alpha_2}+\cdots+2h_{\alpha_{n-2}}+h_{\alpha_{n-1}}+N\right)-h_{\alpha_1}, \\
\cdots\ \cdots\ &\ \ldots\ldots\ldots\ldots\ldots\ldots\ldots\ldots\ldots\ldots\ldots\ldots\ldots\ldots\ldots\ldots\ldots\ldots \\
e_{ii} &= \tfrac{1}{n}\left((n-1)h_{\alpha_1}+(n-2)h_{\alpha_2}+\cdots+2h_{\alpha_{n-2}}+h_{\alpha_{n-1}}+N\right)-\sum_{k=1}^{i-1}h_{\alpha_k}, \\
\cdots\ \cdots\ &\ \ldots\ldots\ldots\ldots\ldots\ldots\ldots\ldots\ldots\ldots\ldots\ldots\ldots\ldots\ldots\ldots\ldots\ldots \\
e_{nn} &= \tfrac{1}{n}\left(-h_{\alpha_1}-2h_{\alpha_2}-\cdots-(n-2)h_{\alpha_{n-2}}-(n-1)h_{\alpha_{n-1}}+N\right).
\end{aligned}
$$

$$(11.3)$$

Here N is a central element of g $(U_q(g))$, which is equal to 0 for the case
$g = sl(n)$ and $N \neq 0$ for $g = gl(n)$. It is easy to see that

$$
\begin{aligned}
h_{\alpha_i} &= e_{ii} - e_{i+1i+1} \qquad (i = 1,\ldots,n-1), \\
N &= e_{11} + e_{22} + \ldots + e_{nn}.
\end{aligned}
$$

$$(11.4)$$

Dual elements to the ones e_{ii} $(i = 1,2,\ldots,n)$ will be denoted by ϵ_i $(i = 1,2,\ldots,n)$: $\epsilon_i(e_{jj}) = (\epsilon_i,\epsilon_j) = \delta_{ij}$. In the terms of ϵ_i the positive root
system Δ_+ of $sl(n)$ is presented as follows

$$\Delta_+ = \{\epsilon_i - \epsilon_j \,|\, 1 \leq i < j \leq n\},\qquad (11.5)$$

where $\epsilon_i - \epsilon_{i+1}$ are the simple roots: $\alpha_i = \epsilon_i - \epsilon_{i+1}$ $(i=1,2,\ldots,n-1)$. For
the root vectors $e_{\epsilon_i-\epsilon_j}$ $(i \neq j)$ the another standard notations are also used

$$e_{ij} := e_{\epsilon_i-\epsilon_j},\qquad e_{ji} := e_{\epsilon_j-\epsilon_i}\qquad (1 \leq i < j \leq n).$$

$$(11.6)$$

In particular, the elements e_{ii+1}, e_{i+1i} are the Chevalley generators: $e_{ii+1} = e_{\alpha_i}$, $e_{i+1i} = e_{-\alpha_i}$ $(i = 1,\ldots,n-1)$.

For construction of the composite root vectors e_{ij} $(j \neq i \pm 1)$ we fix the
following normal ordering of the positive root system Δ_+ (see [26])

$$(\epsilon_1-\epsilon_2),(\epsilon_1-\epsilon_3,\epsilon_2-\epsilon_3),\ldots,(\epsilon_1-\epsilon_i,\ldots,\epsilon_{i-1}-\epsilon_i),\ldots,(\epsilon_1-\epsilon_n,\ldots,\epsilon_1-\epsilon_n).$$

$$(11.7)$$

According to this ordering we set

$$e_{ij} := [e_{ik},e_{kj}]_{q^{-1}},\qquad e_{ji} := [e_{jk},e_{ki}]_q\qquad (1 \leq i < k < j \leq n).$$

$$(11.8)$$

It should be stressed that the structure of the composite root vectors (11.8) is independent of the choice of the index k in the r.h.s. of the definition (11.8). In particular, one has

$$e_{ij} := [e_{ii+1}, e_{i+1j}]_{q^{-1}} = [e_{ij-1}, e_{j-1j}]_{q-1} \quad (1 \le i < j \le n),$$

$$e_{ji} := [e_{ji+1}, e_{i+1i}]_q = [e_{jj-1}, e_{j-1i}]_q \quad (1 \le i < j \le n). \tag{11.9}$$

The explicit formula for the extremal projector (7.2) specialized to the case of $U_q(sl(n))$ has the form

$$p(U_q(sl(n))) = p(U_q(sl(n-1)))(p_{1n}p_{2n}\cdots p_{n-2n}p_{n-1n})$$

$$= p_{12}(p_{13}p_{23})\cdots(p_{1i}\cdots p_{ii+1})\cdots(p_{1n}\cdots p_{n-1n}) \tag{11.10}$$

where the elements p_{ij} are given by the formulas (10.5) with $(1 \le i < j \le n)$. The extremal projector $p := p(U_q(sl(n)))$ satisfies the eqs. (10.6) with $(1 \le i < j \le n)$.

The quantum algebra $U_q(su(n))$ can be considered as the quantum algebra $U_q(sl(n))$ endowed with the additional Cartan involution $*$:

$$e_{ij}^* = e_{ji}, \qquad q^* = q \text{ (or } q^{-1}). \tag{11.11}$$

Since the quantum algebra $U_q(su(n))$ can be interpreted as the algebra $U_q(u(n))$ with the central element $N = 0$ and the inner structure of its representations is more easily described in terms of $U_q(u(n))$, we shall consider the quantum algebra $U_q(u(n))$.

Let V^{λ_n} be a finite-dimensional IR of $U_q(u(n))$ with the highest weight $\lambda_n := (\lambda_{1n}, \lambda_{2n}, \ldots, \lambda_{nn})$ where $\lambda_{in} - \lambda_{i+1n}$ $(i = 1, \ldots, n-1)$ are nonnegative integers. The vector of the highest weight, denoted by the symbol $|\lambda_n\rangle$, satisfy the relations

$$q^{h_{ii}}|\lambda_n\rangle = q^{\lambda_{in}}|\lambda_n\rangle \quad (1 \le i \le n),$$

$$e_{ij}|\lambda_n\rangle = 0 \quad (i < j). \tag{11.12}$$

Labeling of another basis vectors in IR V^{λ_n} depends on choice of subalgebras of $U_q(u(n))$ (in other words which reduction chain from $U_q(u(n))$ to subalgebras is chosen). Here we use the 'so called' Gelfand-Tsetlin reduction chain:

$$U_q(u(n)) \supset U_q(u(n-1)) \supset \ldots \supset U_q(u(k)) \supset \ldots U_q(u(1)), \tag{11.13}$$

where the subalgebra $U_q(u(k))$ is generated by e_{ij} with $i, j = 1, 2, \ldots, k$.
The following theorem can be proved.

Theorem 11.1 *In the $U_q(u(n))$-module V^{λ_n} there is the orthogonal Gelfand-Tsetlin basis consisting of all vectors of the form*

$$
|\lambda\rangle := \left|\begin{array}{ccccc}
\lambda_{1n} & \lambda_{2n} & \cdots & \lambda_{n-1n} & \lambda_{nn} \\
\lambda_{1n-1} & \lambda_{2n-1} & \cdots & \lambda_{n-1n-1} & \\
\cdots & \cdots & \cdots & & \\
\lambda_{12} & \lambda_{22} & & & \\
\lambda_{11} & & & &
\end{array}\right\rangle \tag{11.14}
$$

$$
= F_-(\lambda_1;\lambda_2)F_-(\lambda_1;\lambda_2)\cdots F_-(\lambda_{n-1};\lambda_n)|\lambda_n\rangle ,
$$

where the numbers λ_{ij} satisfy the standard inequalities ('between conditions') for the Lie algebra $u(n)$, i.e.

$$
\lambda_{ij+1} \geq \lambda_{ij} \geq \lambda_{i+1j+1} \qquad \text{for } 1 \leq i \leq j \leq n-1 . \tag{11.15}
$$

The lowering operators $F_-(\lambda_k;\lambda_{k+1})$, $(k = 1, 2 \ldots, n-1)$, are given by

$$
F_-(\lambda_k;\lambda_{k+1}) = \mathcal{N}(\lambda_k;\lambda_{k+1})\, p(U_q(u(k))) \prod_{i=1}^{k} (e_{k+1i})^{\lambda_{ik+1}-\lambda_{ik}} , \tag{11.16}
$$

$$
\mathcal{N}(\lambda_k;\lambda_{k+1}) = \left\{ \prod_{i=1}^{k} \frac{[l_{ik}-l_{k+1k+1}-1]!}{[l_{ik+1}-l_{ik}]![l_{ik+1}-l_{k+1k+1}-1]!} \times \right.
$$
$$
\left. \times \prod_{1\leq i<j\leq k} \frac{[l_{ik+1}-l_{jk}]![l_{ik}-l_{jk+1}-1]!}{[l_{ik}-l_{jk}]![l_{ik+1}-l_{jk+1}-1]!} \right\}^{\frac{1}{2}} \tag{11.17}
$$

where $l_{ij}:=\lambda_{ij}-i$.

The explicit form of the basis vectors (11.14) allows to calculate the actions of the Cartan-Weyl generators on these vectors.

For the Cartan elements $q^{e_{ii}}$, $(i=1,2,\ldots,n)$, we easy find that

$$
q^{e_{ii}}|\lambda\rangle = q^{S_i-S_{i-1}}|\lambda\rangle . \tag{11.18}
$$

Here $S_i=\sum_{j=1}^{i}\lambda_{ji}$, where the numbers $\lambda_{1,i}, \lambda_{2i}, \ldots, \lambda_{ii}$ are ones in the i-th row of the pattern λ of the vector (11.14), $S_0=0$. The computation of the action of the generators e_{ij} for $i \neq j$ is more difficult. The procedure of this computation goes as follows. First of all, using the explicit expression (11.14) for the basis vector $|\lambda\rangle$, we determine the transformation of the basis under the action of the generators e_{ii+1} and e_{i+1i} for all $i=1,2,\ldots,n-1$. Then using the inductive definition (11.8) for the generators e_{ij} we find their action for all $i \neq j$. The result reads as follows.

Theorem 11.2 *The Cartan-Weyl generators e_{kk+s} and e_{k+sk} $(s = 1, 2, \ldots, n - k)$ of the quantum algebra $U_q(u(n))$ act on the Gelfand-Tsetlin basis (11.14) according to relations:*

$$e_{kk+s}|\lambda\rangle = \sum_{j_i,j_2,\ldots,j_s} \langle\lambda + \sum_{p=1}^{s} \epsilon_{j_p,k+p-1}|e_{kk+s}|\lambda\rangle \, |\lambda + \sum_{p=1}^{s} \epsilon_{j_p,k+p-1}\rangle,$$

(11.19)

$$e_{k+sk}|\lambda\rangle = \sum_{j_i,j_2,\ldots,j_s} \langle\lambda - \sum_{p=1}^{s} \epsilon_{j_p,k+p-1}|e_{k+sk}|\lambda\rangle \, |\lambda - \sum_{p=1}^{s} \epsilon_{j_p,k+p-1}\rangle.$$

(11.20)

Here

$$\langle\lambda + \sum_{p=1}^{s} \epsilon_{j_p,k+p-1}|e_{kk+s}|\lambda\rangle = A_s(\lambda)q^{l_{j_1k} - l_{j_sk+s-1}}$$
$$\times \prod_{r=1}^{s} \langle\lambda + \epsilon_{j_r,k+r-1}|e_{k+r-1k+r}|\lambda\rangle,$$

(11.21)

$$\langle\lambda - \sum_{p=1}^{s} \epsilon_{j_p,k+p-1}|e_{k+sk}|\lambda\rangle = A_{s-1}(\lambda)q^{l_{j_sk+s-1} - l_{j_1k}}$$
$$\times \prod_{r=1}^{s} \langle\lambda - \epsilon_{j_r,k+r-1}|e_{k+rk+r-1}|\lambda\rangle,$$

(11.22)

where

$$\langle\lambda + \epsilon_{j_r,k+r-1}|e_{k+r-1k+r}|\lambda\rangle = \left\{ -\frac{\prod\limits_{i=1}^{k+r}[l_{ik+r} - l_{j_rk+r-1}]\prod\limits_{i=1}^{k+r-2}[l_{ik+r-2} - l_{j_rk+r-1} - 1]}{\prod\limits_{\substack{i=1 \\ i \neq j_r}}^{k+r-1}[l_{ik+r-1} - l_{j_rk+r-1}][l_{ik+r-1} - l_{j_rk+r-1} - 1]} \right\}^{\frac{1}{2}} \quad (11.23)$$

$$\langle\lambda + \epsilon_{j_r,k+r-1}|e_{k+r-1k+r}|\lambda\rangle = \left\{ -\frac{\prod\limits_{i=1}^{k+r}[l_{ik+r} - l_{j_rk+r-1} + 1]\prod\limits_{i=1}^{k+r-2}[l_{ik+r-2} - l_{j_rk+r-1}]}{\prod\limits_{\substack{i=1 \\ i \neq j_r}}^{k+r-1}[l_{ik+r-1} - l_{j_rk+r-1} + 1][l_{ik+r-1} - l_{j_rk+r-1}]} \right\}^{\frac{1}{2}} \quad (11.24)$$

$$A_s(\lambda) = \prod_{r=1}^{s} \frac{\text{sign}(l_{j_{r+1}k+r} - l_{j_rk+r-1})}{\{[l_{j_{r+1}k+r} - l_{j_rk+r-1}][l_{j_{r+1}k+r} - l_{j_rk+r-1} - 1]\}^{\frac{1}{2}}}, \quad (11.25)$$

$\text{sign}(x) = 1$ *for* $x \geq 0$ *and* $\text{sign}(x) = -1$ *for* $x < 0$.

In (11.19), (11.20) each summation index j_r runs over integers $1, 2, \ldots, k + r - 1$. The symbol $|\epsilon_{ij}\rangle$ means the Gelfand-Tsetlin patter, which has zeros everywhere, except 1 on the place (ij). The sum of the Gelfand-Tsetlin patterns is given with the sums of the corresponding labels, as a sum of matrices.

In the case $s = 1$ the formulas (11.19)-(11.25) coincide with the results of [8], where they have been given for the first time, however without a proof.

At the limit $q \to 1$ the formulas (11.14)-(11.25) coincides with the results of the paper [3].

Now we consider some generalizations of the extremal projector for the case of the quantum algebra $U_q(sl(2, \mathbb{C}))$.

12. 'Adjoint Extremal Projectors' for $U_q(sl(2, \mathbb{C}))$

Let $\mathbf{J}^2$ be the Casimir invariant for $U_q(sl(2, \mathbb{C}))$:

$$\mathbf{J}^2 = \tfrac{1}{2}(J_+ J_- + J_- J_+ + [2][J_0]^2) \ . \tag{12.1}$$

One easily verifies that:

$$\mathbf{J}^2 = X + [J_0][J_0 + 1] \ , \tag{12.2}$$

where we use the notation

$$X := J_- J_+ \ . \tag{12.3}$$

It is evident that the first equation (9.11) for the extremal projector (9.9) can be rewritten in the form

$$Xp = pX = 0 \ . \tag{12.4}$$

By using the Casimir operator $\mathbf{J}^2$ we can rewrite this equation as follows

$$\mathbf{J}^2 p = p\mathbf{J}^2 = [J_0][J_0 + 1]p \ . \tag{12.5}$$

Thus the extremal projector p is an eigenvector for the Casimir operator $\mathbf{J}^2$ with the eigenvalue $[J_0][J_0 + 1]$. The equation (12.4) is similar to the algebraic equation for the δ-function:

$$x\delta(x) = 0 \ . \tag{12.6}$$

Let us continue this analogy. It is known that $\delta(x + \epsilon)$ is the generating function for derivatives of δ-function, i.e.

$$\delta^{(n)}(x) = \left. \frac{d^n \delta(x+\epsilon)}{d\epsilon^n} \right|_{\epsilon=0} \ . \tag{12.7}$$

Introduce an element $p(\epsilon)$ of the form

$$p(\epsilon) = \sum_{n=-\infty}^{\infty} \frac{(-1)^n}{\overline{\Gamma}(n+1-\epsilon)\overline{\Gamma}(2J_0+2+\epsilon)} J_-^n J_+^n . \tag{12.8}$$

The element $p(\epsilon)$ is an analog of the generating function $\delta(x+\epsilon)$. Its properties are described in the proposition.

Proposition 12.1 *The element $p(\epsilon)$ satisfies the equation*

$$Xp(\epsilon) = p(\epsilon)X = [\epsilon][2J_0 + 1 + \epsilon]p(\epsilon) \tag{12.9}$$

or

$$\mathbf{J}^2 p(\epsilon) = p(\epsilon)\mathbf{J}^2 = [J_0 + \epsilon][J_0 + 1 + \epsilon]p(\epsilon) . \tag{12.10}$$

Proof. The equation (12.9) is easily verified by direct calculation. The equation (12.10) is a consequence of (12.9).

We see from eq. (12.10) that the element $p(\epsilon)$ is a eigenvector of the Casimir operator $\mathbf{J}^2$ with the eigenvalue $[J_0+\epsilon][J_0+ \epsilon+1]$.

Let us introduce a scaling–derivative $\tilde{D}_x f(x)$ of a function $f(x)$ depending on a variable x as follows

$$\tilde{D}_x f(x) = \lim_{\Delta x \to 0} \frac{f(x+\Delta x)-f(x)}{[\Delta x]} = (q^{\frac{1}{2}} - q^{-\frac{1}{2}})(\ln q)^{-1} f_x'(x)) , \tag{12.11}$$

where f_x' is the usual derivative of the function $f(x)$.

Let $\tilde{p}^{(n)}$, $n=0,1,2,\ldots$, be scaling-derivatives of $p(\epsilon)$ at $\epsilon=0$, i.e.

$$\tilde{p}^{(n)} = (\tilde{D}_\epsilon)^n p(\epsilon) \,|_{\epsilon=0} . \tag{12.12}$$

Proposition 12.2 *The elements $\tilde{p}^{(n)}$, $n=0,1,2,\ldots$, satisfy the algebraic equations*

$$X\tilde{p}^{(n)} = \tilde{p}^{(n)}X = \sum_{l=0}^{n-1} a_l^n \tilde{p}^{(l)} , \qquad \text{for } n = 0,1,2,\ldots , \tag{12.13}$$

where

$$a_l^n = (q^{\frac{1}{2}} - q^{-\frac{1}{2}})^{n-l-2}(q^{\frac{2J_0+1}{2}} + (-1)^{n-l}q^{-\frac{2J_0+1}{2}})\frac{n!}{l!(n-l)!} . \tag{12.14}$$

The elements $\tilde{p}^{(n)}$ can be redefined to $p^{(n)}$ such that they will satisfy the simple equations

$$Xp^{(n)} = p^{(n)}X = p^{(n-1)} , \qquad \text{for } n = 0,2,1,\ldots \tag{12.15}$$

or

$$\mathbf{J}^2 p^n = p^{(n)}\mathbf{J}^2 = [J_0][J_0 + 1]p^{(n)} + p^{(n-1)} , \qquad \text{for } n = 0, 1, 2, \ldots .$$
(12.16)

Proof. Applying the scaling–differentiation operator $(\tilde{D}_\epsilon)^n$ to eq. (12.9) and putting $\epsilon = 0$ we obtain the equations (12.13). Let

$$p^n = \sum_{l=0}^{n} d_l^m \tilde{p}^{(l)} .$$
(12.17)

Substituting this in eq. (12.15) we obtain the system of equations

$$\sum_{k=l+1}^{n} d_k^m a_l^k = d_l^{n-1} , \qquad \text{for } l = 0, 1, 2, \ldots, n-1 .$$
(12.18)

This system has a unique solution if $d_1^1 = [2J_0 + 1]^{-1}$ and $d_0^m = 0$ for $m = 1, 2, \ldots, n-1$. We shall not present the solution here, since it has a cumbersome form.

Remark (i) The elements $p^{(n)}$, $n = 1, 2, \ldots$, are adjoint-vectors of the Casimir operator $\mathbf{J}^2$ with eigenvalue $[J_0][J_0 + 1]$. They are joined to the eigenvector p of $\mathbf{J}^2$. In this connection the element $p^{(n)}$ is called the 'adjoint extremal projector' of the n-th order.

(ii) The elements $(-1)^n (n!) p^{(n)}$, $n = 0, 1, 2, \ldots$, are analogs of the δ–function and its derivatives, since they satisfy the same algebraic equations.

(iii) In limit $q \to 1$ the elements $p^{(n)} (\tilde{p}^{(n)})$, $n = 0, 1, 2, \ldots$, turn into the corresponding elements of $sl(2, \mathbf{C})$ [27].

It was found that the 'adjoint extremal projectors, $p^{(n)} (U_q(sl(2))$, $n = 1, 2, \ldots$, are closely connected with a special class of decomposable representations for quantum algebra $U_q(sl(2))$ (details see [6]).

Acknowledgements. The author is thankful to the NATO Advance Study Institute SF2000 for the support of his visit. This work was supported by the Russian Foundation for Fundamental Research, grant No. 99-01-01163, and by the program of French-Russian scientific cooperation (CNRS grant PICS-608 and grant RFBR-98-01-22033).

References

1. R. M. Asherova, Yu. F. Smirnov and V. N. Tolstoy, *Projection operators for the simple Lie groups*, (Russian) Teoret. Mat. Fiz. **8** (1971), 255–271.
2. R. M. Asherova, Yu. F. Smirnov and V. N. Tolstoy, *A description of some class of projection operators for semisimple complex Lie algebras*, (Russian) Matem. Zametki **26** (1979), 15–25.

3. R. M. Asherova, Yu. F. Smirnov and V. N. Tolstoy, *Projection operators for the simple Lie groups. II. General scheme for construction of lowering operators. The case of the group* SU(n), (Russian) Teoret. Mat. Fiz. **15** (1973), 107–119.

4. R. M. Asherova, Yu. F. Smirnov and V. N. Tolstoy, *Weyl q-coefficients for* $U_q(3)$ *and Racah q-coefficients, for* SU$_q$(2), Phys. Atomic Nuclei **59** (1996), 1795–1807.

5. R. M. Asherova, Yu. F. Smirnov and V. N. Tolstoy, *On the general analytical formula for the* $SU_q(3)$ *- Clebsch-Gordan coefficients*, Proc. of Group theoretical methods in physics (Dubna, 2000), (Russian) Yad. Fiz.; math.QA/0103187 2001.

6. H.-D. Doebner and V. N. Tolstoy, *Adjoint "extremal projectors" and indecomposable representations for* $U_q(sl(2, \mathbf{C}))$, in: "Quantum symmetries", H. -D. Doebner and V. K. Dobrev, eds., World Sci. Publishing, River Edge, NJ, 1993, pp. 229–245.

7. V. G. Drinfeld, *Quantum groups*, Proc. ICM-86 (Berkely USA) Vol. 1, Amer. Math. Soc., 1987, pp. 798–820.

8. M. Jimbo, *A q-analogue of* $U_q(gl(n+1))$, *Hecke algebras and Yang-Baxter equation*, Lett. Math. Phys. **11** (1986), 247–252.

9. V. G. Kac, *Lie superalgebras*, Advances in Math. **26** (1977), 8–96.

10. S. M. Khoroshkin, J. Lukierski and V. N. Tolstoy, *Quantum affine (super)algebras* $U_q(A_1^{(1)})$ *and* $U_q(C(2)^{(2)})$; math.QA/0005145, 2000.

11. S. M. Khoroshkin and V. N. Tolstoy, *Universal R-matrix for quantized (super)algebras*, Commun. Math. Phys. **141** (1991), 599–617.

12. S. M. Khoroshkin and V. N. Tolstoy, *Extremal projector and universal R-matrix for quantum contragredient Lie (super)algebras*, in: "Quantum groups and related topics", R. Gelerak et al., eds., Kluwer Acad. Publ., Dordrecht, 1992.

13. S. M. Khoroshkin and V. N. Tolstoy, *The uniqueness theorem for the universal R-matrix*, Lett. Math. Phys. **24** (1992), 231–244.

14. S. M. Khoroshkin and V. N. Tolstoy, *The Cartan-Weyl basis and the universal R-matrix for quantum Kac-Moody algebras and superalgebras*, in: "Quantum Symmetries" (Clausthal 1991), H.-D. Doebner and V. K. Dobrev, eds., World Sci. Publishing, River Edge, NJ, 1993, pp. 336–351.

15. S. M. Khoroshkin and V. N. Tolstoy, *Twisting of quantum (super)algebras. Connection of Drinfeld's and Cartan-Weyl realizations for quantum affine algebras*, MPIM preprint, MPI/94-23, Bonn, 1994; hep-th/9404036.

16. A. N. Kirillov and N. Yu. Reshetikhin, *Representations of the algebra* $u_q(2)$, *q-orthogonal polynomials and invariants of links*, LOMI Preprint E-9-88, Leningrad, 1988.

17. G. Lusztig, *Canonical bases arising from quantized enveloping algebras*, J. Amer. Math. Soc. **3** (1990), 447–498.

18. Z. Pluhar, Yu. F. Smirnov and V. N. Tolstoy, *A novel approach to the* SU(3) *symmetry calculus*, Charles University preprint, Prague (Russian), 1981.

19. Z. Pluhar, Yu. F. Smirnov and V. N. Tolstoy, *Clebsch-Gordan coefficients of* SU(3) *with simple symmetry properties*, J. Phys. A: Math. Gen. **19** (1986), 21–28.

20. Yu. F. Smirnov, V. N. Tolstoy and Yu. I. Kharitonov, *Method of the projection operators and q-analog of the quantum angular momentum theory. Clebsch-Gordan coefficients and irreducible tensor operators*, Soviet J. Nuclear Phys. **53** (1991), 593–605.

21. Yu. F. Smirnov, V. N. Tolstoy and Yu. I. Kharitonov, *Projection-operator method and the q-analog of the quantum theory of angular momentum. Racah coefficients, 3j- and 6j-symbols, and their symmetry properties*, Soviet J. Nuclear Phys. **53** (1991), 1068–1086.

22. Yu. F. Smirnov, V. N. Tolstoy and Yu. I. Kharitonov, *Projection operator method and q-analog of angular momentum theory*, in: "Symmetry in Science" V, Plenum, New York, 1991, pp. 487–518.

23. Yu. F. Smirnov, V. N. Tolstoy and Yu. I. Kharitonov, *Tree technique and irreducible tensor operators for the* $su_q(2)$ *quantum algebra. 9j-symbols*, Soviet J. Nuclear Phys.

55 (1992), 1599–1604.

24. V. N. Tolstoy, *Extremal projectors for reductive classical Lie superalgebras with non-degenerate generalized Killing form*, (Russian) Uspekhi Mat. Nauk **40** (1985), 225–226.

25. V. N. Tolstoy, *Extremal projectors for contragredient Lie algebras and superalgebras of finite growth*, Russian Math. Surveys **44** (1989), 257–258.

26. V. N. Tolstoy, *Extremal projectors for quantized Kac-Moody superalgebras and some of their applications*, Lectures Notes in Phys. **370**, Springer, Berlin, 1990, pp. 118–125.

27. V. N. Tolstoy, *Fundamental system of the extremal projectors for sl(2)*, in: "Group theoretical methods in fundamental and applied physics" (Russian), Nauka, Moscow, 1988, pp. 258–259.

28. V. N. Tolstoy, *Combinatorial structure of root systems for Lie (super)algebras*, JINR, Dubna, 2001.

29. V. N. Tolstoy and S. M. Khoroshkin, *The Universal R-matrix for quantum non-twisted affine Lie algebras*, Functional Anal. Appl. **26** (1992), 69–71.

30. H. Yamane, *A Serre type theorem for affine Lie superalgebras and their quantized universal enveloping superalgebras*, Proc. Japan. Acad. **70** (1994), 31.

UNIFORM ASYMPTOTIC EXPANSIONS

R. WONG
Department of Mathematics
City University of Hong Kong
Tat Chee Avenue
Kowloon Hong Kong

Abstract. In this lecture, I first review an extension of the method of steepest descents given by Chester, Friedman and Ursell (1957). This extension allows one to derive asymptotic expansions which hold uniformly in regions where two saddle points may coalesce at a single point. As an illustration, I give an outline of a derivation of a uniform asymptotic expansion of the Hermite polynomial $H_n(\sqrt{2n+1}t)$. Next, I present a modification of the method of Chester, Friedman and Ursell, which can handle situations where two saddle points may coalesce at *two* distinct locations. Such situations occur in the cases of Meixner, Meixner–Pollaczek, and Krawtchouk polynomials.

1. Introduction

The method of steepest descent is probably the best known procedure for finding asymptotic behavior of integrals of the form

$$I(\lambda) = \int_C g(z)\, e^{\lambda f(z)} dz, \qquad (1.1)$$

where $f(z)$ and $g(z)$ are analytic functions, λ is a large positive parameter, and C is a contour in the z-plane. It was introduced by Debye (1909) in a paper concerning Bessel functions of large order. Debye's basic idea is to deform the contour C into a new path of integration C' so that the following conditions hold:

(a) C' passes through one or more zeros of $f'(z)$.
(b) the imaginary part of $f(z)$ is constant on C'.

J. Bustoz et al. (eds.), Special Functions 2000, 489–503.

If we write $z = x + iy$ and

$$f(z) = u(x, y) + iv(x, y)$$

and suppose that $z_0 = x_0 + iy_0$ is a zero of $f'(z)$, then it is known that (x_0, y_0) is a saddle point of $u(x, y)$ and the new curve $v(x, y) = v(x_0, y_0)$ gives the steepest paths on the surface $u = u(x, y)$ in the Cartesian space (x, y, u). For simplicity, we shall assume that z_0 is a simple zero of $f'(z)$ so that $f''(z_0) \neq 0$. On the steepest path C', we have

$$f(z) = f(z_0) - t^2, \tag{1.2}$$

where t is real and usually increases monotonically from $-\infty$ to $+\infty$. Changing variable from z to t gives

$$I(\lambda) = e^{\lambda f(z_0)} \int_{-\infty}^{\infty} g(z) \frac{dz}{dt} e^{-\lambda t^2} dt.$$

By expanding $f(z)$ into a Taylor series at z_0 and substituting it into (1.2), we have by inversion

$$z - z_0 = \sqrt{\frac{2}{-f''(z_0)}} \, t + c_2 \, t^2 + \cdots$$

Thus, as a first approximation, we obtain

$$I(\lambda) \sim g(z_0) e^{\lambda f(z_0)} \sqrt{\frac{2}{-f''(z_0)}} \int_{-\infty}^{\infty} e^{-\lambda t^2} dt,$$

which in turn yields

$$I(\lambda) \sim g(z_0) e^{\lambda f(z_0)} \sqrt{\frac{-2\pi}{\lambda f''(z_0)}}. \tag{1.3}$$

If $f(z)$ has more than one saddle point, then the full contribution to the asymptotic behavior of the integral $I(\lambda)$ can be obtained by adding the contributions from all relevant saddle points. For instance, if $f(z)$ has two simple saddle points, say z_+ and z_-, then the asymptotic behavior of $I(\lambda)$ is given by

$$I(\lambda) \sim g(z_+) e^{\lambda f(z_+)} \sqrt{\frac{-2\pi}{\lambda f''(z_+)}} + g(z_-) e^{\lambda f(z_-)} \sqrt{\frac{-2\pi}{\lambda f''(z_-)}}. \tag{1.4}$$

We shall assume that $\mathrm{Re} f(z_+) = \mathrm{Re} f(z_-)$, for otherwise one of the terms on the right-hand side of formula (1.4) will dominate the other. For a detailed discussion of the steepest descent method, we refer to Copson [4, Chapter 7] or Wong [10, Chapter II, Sec. 4].

The above situation is completely changed when the function $f(z)$ is allowed to depend on an auxiliary parameter α; the very form of the asymptotic approximation in (1.4) changes when the two saddle points z_+ and z_- coalesce. To be more specific, we consider the integral

$$I(\lambda, \alpha) = \int_C g(z)\, e^{\lambda f(z, \alpha)} dz, \qquad (1.5)$$

and suppose that the saddle points z_+ and z_- depend on α. If there is only one value of α, say α_0, for which the two saddle points coincide, then Chester, Friedman and Ursell [3] have introduced a method in 1957 which will give an asymptotic expansion for the integral in (1.5) as $\lambda \to \infty$, holding uniformly with respect to α in a neighborhood of α_0. A brief discussion of their method is presented in §2 where, as an illustration, we also outline a derivation of a uniform asymptotic expansion of the Hermite polynomial $H_n(\sqrt{2n+1}\, t)$. If there are two values of α, say α_+ and α_-, for which the saddle points z_+ and z_- coincide, then a new method has been devised to yield an asymptotic expansion for the integral $I(\lambda, \alpha)$, which holds uniformly with respect to α in a region containing both α_+ and α_-. This problem was motivated by some recent investigations of the asymptotic behavior of the Meixner, Meixner-Pollaczek, and Krawtchouk polynomials; see [6], [7] and [8]. A brief outline of this new method is given in §3. As an application of uniform asymptotic expansions, we mention in §4 a specific result concerning the behavior of the zeros of the Meixner-Pollaczek polynomials.

2. Airy-Type Expansion

To properly formulate the coalescence of two saddle points, we suppose that there exists a critical value of α, say $\alpha = \alpha_0$, such that for $\alpha \neq \alpha_0$, the two distinct saddle points z_+ and z_- in (1.4) are of multiplicity 1, but at $\alpha = \alpha_0$, these two points coincide and give a single saddle point z_0 of multiplicity 2. Thus

$$f_z(z_0, \alpha_0) = f_{zz}(z_0, \alpha_0) = 0, \qquad\qquad f_{zzz}(z_0, \alpha_0) \neq 0,$$

and

$$f_z(z_+, \alpha) = f_z(z_-, \alpha) = 0, \qquad\qquad f_{zz}(z_\pm, \alpha) \neq 0$$

for $\alpha \neq \alpha_0$. Since $z_\pm \to z_0$ and hence $f_{zz}(z_\pm, \alpha) \to 0$ as $\alpha \to \alpha_0$, approximation (1.4) is not valid in a neighborhood of α_0.

To obtain an asymptotic expansion for $I(\lambda, \alpha)$ as $\lambda \to \infty$, which holds uniformly for α in a neighborhood of α_0, Chester, Friedman and Ursell introduced in their classic paper [3], the cubic transformation

$$f(z, \alpha) = \frac{1}{3}u^3 - \zeta u + \eta, \tag{2.1}$$

where ζ and η are functions of α. These functions are determined by the condition that the transformation $z \to u$ is one-to-one and analytic in a neighborhood of z_0 for all α in a neighborhood of α_0, i.e., in a neighborhood containing the two saddle points. Making the transformation (2.1), the integral in (1.5) is reduced to the canonical form

$$I(\lambda, \alpha) = e^{\lambda \eta} \int_{C^*} \varphi_0(u) \, e^{\lambda(u^3/3 - \zeta u)} du, \tag{2.2}$$

where C^* is the image of C and

$$\varphi_0(u) = g(z) \frac{dz}{du}.$$

To obtain an asymptotic expansion for the last integral, we use a method of Bleistein [1] and write

$$\varphi_0(u) = a_0 + b_0 u + (u^2 - \zeta)\psi_0(u), \tag{2.3}$$

where the coefficients a_0 and b_0 can be determined by setting $u = \pm\sqrt{\zeta}$ on two sides of the equation. Inserting (2.3) in (2.2) gives

$$I(\lambda, \alpha) = e^{\lambda \eta} \left[V(\lambda^{2/3}\zeta) \frac{a_0}{\lambda^{1/3}} + V'(\lambda^{2/3}\zeta) \frac{b_0}{\lambda^{2/3}} + I_1(\lambda, \alpha) \right], \tag{2.4}$$

where

$$V(\lambda) = \int_{C^*} e^{v^3/3 - \lambda v} dv \tag{2.5}$$

and

$$I_1(\lambda, \alpha) = \int_{C^*} (u^2 - \zeta) \, e^{\lambda(u^3/3 - \zeta u)} \psi_0(u) du. \tag{2.6}$$

For simplicity, let us assume that the coefficient ζ in (2.1) is real, and that the contour C^* can be deformed into one which begins at $\infty e^{-\pi i/3}$, passes through $\sqrt{\zeta}$, and ends at $\infty e^{\pi i/3}$. Thus, we have

$$V(\lambda) = 2\pi i \, \text{Ai} \, (\lambda), \tag{2.7}$$

where Ai(λ) is the Airy function. To the integral $I_1(\lambda, \alpha)$ we apply an integration by parts, and the result is

$$I_1(\lambda, \alpha) = \frac{1}{\lambda} \int_{C^*} \varphi_1(u) \, e^{\lambda(u^3/3 - \zeta u)} du, \tag{2.8}$$

where $\varphi_1(u) = \psi_0'(u)$. In view of the factor $\frac{1}{\lambda}$ in (2.8), it is anticipated that the integral $I_1(\lambda, \alpha)$ is of a lower asymptotic order than the first two terms on the right-hand side of equation (2.4). Hence, as a first approximation, we obtain from (2.4) and (2.7)

$$I(\lambda, \alpha) \sim 2\pi i \, e^{\lambda \eta} \left[\text{Ai} \, (\lambda^{2/3} \zeta) \frac{a_0}{\lambda^{1/3}} + \text{Ai}' \, (\lambda^{2/3} \zeta) \frac{b_0}{\lambda^{2/3}} \right]. \tag{2.9}$$

The integral $I_1(\lambda, \alpha)$ is exactly of the same form as the one in (2.2). Hence, the above procedure can be repeated, and will lead to an infinite asymptotic expansion. A detailed discussion of this method can be found in Bleistein and Handelsman [2, Chapter 9] or Wong [10, Chapter VII].

Example 1. The Hermite polynomials $H_n(x)$ have the generating function

$$e^{2xw - w^2} = \sum_{n=0}^{\infty} H_n(x) \frac{w^n}{n!}.$$

By the Cauchy integral formula, we have

$$H_n(x) = \frac{n!}{2\pi i} \int_{C_0} \frac{e^{2xw - w^2}}{w^{n+1}} dw, \tag{2.10}$$

where C_0 is any closed contour encircling the origin in the positive sense. By adding two infinite line segments along the negative real axis, we may replace C_0 by an infinite loop Γ which starts from $-\infty$ in the lower half plane, turns around at the origin, and ends at $-\infty$ in the upper half plane. In (2.10), we let $x = \sqrt{2n+1} \, \alpha$ and make the change of variable $w = \sqrt{2n+1} \, z$. The result is

$$H_n(\sqrt{2n+1} \, \alpha) = \frac{n!}{2\pi i} (2n+1)^{-n/2} \int_{\Gamma} g(z) e^{\lambda f(z, \alpha)} dz, \tag{2.11}$$

where $g(z) = z^{-1/2}, f(z, \alpha) = 2\alpha z - z^2 - \frac{1}{2} \log z$ and $\lambda = 2n+1$. The saddle points are located at

$$z = z_\pm = \frac{1}{2}(\alpha \pm \sqrt{\alpha^2 - 1});$$

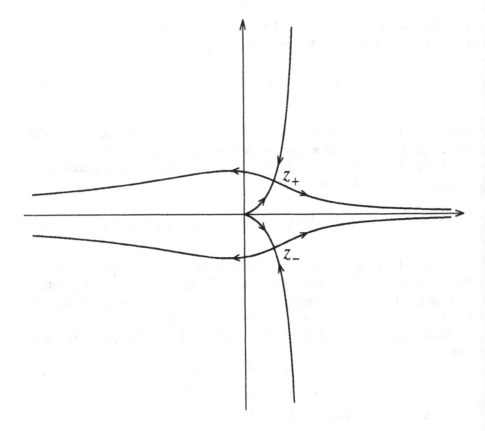

Figure 1. Steepest paths; $0 \leq \alpha < 1$

they coalesce at $z = \frac{1}{2}$ when $\alpha = 1$. For convenience, we consider two separate cases: (i) $0 \leq \alpha < 1$ and (ii) $\alpha \geq 1$. The saddle points and their associated steepest paths are depicted in Figures 1 and 2. In these figures, we have used arrows to indicate directions of descent.

Suggested by (2.1), we set

$$2\alpha z - z^2 - \frac{1}{2}\log z = \frac{1}{3}u^3 - \zeta u + \eta. \qquad (2.12)$$

Upon differentiation on both sides, we get

$$-\frac{2(z - z_-)(z - z_+)}{z}\frac{dz}{du} = u^2 - \zeta.$$

For the transformation to be one-to-one and analytic, we must make $z = z_+$ and $z = z_-$ correspond to $u = -\sqrt{\zeta}$ and $u = \sqrt{\zeta}$, respectively. Substituting

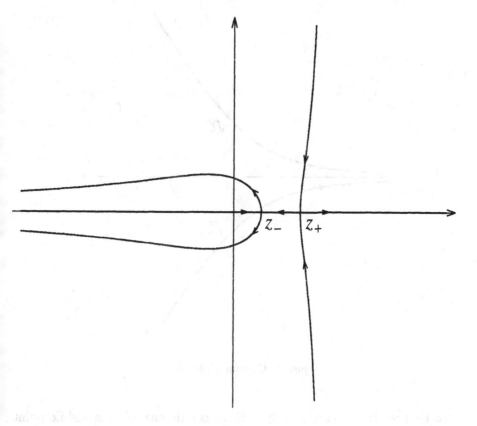

Figure 2. Steepest paths; $\alpha \geq 1$

these values into (2.12) gives

$$
\zeta(\alpha) = \begin{cases} \left[\frac{3}{4}(\alpha\sqrt{\alpha^2 - 1} - \cosh^{-1}\alpha)\right]^{2/3}, & \alpha \geq 1, \\[3mm] -\left[\frac{3}{4}(\cos^{-1}\alpha - \alpha\sqrt{1 - \alpha^2})\right]^{2/3}, & 0 \leq \alpha < 1, \end{cases} \quad (2.13)
$$

and

$$
\eta(\alpha) = \frac{1}{2}\alpha^2 + \frac{1}{4} + \frac{1}{2}\log 2. \tag{2.14}
$$

Note that $\zeta(\alpha) \to 0$ as $\alpha \to 1$.

We now deform the contour Γ in (2.11) into the steepest descent paths shown in Figures 1 and 2. In the case of $0 \leq \alpha < 1$, there are two relevant saddle points z_+ and z_-, and they are complex conjugates. The final integration path in (2.11) consists of two infinite curves, each extending from

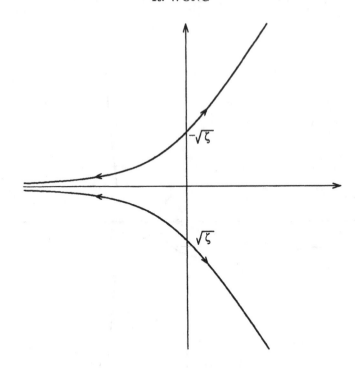

Figure 3. Contour $\Gamma'; 0 \leq \alpha < 1$.

$-\infty$ to $+\infty$. In the case of $\alpha \geq 1$, there is only one relevant saddle point and its associated steepest descent path is an infinite loop.

Changing the integration variable in (2.11) from z to u, we obtain

$$H_n(\sqrt{2n+1}\,\alpha) = \frac{n!}{2\pi i}(2n+1)^{-n/2}\,e^{\lambda\eta}\int_{\Gamma'}\varphi_0(u)e^{\lambda(u^3/\lambda-\zeta u)}du, \tag{2.15}$$

where

$$\varphi_0(u) = z^{-1/2}\frac{dz}{du} \tag{2.16}$$

and Γ' is the image of the steepest descent paths shown in Figures 1 and 2. Note that a conformal map takes steepest descent paths into steepest descent paths. This fact is reassured by the graph of Γ' shown in Figures 3 and 4. The general procedure outlined above (namely, equations (2.3)–(2.9)) can now be applied, and it will lead to the expansion

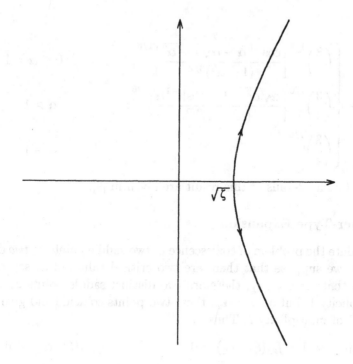

Figure 4. Contour Γ'; $\alpha \geq 1$.

$$\frac{1}{n!}H_n(\sqrt{2n+1}\,\alpha) \sim (2n+1)^{-n/2}e^{(2n+1)\eta(\alpha)}$$

$$\times \left\{ \mathrm{Ai}[(2n+1)^{2/3}\zeta(\alpha)] \sum_{m=0}^{\infty} \frac{(-1)^m a_m}{(2n+1)^{m+1/3}} \right.$$

$$\left. - \mathrm{Ai}'[(2n+1)^{2/3}\zeta(\alpha)] \sum_{m=0}^{\infty} \frac{(-1)^m b_m}{(2n+1)^{m+2/3}} \right\},$$

holding uniformly for all bounded $\alpha \geq 0$. The coefficients a_m and b_m are defined recursively by the following equations

$$\varphi_m(u) = a_m + b_m u + (u^2 - \zeta)\psi_m(u),$$
$$\varphi_{m+1}(u) = \psi'_m(u),$$
$$a_m = \frac{1}{2}[\varphi_m(\sqrt{\zeta}) + \varphi_m(-\sqrt{\zeta})],$$
$$b_m = \frac{1}{2\sqrt{\zeta}}[\varphi_m(\sqrt{\zeta}) - \varphi_m(-\sqrt{\zeta})]$$

with

$$
a_0 = \begin{cases} \left(\dfrac{3}{4}\right)^{1/6}\left[\dfrac{\cos^{-1}\alpha - \alpha\sqrt{1-\alpha^2}}{(1-\alpha^2)^{3/2}}\right]^{1/6}, & 0 \le \alpha < 1 \\[4ex] \left(\dfrac{3}{4}\right)^{1/6}\left[\dfrac{\alpha\sqrt{\alpha^2-1} - \cosh^{-1}\alpha}{(\alpha^2-1)^{3/2}}\right]^{1/6}, & \alpha > 1 \\[4ex] \left(\dfrac{3}{4}\right)^{1/6}, & \alpha = 1 \end{cases}
$$

and $b_0 = 0$. The details of this result are given in [9].

3. Weber-Type Expansion

To formulate the problem of coalescence of two saddle points at *two* distinct locations, we suppose that there are two critical values of α, say α_+ and α_-, such that for $\alpha \ne \alpha_\pm$, there are two distinct saddle points z_+ and z_- of multiplicity 1, but at $\alpha = \alpha_\pm$, these two points coincide and give saddle points $\xi^\pm$ of multiplicity 2. Thus,

$$
f_z(\xi^\pm, \alpha_\pm) = f_{zz}(\xi^\pm, \alpha_\pm) = 0, \qquad f_{zzz}(\xi^\pm, \alpha_\pm) \ne 0,
$$

(3.1)

and

$$
f_z(z_+, \alpha) = f_z(z_-, \alpha) = 0, \qquad f_{zz}(z_\pm, \alpha) \ne 0 \qquad (3.2)
$$

for $\alpha \ne \alpha_\pm$. The following examples provide concrete illustrations of such situations.

Example 2. *Meixner polynomials* $m_n(x; \beta, c)$; see [6]. From their generating function, one can get the integral representation

$$
\frac{1}{n!} m_n(x; \beta, c) = \frac{1}{2\pi i} \int_\Gamma e^{nf(z,\alpha)} \frac{dz}{z(1-z)^\beta},
$$

where $x = n\alpha, \alpha \in (0, \infty)$,

$$
f(z, \alpha) = \alpha \log\left(1 - \frac{z}{c}\right) - \alpha \log(1-z) - \log z \qquad (3.3)
$$

and Γ is a circle centered at the origin with radius less than $\min(1, |c|)$.

Example 3. *Meixner-Pollaczek polynomials* $M_n(x; \delta, \eta)$; see [7]. By the same argument, one also has the integral representation

$$
\frac{1}{n!} M_n(x; \delta, \eta) = \frac{1}{2\pi i} \int_C [(1 + \delta z)^2 + z^2]^{-\eta/2} e^{nf(z,\alpha)} \frac{dz}{z},
$$

where $x = n\alpha, \alpha \in (-\infty, \infty)$,

$$f(z, \alpha) = \alpha \tan^{-1}\left(\frac{z}{1 + \delta z}\right) - \log z \qquad (3.4)$$

and C is a circle centered at the origin with radius $1/\sqrt{1 + \delta^2}$. If we put

$$z_0 = \frac{-\delta + i}{1 + \delta^2}$$

and

$$z_0 = r_0\, e^{i\theta_0} \qquad \text{with} \qquad r_0 = \frac{1}{\sqrt{1 + \delta^2}},$$

then (3.4) can also be written as

$$f(z, \alpha) = \frac{\alpha}{2i} \log (z - z_0) - \frac{\alpha}{2i} \log(z - \bar{z}_0) - \log z + \alpha(\pi - \theta_0). \qquad (3.5)$$

Example 4. *Krawtchouk polynomials* $K_n^{(N)}(x; p, q)$; see [8]. They have the integral representation

$$K_n^N(x; p, q) = \frac{p^n \sigma^{-\alpha N}}{2\pi i} \int_C e^{Nf(z, \alpha)} \frac{dz}{z},$$

where $\sigma \equiv p/q, \alpha \equiv x/N, \nu \equiv n/N$,

$$f(z, \alpha) = (1 - \alpha) \log(1 - z) + \alpha \log(\sigma + z) - \nu \log z \qquad (3.6)$$

and C is a small closed contour surrounding $z = 0$.

In all three examples above, the large variable is n (or N). A simple function which exhibits two saddle points coalescing at two distinct places is given by

$$\psi(u, \eta) = -\rho \log u + \eta u - \frac{u^2}{2}. \qquad (3.7)$$

The saddle points occur at

$$u_\pm = \frac{\eta \pm \sqrt{\eta^2 - 4\rho}}{2}, \qquad (3.8)$$

and they coincide when $\eta = \pm 2\sqrt{\rho}$. If we put $d = -\lambda\rho - \frac{1}{2}(\rho > 0)$ and $z = \sqrt{\lambda}\,\eta$ in the integral representation of the parabolic cylinder function

$$U(d, z) = \frac{\Gamma(\frac{1}{2} - d)}{2\pi i}\, e^{-z^2/4} \int_{-\infty}^{(0+)} e^{zu - \frac{1}{2}u^2}\, u^{d - \frac{1}{2}}\, du, \qquad (3.9)$$

we obtain

$$U(-\lambda\rho - \tfrac{1}{2}, \sqrt{\lambda}\,\eta) = \frac{\Gamma(\lambda\rho+1)}{2\pi i}\, e^{-\lambda\eta^2/4}\, \lambda^{-\lambda\rho/2} \int_{-\infty}^{(0^+)} e^{\lambda\psi(u,\eta)}\frac{du}{u}. \tag{3.10}$$

We now return to the integral $I(\lambda, \alpha)$ in equation (1.5), and suppose that $f(z, \alpha)$ satisfies the conditions in (3.1) and (3.2). To derive an asymptotic expansion for $I(\lambda, \alpha)$, as $\lambda \to \infty$, which holds uniformly in a region containing both α_+ and α_-, we compare it with the integral in (3.10). This suggests that we make the transformation $z \leftrightarrow u(z)$ defined by

$$f(z, \alpha) = \psi(u, \eta) + \gamma, \tag{3.11}$$

where γ is a constant to be determined, and require $u(0) = 0$. Changing variable from z to u, the integral in (1.5) becomes

$$I(\lambda, \alpha) = e^{\lambda\gamma} \int_{-\infty}^{(0^+)} \phi(u)\, e^{\lambda\psi(u,\eta)}\frac{du}{u}, \tag{3.12}$$

where

$$\phi(u) = g(z)\,\frac{\psi_u(u,\eta)}{f_z(z,\alpha)}\, u. \tag{3.13}$$

The contour C in the z-plane should first be deformed into a steepest descent path; it will then be mapped into the loop path shown in (3.12) in the u-plane. Put $\phi_0(u) = \phi(u)$, and write

$$\phi_0(u) = a_0 + b_0 u + (u - u_+)(u - u_-)h_0(u), \tag{3.14}$$

where u_+ and u_- are given in (3.8). By setting $u = u_+$ and $u = u_-$ on two sides of the equation, one finds that the coefficients a_0 and b_0 can be expressed in terms of $\phi_0(u_+)$ and $\phi_0(u_-)$. For simplicity, let us define the new function

$$W(x, \lambda) \equiv e^{x^2/4}U(-\lambda\rho - \tfrac{1}{2}, x). \tag{3.15}$$

Clearly

$$W(\sqrt{\lambda}\,\eta, \lambda) = \frac{\Gamma(\lambda\rho + 1)}{2\pi i}\, \lambda^{-\lambda\rho/2} \int_{-\infty}^{(0^+)} e^{\lambda\psi(u,\eta)}\frac{du}{u},$$

and from (3.15)

$$W_x(\sqrt{\lambda}\,\eta, \lambda) = \frac{\Gamma(\lambda\rho + 1)}{2\pi i}\, \lambda^{1/2 - \lambda\rho/2} \int_{-\infty}^{(0^+)} e^{\lambda\psi(u,\eta)}\, du.$$

Substituting (3.14) in (3.12) gives

$$I(\lambda, \alpha) = \frac{2\pi i}{\Gamma(\lambda\rho + 1)} \lambda^{\lambda\rho/2} e^{\lambda\gamma} \left[a_0 W(\sqrt{\lambda}\,\eta, \lambda) + \frac{b_0}{\sqrt{\lambda}} W_x(\sqrt{\lambda}\,\eta, \lambda) + \varepsilon_1 \right],$$

$$(3.16)$$

where

$$\varepsilon_1 = \frac{\Gamma(\lambda\rho + 1)}{2\pi i} \lambda^{-\lambda\rho/2} \int_{-\infty}^{(0^+)} e^{\lambda\psi(u,\eta)} (u - u_+)(u - u_-) h_0(u) \frac{du}{u}.$$

An integration by parts gives

$$\varepsilon_1 = \frac{1}{\lambda} \frac{\Gamma(\lambda\rho + 1)}{2\pi i} \lambda^{-\lambda\rho/2} \int_{-\infty}^{(0^+)} e^{\lambda\psi(u,\eta)} \phi_1(u) \frac{du}{u},$$

where $\phi_1(u) = uh_0'(u)$. Neglecting the error term ε_1, we have from (3.16), as a first approximation,

$$I(\lambda, \alpha) \sim \frac{2\pi i}{\Gamma(\lambda\rho + 1)} \lambda^{\lambda\rho/2} e^{\lambda\gamma} \left[a_0 W(\sqrt{\lambda}\,\eta, \lambda) + \frac{b_0}{\sqrt{\lambda}} W_x(\sqrt{\lambda}\,\eta, \lambda) \right].$$
$$(3.17)$$

This process can be repeated to yield an infinite asymptotic expansion of the form

$$I(\lambda, \alpha) \sim \frac{2\pi i}{\Gamma(\lambda\rho + 1)} \lambda^{\lambda\rho/2} e^{\lambda\gamma} \left\{ W(\sqrt{\lambda}\,\eta, \lambda) \sum_{s=0}^{\infty} \frac{a_s}{\lambda^s} \right.$$

$$\left. + W_x(\sqrt{\lambda}\,\eta, \lambda) \sum_{s=0}^{\infty} \frac{b_s}{\lambda^{s+1/2}} \right\}, \quad (3.18)$$

where the coefficients a_s and b_s are determined recursively.

4. Zeros

An important application of uniform approximations/expansions such as (2.9) and (3.18) is to provide asymptotic formulas for the zeros of orthogonal polynomials and special functions, and the basic tool used in this application is the following simple result (see [5]).

Lemma. *In the interval* $[a - \rho, a + \rho]$, *suppose* $f(t) = g(t) + \varepsilon(t)$, *where* $f(t)$ *is continuous,* $g(t)$ *is differentiable,* $g(a) = 0, m = \min |g'(t)| > 0$, *and*

$$E = \max |\varepsilon(t)| < \min\{|g(a - \rho)|, |g(a + \rho)|\}.$$

Then there exists a zero c of f(t) in the interval such that

$$|c - a| \le \frac{E}{m}.$$

As an illustration, we take the Meixner-Pollaczek polynomial $M_n(n\alpha; \delta, \eta)$ mentioned in §3. Here, the phase function $f(z, \alpha)$ in (3.5) has two saddle points

$$z_{\pm} = \frac{(\alpha - 2\delta) \pm \sqrt{(\alpha - 2\delta)^2 - 4(1 + \delta^2)}}{2(1 + \delta^2)}, \qquad (4.1)$$

which coalesce at $z = r_0$ and $z = -r_0$, when α assumes the values α_+ and α_-, respectively, where

$$\alpha_{\pm} = 2\delta \pm 2\sqrt{1 + \delta^2} \qquad (4.2)$$

and $r_0 = 1/\sqrt{1 + \delta^2}$ is as defined in Example 3. Since δ is a real number, we have $\alpha_- < 0 < \alpha_+$. From the asymptotic approximations given in [7], it is clear that $M_n(n\alpha; \delta, \eta)$ does not have zeros when $\alpha \ge \alpha_+$ or $\alpha \le \alpha_-$. Hence we may restrict ourselves to the interval $\alpha_- < \alpha < \alpha_+$. Let $\alpha_{n,s}$ denote the sth zero, counted in descending order

$$\alpha_- < \alpha_{n,n} < \alpha_{n,n-1} < \cdots < \alpha_{n,2} < \alpha_{n,1} < \alpha_+. \qquad (4.3)$$

In [7], we have also shown that when $n^{2/3}(\alpha_+ - \alpha)$ is bounded,

$$\frac{1}{n!} M_n(n\alpha; \delta, \eta) = n^{-1/3} e^{n\alpha(\pi - \theta_0)/2} r_0^{-n} (r_0\alpha_+)^{1/3 - \eta/2} \qquad (4.4)$$
$$\times \ \{\mathrm{Ai}\,[ar_0^{1/3}\alpha_+^{-2/3} + \frac{1}{2}(1 - \eta r_0^{1/3}\alpha_+^{1/3})\,n^{-1/3}] + O(n^{-2/3})\},$$

where $a = n^{2/3}(\alpha - \alpha_+)$ and θ_0 is the argument of $z_0 \equiv (-\delta + i)/(1 + \delta^2)$; see Example 3.

To apply the above lemma, we let

$$f(\alpha) \equiv \frac{1}{n!}\, n^{1/3}\, r_0^n\, e^{-n\alpha(\pi - \theta_0)/2} (r_0\alpha_+)^{\eta/2 - 1/3} M_n(n\alpha; \delta, \eta),$$

$$g(\alpha) \equiv \mathrm{Ai}\,[ar_0^{1/3}\alpha_+^{-2/3} + \frac{1}{2}(1 - \eta r_0^{1/3}\alpha_+^{1/3})n^{-1/3}] \qquad (4.5)$$

and

$$\varepsilon(\alpha) \equiv O(n^{-2/3}).$$

If g_s is a zero of $g(\alpha)$, then

$$g_s = \alpha_+ + a_s r_0^{-1/3} \alpha_+^{2/3} n^{-2/3} + \frac{1}{2}(\eta \alpha_+ - r_0^{-1/3} \alpha_+^{2/3}) n^{-1}$$

since $a = n^{2/3}(\alpha - \alpha_+)$, where a_s is the sth negative zero of the Airy function $\mathrm{Ai}(\alpha)$. Furthermore, since

$$|g'(\alpha)| = |\mathrm{Ai}'(\cdots)| r_0^{1/3} \alpha_+^{-2/3} n^{2/3} \geq G n^{2/3}$$

for some positive constant G, where the dots inside the function Ai' represents the same quantity as that in (4.5), by the lemma $f(\alpha)$ has a zero $\alpha_{n,s}$ such that $\alpha_{n,s} = g_s + \varepsilon_{n,s}$, i.e.,

$$\alpha_{n,s} = \alpha_+ + a_s (r_0^{-1} \alpha_+^2)^{1/3} n^{-2/3} + \frac{1}{2}(\eta \alpha_+ - (r_0^{-1} \alpha_+^2)^{1/3}) n^{-1} + \varepsilon_{n,s}, \tag{4.6}$$

where $\varepsilon_{n,s} = O(n^{-4/3})$. A corresponding result can be found in [7] for the zeros arranged in the ascending order.

References

1. N. Bleistein, *Uniform asymptotic expansions of integrals with many nearby stationary points and algebraic singularities*, J. Math. Mech. **17** (1967), 533-559.
2. N. Bleistein and R. A. Handelsman, *Asymptotic Expansions of Integrals*, Holt, Rinehart and Winston, New York, 1975. (Reprinted in 1986 by Dover Publications, New York.)
3. C. Chester, B. Friedman and F. Ursell, *An extension of the method of steepest descents*, Philos. Soc. **53** (1957), 599-611.
4. E. T. Copson, *Asymptotic Expansions*, Cambridge Tracts in Math. and Math. Phys. No. 55, Cambridge University Press, London, 1965.
5. H. W. Hethcote, *Error bounds for asymptotic approximations of zeros of transcendental functions*, Proc. Roy. Soc. Lond. A SIAM J. Math. Anal. **1** (1970), 147-152.
6. X. -S. Jin and R. Wong, *Uniform asymptotic expansions for Meixner polynomials*, Constr. Approx. **14** (1998), 113-150.
7. X. -C. Li and R. Wong, *Asymptotic Expansions: Their Derivations and Interpretation*, Constr. Approx., to appear.
8. X. -C. Li and R. Wong, *On the asymptotics of the Meixner-Pollaczek polynomials and their zeros*, Constr. Approx., to appear.
9. W.-F. Sun, Uniform Asymptotic Expansions of Hermite Polynomials, M.Phil. thesis, City University of Hong Kong, 1996.
10. R. Wong, *Asymptotic Approximations of Integrals*, Academic Press, Boston, 1989.

EXPONENTIAL ASYMPTOTICS

R. WONG
Department of Mathematics
City University of Hong Kong
Tat Chee Avenue
Kowloon Hong Kong

Abstract. Significant developments have occurred recently in the general theory of asymptotic expansions. These developments include smoothing of the Stokes phenomenon (M. V. Berry), uniform exponentially-improved asymptotic expansions (F. W. J. Olver), superasymptotics and hyperasymptotics (M. V. Berry and C. J. Howls). In this lecture, I will use the Airy function to give a flavor of each of these new concepts in what is now known as "exponential asymptotics" or "asymptotics beyond all orders" (M. D. Kruskal and H. Segur). My approach is based on a modified version of the steepest-descent method introduced by Berry and Howls in 1991.

1. Introduction

Stokes phenomenon occurs typically in compound asymptotic expansions. It concerns the abrupt change in the coefficients of these expansions, when the variable crosses certain lines in the complex plane. To illustrate this phenomenon, we consider the Airy integral

$$\mathrm{Ai}(z) = \frac{1}{2\pi i} \int_L \exp(\tfrac{1}{3}t^3 - zt)dt, \qquad (1.1)$$

where L is any contour which begins at infinity in the sector $-\frac{1}{2}\pi < \mathrm{art}\ t < -\frac{1}{6}\pi$ and ends at infinity in the sector $\frac{1}{6}\pi < \arg t < \frac{1}{2}\pi$. From (1.1), it can be easily verified that $\mathrm{Ai}(z)$ has the Maclaurin expansion

$$\mathrm{Ai}(z) = 3^{-2/3} \sum_{n=0}^{\infty} \frac{z^{3n}}{3^{2n} n! \Gamma(n + \frac{2}{3})} - 3^{-4/3} \sum_{n=0}^{\infty} \frac{z^{3n+1}}{3^{2n} n! \Gamma(n + \frac{4}{3})}.$$

J. Bustoz et al. (eds.), Special Functions 2000, 505–518.

Thus, Ai(z) is an entire function. If we denote by $u_+(z)$ and $u_-(z)$ the formal series

$$u_+(z) = \frac{1}{2\pi z^{1/4}} \exp\left(\tfrac{2}{3}z^{3/2}\right) \sum_{s=0}^{\infty} \frac{\Gamma(3s+\tfrac{1}{2})}{(2s)!9^s} z^{-3s/2} \tag{1.2}$$

and

$$u_-(z) = \frac{1}{2\pi z^{1/4}} \exp\left(-\tfrac{2}{3}z^{3/2}\right) \sum_{s=0}^{\infty} \frac{(-1)^s \Gamma(3s+\tfrac{1}{2})}{(2s)!9^s} z^{-3s/2}, \tag{1.3}$$

then the asymptotic behavior of Ai(z) is given by

$$\text{Ai}(z) \sim u_-(z), \tag{1.4}$$

as $z \to \infty$ in $-\pi < \arg z < \pi$; see [12, p. 93]. (Clearly, this result can not be valid in a wider sector, since Ai(z) is a single-valued function and the factor multiplying the infinite series in (1.3) is not.) On the other hand, as $z \to \infty$ in $\tfrac{1}{3}\pi < \arg z < \tfrac{5}{3}\pi$, we have the compound expansion

$$\text{Ai}(z) \sim u_-(z) + iu_+(z). \tag{1.5}$$

In the common sector of validity, given by $\tfrac{1}{3}\pi < \arg z < \pi$, either (1.4) or (1.5) can be used since the contribution of $u_+(z)$ is exponentially small compared with that of $u_-(z)$. However, in the sector $-\tfrac{1}{3}\pi < \arg z < \tfrac{1}{3}\pi$, $u_+(z)$ dominates $u_-(z)$; hence it is mandatory to drop $u_+(z)$ from the asymptotic expansion of Ai(z). By introducing a constant (coefficient) C, which is 0 for $\arg z \in (-\tfrac{1}{3}\pi, \tfrac{1}{3}\pi)$ and i for $\arg z \in (\tfrac{1}{3}\pi, \tfrac{5}{3}\pi)$, to the asymptotic series $u_+(z)$, the two results (1.4) and (1.5) can be combined into one

$$\text{Ai}(z) \sim u_-(z) + Cu_+(z), \tag{1.6}$$

as $z \to \infty$ in $\arg z \in (-\pi, \tfrac{5}{3}\pi)$. The coefficient C is called a *Stokes multiplier*, and is domain dependent. The discontinuous change of the coefficient C, when the argument of z changes in a continuous manner, is known as *Stokes' phenomenon*.

Returning to the series $u_+(z)$ and $u_-(z)$ in (1.2) and (1.3), we let

$$S_+(z) = \tfrac{2}{3} z^{3/2} \qquad \text{and} \qquad S_-(z) = -\tfrac{2}{3} z^{3/2}.$$

Note that the behavior of $u_+(z)$ and $u_-(z)$ are most unequal on the curves given by

$$\text{Im}\{S_+(z) - S_-(z)\} = 0, \tag{1.7}$$

and they are nearly equal on the curves given by

$$\text{Re}\{S_+(z) - S_-(z)\} = 0. \tag{1.8}$$

The curves given in (1.7) and (1.8) are known, respectively, as the Stokes and anti-Stokes lines. In the case of Airy function, it is easily seen that the rays $\arg z = 0, \pm\frac{2}{3}\pi$ are the Stokes lines and the rays $\arg z = \pm\frac{\pi}{3}, \pm\pi$ are the anti-Stokes lines.

Since we are dealing with a continuous (in fact, analytic) function, it is rather unsatisfactory to have a discontinuous coefficient in the asymptotic expansion (1.6). In 1989, Berry wrote an innovative paper [2], in which he adopted a different interpretation of the Stokes phenomenon. In his view, the coefficient of the series $u_+(z)$ should be a continuous function of arg z, instead of a discontinuous constant. He gave a beautiful, although not mathematically rigorous, demonstration of this theory by truncating the series $u_-(z)$ at its optimal number of terms and the series $u_+(z)$ at its first term. Berry's theory has since become known variously as "exponential asymptotics"or "superasymptotics", and has been successfully applied to several known asymptotic expansions in a mathematically rigorous manner (see, for example, Boyd [5]; Olver [9, 10]; Olde Daalhuis and Olver [8]; Wong and Zhao [13, 14]).

In this lecture, we shall illustrate Berry's theory with the simple Airy function given in (1.1). Our approach is based on a modified version of the steepest-descent method introduced by Berry and Howls [4].

2. Steepest Descent Method

In (1.1), we make the change of variable $t = z^{1/2}u$. Assuming temporarily that z is real and positive, the integral in (1.1) becomes

$$\mathrm{Ai}(z) = \frac{z^{1/2}}{2\pi i} \int_L e^{z^{3/2}(\frac{1}{3}u^3 - u)} du.$$

Once this identity has been established, the restriction on z can be removed by using analytic continuation. Put

$$\xi := \frac{2}{3} z^{3/2} \qquad \text{and} \qquad f(u) = \frac{1}{2}(u^3 - 3u) \qquad (2.1)$$

so that

$$\mathrm{Ai}(z) = \frac{1}{2\pi i} \left(\frac{3}{2}\xi\right)^{1/3} \int_L e^{\xi f(u)} du. \qquad (2.2)$$

The saddle points of $f(u)$ are located at $u = \pm 1$. Clearly, $f(\pm 1) = \mp 1$. Let

$$\theta := \arg \xi, \qquad (2.3)$$

and consider the steepest descent curves

$$\Gamma_{\pm 1}(\theta): \quad \arg\{\xi[f(\pm 1) - f(u)]\} = \arg\{e^{i\theta}[\mp 1 - f(u)] = 0. \tag{2.4}$$

From [12, p. 92], we know that only the saddle point at $u = 1$ is relevant. Deform L into $\Gamma_1(\theta)$, and write

$$\text{Ai}(z) = \frac{e^{-\xi}}{2\pi i}\left(\frac{3}{2}\xi\right)^{1/3}\int_{\Gamma_1(\theta)} e^{\xi[f(u)+1]}du. \tag{2.5}$$

Furthermore, we introduce the notations

$$I^{(\pm 1)}(\xi) = \xi^{1/2}\int_{\Gamma_1(\theta)} e^{\xi[f(u)\pm 1]}du. \tag{2.6}$$

In terms of $I^{(1)}(\xi)$, (2.5) can be written as

$$\text{Ai}(z) = \frac{1}{2\pi i}\left(\frac{3}{2}\right)^{1/3}\xi^{-1/6}e^{-\xi}I^{(1)}(\xi). \tag{2.7}$$

In (2.6), we make the change of variable

$$\tau = \xi[-1 - f(u)]. \tag{2.8}$$

For $u \in \Gamma_1(\theta), \tau$ is real and positive; see (2.4). Expanding $f(u)$ into a Taylor series at $u = 1$, we have from (2.8)

$$-\tau = \frac{3}{2}\xi\left[(u-1)^2 + \frac{1}{3}(u-1)^3\right],$$

from which it follows that

$$\pm i\left(\frac{2}{3}\frac{\tau}{\xi}\right)^{1/2} = (u-1)\left[1 + \frac{1}{3}(u-1)\right]^{1/2}$$

where proper branches of square roots have to be chosen and $[\cdots]^{1/2}$ is that branch which reduces to 1 at $u = 1$. By Lagrange's formula for the reversion of series,

$$u^{\pm} = 1 + \sum_{n=0}^{\infty} \alpha_n\left(\pm i\sqrt{\frac{2\tau}{3\xi}}\right)^n. \tag{2.9}$$

Breaking the integration path $\Gamma_1(\theta)$ in (2.6) at $u = 1$, we can rewrite the integral $I^{(1)}(\xi)$ as

$$I^{(1)}(\xi) = \xi^{1/2} \int_0^\infty \left[\frac{du^+}{d\tau} - \frac{du^-}{d\tau} \right] e^{-\tau} d\tau. \qquad (2.10)$$

Coupling (2.9) and (2.10) gives the asymptotic expansion

$$I^{(1)}(\xi) \sim \sum_{n=0}^\infty \frac{d_n}{\xi^n}, \qquad (2.11)$$

where the coefficients d_n can be given explicitly. For details, we refer to [12, p. 93].

3. A Modification by Berry and Howls

Instead of (2.10), Berry and Howls [4] used the equivalent representation

$$I^{(1)}(\xi) = \xi^{-1/2} \int_0^\infty \left[\frac{1}{f'(u^-(\tau))} - \frac{1}{f'(u^+(\tau))} \right] e^{-\tau} d\tau. \qquad (3.1)$$

They further observed that the integrand in (3.1) can be expressed as a residue. Indeed, they gave

$$\frac{1}{f'(u^-(\tau))} - \frac{1}{f'(u^+(\tau))} = \frac{1}{2\pi i} \frac{\xi^{3/2}}{\tau^{1/2}} \int_{C_1(\theta)} \frac{[-1 - f(u)]^{1/2}}{\xi[-1 - f(u)] - \tau} du, \qquad (3.2)$$

where $C_1(\theta)$ is a positively oriented curve surrounding the steepest-descent path $\Gamma_1(\theta)$. (Since $\Gamma_1(\theta)$ is an infinite contour, $C_1(\theta)$ actually consists of two infinite curves embracing $\Gamma_1(\theta)$; see Figure 1.) To establish (3.2), we recall the formula

$$\text{Res}\left\{ \frac{Q(u)}{P(u)}; u_0 \right\} = \frac{Q(u_0)}{P'(u_0)},$$

where $P(u)$ and $Q(u)$ are analytic functions with $P(u_0) = 0$, $P'(u_0) \neq 0$ and $Q(u_0) \neq 0$. Take $P(u) \equiv \xi[-1 - f(u)] - \tau, Q(u) \equiv [-1 - f(u)]^{1/2}$ and $u_0 = u^\pm(\tau)$. Then

$$\frac{Q(u^\pm(\tau))}{P'(u^\pm(\tau))} = \text{Res}\left\{ \frac{Q(u)}{P(u)}; u^\pm(\tau) \right\} = \frac{1}{2\pi i} \int_{C^\pm} \frac{Q(u)}{P(u)} du, \qquad (3.3)$$

where the two contours C^+ and C^- are shown in Figure 1. The desired result (3.2) now follows from (3.3) and (2.8). Inserting (3.2) into (3.1) gives

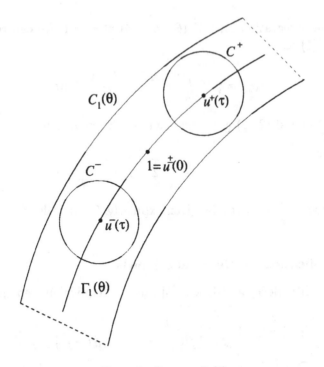

Figure 1. Contour $C_1(\theta)$.

a double integral, and, upon interchanging the order of integration, one obtains

$$I^{(1)}(\xi) = \frac{1}{2\pi i} \int_{C_1(\theta)} [-1 - f(u)]^{-\frac{1}{2}} \int_0^\infty \frac{e^{-\tau}\tau^{-\frac{1}{2}}}{1 - \{\tau/\xi[-1 - f(u)]\}} d\tau du. \tag{3.4}$$

The geometric series

$$\frac{1}{1-x} = \sum_{s=0}^{N-1} x^s + \frac{x^N}{1-x}$$

then gives the asymptotic expansion

$$I^{(1)}(\xi) = \sum_{s=0}^{N-1} c_s\, \xi^{-s} + R_N(\xi), \tag{3.5}$$

where

$$c_s = \frac{\Gamma(s + \frac{1}{2})}{2\pi i} \int_{C_1(\theta)} [-1 - f(u)]^{-s-\frac{1}{2}} du \tag{3.6}$$

and

$$R_N(\xi) = \frac{\xi^{-N}}{2\pi i} \int_0^\infty e^{-\tau} \tau^{N-\frac{1}{2}} \tag{3.7}$$

$$\times \int_{C_1(\theta)} [-1 - f(u)]^{-N-\frac{1}{2}} \frac{1}{1 - \{\tau/\xi[-1 - f(u)]\}} du d\tau.$$

The coefficients c_s can be evaluated exactly, and we have

$$c_s = i \frac{(-1)^s \Gamma(3s + \frac{1}{2})}{(2s)! 9^s} \left(\frac{3}{2}\right)^{-s-\frac{1}{2}}.$$

Now consider all steepest descent paths $\Gamma_1(\theta)$ passing through $u = 1$ for different values of θ; see Figure 2. Since $f(1) - f(-1) = -2$, the path

$$\Gamma_1(\pi): \qquad \arg\{e^{i\pi}[f(1) - f(u)]\} = 0 \tag{3.8}$$

runs into the saddle point $u = -1$. Berry and Howls called $u = -1$ an *adjacent saddle* of $u = 1$, and the steepest-descent path

$$\Gamma_{-1}(\pi): \qquad \arg\{e^{i\pi}[f(-1) - f(u)]\} = 0 \tag{3.9}$$

an *adjacent contour*; see also [6]. The next step is to deform $C_1(\theta)$ into $\Gamma_{-1}(\pi)$ so that (3.7) becomes

$$R_N(\xi) = \frac{\xi^{-N}}{2\pi i} \int_0^\infty \qquad\qquad e^{-\tau} \tau^{N-\frac{1}{2}} \tag{3.10}$$

$$\times \int_{\Gamma_{-1}(\pi)} [-1 - f(u)]^{-N-\frac{1}{2}} \frac{1}{1 - \{\tau/\xi[-1 - f(u)]\}} du d\tau,$$

In the last equation, we make the change of variable

$$\tau = t \frac{f(1) - f(u)}{f(1) - f(-1)}; \tag{3.11}$$

recall that $f(1) = -1$. Since

$$\frac{f(1) - f(u)}{f(1) - f(-1)} = 1 + \frac{f(-1) - f(u)}{f(1) - f(-1)}, \tag{3.12}$$

and since the quotient on the right-hand side is real and positive when $u \in \Gamma_{-1}(\pi)$, the quotient on the left-hand side is also real and positive for $u \in \Gamma_{-1}(\pi)$. Dingle [7] called the quantity $f(1) - f(-1)$ a *singlant*; see also [4]. Substituting (3.11) in (3.10), and making use of (3.12), we obtain

$$R_N(\xi) = \frac{\xi^{-N}}{2\pi i} (-2)^{-N-\frac{1}{2}} \int_0^\infty e^{-t} t^{N-\frac{1}{2}} \left(1 + \frac{t}{2\xi}\right)^{-1} \tag{3.13}$$

$$\times \int_{\Gamma_{-1}(\pi)} \exp\left\{-t \frac{f(-1) - f(u)}{f(1) - f(-1)}\right\} du \, dt.$$

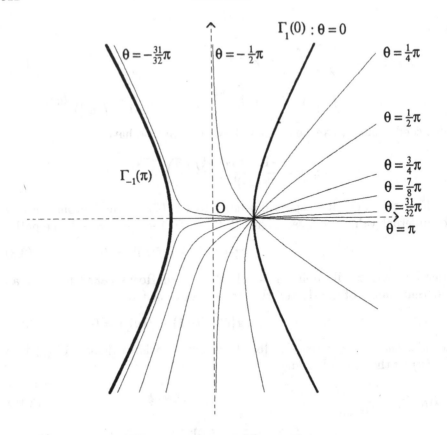

Figure 2. Contours $\Gamma_1(\theta)$, $-\pi < \theta < \pi$.

In the inner integral, we write $u = -w$. Since $f(u)$ is an odd function and $f(1) - f(-1) = -2$, it follows that

$$\int_{\Gamma_{-1}(\pi)} \exp\left\{-t\,\frac{f(-1)-f(u)}{f(1)-f(-1)}\right\}du = -\int_{\Gamma_1(0)} \exp\left\{\frac{t}{2}[f(w)-f(1)]\right\}dw.$$

The last integral can be expressed in terms of the integral $I^{(1)}(\xi)$ given in (2.6). Indeed, we have

$$\int_{\Gamma_{-1}(\pi)} \exp\left\{-t\,\frac{f(-1)-f(u)}{f(1)-f(-1)}\right\}du = -\left(\frac{t}{2}\right)^{-1/2} I^{(1)}\left(\frac{t}{2}\right). \tag{3.14}$$

Inserting (3.14) into (3.13) gives

$$R_N(\xi) = \frac{1}{2\pi}(-2\xi)^{-N}\int_0^\infty e^{-t}\,t^{N-1}\left(1+\frac{t}{2\xi}\right)^{-1} I^{(1)}\left(\frac{t}{2}\right)dt. \tag{3.15}$$

Equations (3.5) and (3.15) coupled together is known as a *resurgence formula*, since the integral $I^{(1)}(\xi)$ on the left-hand side of (3.5) appears again in the remainder term $R_N(\xi)$ given in (3.15).

4. Exponentially Improved Estimates

To estimate the remainder $R_N(\xi)$ in (3.10), we first observe that if $\zeta = |\zeta|e^{i\varphi}$ then

$$|1 - \zeta| \geq \begin{cases} |\sin \varphi| & \text{if } |\varphi| < \frac{\pi}{2}, \\ 1 & \text{if } \frac{\pi}{2} \leq |\varphi| \leq \pi. \end{cases} \tag{4.1}$$

For convenience, let us introduce the function

$$\csc_1(\varphi) = \begin{cases} |\csc \varphi| & \text{if } 0 < |\varphi| < \frac{\pi}{2} \\ 1 & \text{if } \frac{\pi}{2} \leq |\varphi| \leq \pi. \end{cases} \tag{4.2}$$

In terms of this function, (4.1) can be written as

$$\left| \frac{1}{1 - \zeta} \right| \leq \csc_1(\varphi). \tag{4.3}$$

We shall extend $\csc_1(\varphi)$ into a 2π-periodic function by defining

$$\csc_1(\varphi + 2\pi) = \csc_1(\varphi). \tag{4.4}$$

Note that this is an even function, and that it is unbounded at $0, \pm 2\pi, \pm 4\pi$, $\cdots$. For $u \in \Gamma_{-1}(\pi)$, we know that the quotient on the left-hand side of (3.12) is real and positive. Since $f(\pm 1) = \mp 1$, it follows that $\arg\{e^{-i\pi}[-1 - f(u)]\} = 0$ or, equivalently, $\arg[-1 - f(u)] = \pi$. If we take $\zeta = \tau/\xi[-1 - f(u)]$, then $\arg \zeta = -\pi - \theta$, and we have from (4.3)

$$\left| 1 - \frac{\tau}{\xi[-1 - f(u)]} \right|^{-1} \leq \csc_1(\pi + \theta) \tag{4.5}$$

for $u \in \Gamma_{-1}(\pi)$. A direct application of (4.5) to (3.10) gives

$$|R_N(\xi)| \leq \frac{A_N}{|\xi|^N} \csc_1(\theta + \pi), \tag{4.6}$$

where A_N is a constant given by

$$A_N = \frac{\Gamma(N + \frac{1}{2})}{2\pi} \int_{\Gamma_{-1}(\pi)} \left| [-1 - f(u)]^{-N - \frac{1}{2}} \, du \right|. \tag{4.7}$$

There is another error estimate which is probably easier to compute. To get that estimate, we first note that $I^{(1)}(\xi) = R_0(\xi)$; see (3.5). Hence, by (4.6),

$$\left| I^{(1)}\left(\frac{t}{2}\right) \right| \le A_0 \csc_1(\pi). \tag{4.8}$$

Next, we return to (3.15). With $\zeta = -t/2\xi$, it follows from (4.3)

$$|R_N(\xi)| \le \frac{B_N}{|\xi|^N} \csc_1(\pi - \theta), \tag{4.9}$$

where

$$B_N = \frac{A_0 \Gamma(N)}{2^{N+1}\pi}. \tag{4.10}$$

Since the function $\csc_1(\varphi)$ becomes unbounded at $\varphi = 0, \pm 2\pi, \pm 4\pi, \cdots$, both bounds in (4.6) and (4.8) break down when $\theta = \pm\pi, \pm 3\pi, \pm 5\pi, \cdots$.

To obtain an error estimate near the Stokes lines $\theta = \pm\pi$ or, equivalently, $\arg z = \pm\frac{2}{3}\pi$, we use a device suggested by Boyd [5]. Rotating the path of integration in (3.15) by an angle $\eta \in (-\frac{\pi}{2}, 0)$ gives

$$R_N(\xi) = \frac{1}{2\pi}(-2\xi)^{-N} \int_0^{\infty e^{i\eta}} e^{-t} t^{N-1} \left(1 + \frac{t}{2\xi}\right)^{-1} I^{(1)}\left(\frac{t}{2}\right) dt. \tag{4.11}$$

On the path of integration in (4.10), we have $\arg t = \eta$. Hence, it follows from (4.6)

$$\left| I^{(1)}\left(\frac{t}{2}\right) \right| \le A_0 \csc_1(\eta + \pi) = A_0; \tag{4.12}$$

see (4.8). The equality in (4.12) follows from the fact that $\frac{\pi}{2} < \eta + \pi < \pi$. On the other hand, we also have from (4.3)

$$\left| 1 + \frac{t}{2\xi} \right|^{-1} \le \csc_1(\pi + \eta - \theta). \tag{4.13}$$

A combination of (4.11), (4.12) and (4.13) yields

$$|R_N(\xi)| \le \frac{C}{(2|\xi|)^N} \csc_1(\pi + \eta - \theta) \frac{\Gamma(\eta)}{(\cos\eta)^N}, \tag{4.14}$$

where $C = A_0/2\pi$.

If $0 < \theta < \pi$ and θ is close to π, then for any sufficiently small $\eta \in (-\frac{\pi}{2}, 0)$ we have $-\frac{\pi}{2} < \pi + \eta - \theta < 0$. Similarly, if $-\pi < \theta < 0$ and θ is close to $-\pi$, then for any sufficiently small $\eta \in (-\frac{\pi}{2}, 0)$ we have $\frac{3\pi}{2} < \pi + \eta - \theta < 2\pi$. Since the function $\csc_1(\varphi)$ defined in (4.2) and (4.3) is 2π-periodic, from (4.14) we get

$$|R_N(\xi)| \leq \frac{C}{(2|\xi|)^N} \frac{1}{|\sin(\pm\pi + \eta - \theta)|} \frac{\Gamma(N)}{(\cos\eta)^N}, \qquad (4.15)$$

where $+$ sign is taken when $\theta \nearrow \pi$ and $-$ sign is taken when $\theta \searrow -\pi$. As $\theta \to \pm\pi$, we have $\sin(\pm\pi + \eta - \theta) \sim \sin\eta$. As η is arbitrary, we may choose it so that

$$\cos\eta \sim \left(1 - \frac{1}{N}\right) \qquad \text{and} \qquad \sin\eta \sim \sqrt{\frac{2}{N}}.$$

Thus, for θ near $\pm\pi$, we can find a constant C_1 such that

$$|R_N(\xi)| \leq \frac{C_1}{(2|\xi|)^N} \sqrt{\frac{N}{2}} \left(1 - \frac{1}{N}\right)^{-N} \Gamma(N).$$

(An explicit value for C_1 can be given if desired.) Since $\frac{1}{x}\log(1 + x) \leq \frac{1}{1+x}$ for negative values of x, we have $(1 - \frac{1}{N})^{-N} \leq e^{N/(N-1)}$. This together with the inequality [1]

$$\Gamma(N) < \sqrt{2\pi}\, N^{N-\frac{1}{2}} e^{-N+(1/12N)} \qquad (4.16)$$

yields the desired estimate

$$|R_N(\xi)| \leq C_2\, e^{N(-1+\log N - \log|2\xi|)}, \qquad (4.17)$$

where $C_2 = \sqrt{\pi}e^2 C_1$. The minimum value of the exponential function on the right-hand side of (4.16) is attained when

$$\frac{d}{dN}(-N + N\log N - N\log|2\xi|) = \log N - \log(2|\xi|) = 0.$$

Therefore, optimal truncation occurs near

$$N = N^* = 2|\xi|.$$

With N given by this value, we obtain from (4.16)

$$|R_N(\xi)| \leq C_2\, e^{-2|\xi|}, \qquad (4.18)$$

as $|\xi| \to \infty$, when $\arg \xi = \theta$ is near the Stokes lines $\theta = \pm\pi$. Applying (4.16) to the error estimate in (4.9), one readily sees that inequality (4.18) also holds for ξ away from the Stokes lines. Olver [9] called the expansion (3.5) with error given by (4.18) a *uniform, exponentially improved,* asymptotic expansion in the sector $|\theta| \leq \pi$. Optimally truncated asymptotic expansions have now been also called *superasymptotic expansions* by Berry and Howls [3].

5. Hyperasymptotics

Returning to (3.15), we now replace the function $I^{(1)}(t/2)$ by its asymptotic expansion (3.5). Termwise integration gives a series of integrals which can be expressed in terms of Dingle's *terminant function*

$$\int_0^\infty \frac{t^{k-1}e^{-t}}{1+t/\zeta}dt := 2\pi i(-\zeta)^k e^\zeta T_k(\zeta); \qquad (5.1)$$

see Olver [9]. More precisely, we have

$$R_N(\xi) = ie^{2\xi} \sum_{r=0}^{N'-1} (-1)^r \frac{c_r}{\xi^r} T_{N-r}(2\xi) + R_{N,N'}(\xi), \qquad (5.2)$$

where

$$R_{N,N'}(\xi) = \frac{1}{2\pi}(-2|\xi|)^{-N} \int_0^\infty e^{-t}t^{N-1}\left(1+\frac{t}{2\xi}\right)^{-1} R_{N'}\left(\frac{t}{2}\right)dt. \qquad (5.3)$$

The idea of re-expanding the remainder terms in optimally truncated asymptotic series was introduced by Berry and Howls [3]. They called this theory *hyperasymptotics*; see also Olver [11] and Olde Daalhuis & Olver [8].

With $\arg \zeta = \phi, u = t/|\zeta|$, and $k = |\zeta| + \alpha, \alpha$ being a positive and bounded quantity, the integral in (5.1) can be written as

$$\int_0^\infty \frac{t^{k-1}e^{-t}}{1+t/\zeta}dt = |\zeta|^k \int_0^\infty \frac{u^{\alpha-1}e^{-|\zeta|(u-\log u)}}{1+e^{-i\phi}u}du.$$

The integrand on the right-hand side has a pole at $u = -e^{i\phi}$, which coalesces with the saddle point at $u = 1$ where $\phi = \pm\pi$. An existing theory on uniform asymptotic expansions (see [13, pp. 356 – 358) can now be used to show that as $|\zeta| \to \infty$,

$$T_k(\zeta) \sim \frac{1}{2}\,\mathrm{erfc}\,(Z) - \frac{i}{\sqrt{2\pi|\zeta|}}\,e^{-Z^2} \sum_{s=0}^\infty \left(\frac{1}{2}\right)_s g_{2s}(\phi,\alpha)\left(\frac{2}{|\zeta|}\right)^s \qquad (5.4)$$

uniformly with respect to $\phi \in [-\pi + \delta, 3\pi - \delta]$, where erfc is the complementary error function, $Z := c(\phi)\sqrt{|\zeta|/2}$, and

$$\frac{1}{2}[c(\phi)]^2 := -e^{i(\phi-\pi)} + i(\phi - \pi) + 1.$$

Near $\phi = \pi$, the Taylor series expansion of $c(\phi)$ begins

$$c(\phi) = -(\phi - \pi) - \frac{i}{6}(\phi - \pi)^2 + \frac{1}{36}(\phi - \pi)^3 + \cdots.$$

The coefficients $g_{2s}(\phi, \alpha)$ can be given explicitly and the first one is

$$g_0(\phi, \alpha) = \frac{e^{i\alpha(\pi-\phi)}}{1 + e^{-i\phi}} - \frac{i}{c(\phi)}.$$

This result is taken from Olver [9]. He has also shown that Z lies in the sector $-\frac{1}{4}\pi \leq \arg Z \leq 0$ when $-\pi \leq \phi \leq \pi$, and in the sector $0 \leq \arg(-Z) \leq \frac{1}{4}\pi$ when $\pi \leq \phi \leq 3\pi$. As ϕ increases from values below π to values above π, Z moves from the first sector to the second sector through the origin. Since $\text{erfc}(Z) = O(e^{-Z^2})$ uniformly throughout the first sector and $\text{erfc}(Z) = 2 + O(e^{-Z^2})$ uniformly throughout the second sector (see [13, p. 42]), if $|\zeta|$ is large and fixed and ϕ increases continuously from $\pi-$ to $\pi+$, then (5.4) shows that $T_k(\zeta)$ changes rapidly, but smoothly, from being exponentially small to being exponentially close to one.

Coupling (3.5) and (5.2) gives

$$I^{(1)}(\xi) = \sum_{s=0}^{N-1} c_s\, \xi^{-s} + i e^{2\xi} \sum_{r=0}^{N'-1} (-1)^r c_r\, \xi^{-r} T_{N-r}(2\xi) + R_{N,N'}(\xi). \tag{5.5}$$

The remainder $R_{N,N'}(\xi)$ is given in (5.3), and can be estimated as before. Of course, it is expected to be of lower order of magnitude, and hence can be neglected. Inserting (5.4) into (5.5), we obtain from (2.7)

$$\text{Ai}(z) \sim \frac{(3/2)^{1/2}}{2\pi i} z^{-1/4}\left[e^{-\xi} \sum_{s=0}^{2|\xi|-1} c_s\, \xi^{-s} \right.$$
$$\left. + \frac{i}{2}\,\text{erfc}\{c(\theta)|\xi|^{1/2}\}e^{\xi} \sum_{r=0}^{N'-1} (-1)^r c_r\, \xi^{-r} \right] \tag{5.6}$$

where $\xi = \frac{2}{3}z^{3/2}$. Note that in (5.6), we have truncated the first series at an optimal place. When θ is near π, $\text{erfc}\{c(\theta)|\xi|^{1/2}\}$ will have an abrupt but smooth change. In Berry's terminology, this function is called a *Stokes multiplier*. A similar result holds for θ near $-\pi$.

References

1. E. Artin, *The Gamma Function*, Holt, Reinehart and Winston, New York, 1964.
2. M. V. Berry, *Uniform asymptotic smoothing of Stokes' discontinuities*, Proc. Roy. Soc. Lond. A **422** (1989), 7–21.
3. M. V. Berry and C. J. Howls, *Hyperasymptotics*, Proc. Roy. Soc. Lond. A **430** (1990), 653 – 668.
4. M. V. Berry and C. J. Howls, *Hyperasymptotics for integrals with saddles*, Proc. Roy. Soc. Lond. A **434** (1991), 657–675.
5. W. G. C. Boyd, *Stieltjes transforms and the Stokes phenomenon*, Proc. Roy. Soc. Lond. A **429** (1990), 227–246.
6. W. G. C. Boyd, *Error bounds for the method of steepest descents*, Proc. Roy. Soc. Lond. A **440** (1993), 493–518.
7. R. B. Dingle, *Asymptotic Expansions: Their Derivations and Interpretation*, Academic Press, New York, 1973.
8. A. B. Olde Daalhuis and F. W. J. Olver, *Exponentially improved asymptotic solutions of ordinary differential equations, II. Irregular singularities of rank one*, Proc. Roy. Soc. Lond. A **445** (1994), 39–56.
9. F. W. J. Olver, *Uniform, exponentially improved, asymptotic expansions for the generalized exponential integrals*, SIAM J. Math. Anal. **22** (1991), 1460–1474.
10. F. W. J. Olver, *Uniform, exponentially improved, asymptotic expansions for the confluent hypergeometric function and other integral transforms*, SIAM J. Math. Anal. **22** (1991), 1475–1489.
11. F. W. J. Olver, *Exponentially improved asymptotic solutions of ordinary differential equations I, The confluent hypergeometric function*, SIAM J. Math. Anal. **24** (1993), 756–767.
12. R. Wong, *Asymptotic Approximations of Integrals*, Academic Press, Boston, 1989.
13. R. Wong and Y. -Q. Zhao, *Smoothing of Stokes' discontinuity for the generalized Bessel function*, Proc. Roy. Soc. Lond. A **455** (1999), 1381–1400.
14. R. Wong and Y. -Q. Zhao, *Smoothing of Stokes' discontinuity for the generalized Bessel function, II*, Proc. Roy. Soc. Lond. A **455** (1999), 3065–3084.

INDEX